Ane-Marie

Methods of Social Research

Methods of Social Research

Thomas J. Sullivan
Northern Michigan University

Harcourt College Publishers

Fort Worth Philadelphia San Diego New York Orlando Austin San Antonio

Toronto Montreal London Sydney Tokyo

Publisher   Earl McPeek
Developmental Editor   Peggy Howell
Project Editor   Claudia Gravier
Art Director   David Beard
Production Manager   Suzie Wurzer

ISBN: 0-15-507463-6
Library of Congress Catalog Card Number: 00-101063

Address for Domestic Orders
Harcourt College Publishers, 6277 Sea Harbor Drive, Orlando, FL 32887-6777
800-782-4479

Address for International Orders
International Customer Service
Harcourt Inc., 6277 Sea Harbor Drive, Orlando, FL 32887-6777
407-345-3800
(fax) 407-345-4060
(e-mail) hbintl@harcourtbrace.com

Address for Editorial Correspondence
Harcourt College Publishers, 301 Commerce Street, Suite 3700, Fort Worth, TX 76102

Web Site Address
http://www.harcourtcollege.com

Harcourt College Publishers will provide complimentary supplements or supplement packages to those adopters qualified under our adoption policy. Please contact your sales representative to learn how you qualify. If as an adopter or potential user you receive supplements you do not need, please return them to your sales representative or send them to: Attn: Returns Department, Troy Warehouse, 465 South Lincoln Drive, Troy, MO 63379.

Printed in the United States of America

0 1 2 3 4 5 6 7 8 9 043 9 8 7 6 5 4 3 2 1

To Nancy and Ginnie

Preface

When I first started teaching courses in sociology and the social sciences, I recall lecturing one day on the topic of social inequality. I was presenting research showing how achievement opportunities are inequitably distributed over the social class hierarchy when a student expressed disagreement with some of the conclusions that I was drawing. The two of us went back and forth for a few minutes, expressing diametrically opposed stances on the issue. Finally, she ended her side of the debate with a comment that was polite but dismissive: "Well, that's your opinion."

As I have learned from my colleagues, an interchange of this nature is not uncommon in introductory courses in the social sciences because students in these courses often don't fully grasp the procedural rules and foundations upon which social scientific knowledge is based. Of course, I do have opinions, as does every other social scientist. But that day in class I was presenting the results of social science research studies, and those results go far beyond just "opinion."

At the other end of the educational journey, my students who have finished taking our research methods courses have a very different response to the positions and issues I present in class: "Where are the data?" "What research methods and procedures were used?" "Let's analyze whether the conclusions are warranted given the data and methods." These students, having taken research methods, have learned and internalized the social science rules and procedures for gaining knowledge about the world. They have learned to think a little more critically about their world. And that is one of the things that gives me great joy and satisfaction in teaching research methods: opening students' eyes and providing them with tools that help them to better understand their world.

So, this book is a comprehensive introduction to the research methods used in sociology and the other social sciences. The book explores the rules and procedures that social scientists use in assessing information and drawing conclusions. You will see that these research methods are not cut-and-dried. There is a lot of controversy and disagreement among social scientists about what are the best ways to study particular social phenomena, and there are many areas where researchers must exercise judgment in designing research. However, despite this disagreement and controversy, the research methods laid out in this book provide the most rigorous set of procedures for gaining knowledge about the world.

This book is designed to achieve a number of goals:

1. to provide a complete introduction to social science research methods,

2. to equip the student to critically analyze research reports, to identify their strengths and weaknesses, and to assess the legitimacy of their conclusions,

3. to prepare the student to participate in the design of research projects in the social sciences, and

4. to encourage the student to recognize how research logic and procedures can translate into critical thinking skills in one's everyday life.

Special Features

A number of features have been incorporated into this book to achieve these goals.

1. *Comprehensive Coverage:* The book provides a thorough coverage of all the topics introduced in a basic research methods course. Issues and procedures are illustrated through the presentation of many interesting classical and contemporary research projects.

2. *Emphasis on Both Qualitative and Quantitative Approaches:* The book provides thorough coverage of the important paradigms in social science research: the positivist, interpretivist, and critical approaches. These are introduced in Chapter 2 and addressed in a number of chapters where appropriate. The paradigms are not presented as mutually exclusive alternatives but in terms of their usefulness for understanding some aspects of social reality. As an outgrowth of this, thorough treatment is given to both qualitative and quantitative research. Many books of this sort stress quantitative research or relegate the discussion of qualitative issues to a single chapter. Instead, I have incorporated discussions of qualitative research throughout the book. Some chapters (11 and 15) are devoted especially to qualitative research. In other chapters, special attention is devoted to qualitative research, such as an assessment of non-positivist approaches to measurement in Chapter 5 and a discussion of qualitative approaches to evaluation research in Chapter 12. In this way, qualitative research is presented as an equal partner with quantitative research, and this approach encourages students to consider qualitative approaches to all research issues.

3. *Eye on Diversity:* Special attention is devoted to conducting research in a socially and culturally diverse world. In almost every chapter, there is a separate insert titled "Eye on Diversity" that addresses the issues encountered when conducting research on groups that are diverse in terms of race, culture, nationality, gender, disability, or sexual orientation. I discuss special considerations and techniques that enable us to do more effective research on diverse groups.

4. *Social Policy and Applied Research:* Throughout the book, special attention is given to the many forms of applied research and its impact on social policy. The distinction between basic and applied research is made in Chapter 1, and then various applied research techniques are introduced in appropriate chapters, such as focus groups in Chapters 9 and 11. Chapter 12 on evaluation research discusses many forms of applied research, and the book concludes in Chapter 17 with applied research used in marketing, social policy development, public opinion polling, and other types of applications. "Applied Scenarios" appear in most chapters, elaborating on a particular applied research project. Each "Applied Scenario" ends with some exercises for the students to carry out and bring back to the class. So, although the book does focus on basic social science research methods, theory development, and hypothesis testing, it also appeals to those professors who want to develop the applications of research.

5. *Computers and the Internet:* Special attention is given to the use of computers and the Internet in research. In some cases, this is done in the body of the chapter, where particular computer hardware, software, or capabilities are discussed along with research applications. This will not teach the student how to use a particular piece of hardware or software, since there are far too many variations available to make this feasible; instead, it will inform the student about the kinds of computer applications that are commonly used. For example, in Chapter 9 on survey research, the general features of computer assisted telephone interviewing (CATI) are discussed, but a description of specific CATI software will

be avoided. Likewise, Chapter 15 discusses the general process of computer assisted qualitative data analysis (CAQDA) but avoids an analysis of specific software. All chapters end with a special insert titled "Computers and the Internet," and there are additional computer applications throughout. Each presents ways to use the Internet to further explore issues addressed in the chapter. The students are encouraged to explore the Internet. This discussion is tied into my own Web page where URLs are presented and updated and new Internet material will be made available to students.

6. *Comprehensive Ethics Content:* Ethical issues in research are covered extensively in Chapter 3 but are also incorporated as a separate section titled "Eye on Ethics" in other chapters. In this way, ethical considerations are integrated into the design of research.

7. *Review and Critical Thinking:* At the end of each chapter, a section titled "Review and Critical Thinking" presents a series of questions that enable students to develop critical thinking skills; students are asked to use what they learn about research methods as a foundation for thinking critically, about any issues and information they encounter in their everyday lives. Chapter 17 reinforces the development of critical thinking skills and their application to social research.

8. *Pedagogy:* At the end of each chapter, there is a summary of the chapter, a listing of the boldfaced terms introduced in the chapter, and a list of suggested readings.

9. *Appendix on Library Usage:* The book contains an appendix on how to use the library, including discussion on how to use the Internet and various computerized and online information-retrieval systems. Although many students will be familiar with this by the time they take a research methods course, I have learned from experience that some students at this level will be less familiar with it and others can benefit from a review.

Ancillary Package:
For the Instructor

- An Instructor's Manual/Test Bank has been prepared with multiple-choice and essay questions as well as suggestions for lecture and discussion topics.

- EXAMaster software lets you create a test using a single screen or using the full test feature editing items. It comes with a customized grade book software program.

- PowerPoint Slides, available on the Web site, are downloadable to make a great slide presentation of the main concepts in the book.

- Web site links the book through special inserts to the Internet and the author's Web page where URLs are presented and updated constantly.

- Solutions Manual for Student Workbook will make practical application assignments easier.

For the Student

- Workbook, written by Gregg Lee Carter as a companion for the text, provides students with real world applications to strengthen their research skills.

- Data set provided to support the test book and the workbook.

- Student version of SPSS software (10.0) is available on CD with the text for those who want easy access.

Organization of the Book

In many ways, this is a difficult book to organize. Because there are different kinds of research methods and special circumstances might call for a novel arrangement of the elements of a research project, it is impossible to define the proper sequence of topics about research methods. Each instructor who teaches this course may prefer some particular sequencing of topics that others would not. So, what I have done is to, first, make sure that all the necessary elements are covered in this book, and second, use a logical and fairly common sequencing of topics, recognizing that many instructors will rearrange the order in which they use some of this material in their courses. As one example of this, I have placed the chapter on designing questions, indexes, and scales (Chapter 6) after the chapter on measurement (Chapter 5). I did this because, along with many others, I see the designing of questions and scales as fundamentally a matter of how to measure some variables. Other instructors teach the content of Chapter 6 as an integral part of survey research (Chapter 9). Such instructors can easily change the order in which they cover the chapters to meet their special considerations.

Part I of the book contains four chapters that introduce the topic of social research. Chapter 1 describes what social research is, what its goals are, and how it is distinguished from other ways of gaining knowledge about the world. Chapter 2 describes the basic building blocks of research—theories, variables, and hypotheses—and provides an overview of three prominent paradigms in science: positivism, interpretivism, and the critical approach. The major ethical issues confronted in social research are discussed in Chapter 3. Chapter 4 assesses causality and addresses issues of feasibility in research; it also delineates how an issue or concern is shaped into a feasible and researchable problem.

Part II of the book describes two critical issues in research, measurement and sampling. Chapters 5 and 6 focus on the measurement process: the procedures social scientists use to determine whether something exists and how much of it exists. Their treatment here shows the importance of these fundamental observations upon which scientific knowledge rests.

Chapter 5 explains generally how scientists measure and how they evaluate their measures. Chapter 6 covers in more detail a common type of measurement: The use of verbal questions and statements to measure social phenomena. The chapter addresses how to construct those questions and statements and how to combine them into measurement instruments called indexes and scales. Chapter 7 focuses on the topic of sampling: selecting elements from a population that will be the focus of observation.

Part III describes the most commonly used research designs in the social sciences. Chapter 8 discusses how to design and conduct experiments. Chapter 9 focuses on survey research. Chapter 10 explores research designs that use unobtrusive or available data, such as documents or organizational records. Chapter 11 deals with qualitative research where the data is in the form of words, narratives, or some other nonnumerical format. While most attention is given to field research, some other qualitative designs are considered. Finally, Chapter 12 focuses on applied research, especially evaluation research.

The three chapters in Part IV focus on data analysis, deriving some meaning from the observations that were made during the research process. The first two chapters address issues of quantitative data analysis. Chapter 13 focuses on preparing quantitative data for analysis, including entering the data into a computer file. This chapter also addresses how to present data to show conclusions, especially in graphs, charts, and tables. Chapter 14 deals with more sophisticated data analysis where statistics are used to summarize data or to draw conclusions about populations. Chapter 15 addresses the analysis of qualitative data, showing how meaning can be extracted from data in the form of narratives and descriptions without quantifying the data.

Part V of the book deals with how research findings are used. Chapter 16 focuses on the role that grants play in the research process and how to seek a grant. Chapter 16 also addresses properly preparing a research report to communicate research findings to various audiences. Chapter 17 discusses the specific applications of social science research and how researchers and others advance knowledge or improve society using basic and applied research.

Acknowledgments

I have taught research methods to countless undergraduate and graduate students over the years, and it is impossible to overestimate their contribution to this book. They always kept me on my toes by demanding that only the most careful and considered material is acceptable in the classroom. They forced me to hone my pedagogy, and if I didn't, they peppered me with questions. The discipline they forced on me helped to make this a much better book.

I am also indebted to a number of colleagues. Rem DeJong and Gregg Carter helped by preparing supplementary materials to accompany this book. DeJong has also provided immeasurable support by being a kindred methodologist who has always been willing to explore issues with me. In addition, a number of other colleagues reviewed some or all of this manuscript and provided invaluable suggestions as to how it might be improved: Lisa Frehill, New Mexico State University; Judith Linneman, Texas A & M University; Craig

T. Robertson, University of North Alabama; Gregg Carter, Bryant College; Chalon Keller, Utah State University; Francis Adeola, University of New Orleans; Teodora Amaloza, Illinois Wesleyan University; Daniel Cervi, University of West Virginia; Richard Dukes, University of Colorado–Colorado Springs; James Bearden, SUNY–Geneseo, Richard Nagasawa, Arizona State University, Sharlene Hesse-Biber, Boston College; and Terry Miethe, University of Nevada–Las Vegas.

I would like to thank the people at Harcourt for their help in producing this book, including Lin Marshall, acquisitions editor; Peggy Boone Howell, developmental editor; Claudia Gravier, project editor; David Beard, art director; and Suzie Wurzer, production manager.

In the end, of course, I take responsibility for the content of this book. I hope that at least some of the students who use this book will become enthralled by the possibilities offered for using scientific observation as a way of gaining knowledge about the social world. I know it has been exciting for me to spend a life as a social science researcher. Possibly some others will recognize that it could also be a path for them.

Thomas J. Sullivan
Marquette, Michigan

Contents in Brief

Contents

Part I

SHAPING A RESEARCH PROBLEM

CHAPTER 1:
Introduction to Social Research

CHAPTER 2:
Concepts, Hypotheses, and Theories in Research

CHAPTER 3:
Ethical Issues in Social Research

CHAPTER 4:
Problem Formulation and Research Design Development

The four chapters in Part I introduce the topic of social research, showing what it is and how it is distinguished from other ways of gaining knowledge about the world. These chapters describe the basic building blocks of research, such as theories, variables, and hypotheses, and delineate how an issue or concern is shaped into a feasible and researchable problem. The major ethical issues confronted in social research are also addressed.

Chapter 1

Introduction to Social Research

In the twenty-first century, the code word seems to be *information*, as in "information superhighway," and "information overload." With computer technology, satellites, and the Internet, we are surrounded by a greater volume of information than human beings ever confronted before. Information is often linked with the word *research*, but the term "research" is applied to many activities: the student who browses in the library for a few hours; the journalist who interviews five men living on the street in order to write a newspaper article about homelessness; the professor who goes online to discover which teaching techniques are used at other universities. All these people might claim to be gathering information or doing "research." Yet the term "research," as it is commonly used in the social sciences, has a considerably more precise meaning according to which none of these activities would be considered scientific research. This is not to say that these activities are unimportant. They may have a variety of uses. However, social research has very specific goals that can be achieved only through utilizing the proper procedures.

Fundamentally, these differing views of information and research have to do with how we gain knowledge about the world. How do we know and learn things about the world? Some things we know because they are obvious or because they are facts, such as "today is Thursday," or "the sun will rise at 7:12 A.M. tomorrow." But most of the world is not so simple. How do we know what causes juvenile delinquency? or race riots? or the ethnic cleansing that plagued the Balkans region for much of the 1990s? These much more complex social phenomena are neither obvious nor easily understood as mere "facts." This book is an exploration of one procedure for gaining knowledge about the world—the scientific method, which rests on the conduct of scientific research.

Social research is the systematic examination (or reexamination) of empirical data, collected by someone firsthand, concerning the social, cultural, or psychological forces operating in a situation. This definition contains three key elements. First, social research is *systematic*—that is, all aspects of the research process are carefully planned in advance, and nothing is done in a casual or haphazard fashion. The systematic nature of research is at the core of the scientific method, which is discussed in more detail in Chapter 2. Second, social research involves the collection of *empirical data*—that is, information or facts about the world based on sensory experiences. As such, it should not be confused with philosophizing or speculating, which lack the empirical base of research. Third, social research studies *social, cultural,* and *psychological factors* that affect human behavior. Biological, physiological, nutritional, or other such factors would be a part of social research only to the extent that they affect, or are affected by, social and cultural factors. However, the basic logic underlying these other sciences is identical with the logic of the social sciences, although some of the specific procedures that are used are quite different.

This chapter provides an introduction to the nature of social research, including what sets it apart from other ways of knowing about the world. Chapter 1 also gives an overview of some of the key issues in conducting research and the steps in the research process.

SOURCES OF KNOWLEDGE

There are numerous ways of gaining knowledge about the world, and all these sources of knowledge contain both benefits and pitfalls. This book argues that science is the preferred way of gaining knowledge about the world that we can observe with our senses. This does not mean that science is infallible; rather, it means that science has advantages as a source of knowledge that make it superior to other ways of gaining knowledge. To see why this is the case, I will begin by contrasting science with five other common sources of knowledge: tradition, experience, authority, common sense, and journalism (O'Hear, 1989; Wallace, 1971).

Tradition

Traditional knowledge is based on custom, habit, and repetition. It is founded on a belief in the sanctity of ancient wisdom and the ways of our forebears. People familiar with the musical *Fiddler on the Roof* will recall how the delightful character Tevye, a dairyman in the village of Anatevka, sang the praises of tradition:

> Because of our traditions, we've kept our balance for many, many years. Here in Anatevka we have traditions for everything—how to eat, how to sleep, how to wear clothes . . . You may ask, how did this tradition start! I'11 tell you—I don't know! But it's a tradition. Because of our traditions, everyone knows who he is and what God expects him to do. Tradition. Without our traditions, our lives would be as shaky as—as a fiddler on the roof (Stein, 1964, pp. 1, 61)!

For Tevye and the villagers of Anatevka, it is unimportant where traditions come from. Traditions provide guidance; they offer "truth"; they are the final word. Tradition tells us that some social practice is correct because it has always been done that way.

Traditional knowledge is widespread in all societies. Many people, for example, believe that the two-parent family is preferable to the single-parent family in that the former provides a more stable and effective socializing experience for children and reduces the likelihood of maladjustment. In some cases, these beliefs are grounded in religious traditions, whereas in other cases they are accepted because "everybody knows" how important two parents are to a child's development. In fact, some people maintain these beliefs about the traditional two-parent family despite the existence of considerable social science research suggesting that the two-parent family may not

always be indispensable for high-quality parenting. Some research, for example, suggests that children can actually gain some benefits from living in a single-parent family and that adoptions by single parents can be good for children in some cases (Amato, 1987; Goldscheider & Waite, 1991; Groze, 1991). However, tradition often leads people to conclusions that are at variance with the results of social science research.

Tradition can be an important source of knowledge, especially in areas such as moral judgments or value decisions, but it does have some major disadvantages. First, it is extremely resistant to change, even in those cases where change may be necessary because of the surfacing of new information or developments. Second, traditional knowledge sometimes confuses knowledge (an understanding of what *is*) with values (a preference for what *ought to be*). For many people, the traditional emphasis on the two-parent family is actually based on a value regarding the preferred family form rather than on knowledge of the effect of family on child development.

Experience

Experiential knowledge refers to firsthand, personal observations of events. It is based on the assumption that truth and understanding can be achieved through personal experience and that witnessing events will lead to an accurate comprehension of those events.

Experience is a common source of knowledge for all people because we all have numerous opportunities to make firsthand observations of a whole variety of human social behaviors. From these observations, people develop an understanding—not necessarily accurate—of what motivates themselves and others and what social or psychological processes have influenced them. For example, a person who works with welfare recipients at a particular agency may note that a seemingly large number of the recipients have children and are not married—a focus on single parenthood again. These observations, based on experience, may lead the person to conclude that welfare encourages women to have children and avoid marriage, possibly as a means of obtaining more benefits from the welfare system. Yet social science research has taken a more careful look at the impact of welfare, looking at how welfare works in different locales and at different times and comparing women receiving welfare to comparable women not receiving welfare. The conclusion of this research is that welfare seems neither to discourage marriage nor encourage single parenthood (Jencks, 1992; Rank, 1994). One finding on which this conclusion is based is the observation that women on welfare actually have a birthrate that is no higher than the birthrate among nonwelfare women of comparable age and social standing.

So experiential knowledge should be relied on only with great caution because it has some severe limitations that can lead to erroneous conclusions. One limitation is that human perceptions are notoriously unreliable. Perception is affected by many factors, including the cultural background and the mood of the observer, the conditions under which something is observed, and the nature of what is being observed. Even under the best conditions, some misperception is likely and thus knowledge based on experience is often inaccurate.

Second, human knowledge and understanding do not result from direct perception but rather from inferences made from those perceptions. In the example above, the conclusion that the benefits of welfare encourage women to have children is an inference—it is not directly observed. All that has been observed is that these women on welfare often have children while not married. There is no direct observation of the effect of the benefits or even that the rate of single parenthood is higher among women receiving welfare. (Chapter 4 contains a more detailed discussion of the problem of drawing inferences from observations when I address the issue of causality.)

A third limitation of experiential knowledge is that the very people in a position to experience something directly often have a vested interest in perceiving that thing in a certain way. Teachers, for example, observe that the students who do poorly are the ones who do not pay attention during class. However, teachers have a vested interest in showing that their teaching techniques are not the reason for poor performance among students. Teachers would probably be inclined to attribute students' failings to the students' lack of effort and attentiveness rather than to their own inadequacies as teachers.

A final limitation on experiential knowledge is that it is difficult to know if the people directly available to you are representative of all the people about whom you wish to draw conclusions. If they are not, any conclusions drawn from your observations may be in error. To use my earlier example, are the welfare women observed at a particular agency representative of all women receiving welfare? If not, you cannot generalize conclusions from the women observed to the experiences of all women welfare recipients.

Authority

Authority refers to the unquestioned acceptance of someone's leadership or knowledge because of their social position, expertise, or experience. People may accept a person's conclusions, assessments, or interpretations about the world because that person's authoritative position seems to warrant such acceptance. A person who is particularly knowledgeable about some issue or who has had much experience in some realm might seem to know more about those issues and realms than does the average person. For example, a police officer might be accorded such an authoritative position in considerations of crime or punishment, whereas a physician might be given special authority on issues of health and illness. Sometimes people accord such authority to people who are highly intelligent or educated, more willing to accept judgments from such people than from less intelligent or educated people. For this very reason, scientists are often accorded just such an authoritative position. Authority, then, can lead people to accept one person's knowledge as more accurate than or superior to another's knowledge.

As with tradition and experience, authority also has some advantages as a source of knowledge. One advantage is that, at times, people in positions of authority do have special knowledge or experience that makes them more knowledgeable about particular matters. Physicians, for example, are often more expert about health matters (at least some health matters) than is the average person. It is, therefore, warranted to pay special attention to them on these matters. Another advantage of authoritative knowledge is that it is often relatively easy to obtain, requiring only that we solicit it from the person in a position of authority.

However, authority also has its weaknesses. One major weakness is the tendency to extend authority into realms where it is unwarranted. A physician, for instance, may be authoritative in health matters, but people may be inclined to extend that authority into realms of politics or religion, giving physicians' pronouncements on these issues more weight than the pronouncements of other people. As another example, a physician's authority might be quite appropriate in her specialty field, such as cardiac surgery, but her knowledge may be no more valid than other people's knowledge on such health fields as acupuncture. We see this overextension of authority repeatedly in advertising where the testimonials of physicians, star athletes, and successful actors are used to sell us all manner of products about which the physicians, athletes, and actors have little actual expertise.

Another weakness of authoritative knowledge is that it does not include procedures for detecting truth from falsity. Instead, it relies on an individual's expertise, intelligence, or experience, whose appropriateness is often difficult to judge. Authoritative knowledge by itself does not include procedures for assessing the accuracy of an authority's pronouncements.

Common Sense

The accumulation of knowledge from tradition, experience, and authority often blends to form what people call **common sense**: practical judgments based on the experiences, wisdom, and prejudices of a people. People with common sense are presumed to be able to make sound decisions even though they lack any specialized training and knowledge. Yet, is common sense an accurate source of knowledge? Consider the following contradictory examples. Common sense tells us that people with similar interests and inclinations will likely associate with one another. And when we see a youngster who smokes marijuana associating with others who do the same, we sagely comment that "Birds of a feather flock together." Then we see an athletic woman become involved with a bookish, cerebral man, and we say "Opposites attract."

In other words, common sense often explains everything—even when those explanations contradict one another. Not that common sense is unimportant or always useless. Common sense can be valuable and accurate, which is not surprising because people need sound information as a basis for interacting with others and functioning in society. However, common sense does not normally involve a rigorous and systematic attempt to distinguish reality from fiction. Rather, it tends to accept what "everyone knows" to be true and to reject contradictory information. Further, common sense is often considered something people either do or do not possess because it is not "teachable." In fact, it is often contrasted with "book learning." This discourages people from critically assessing their commonsense knowledge and tempering it with knowledge acquired from other sources. For this reason, commonsense knowledge should be accepted and used cautiously.

Journalism

Another important source of knowledge about the world for most people is the material written by journalists for newspapers, magazines, television, or other media. With the explosion of sources available on cable television and the Internet, people now have access to vast amounts of information produced by journalists. While some journalism consists of opinion pieces based on the speculations and inferences of the journalist, much of journalism, like science, is grounded in observation: Reporters interview people or observe events and write their reports based on those observations. Additionally, using video equipment, journalists are often in a position to provide an audio and/or video record of what happened at a scene.

It may seem at first glance that science and journalism share much in common as sources of knowledge, and that significant similarities between the two endeavors can be identified—both use observation to seek out and record accurate knowledge about the world. In fact, journalism can at times take on many of the characteristics of social science research. Some journalistic output, for example, can look like the in-depth interviews and case studies that are discussed in Chapters 11 and 15. However, while scientific standards require that scientists use the systematic procedures discussed in this book, journalism can and often does fall far short of meeting these standards.

One key difference between science and journalism is that the observations of scientists are far more systematic in nature; scientists utilize careful procedures to reduce the chances that their conclusions will be inaccurate. For example, a journalist interested in the experiences of prison inmates will likely interview a few inmates that are made available to him or her by prison authorities and then use these interviews to draw conclusions, at least implicitly, about the experiences of all prisoners. Social scientists, on the other hand, recognize that the prisoners selected by the authorities are likely to differ in some important ways from other prisoners. They may have been selected because they committed less serious offenses or were model prisoners. Their experience of prison is likely to be very different from that of a more serious offender or someone who has chronic confrontations with prison authorities. Recognizing this, social scientists would—before making observations and collecting data—be very careful about how they selected inmates. They would not usually accept a sample selected by prison authorities. The best sampling procedures (discussed in Chapter 7) would be those that ensure that *all* types of prisoners have a chance to appear in the sample. This could be done, for example, by interviewing all prisoners or, if that was not feasible, interviewing a group of prisoners selected randomly. If sampling procedures fall short of these standards, then social scientists have reduced confidence in the conclusions arrived at, and journalists do not often use such rigorous sampling procedures.

A second key difference between science and journalism is that journalism is not concerned with theory building and theory verification as a way of developing an abstract explanation of people's behavior. Journalists are much more focused on, as the saying goes, "just the facts." Scientists, on the other hand, recognize that facts don't often speak for themselves—they need to be interpreted in the context of a theoretical understanding in order to fully comprehend what the facts signify.

When people depend on the sources of knowledge discussed to this point—tradition, experience, authority, common sense, journalism—they can engage in erroneous modes of reasoning that can produce an inaccurate understanding of the world. Table 1.1 summarizes some of these errors in reasoning.

TABLE 1.1 Fallacies in Reasoning

Fallacy:	Description:
Selective perception	Noticing only evidence that supports your position and not evidence that refutes it
Inaccurate perception	Misperceiving evidence that actually refutes your position as being evidence that supports your position
Overgeneralization	Using evidence from a very limited number of cases of a phenomenon as evidence for all cases of a phenomenon
Illogical inference	Drawing conclusions from evidence when the same evidence could lead to different conclusions
Resistance to change	An unwillingness to entertain new ideas or conclusions even in the face of evidence supporting them

The scientific method has built-in procedures that help overcome these errors.

Science

Winston Churchill, Prime Minister of Britain during World War II, is reported to have said that democracy is an imperfect form of government but that it is far superior to all other forms of government. Many scientists have a similar view of science. They realize that science is imperfect and limited, but also recognize that it is far superior to other sources of knowledge for gaining an understanding of the world. **Science** is a method of obtaining objective knowledge about the world through systematic observation. (The term "science" is also sometimes used to refer to the accumulated body of knowledge that results from scientific inquiry.) Science has five distinguishing characteristics that, taken together, set it apart from other sources of knowledge.

First, science is *empirical*, which means simply that science is based on direct observation of the world. Science is not, as some people mistakenly believe, founded in theorizing, philosophizing, or speculating. Although scientists at times do all these things, they must eventually observe the world to see whether their theories or speculations agree with the facts. Because of this, the number of topics that can be subjected to scientific scrutiny are limited; any issue that cannot be resolved through observation is not within the scope of science. For example, the questions of whether God exists, or which values should underlie a moral life, are not scientific issues because to determine their truth or falsity through observation is impossible. These are matters of faith or preference, not of science.

Second, science is *systematic*, meaning that the procedures used by scientists are organized, methodical, public, and recognized by other scientists. One dimension of the systematic nature of science is that scientists report in detail all the procedures used in coming to a conclusion. This enables other scientists to assess whether inferences and conclusions drawn are warranted given the observations that were made. A second dimension of the systematic nature of science is *replication*—repeating studies numerous times by different scientists to determine if the same results will be obtained. We gain confidence in scientific findings only after they pass this test of reproducibility. Scientists are very cautious about drawing hard-and-fast conclusions from a single observation or an investigation conducted by a single researcher. In fact, quite at variance with experiential knowledge, scientists assume that a single, direct observation is as likely to be incorrect as correct. Only repeated observations can reduce the chance of error and misinterpretation (Rosenthal, 1991).

Third, science is the *search for causes*. Scientists assume that there is order in the universe, that there are ascertainable reasons for the occurrence of all events, and that science can be used to discover the orderly nature of the world. If we assume there is no order, no pattern, then there would be no reason to search for it. We could write off events as due to chance or the intervention of some benevolent or malevolent otherworldly force that can never be understood.

Fourth, science is provisional, which means that scientific conclusions are always accepted as tentative and subject to question and possible refutation. There are no ultimate, untouchable, irrevocable truths in science. There are no scientists whose work is held in such esteem that it cannot be criticized or rejected. As philosopher Jacob Bronowski put it: "Science is not a finished enterprise. . . . The truth is [not] a thing, that you could find . . . the way you could find your hat or your umbrella." Science is a process of continuous movement toward a more accurate picture of the world, and scientists fully realize that the ultimate and final picture may never be achieved (1978, pp. 121–122).

Finally, science strives for *objectivity*, which means that scientists attempt to avoid having their personal biases and values influence their scientific conclusions. This is a controversial and complicated characteristic of science because many social scientists would argue that true objectivity is impossible for human beings ever to achieve. This issue will be discussed at a number of points in this book, but it is sufficient to say at this point that all scientists are concerned that their scientific conclusions will be solely or merely a product of their own personal biases and values. This doesn't mean that scientists are devoid of values. Quite the contrary, they can be as passionate, concerned, and involved as any other group of citizens. They realize, however, that their values and biases can and probably will lead to erroneous scientific conclusions. To address this problem, science incorporates mechanisms to reduce the likelihood of biased observations becoming an accepted part of the body of scientific knowledge. For example, publicizing all research procedures enables others to assess whether the research was conducted in a way that justifies the conclusions reached. Further, such detailed reporting permits replication, so that other researchers, with differing values, can see if they arrive at the same conclusions regarding a specific set of observations.

Despite these checks, values and biases still, of course, impact research. The very decision of which topics to investigate, for example, is often shaped by the researcher's personal values. One person studies family violence because a close friend was the victim of spouse abuse, and another studies factors contributing to job satisfaction because of a personal belief that work is central to a person's identity. Values and biases also enter research through the interpretation of observations. For personal reasons, one researcher may want to show that the criminal justice system rehabilitates, whereas another researcher would like to show the opposite. This may well influence how each researcher goes about conducting research and interpreting the results. There are even a few cases, most common in biomedical research, of outright falsification of data to show a certain conclusion. The point is that values and biases commonly intrude on scientific research, but the overall scientific enterprise is organized to reduce their impact on the body of scientific knowledge.

The scientific method, then, with the characteristics just described, is viewed by scientists as preferable to other ways of gaining knowledge because it is more likely to lead to accurate knowledge of the world. In particular, scientific knowledge is based on careful, systematic, and repeated observations that enable one to accumulate accurate information despite the personal biases of individual researchers. These positive attributes of science do not mean, however, that science is perfect. Scientists make errors. But, as Bronowski so aptly put it, "Science is essentially a self-correcting activity." If proper scientific procedures are followed, today's errors will be corrected by researchers in the future, whose errors in turn will be corrected by yet further research (1978, p. 122).

THE PROCESS OF SOCIAL SCIENCE RESEARCH

This book is dedicated to the exploration of the "process" of social science research. But there are some very basic foundations upon which all of the later material rests. These foundations have to do with how the scientific community is organized and the role of theories in the research process.

The Scientific Community

Being a scientist is like being a soldier, a hockey player, or an astronaut: Each is involved in a special social world or community that shapes the attitudes and behaviors of the people who are part of it. Each of these social worlds is a subculture that, like any subculture, consists of values, norms, material objects, and a special language. As with any subculture, the scientific subculture promotes attitudes and behaviors that encourage sound scientific work whose ultimate goal is an accurate description of the world. Ergo, the **scientific community** consists of the people engaged in scientific work along with the values, norms, attitudes, behaviors, and language that guide and direct their work. The members of this community share a culture and social life that connects them to one another and gives meaning and direction to their actions. I want to spend a little time talking about this community because, while it may seem subtle or even intangible to nonscientists, it is an important and powerful force in the lives of people who become part of it.

As with any other human community, the scientific community contains a set of values or norms that serve to guide and regulate behavior. These values are, in a sense, ideal guidelines, in that they describe how people *ought to* behave rather than how they actually *do* behave. This same thing happens in all cultures. The culture of the United States, for example, might direct its members to tell the truth or help people in distress when in reality people

do violate these guidelines. However, the promotion of these values puts pressure on people to behave in certain socially acceptable or ethical ways. Likewise, the scientific culture describes and promotes ideals for behavior even though particular scientists may not live up to those ideals constantly. Different lists of the values of the scientific community exist, but they all can be distilled into five general guidelines, which sociologist Robert Merton called "the ethos of modern science" (Hagstrom, 1965; Merton, 1973; Mitroff, 1974):

1. *Organized skepticism.* All knowledge, whether based on scientific research or other sources, is open to criticism and questioning. No scientist or researcher has such high standing or reputation that his or her research cannot be criticized. There is no ruler in the scientific world who is beyond dethroning. A scientist is therefore always skeptical of claims of truth or knowledge, but it is organized skepticism—it evaluates claims on the basis of the scientific procedures used to create the knowledge. Being skeptical doesn't mean one rejects all claims; neither is it a cynical stance that holds that nothing is to be believed. Rather, it means one evaluates all claims using certain established and agreed upon criteria. So science is very much about critical thinking based on scientific procedures and practices.

2. *Universalism.* The same criteria are used in judging all research, irrespective of who does the research or when or where it is done. These criteria (explained in this book) are the standards of research design that are widely accepted by most scientists. The reputation or previous success in research of the researcher is irrelevant to judging the merits of the current research under scrutiny.

3. *Disinterestedness.* Scientists should not be bound to any particular research outcome or finding but should be open to any possibility. The scientist's stance should be one

of neutrality and impartiality. If proper scientific procedures are followed, then any outcome should be acceptable to a scientist. The point of science is to produce accurate and valid knowledge—not to advance a particular viewpoint or outcome.

4. *Communalism.* Scientific research is engaged in to advance the interests of society as a whole and, thus, scientific knowledge should be made available to everyone. Scientific communities have built-in procedures for sharing the results of science within the scientific community and with society at large. The results of scientific work are often made available in public libraries, or increasingly on Internet Web sites that are accessible to all who own or have access to a computer. The point is that scientific knowledge is not proprietary; it is not the private property of those who produce it. It is the property of the community.

5. *Honesty.* Scientists must be honest about how they conduct their research and how they draw their conclusions. Without this foundation of honesty, the accuracy of scientific knowledge could not be trusted. If a corporation says that its product is good for you, we might understandably question the truth of the statement because of the self-interest of the corporation. If we also had to assume that the statements of scientists are equally self-serving and suspect, then the body of scientific knowledge could not be relied on as a source of knowledge for understanding the world.

These are the values that serve to organize the scientific community. In addition, procedures and mechanisms exist to encourage scientists to live up to these values and increase the likelihood that the scientific endeavor will reflect these values rather than their opposite (such as dishonesty or proprietariness). One such procedure is the socialization that scientists go through as a part of their education and training. That socialization occurs during their collegiate and post-collegiate educational

experiences. Depending on their discipline, scientists will have completed a college major in their scientific discipline, received graduate training at the master's or doctoral level and, in some cases, done postdoctoral research internships. The course in which you are reading this text is a part of that socialization process. As you read this text and take this course, you will see the extent to which these values are promoted as the proper way to conduct scientific research. Throughout the educational experiences of scientists, those who follow these guidelines are portrayed as role models, whereas researchers who violate these norms are portrayed as poor scientists. For those who go on to obtain advanced degrees and become sociologists, psychologists, or biologists, this socialization process continues for many years, to the point where these values are internalized as the preferred way to conduct their scientific investigations. In short, becoming a scientist does not mean merely gaining and retaining a lot of information and learning scientific techniques; it involves the development of new attitudes—especially as they relate to assessing information and making critical judgments about issues.

The Peer Review Process

In addition to the socialization that occurs during the educational process, the norms of the scientific community are also promoted through a process of peer review and evaluation that is central to scientific work. When scientists conduct research, they write up their results in a report or paper that is either (1) presented at a professional conference to other scientists knowledgeable about the topic of the research, and/or (2) submitted for publication in a scientific journal or book in the field. The procedures for scientific report preparation are discussed in more detail in Chapter 16, but the point here is that professional conferences and scientific journals and books only accept papers after they have been reviewed by scientific peers—researchers who are expert on a particular topic—and judged to be acceptable in terms of the research methods used, the data analysis techniques applied, and the conclusions drawn. Since prestige, jobs, and income for scientists depend on their success in presenting papers and publishing research, scientists are highly motivated to see their research in such forums. However, the only way to accomplish this is to produce research that can stand up to the intense scrutiny of one's scientific peers. If a scientist publishes his or her research in a journal or magazine that does not use peer review, then it is considered a low-quality publication and will carry less weight (or possibly no weight) in decisions about employment, promotion, or tenure.

Just as other cultures sanction their members through rewards and punishment in order to encourage acceptable behavior, so too does the scientific community. When laypeople learn of this peer review process, they are sometimes inclined to view it as a system of censorship where those who control the journals' content (editors) determine what is scientific knowledge at a given moment. And I would be naive to say that something like this never occurs. Certainly, control of the peer review process could enable some scientists to thwart the emergence of new ideas. Intense controversy sometimes emerges among scientists over what are to be considered acceptable scientific techniques and research findings, and I will discuss these controversies repeatedly in this book. But scientists view the peer review process not as censorship but rather as a demand for high standards. Scientists agree that there are certain procedures—the topic of this book—more likely to produce accurate and valid knowledge. Peer reviewed journals in science demand that authors use these procedures in order to get published in the journals (or make valid arguments for deviating from the procedures) because this will result in a more accurate body of scientific knowledge.

Theories in Research

If two themes that are most central to what science is all about had to be identified, those two themes would probably be "observation" and "theory." As we have seen, science is empirical—it is based on the observation of things that occur in the world. Science is not just speculation, opinion, or philosophizing, although each of these might come into play at some stage in the scientific process. Ultimately, however, scientists must prove what they claim by making observations that will stand as evidence of those claims. In the absence of such confirming observations, scientific knowledge is at best incomplete, or at worst no better than knowledge based on experience, intuition, or speculation.

However, science goes beyond observation to develop theories, which represent a more abstract and general level of knowledge than that gained from observation. Each observation is a unique, one-time occurrence. Yet we don't experience the world as a series of unique occurrences; instead we categorically lump together things or patterns that seem to go together for some reason. For example, an observation might be "Jim slapped Mary at 10:30 on Friday night." We could then generalize by saying that that slap is an instance of a more broad and abstract category or phenomenon called "spouse abuse." Also included in this spouse abuse category would be other observations, such as "Bill twists Mary's arm," "Nancy stabs Fred," and "Jim slaps Mary again on Saturday night." Being able to place these observations into a more general category, such as "spouse abuse," tells us much more about the event; it is no longer a unique event but an event of a certain type with a certain meaning. We could take these instances of spouse abuse and place them into even more general categories, such as "gender stratification" or "family disruption." Thus, gender stratification could manifest itself in the form of spouse abuse, but it could also manifest in other forms, such as income inequities or other forms of economic discrimination (see Figure 1.1). When we take the original observations and place them into the more abstract context of spouse abuse and gender stratification, we have a far more complicated and sophisticated

FIGURE 1.1   Moving from Observation to General Explanations

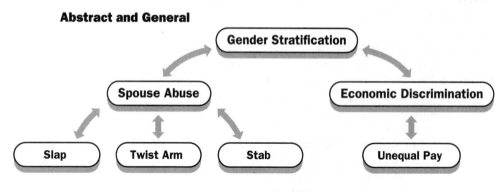

Figure 1.1 shows an inverted-tree diagram labeled "Abstract and General" at the top and "Observations/Unique Events" at the bottom, with boxes "Gender Stratification," "Spouse Abuse," "Economic Discrimination," "Slap," "Twist Arm," "Stab," and "Unequal Pay."

understanding of social reality. We can see that reality is like an onion, with many layers of understanding to it. As we move from layer to layer, we gain a more general and abstract understanding of reality. In other words, we develop a **theory**: a set of interrelated, abstract concepts and statements that offer an explanation of some phenomenon. An observation tells us only that something occurred; as an explanation, the theory tells us why it occurred. Jim slapped Mary (the observation) because he was engaging in spouse abuse (a more general explanation), and he engaged in spouse abuse because his sociocultural context was one of gender stratification (a still more general explanation) where men are permitted, perhaps even encouraged, to dominate their wives.

Nonscientists often misunderstand what scientists mean by the term "theory." Some people, for example, believe that theory refers to speculation or opinion, as when one says, "Oh, that's just a theory," as a way of suggesting that someone's statement need not be taken very seriously. Related to this is the misunderstanding that a theory is something that has not yet been proved to be true. A theory, especially when first formulated, might be based—in part or in whole—on speculation, and might be unverified. However, other theories have been extensively verified by research and contain little, if anything, speculative about them. What makes all of these things theories—no matter how speculative or verified they may be—is that they are abstract explanations of phenomena.

Another misunderstanding about theories is that they describe what the world should be like, as in a preferred or utopian way to organize things. In actuality, theories are efforts to describe what the world is like—how it actually operates. Philosophy or theology may entertain issues of what is preferred; science is an effort to develop an accurate description of things as they are. Of course, it is not until the theory is verified that we know if the description proposed by the theory is how the world actually operates.

One of the major outcomes of scientific research, then, is to develop abstract theories and verify them. In Chapter 2 we begin to describe this process in detail.

ISSUES IN RESEARCH IN THE SOCIAL SCIENCES

Because social phenomena come in many different forms, social researchers need many different types of research methods in their arsenal. Researchers must also make a variety of choices about what kind of research methods are most appropriate to a particular research problem. This section explores some of these basic choices in terms of three ways in which research can be classified: (1) the goals of the research, (2) the applications of the research, and (3) the types of observations that are made.

This book describes the criteria researchers apply in making these choices. These decisions are not always easy and are sometimes controversial because researchers may disagree among themselves about what is appropriate for a particular research problem.

Goals of Research

One of the most basic decisions a researcher must make is about what the goal of his or her research is going to be. Research in the social sciences generally focuses on one or more of the following goals: description, prediction, explanation, and/or evaluation.

Descriptive research has as its goal *description*, or the attempt to discover facts or describe reality. It is a picture or account of what exists, sometimes summarized in numbers, percentages, or some other statistics. Descriptive research, for example, might deal with such questions as: What are people's attitudes toward welfare? How widespread is child abuse? What are the attitudes of people in a particular community toward Asians, Latinos, or other minorities? Some descriptive research efforts are quite extensive. For example, the National Center for Health Statistics and The Centers for Disease Control and Prevention collect voluminous amounts of data each year on such things as the death rate and the infant mortality rate for purposes of describing the health status of people in the United States.

Predictive research focuses on *prediction*, or making projections about what may occur in the future or in other settings. It uses present knowledge to predict what we would expect to find at another time or place. For example, if a river is to be dammed to produce electrical power, we can try to predict the social impacts this will have on adjacent communities: Increased jobs lead to increased population, putting additional pressure on the school system; if new workers migrate to the area to build the dam, they may be young and unmarried, and crime rates may increase; and the newly created lake will provide enhanced recreational opportunities. Social impact assessments, conducted by social scientists, are an attempt to predict the impacts on the social and cultural environment of establishing a program in order to decide whether the benefits outweigh the costs or to see how the costs can be reduced.

Explanatory research involves *explanation*, or determining why or how something occurred. It goes beyond description by trying to find the causes of some phenomenon, and it strives to develop and verify theories as explanatory systems, which are described in the preceding section. Explanatory research, for example, would go beyond describing rates of juvenile delinquency or even predicting who will engage in delinquent acts; it would also focus on determining *why* certain people become delinquents and others do not. The goal of explanation may appear to be quite similar to that of prediction, but there is a difference: One can make predictions without an accompanying explanation. Actuaries at insurance companies do this all the time by calculating risks and premium levels for insurance policies. From data on past automobile accidents, for example, they predict that young males will—for whatever reasons—have more accidents than older people and females, and their insurance rates reflect this fact.

Evaluation research focuses on *evaluation*, or the use of scientific research methods to plan and assess social policies, social interventions, and social programs. Evaluation research can also determine whether a program has unintended consequences that are either desirable or undesirable. For example, social programs that attempt to reduce poverty, promote positive child development, or fight crime or alcoholism need to be evaluated in terms of how well they achieve their goals. With the intense competition for funds for such programs, program managers are routinely required to justify and defend their programs in terms of cost effectiveness. Social scientists conduct evaluation research to provide scientifically based answers to these questions. Evaluation research is, in a sense, an extension of the three previous goals because it typically uses some combination of description, prediction, and explanation in order to achieve the goal of evaluation. Chapter 12 is devoted entirely to the special considerations involved in doing evaluation research.

Applications of Research

The decision about the goals of research leads directly to the issue of the applications to which research is put. At a general level, research has two fundamental uses: to advance our understanding and to solve some practical problem. In fact, reflecting these two uses, all sciences—including the social sciences—make a distinction between basic (or pure) science and applied science. Both basic and applied science are based on observation, but what each does with these observations is quite distinct (Larson, 1995; Sullivan, 1992).

Basic research refers to research whose purpose is to advance our knowledge about human social behavior with little concern for any immediate practical benefits that might result. The primary focus of basic research is on developing theories of human social behavior and testing hypotheses derived from those theories. Basic social science research is an attempt to develop theories that explain how societies work and why people behave the way they do; the goal of basic research is to subject theories to the process of theory verification (described in more detail in Chapter 2). Basic research could focus on one or all of the goals of research described in the preceding section, but

explanation is usually seen as the ultimate or long-term reason why research is conducted. Lest you conclude that basic research is impractical or useless, keep in mind that practical applications for basic research are often found by people other than the researchers themselves. In some cases, it takes years for this to happen, but eventually someone will discover that the knowledge base developed in earlier basic research will enable them to find solutions for real-world problems.

Applied research consists of research designed to focus scientific research tools on a practical, real-world problem identified by some client with some practical outcome in mind. The primary focus of applied research is on making recommendations about programs or policies that might be implemented or changed or about how to change behaviors. Applied social science research uses social science knowledge and research tools to tackle some particular problem that somebody wants solved. It addresses questions like: How well does this program work? Does it achieve its goals? What impact does it have on other people? groups? institutions? The overall focus is to direct research tools toward the end of social intervention. We may learn something new about human behavior while doing the research but that is not its main point. As for goals, applied research could focus on any of the four goals, but it would more likely focus on evaluation and description and less likely on explanation than would basic research.

Another important difference between basic and applied research has to do with how researchers select problems for study (DeMartini, 1982). Basic researchers usually decide themselves which research questions are most important based on their judgment of the state of knowledge. Researchers assess the extent to which particular theories have been verified and whether further development and verification of a theory is warranted. In doing this, they are strongly influenced by evaluations of their work by their scientific peers. In applied

TABLE 1.2   Types of Applied Social Research

Type	Definition	Example
Program Evaluation	An assessment of how well social programs or social interventions operate and whether they achieve their goals	Do teenage mothers in a pregnancy prevention program avoid repeat pregnancies and stay in school?
Needs Assessment	An evaluation of how extensive some problem is, what resources exist for meeting the problem, or whether some goods or services are needed by a particular group	How many teenagers in the community would avail themselves of the services or a pregnancy prevention program?
Social Impact Assessment	A look at the effects of some program or practice on the social and cultural environment of a community	Does the teenage pregnancy program place additional demands on schools or social agencies?
Social Indicators	An effort to devise quantitative measures of significant social phenomena	Can the rate of teenage pregnancy be used as part of a broader measure of whether a healthy environment for teens exists in a community?
Cost–Benefit Analysis	A quantitative assessment of the costs and the benefits of a program or practice in order to decide its future	What are the costs of operating a pregnancy prevention program? Could other alternatives achieve the same goals with lower costs?

research, on the other hand, the client who is paying to have the research done plays a leading role in deciding which research problems will be addressed. The researcher's task is to develop scientifically sound procedures for assessing the problem.

While the distinction between basic and applied science can be made in the abstract, it is sometimes difficult in reality to draw a hard and fast line between the two. A single research project may have both basic and applied dimensions to it. Nevertheless, even in a project that addresses both issues, one of the two uses of research—either theory building or practical problem solving—will usually predominate. This book will focus on the research methods and activities used both by basic and by applied researchers.

As a final illustration, at this point, of the kinds of issues addressed in applied research, Table 1.2 displays definitions of five types of applied research. For each type, an example is given of a research question it would address if

an evaluation were being done of a program whose goal was to reduce repeat pregnancies among teenage mothers and to keep those mothers in school. Chapter 12 and Chapter 17 will focus greater attention on the different kinds of applied research and the special consideration that must be taken into account in doing good applied research. In addition, a number of other chapters will contain boxes titled Applied Scenario, which will describe some particular applied research challenges with some suggested exercises for the student.

While important differences can be found between basic and applied research, they do share one important characteristic: Both types of research strive to maintain the most rigorous scientific standards possible for conducting their research and both are open to criticism and evaluation by other scientists. In fact, much of what you will learn in this book about research methods applies to both basic and applied research. Sometimes applied researchers confront some practical problems

EYE ON ETHICS Ethics and Values in Research

Social research affects people's lives and livelihoods. It can have both beneficial impacts and detrimental ones. Sometimes the effect is direct and obvious, as when an evaluation research project shows that a social service intervention program doesn't achieve its goals; the people who run the program may lose their jobs. In other cases, the impact is less direct but can still be important. For example, in a now famous study, sociologist James Coleman and his colleagues (1966) documented how segregated schools have a negative impact on the academic performance of minority children. These research results played a key part in bringing about policies of school busing across the country in order to promote integration in the schools, and busing has had both positive and negative consequences. So, millions of school children over many decades had their lives changed in substantial ways due in part to research conducted by social scientists.

The conduct of social research carries with it serious and heavy responsibilities. Therefore, social researchers are very careful about how they do their research, and they are guided by certain ethical standards in the conduct of research. Chapter 3 is devoted to a detailed discussion of ethics in research; in addition, specific ethical considerations will be addressed in Eye on Ethics boxes in other chapters. However, it is worth a brief mention of some key ethical issues at this point to show how every step in the research process is suffused with considerations about ethics.

Honesty and Accuracy. Although it would seem obvious, it is fundamental that research be conducted and reported with honesty and accuracy in the forefront. Researchers take great care to design their research so that they produce an accurate vision of reality, and they analyze and report their results to give an accurate and honest portrayal of the outcome. Any kind of dishonesty or fraud is considered deeply offensive to scientists, and as we have seen, the scientific community discourages such behavior through sanctions against people who violate these ethical standards.

As you learn more about designing research in this book, you will begin to recognize the many ways in which dishonesty could be perpetrated by researchers. Some are obvious, such as not reporting some of the results when you are dissatisfied with them. But dishonesty can also be much subtler, as when researchers choose a way to measure a phenomenon that will sway the results in a direction they prefer. (Chapters 5 and 6 examine measurement issues.) In some cases, researchers may knowingly carry out deceptive research, but more often researchers may not even realize that the design they have chosen is producing misleading results. As you learn more about research, you will become more adept at identifying possible dishonesty or fraud in research.

Do No Harm. Most social researchers are motivated by the desire to use their research to help people and make the world a better place, but they are also guided by the principle of doing no harm to people. Most social research does not bring physical harm to people. However, social research in some cases can threaten people with psychological harm. For example, asking people to report on their sexual behavior could be embarrassing, while interviewing people about their past misdeeds might produce bouts of depression or remorse in them. In other cases, the results of research may lead to changes in policies that could negatively affect people, as with the research on school integration that produced policies of school busing.

The ideal would be to design research that does no harm. If that is not possible and the research is still considered sufficiently important to proceed despite the possible harm, then the researchers have the obligation to alleviate the harm. If the harm is psychological in nature, then it may be necessary to counsel or debrief the research subjects, possibly using professional counseling assistance, to help them overcome the difficulties.

Privacy and Self Determination. The cultural heritage of the United States places great value on people having a right to privacy and self determination. In other words, people should be able to control what is known about them and what happens to them, and this should be true in research as it is elsewhere. In the research context, this means that participation in research should always be voluntary. People should be informed of all aspects of the research before they make a decision to join; they should be able to withdraw from participation at any point; and they should be able to withhold whatever information from the researcher that they wish. Anything less would violate their fundamental human right to privacy and self determination, and these rights supercede any benefits that might accrue from a research project.

All research projects are designed specifically with these issues of privacy and self determination in mind. Potential participants are told about all aspects of the research project that might influence their decision to participate. They may also be required to sign a consent form indicating that they were so informed and volunteered freely. Researchers go to great lengths to ensure that information about participants is kept private or that participants know ahead of time — before they agree to participate — whether any information about them will be made public. However, the issue gets complicated: One sociologist we will meet in Chapter 3 conducted his research by observing men having quick and impersonal sex in public bathrooms. He did not inform the men that he was a sociologist who would use his observations of their behavior as part of his research findings. Was he being ethical? Both sides of that issue will be examined in Chapter 3.

You can begin to see that ethical considerations are integral to the research process, and they are among the choices and decisions that all researchers must make. If ethical difficulties cannot be resolved, then researchers may abandon a research idea or change it so that it can be accomplished without violating ethical guidelines. And to reiterate: these issues are important because research impacts on people's lives, and researchers have responsibilities to people that go beyond the goal of getting research results.

that make it difficult or impossible for them to use the same procedures that basic researchers might prefer. However, when this happens, applied researchers are not absolved from the demands of science: If they adopt less rigorous procedures, then their work will be criticized and their results accorded a lower level of confidence than if they had used preferred procedures. The standards of science are the same for both basic and applied research. In some cases, basic researchers may also find it impossible to maintain those standards, and they will be criticized for not doing so.

Qualitative and Quantitative Research

When researchers make observations, or gather data, they must decide whether to use one or both of two broad categories of data: qualitative or quantitative (Berg, 1998). **Qualitative research** basically involves data in the form of words, pictures, descriptions, or narratives. **Quantitative research** uses numbers, counts, and measures of things. The decision as to which general orientation to follow in a given research project depends primarily on two factors: the state of one's knowledge on a particular research topic, and the individual researcher's position regarding the nature of human social behavior. When knowledge about some phenomenon is sketchy or when there is little theoretical understanding of the phenomenon, it may be not be possible to develop precise statements of concepts or quantitative ways to measure them. In such cases, researchers often turn to qualitative research that is more exploratory in nature. The research can be very descriptive, possibly resulting in the development of concepts and theories rather than the verification of them. As previous research on a topic accumulates, it may be more feasible to precisely state theories and derive testable, quantitative predictions from them.

The second consideration in choosing between quantitative and qualitative research stems from a more fundamental controversy over the nature of human social behavior and social research. Basically, it involves a debate over whether the human experience can be meaningfully reduced to numbers and counts, as is called for in quantitative research. Quantitative researchers argue that quantitative observations are precise ways of discovering and describing social phenomena, the equivalent of what is done in the natural sciences such as physics and chemistry. As in those fields, social scientists can gain an objective and precise assessment of social life through numbers and counts involving how often things occur.

Other social scientists dispute this position, arguing that the human experience has a subjective dimension to it—the very personal meanings and feelings that people have about themselves and what they do. These meanings or feelings cannot be captured very well through numbers or counts. They are better understood through narrative descriptions of people going about their daily routines or through lengthy and broad-ranging interviews with them. Descriptions and interviews are better able to capture the very critical subjective meanings that are an essential element of understanding human behavior. Proponents of this view argue that, while quantitative research appears more precise, it is actually a distortion of social reality that misses an important aspect of human social behavior.

I mentioned earlier in this chapter that social scientists sometimes clash with one another over significant controversies, and the issue of quantitative versus qualitative research is one of these key controversies. It is a complicated issue that is discussed in more detail at a number of points in this book. However, you need to understand that the line between qualitative and quantitative approaches is not always completely clear, and the choice between the two can be difficult. Many research projects incorporate both approaches in order to take advantage of the benefits that each has (Newman & Benz, 1998). Your task is to learn the advantages and disadvantages of each, how each is conducted, and the criteria used to decide when each is appropriate.

A final set of issues that researchers must make decisions about are maintaining appropriate ethical standards. In this and a number of other chapters, Eye on Ethics boxes will address the very important issue of ethics in research.

The Limitations of Science

Although the methods of science are very useful tools for acquiring valid knowledge about the physical and social worlds, science does have its limitations. One limitation is that it cannot resolve fundamental moral issues. Because science deals with matters of fact, there are many conflicts that science cannot settle for us. When the conflict is over competing values, ideologies, or religious beliefs, science cannot settle the dispute because the dispute is typically not over matters of fact. So science cannot tell you whether you ought to marry, go to college, be a Catholic, Buddhist, or atheist, or whether the fall of communism was a good thing—unless we can agree on certain factual criteria that will settle these disputes. If you want to live in social circumstances that will reduce the likelihood of suicide, then you should marry because married people have lower suicide rates than single people; if you want to increase your chances of earning a high income, then go to college because college graduates, on average, earn higher incomes than those who have less than a college degree. Science can help us with these disputes only if we can reduce the choice to a set of agreed-upon facts that can be observed by researchers in the world. Even then, science only tells you the facts (married people have lower suicide rates); people then have to make the choice (I will marry rather than remain single). In reality, of course, complex questions of morals or values can rarely be reduced to one, or even a few, factual questions that all parties to the dispute will agree to. To resolve these moral issues, in most cases, we need to rely on basic values or religious, ethical, or philosophical

principles. Science may be able to inform us about the consequences of particular choices, and this could help us make a decision. However, the decision itself is still, fundamentally, a moral one.

Science also cannot help us with realities that are not observable. The spiritual or transcendent realities posited by many religions, for example, are beyond the realm of science. Science cannot tell us whether any gods or spirits exist or whether there is a heaven or hell. These matters involve a reality different from the physical reality that is available to us through our senses. Science can tell us how many people go to church or how many people believe in God—these issues can be determined through observation—but not whether the spiritual beliefs of a church are correct or whether some god or spirit actually exists.

It is therefore important to be clear about the limitations of science: It can deal with issues that can be resolved through observation of the world available to us through our senses. Anything beyond that must be approached with other ways of knowing, such as tradition, authority, or common sense.

STEPS IN CONDUCTING RESEARCH

I want to conclude this chapter by providing a little more concrete detail to the research process by describing the steps that scientists go through when conducting research. Although each research project is unique to some degree, some general steps characterize virtually every project. The research process can be divided into six identifiable stages, as summarized in Figure 1.2.

Problem Formulation

The first step in conducting social research is to decide on the problem that will be researched. When first encountering the issue of problem formulation, students commonly question its importance. So many problems exist that it

FIGURE 1.2   Stages in the Process of Social Research

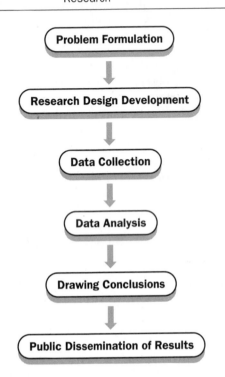

social development than children raised in two-parent families? As the initial step in research, developing a researchable problem is highly important. Part I of this book outlines the many issues involved in this process.

Research Design Development

Having successfully established a researchable problem, the next step is to develop a **research design,** a detailed plan outlining how observations will be made. The plan is followed by the researcher as the project is carried out. Research designs always address certain key issues, such as who will be studied, how these people will be selected, and what information will be gathered from or about them. In fact, the research design spells out in considerable detail what will occur in the other stages of the research process. Part II and Part III describe the different kinds of research designs and issues that must be considered in their development. The Eye on Diversity box discusses the importance of considering human diversity in the design of social science research.

Data Collection

A part of any research design is a description of the kinds of data to be collected and how this will be done. The data collected at this stage constitute the basic observations from which conclusions will be drawn, so great care must be exercised. Two aspects of data collection—*pretests* and *pilot studies*—illustrate just how careful scientists are about this. The **pretest,** as the name implies, is a preliminary application of the data-gathering technique for the purpose of determining its adequacy. It certainly would be risky and unwise to jump prematurely into data gathering without first knowing that all the data collection procedures are sound. For example, if our study involved interviews with professional hockey players in the National Hockey League, we would choose a small group of

would appear to be a simple matter to select one on which to conduct research. However, such a casual view of scientific problem formulation is erroneous. For example, some problems about which we might desire answers are not scientific questions at all, and no amount of research will answer them. Other problems, though possibly interesting and intriguing, might prove impractical from a methodological, ethical, or financial standpoint.

Another element of problem formulation is to shape a concern into a specific researchable question. Such global concerns as "the state of the modern family" are far too broad to be considered research problems. They need to be narrowed down to specific problems for which empirical data can be gathered, such as: What is the divorce rate? How does it compare to the divorce rate of past years? Do children raised in single-parent families exhibit poorer

Social diversity can be based in culture, ethnicity, race, gender, sexual orientation, disability, or other characteristics, and it has been an abiding concern in the social sciences since their inception. One reason for this concern is that diversity is often a pretext for discrimination and the inequitable treatment of some groups. In fact, it is possible for social science research, if not carefully done, to produce inaccurate knowledge of diverse groups that may have negative consequences for them. For this reason, social scientists make special efforts to consider the issue of diversity when designing research.

In some cases, social scientists focus on how diversity affects behavior, such as studies on whether the race of a defendant plays a part in sentencing decisions in criminal courts. In other cases, diversity enters the research process because the characteristics of either the researchers or the groups being studied may be influencing the research outcome, even though those characteristics are not an explicit focus of the research. As one example of this, a researcher may assume that an Asian American respondent will give the same information to an interviewer who is an Anglo American, African American, or Asian American. Yet, Asian respondents may, without realizing it, feel more rapport with an interviewer of the same race, and give more complete answers, than they would to an African American or Anglo interviewer. If this occurs, it means that race is influencing the data collected, and because of it, the interviews may produce distorted or inaccurate data. Keep in mind that this effect can occur even though race is not a part of the research and even though no one, including the researchers, even notices that it is occurring. Conducting research that is accurate and complete in the context of diverse populations can be challenging.

Therefore, issues of diversity are important to consider in conducting social science research. In fact, in certain circumstances, the standard research methods of the social sciences can produce misleading and, in some cases, outright false conclusions if diversity issues are not taken into consideration. This problem is sufficiently important that special attention is devoted to it throughout this book. In each chapter, particular ways are pointed to in which problems or biases in research can occur if diversity is not addressed. Some chapters have a special Eye on Diversity box. The goals of this special attention are to teach techniques for overcoming common research problems, to encourage the development of research designs that provide accurate pictures of diverse groups, and to create a sensitivity to the importance of diversity issues.

such hockey players and collect the same data from them that we plan to collect in the final project. It is, in a sense, a "trial run." And, unless we are very good or very lucky, some modifications in the data collection technique will likely be required based on the results of the pretest. After these modifications are made, the technique is pretested again. Additional pretests are always desirable after any modifications in the data-gathering technique, in order to determine whether the modifications handle the problems encountered in the previous pretest.

In some cases, it may even be necessary to do a **pilot study**, which is a small-scale "trial run" of all the procedures planned for use in the main study. In addition to administering the data-gathering instrument, a pilot study might include such steps as a test of the procedures for selecting the sample and an application of the statistical procedures to be used in the data analysis stage.

It is this kind of care in data collection that improves the validity of the data collected and bolsters confidence in the conclusions drawn. Issues of data collection are covered at various points in Part II and Part III.

Data Analysis

As with data collection, data analysis is spelled out in the research design, and it can be the most challenging and interesting aspect of a research project. It is challenging because data in raw form can be quite unrevealing. Data analysis is what unlocks the information hidden in the raw data and transforms it into something useful and meaningful. During data analysis, you learn whether your ideas are confirmed or refuted by empirical reality. During the course of data analysis, researchers often make use of statistical tools that can range from simple percentages to very complex statistical tests that require much training to understand and master. These statistics aid in communicating the findings of research to others. Once you learn the special language and interpretations of statistics, you can be more effective at communicating research findings in a clear, concise manner than when using conventional English. Part IV reviews some of the basic data manipulation and data analysis techniques that are used in both quantitative and qualitative social research. We will see that modern data analysis is typically done via computers, so a research design must specify data collection procedures that are compatible with the current computer hardware and software.

Drawing Conclusions

The next step in conducting social research is to draw some conclusions from the data analy-

sis. The form this takes depends in part on the goals of the research project. A descriptive study, for example, would simply present what was found, possibly in a summarized form to make it more easily understood. Predictive and explanatory research, on the other hand, usually have hypotheses or statements of what the researchers expect to find, stated before the data are collected. In this case, a major element of drawing conclusions is to assess how much support exists for the hypotheses. The support that data provide for hypotheses can range from strong to weak to none, and researchers have an obligation to those who might use their research to represent accurately the strength of their findings. Finally, in evaluation research, drawing conclusions usually involves making a judgment about the adequacy and effectiveness of programs and changes that might improve conditions.

The findings of all research projects have limitations. Though some questions are answered, others always remain, and new questions may be raised by the findings. Because of this, researchers typically conclude research reports with a discussion of the weaknesses and limitations of their research, including suggestions for future research that follow from the findings that have been presented. In addition, research sometimes discovers the unexpected: things that do not relate directly to any specific hypothesis or things that are completely unanticipated. When drawing conclusions, the researchers should make note of the implications of any such findings that are of sufficient importance to warrant mention. When complete, the conclusions should clearly indicate what has been learned by conducting the research and the impact of this new knowledge. Drawing conclusions is also discussed in Part III and Part IV.

Public Dissemination of Results

Research findings are of little value if they remain the private property of the researchers who produce them. A crucial stage of social research is the public dissemination of the findings by publishing them in a book or professional journal or presenting them to a professional organization. This disseminates the newly created knowledge to those who can put that knowledge to use or who can build on it in future research. In fact, public dissemination of knowledge is a major mechanism for scientific advancement. Public dissemination also makes it possible for others to reanalyze or replicate the research and to confirm the findings or identify cases of error-filled, biased, or fraudulent research. Part V explores many of these issues.

The steps in research described here are not always followed in the exact order just presented, neither does one step necessarily end before the next begins. While research would always begin with some degree of problem formulation, after that the steps may overlap and the order may in some cases be rearranged. In some qualitative research, for example, the results of data analysis may lead to changes in problem formulation, research design development, or the kind of data collected (see Chapter 15). However, in all social research, these six steps can be readily identified.

REVIEW AND CRITICAL THINKING

This book provides an introduction to research in the social sciences, but it is much more than that. It can also provide a new window on your world—a new perspective from which to assess the myriad of information that you confront in the modern world. The logic of social research presented can be adapted for use in thinking about your daily life and everyday world. The term **critical thinking** refers to a mode of assessment or a reflective process that helps you assess information and decide on courses of action (Browne & Keeley, 1997; Paul, 1993). While lessons for critical thinking can be gained from many different realms of life, I mentioned earlier in this chapter that the norms of the scientific community promote critical thinking, and the process of scientific research has lessons about this that are directly transferable to everyday life. Near the end of each chapter in this book, a special section titled Review and Critical Thinking is devoted to exploring the issue of how the content of the chapter has lessons for critical thinking.

One of the issues addressed in Chapter 1 is that the "truth" is not so simple a matter as people often believe, and the procedures that scientists use for detecting the truth are, of course, the topic of this book. The truth can be complex and elusive. In fact, it may not be very useful to talk of "truth" without also talking about the procedures that were used to decide what is true or accurate. It is impossible to assess a claim to truth without also assessing the procedures used to arrive at that claim. By learning how scientists assess information, you not only learn about the scientific process but also some lessons about how to engage in critical thinking in your everyday life.

In terms of critical thinking skills, then, this chapter suggests that, when confronted with information or assertions about truth, you consider the following guidelines:

1. What sources of knowledge serve as the basis for a claim or assertion? What are the strengths or weaknesses of those sources of knowledge?

2. Is there any group or group process that serves to protect the search for knowledge? Is there anything equivalent to the scientific community or the peer review process that can enhance your confidence in the claims to truth?

3. Is the information produced by a systematic inquiry that involves the kinds of goals and steps that are an inherent part of conducting scientific research?

Computers and the Internet

The computer, probably more than any other single invention, stands as a symbol of technological progress at the beginning of the twenty-first century. Computers have come to dominate practically all realms of human life, and this is clearly the case in social science research. This includes the Internet and World Wide Web, which have increasingly become resources for researchers.

The traditional use of computers in social research has been to assist in analyzing data. In later chapters, computer software that can run statistical tests, assist in interviewing respondents, help select a sample to use in data collection, and accomplish a host of other activities related directly to research fundamentals are described. Computers today have many uses beyond the traditional. They assist in the literature review process through computerized database searches available at most libraries (see Appendix A); they help in report preparation as word processors; and graphic display programs have become essential in producing charts and figures to communicate the findings of research. The combined technology of computers and the Internet make it possible to store, transmit, and make very large datasets available even to researchers with only modest resources. Helpful information about the use of computers and the Internet in social science research can be found in the journal *Social Science Computer Review.*

A Computers and the Internet section is included in every chapter. The purpose of this is not to teach you *how* to use computer software because this depends on the computer software and hardware you have available at your school, home, or organization; instead, its purpose is to educate you regarding the range of computer uses in the field of social research today and to stimulate your thinking about ways of using this technology. This will enable you, when confronted with a research problem, to search for the computer tools that can assist in solving the problem at hand.

In every Computers and the Internet section, exercises are presented that will teach you how to use the Internet to enhance your understanding of social science research methods. To begin, I suggest that you explore some of the search engines, such as Altavista or Yahoo. Try entering search terms such as "sociology," "psychology," "research," "social science," "criminal justice," "education," or "health care" in the search engine. Or use some combination of these terms. Print a copy of interesting sites that you locate, or record the Internet addresses and share these with other students in the class. The class may wish to establish a discussion list on the university computer system to share information on research topics as you discover them.

These Internet exercises mention some Internet addresses, but these addresses tend to change periodically and may have changed by the time you read this book. So, many exercises will focus on how you can direct a search yourself. As an adjunct to these Internet exercises, I will place useful and interesting Internet addresses (URLs) at a Web page that I maintain for users of this book. That Web page can be accessed at: **http://members.aol.com/ tsulli3206/research.htm.** I will be updating the addresses at this Web page continuously, so make good use of it and enjoy your exploration of social research!

Main Points

1. There are many ways of gaining knowledge about the world: tradition, experience, authority, common sense, and journalism. Each has its uses, but each also has shortcomings.

2. Science is considered superior to these sources of knowledge when investigating things about the observable world. Science is superior because of its five characteristics: It is empirical, uses systematic observation, searches for causes, is provisional in nature, and strives toward objectivity.

3. The scientific community is like any other culture or society in that its members share a culture and social life that connects them to one another and gives meaning and direction to their actions. The scientific community is guided by these principles: organized skepticism, universalism, disinterestedness, communalism, and honesty.

4. The scientific community uses the peer review process to evaluate the work of scientists to ensure that the highest scientific standards are maintained and that the guidelines of the scientific community are followed.

5. Beginning with observations, scientists strive for a more abstract level of knowledge by developing theories that explain why the events observed have occurred. Theories are abstract explanations, not speculations or statements of preferences as some people believe.

6. In conducting research, scientists have to make many choices. They have to decide whether to conduct descriptive, predictive, explanatory, or evaluation research and whether their research will be basic or applied in nature. Scientists also have to decide whether quantitative or qualitative research is most appropriate to their research questions and what ethical considerations must be taken into account.

7. Science has its limitations. In particular, science cannot resolve moral issues, the questions of "should," and it cannot help us understand worlds that are beyond observation by our physical senses.

8. The research process can be divided into six identifiable stages: problem formulation, research design development, data collection, data analysis, drawing conclusions, and public dissemination of the results.

Important Terms for Review

applied research
authority
basic research
common sense
critical thinking
descriptive research
evaluation research
experiential knowledge
explanatory research
pilot study
predictive research
pretest
qualitative research
quantitative research
research design
science
scientific community
social research
theory
traditional knowledge

For Further Reading

Agnew, Neil M., & Pyke, Sandra W. (1994). *The science game: An introduction to research in the behavioral sciences* (6th ed.). Englewood Cliffs, NJ: Prentice-Hall. Presented as a "consumers guide" to the products of science, *The Science Game* is much more than that. It is a compact and highly readable introduction to the research craft.

Frost, Peter J., & Stablein, Ralph E. (Eds.). (1992). *Doing exemplary research*. Newbury Park, CA: Sage. This is a collection of the reflections of several researchers about their research experiences. Among other things, it illustrates a wide array of research topics and approaches.

Hoover, Kenneth R. (1992). *The elements of social scientific thinking* (5th ed.). New York: St. Martin's. This book is an initiation to social science research intended for those who use the results of research or those just beginning as researchers. Through several editions it has remained to-the-point and up-to-date.

Kemeny, John G. (1959). *A philosopher looks at science*. Princeton, NJ: Van Nostrand. A discussion of the philosophical underpinnings of science and scientific research. To truly understand research, you need some knowledge of the basic logic of science, a field generally referred to as the "philosophy of science."

Rossman, Gretchen B., & Rallis, Sharon F. (1998). *Learning in the field: An introduction to qualitative research*. Thousand Oaks, CA: Sage. This text provides a comprehensive overview of qualitative research, discussing both when it is preferred over quantitative research and how it is conducted.

Sagan, Carl. (1996). *The demon-haunted world*. New York: Random House. In this controversial book, the noted astronomer and author makes the case that science is a kind of savior of humankind: It can save us from believing in the unproven. In making his argument for science, the author pillories many other ways of trying to understand the world, such as astrology, turning to psychics, mysticism, faith healing, and even conventional religion.

Chapter 2

CONCEPTS, HYPOTHESES, AND THEORIES IN RESEARCH

Chapter 1 explored different ways of gaining knowledge about the world, concluding that science is the preferred method for gaining such knowledge. Drawing such a conclusion was, in a sense, the easy part; Chapter 2 explores the more difficult and challenging task of explaining exactly what that scientific method is and how it operates. The world, after all, is a complex place, and understanding it accurately is no minor feat. I sometimes find it useful to turn to literature to make a point. In this case, a segment from Lewis Carroll's *Through the Looking-Glass* helps drive home the point about the challenge of understanding our surroundings. After dashing through the looking-glass house in order to view its garden, Alice says:

> I should see the garden far better . . . if I could get to the top of that hill: and here's a path that leads straight to it—at least, no, it doesn't do that . . . but I suppose it will at last. But how curiously it twists! It's more like a corkscrew than a path! Well this turn goes to the hill, I suppose—no it doesn't! This goes straight back to the house! Well then, I'll try it the other way (Carroll, 1946, pp. 21–22).

Understanding the world—especially human behavior—sometimes bears a striking resemblance to Alice's convoluted and frustrating journey. People, without any apparent rhyme or reason, do what we least expect. A prisoner on parole who appeared to be "making it on the outside" suddenly commits another offense and lands back in jail; a marriage of 25 years that seemed to be quite solid suddenly ends in divorce; a respected and seemingly successful business executive commits suicide. When we see the unusual or seemingly unpredictable, the path to understanding seems to mirror Alice's corkscrew.

Science, however, provides a method for mapping and understanding that corkscrew. In this chapter, the basic logic underlying scientific research is introduced; it analyzes the im-portance of theories in scientific research and describes how theories are constructed and verified. As a part of this, the role of concepts and hypotheses is discussed, showing how hypotheses serve to link theory and research. Finally, an understanding of some of the overarching theories and paradigms in the social sciences is important because a researcher's stance on the issues raised by these theories can influence how research is conducted.

CONCEPTS, HYPOTHESES, AND THEORIES

Theories are a central part of the process of science, so it is essential to begin an exploration of the scientific method with a discussion of theories. Yet, "theory" is a word that is often misunderstood. To many people, theories are associated with the abstract, the impractical, or the unreal. We have all heard the refrain "It's only a theory" or "That's just your theory." Such phrases are often used in the context of deflating an argument that someone has put forth. Actually, these comments, though often intended in a disparaging sense, convey some truth regarding theories. In particular, they point out that theories are, in some cases, unverified (but verifiable) assertions about reality. Yet, other theories have been thoroughly tested and either substantiated or refuted by empirical evidence. So it is not the level of substantiation that is the defining characteristic of a theory. Rather, theories play a critical role in our understanding of reality, and in fact, people commonly use theories in their daily lives without recognizing it. So, to understand science, we first need to understand what theories are.

What Is a Theory?

A *theory* is a set of interrelated, abstract propositions or statements that offers an explanation of some phenomenon (Skidmore, 1979). There are three key elements in this definition that are important to understanding theories. First, theories are made up of *propositions*, which are statements about the relationships between some elements in the theory. Second, theories are *abstract systems*, meaning they link general and abstract propositions to particular, testable events or phenomena. For a set of propositions to constitute a theory, deducing further relationships between the elements must be possible. The third key aspect of theories is that they provide *explanations* for the phenomena they address. Indeed, the ultimate purpose of a theory is to explain *why* something has occurred.

Science accumulates knowledge through the development and verification of theories. To begin to understand this, we can review the various elements that come together to make up a theory. Science begins, of course, with **observations**, or **facts**: this means that we note or record that some thing or event has happened or is true. In a sense, observations and facts have to do with reality, with what is true: we could observe that today is Tuesday, that the Portuguese writer José Saramago won the Nobel Prize in Literature in 1998, and that a survey respondent circled a number on a survey form indicating that her sex is female. While observations are central to the scientific enterprise, there are two important reasons to be cautious with facts. One reason for care is that what people think are facts are often inferences that go beyond the facts. For example, one might be tempted to say that it is a fact that José Saramago is a great writer. However, what has been observed is that he won the Nobel Prize, which is true. To say that he is a great writer is to make a judgment about the meaning of receiving a Nobel Prize—it is an inference that is made based on the facts. Although science deals both with facts and with inferences, it is critically important to keep the two separate. A second reason for care about facts is that facts themselves are often not very enlightening. It is the meaning of the facts in a particular context, or the relationships between disparate facts, that most often lead to real enlightenment in the sense of understanding how the world operates.

Concepts and Variables

Sociologist Robert Alford calls concepts "the building blocks of theories" (1998, p. 33). The term **concept** refers to a mental construct or image developed to symbolize ideas, persons, things, events, or processes. Human beings conceptualize all the time; it is in our very nature. *Conceptualize* means to form an idea of something or to conceive of it mentally. When you get up in the morning, you conceptualize what your day will be like: where you will go, who you will see, what you will do. You form a mental image of the day. Concepts are also at the core of all scientific work as scientists conceptualize about what they study. Concepts are different from the facts just discussed in that facts refer to things or events in reality whereas concepts are mental images that may or may not be true.

Each concept refers to a group of elements that are presumed to possess some characteristics in common. Examples of concepts used in the social sciences include: social bonds, gender, deviance, crime, poverty, and social class. We could define each of these concepts by specifying what it is that all elements identified by the particular concept share in common. For example, the concept of "gender" refers to people's status in society based on their sexual characteristics. The woman mentioned above who marked "female" on the survey form is one instance of this general category. As another example, the control theory of deviant behavior includes, among other things, the concepts of "social bonds" and "deviant behavior" (Agnew, 1991; Hirschi, 1969). "Social bonds" can be defined as the ties or linkages that connect people to social groups; "deviance" can be defined as behavior or characteristics that violate the norms of a group to the extent that social disapproval results. These concepts are some of the elements of control theory; they are the building blocks

FIGURE 2.1 Elements of Theories and the Process of Research

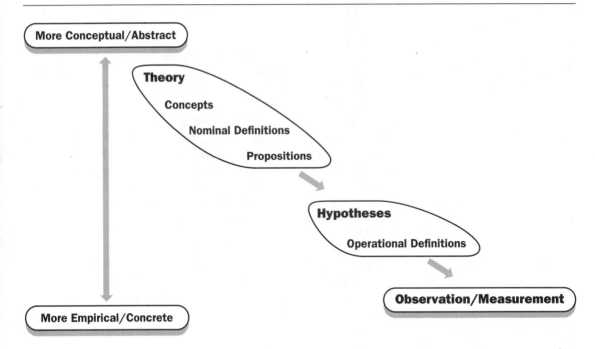

that are interrelated in propositions that form the explanatory statements of the theory (see Figure 2.1).

Some concepts have only a single category or value. The concept "universe," for example, refers to the totality of all things that exist. There is, by definition, only one universe (ignoring the science fiction device of alternate time lines, each with their own universe), and it doesn't make sense to talk about degrees or amounts of universe. It either exists or it doesn't. Most scientific concepts, on the other hand, contain a number of categories or values, and these concepts are called **variables** because their value can vary. Age and sex are two important variables in social science research. Age can take on a whole range of values, such as 5 years old or 52 years old, whereas sex can take on only two values: female or male. The concept of social bonds might include the values of "strong bonds," "weak bonds," or "no social bonds at all." Other illustrations of con-

cepts and their associated categories or values are presented in Table 2.1. It is important to develop an ability to identify variables and their categories because these are central to the scientific process and are widely used in the language of science. (Categories are also called "values" because they sometimes do have a numerical value, as with a variable like age.) Because social scientists almost always deal with concepts that are, at least potentially, variables, the two terms are often used interchangeably by social scientists.

Concepts are similar in function to the words we use in everyday communication. The word *automobile*, for example, is the agreed-on symbol for a particular object that is used as a mode of transportation. The symbol or word is not the object itself but rather something that stands for or represents that object. Scientific concepts, like words in everyday language, are also symbols that can refer to an extremely broad range of referents. They may refer to

TABLE 2.1 Examples of Some Concepts and the Categories or Values that Might Be Associated with Them

Concepts/Variables	Categories/Values
sex	male, female
age	1, 2, 3, 4, 5 . . .
social bonds	none, weak, moderate, strong
rates of crime	low, moderate, high
types of crime	shoplifting, illegal drug use . . .
social class	lower, middle, upper
level of education	less than high school, high school graduate, some college, college graduate, post-baccalaureate

something fairly concrete, such as sex, or something highly abstract, such as social bonds.

Despite the similarities between scientific concepts and ordinary words, some differences between them are critical to the scientific endeavor. In particular, concepts used in scientific research must be defined very carefully. With the words we use for everyday communication, we can get along quite well with only a general idea of how these words are defined. In fact, whether most people could give a dictionary-perfect definition of even the most commonly used words is doubtful. Such imprecision in the use of scientific concepts, however, is inadequate. Scientists, widely scattered both geographically and temporally, carry on research that tests various aspects of theories. For these disconnected research projects to produce information of maximum utility, all the bits of knowledge need to be integrated into an explanatory scheme—a theory. This accumulation of knowledge is difficult—in fact, becomes practically impossible—if isolated scientists use differing definitions to refer to the same concepts. For example, if two independent researchers produce some findings on the relationship between "social class" and "crime," the reader needs to be assured that they both mean the same thing by "social class" and by "crime." If, by "social class," one researcher is referring to people's level of education while the other is referring to their income level, the two are then talking about two quite distinct, although related, phenomena. If their concepts have differing meanings, then their research results are not directly comparable.

Scientific analysis involves a number of different types of definitions of concepts, with each type functioning at a different level of analysis and serving a different purpose. At the theoretical or abstract level, concepts are given **nominal definitions**: verbal definitions in which scientists agree that one set of words or symbols will be used to represent another set of words or symbols (see Figure 2.1). Nominal definitions are directly analogous to the dictionary definitions of ordinary words in which a phrase is designed to give meaning to the word or concept being defined. For example, the definitions of "social bonds" and "deviance" given above are nominal definitions of those concepts.

Theories often consist of concepts at varying levels of abstraction. Less abstract concepts provide a more detailed and precise specification of a phenomenon in comparison to the more abstract concepts. I continue to use control theory to illustrate the ideas in this section of the chapter because it conveys the issues relating to concepts and theories very well. In control theory, the concept of social bonds is very general and abstract, but it has been made more precise through the less abstract concepts of "attachment" and "involvement." These concepts refer to more specific ways in which people can be linked to groups. "Attachment" refers to having connections with specific other people, such as parents, teachers, or clergy; "involvement" refers to taking part in various group activities, such as at school, work, or church. These are nominal definitions of these more concrete concepts. As shown in Figure 2.2, the general concept of deviance can also

FIGURE 2.2   Some Concepts and Propositions of the Social Control Theory of Deviant Behavior

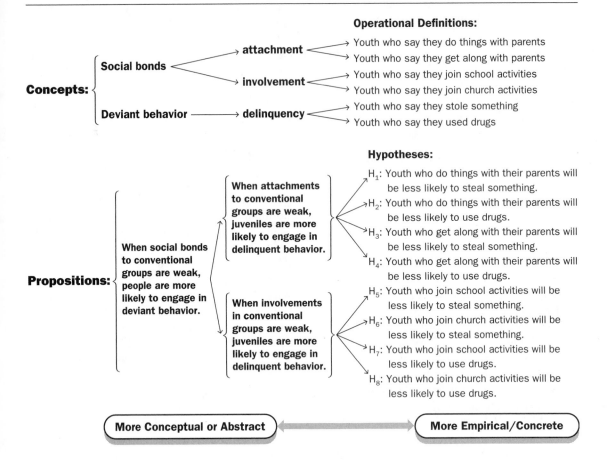

be more specifically identified by the concept of "delinquency."

Propositions and Empirical Generalizations

The concepts that form a theory are cast into propositions. **Propositions** are statements linking concepts or variables, often statements about the relationship between two or more variables. Depending on the concepts involved, propositions can be fairly concrete or quite abstract, and often include information about

why a specified relationship occurs between two variables. Using the example of control theory (see Figure 2.2), a general proposition would be: "When social bonds to conventional groups are weak, people are more likely to engage in deviant behavior" This proposition links the concepts of "social bonds" and "deviance" by stating how they are related. (This proposition also introduces a third concept of "conventional group," or at least it is a specification of the concept of social bonds.) Using the less abstract concepts in control theory, this general proposition can be transformed into two less abstract propositions: "When attachments to conventional groups are weak, juveniles are more likely to engage in delinquent behavior"; and "When involvements in conventional groups are weak, juveniles are more likely to engage in delinquent behavior."

If sufficiently abstract, a proposition might be called an **empirical generalization**: a statement about how two variables will be related in most or all situations of a certain type. Many social scientists consider the proposition relating "social bonds" to "deviant behavior" an empirical generalization because it can be applied to so many different kinds of situations. It has been used to explain, for example, many different kinds of crime and juvenile delinquency as well as suicide and other forms of deviant behavior (Goode, 1994). By definition, such empirical generalizations are very general and abstract in nature.

Operational Definitions

In order to move from the abstract level of theory to the concrete level of research and observation, scientists give concepts *operational definitions* (see Figure 2.1). Unlike the more abstract nominal definitions, **operational definitions** indicate the precise procedures or operations to be followed when measuring a concept. For example, to measure "social class," we could ask people the question listed in Table 2.2. This question, along with a specification of which income levels would place a person in a particular social class, would then be our operational definition of "social class." In this case, the operational definition includes the precise wording of the question, the use of income categories instead of exact income, and the procedure of having people identify an alphabetic character rather than an income level. Anyone using these procedures to measure "social class" will be using the same operational definition.

In the case of control theory, the operational definitions of the concepts often consist

TABLE 2.2   A Question that Can Serve as an Operational Definition of the Concept of Social Class

Did you earn any income from (OCCUPATION DESCRIBED IN PREVIOUS QUESTION) in 1994?

   Yes .... (ASK A)

   No ....

A. IF YES: In which of these groups did your earnings from (OCCUPATION IN PREVIOUS QUESTION) for last year–1994—fall? That is, before taxes or other deductions. Just tell me the letter.

A. Under $1,000

B. $ 1,000 to 2,999

C. $ 3,000 to 3,999

D. $ 4,000 to 4,999

E. $ 5,000 to 5,999

F. $ 6,000 to 6,999

G. $ 7,000 to 7,999

H. $ 8,000 to 9,999

I. $10,000 to 14,999

J. $15,000 to 19,999

K. $20,000 to 24,999

L. $25,000 or over

M. Don't know

Source   *General Social Surveys, 1972–1994*. Produced by the National Opinion Research Center, University of Chicago; Distributed by The Roper Center for Public Opinion Research, University of Connecticut. ⟨**http://www. icpsr.umich.edu/GSS/codebook/ rincome.htm**⟩ October 20, 1998.

of questions that people are asked in a survey (see Figure 2.2). The variable *of attachment to parents*, for example, can be operationally defined as people's agreement with the statements that they "often do things with their parents" or the statement that they "get along with their parents." Involvement can likewise be operationally defined as people's responses to questions about their activities at school or church, and delinquency can be measured by questions that ask whether they have stolen something, used drugs, been involved in a gang fight, or other delinquent acts. In these cases, the operational definitions of the concepts are the precise survey questions or statements that people respond to.

The process of moving from nominal to operational definitions can be complex because concepts are often very general and abstract, and controversy often arises over exactly what they refer to. Additionally, more than one operational definition of a concept can normally be identified. In the case of control theory, for example, there are actually a large number of questions or statements that could be used to measure "attachment," "involvement," and "delinquency." Each separate operational definition of a concept is called an **indicator** of that concept. You can now begin to see that operationalizing concepts can be difficult, complex, and sometimes controversial. This process of moving from the nominal to the operational level is called *measurement* and is treated extensively in Chapter 5 and Chapter 6.

Developing Hypotheses

A common strategy in scientific investigations is to move from a general theory to a specific, researchable problem. A part of this strategy is to develop **hypotheses**, which are testable statements of presumed relationships between two or more concepts. Hypotheses are one type of proposition—a sufficiently concrete statement that can be tested. Hypotheses state what we expect to find rather than what has already been determined to exist. A major purpose of developing hypotheses in research is to test the accuracy of a theory. The concepts and propositions of which theories are composed are usually too broad and too abstract to be tested directly. Such concepts as "social bonds" or "deviant behavior," for example, must be specified empirically through operational definitions before they are amenable to testing.

Hypotheses are stated in terms of how changes in one variable will be associated with changes in another variable. In the case of control theory, one hypothesis might be stated as "Juveniles who do things with their parents will be less likely to use drugs." You can see that this hypothesis is fairly concrete, specifying relationships with "parents" and "using drugs". Useful guidelines to keep in mind for developing hypotheses include the following:

1. *Hypotheses are linked to more abstract theories.* Although generating hypotheses without deriving them from theories is possible, hypotheses are at some point linked to theories because the theories provide explanations for why things occur.

2. *It is important that the independent and dependent variables in hypotheses be clearly specified.* An **independent variable** is the presumed active or causal variable. It is a variable that brings about a change in another variable. The variable that is changed is the **dependent variable**; it is the passive variable. Its value is, in a sense, dependent upon the value of the independent variable. A hypothesis would therefore typically take the following form: Independent variable X has a specified effect on a dependent variable Y. In the hypothesis from control theory, the independent variable *attachments to parents* is linked with the dependent variable *using drugs*.

3. *It is important that the precise nature and direction of the relationship between variables be specified in the hypothesis.* Students are sometimes tempted to state hypotheses such as: "Attachment to parents will have an effect on using drugs." While this statement says that there is a relationship, it doesn't say *exactly what the nature or direction of the relationship is.* A proper hypothesis, as above, would state how changes in one variable will be associated with particular changes in the other:

"As attachment to parents decreases, drug use increases."

4. *Hypotheses should be so stated that they can be verified or refuted.* Hypotheses, after all, are statements about which we can gather empirical evidence to determine whether they are correct or false. A common pitfall is to make statements that involve judgments or values rather than issues of empirical observation. For example, we might hypothesize that juveniles should be attached to their parents in order to reduce the incidence of delinquency. On the surface, this statement might appear to be a hypothesis because it relates *attachment to parents* and *delinquency* and specifies the nature of the relationship. But note that, as stated, it is not a testable hypothesis. The problem is the evaluative phrase "should be." What should or should not be has no place in hypotheses. However, the statement can be modified so that it qualifies as a testable hypothesis: "Increased levels of attachment tend to reduce delinquency." The hypothesis is now making an empirical assertion that can be checked against fact.

5. *All concepts and comparisons in hypotheses must be clearly stated.* For example, consider the hypothesis "Southern Baptists have superior moral standards." The concept of "moral standards" is so abstract and vague that it is impossible to know what it means; it would have to be clearly specified in terms of what is considered a moral standard. Also, to say that someone's standards are superior requires a referent for comparison: superior to whom or to what? It could mean higher than some other religious group, or it could mean above some chosen, absolute standard.

Developing hypotheses from theories is a creative process that depends in part on the insight of the investigator. Because hypotheses link theories to particular concrete settings, the researcher's insight is often the trigger to making such connections. In addition, researchers at times combine two or more theories to develop hypotheses that neither theory alone is capable of generating.

Theory Verification

Theories, then, play an important part in research. But one point needs to be reiterated: The utility of theories must be based on their *demonstrated* effectiveness. Theories should never be allowed to become "sacred cows" whose use is based on tradition or custom. In scientific research, theories are accepted only to the extent that they have been verified. This process of **verification** is diagrammed in Figure 2.3. As that figure shows, theory verification is a long and complex process involving the derivation and testing of many hypotheses from a theory. Any theory involves many different concepts (such as "attachment" and "involvement"), and each concept can be formed into a number of different hypotheses based on different indicators, or operational definitions, of the concept. Figure 2.3 shows the eight hypotheses that were developed in Figure 2.2 based on the operational definitions of the concepts of "attachment," "involvement," and "delinquency" in social control theory. Observations are made to test these hypotheses, and a hypothetical outcome is presented in Figure 2.3. Suppose that the first four hypotheses, which link "attachment" and "delinquency," are confirmed; this outcome would provide strong verification for this aspect of social control theory. On the other hand, only two of the next four hypotheses, which link "involvement" and "delinquency," are confirmed; this provides weaker verification for that part of control theory.

Over many tests of a theory, some hypotheses will be confirmed, others will not. The more hypotheses derived from a particular theory are confirmed, the more confidence we have in the overall theory. If, after a series of tests, some hypotheses are not confirmed, we

FIGURE 2.3   Hypothetical Illustration of the Verification of Social Control Theory

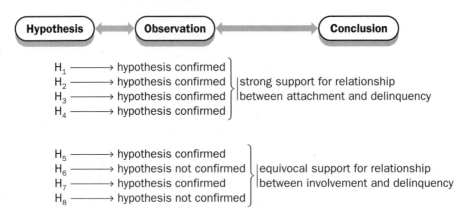

need to consider revising the theory to account for this. However, no theory stands or falls on the basis of one or a few trials; theories are tested over long periods of time by many investigations. Only with the accumulation of research outcomes can we begin to have confidence concerning the validity of a theory. With this understanding of theories and the process of theory verification, let's return to a point made earlier in this section: What characterizes a theory is not whether it is unsubstantiated or substantiated; theories exist at all levels of substantiation or verification. What does characterize a theory is that it is an abstract, explanatory set of propositions.

Theories and Common Sense

In comprehending theories and the role they play, it is helpful to realize that we all use theories in our everyday lives even though we may not call them theories or consciously be aware of using them. Notwithstanding, we base our decisions and behavior on our past experience and what we have learned from others. From these experiences, we generalize that certain physical, psychological, or social processes are operative and will continue to be important in the future with predictable consequences. This

is our "commonsense theory" about how the world operates and forms the basis for our decisions. For example, most people have certain general notions—personal theories—about what causes poverty. Some personal theories emphasize poverty as an individual problem: People are poor because of their individual characteristics, such as laziness, low intelligence, poor education, and/or lack of marketable skills. Others' theories of poverty emphasize the structural features of the economy that dictate that even in times of economic expansion some people will remain impoverished through no fault of their own.

Actually, these so-called theories are more like the scientists' hypotheses than theories in that they are relatively concrete and can be observed in the world. These concrete notions derive from more abstract ideas, or "theories," regarding human motivation and action. The former derives from an individualistic theory that posits that people's behaviors in most or all settings are a function of their personal desires, abilities, and characteristics. The latter derives from a structural theory that posits that people's behavior is constrained, if not determined, by social, organizational, cultural, or physical factors in their environment. Which of these theories people accept most closely will determine, in part, how they react to poor people and which public policy provisions concerning poverty they will support. Advocates of the individualistic theory might oppose programs to aid the poor because they believe the poor are undeserving persons suffering because of their own shortcomings. Supporters of the structural theory may view the poor as victims and tend to be more benevolent toward them.

Personal theories, like the above concerning poverty, may be extreme and misleading because they are based on casual observations, personal experience, or other information lacking the rigorous concern for accuracy found in scientific investigations. People also tend to accept their personal theories as true rather than subject them to critical evaluation and attempt to verify them. Unlike commonsense theories, theories in research are precise, detailed, and explicit. It is, however, important to recognize that a theory is always tentative in nature; that is, any theory is best viewed as a possible explanation for the phenomenon under investigation. By conducting research, scientists gather evidence that either supports or fails to support a theoretical explanation.

The Functions of Theories

Theories are used for particular purposes in research. I can identify three major functions of theories in research:

Explaining Phenomena

As we have seen, theories provide an explanation for phenomena. They not only say what will happen under certain conditions (which is what hypotheses also do, but more concretely) but also why it will happen. This provides a far more powerful understanding of human behavior. In control theory, for example, the phenomenon to be explained is deviant behavior, and the explanation is that, to avoid rejection, people conform to groups to which they have connections. The more connections people have to conventional groups, the more pressure they feel to conform to these groups and

avoid behaviors the groups consider deviant, such as delinquency. Thus control theory provides a broad and abstract explanation for the development of deviant behavior that links it in with processes of conformity and group process.

Guiding Research

Theories serve to guide and direct research. They focus attention on certain phenomena as being relevant to the issues of concern. If we were to dispense with theories altogether, as some would suggest, then what would we study? What data would be collected? Theories aid us in finding answers to these questions.

Imagine that a school district is concerned about the problem of vandalism or shoplifting among teenagers and that the staff decides to study the problem. Where do they begin? Which variables are important? As a first step, it might help to fall back on some theory related to these issues. They might, for example, use control theory, which posits the importance of attachments to groups and involvement in activities. In order to test this theory, they could determine whether shoplifting or vandalism are more common among teenagers who are less attached and involved. They are then in a position to collect data on attachments and involvements and rates of delinquency. If their theory is confirmed, then it supports the idea that social bonds affect deviant behavior.

They could have selected a different theory regarding deviant behavior. For example, some theories posit that deviance arises when juveniles learn from their peers to value deviant activities over conventional ones. Other theories suggest that deviance results from a lack of opportunities to participate in more conventional ways of obtaining money or other gratifications. I am not interested here in which of these is the more accurate theory—research will settle that issue. The point is that the theories used by researchers serve to guide their approaches and focus their attention on particular phenomena.

Integrating Multiple Observations

Theories serve to integrate and explain the many observations made in diverse settings by researchers. They tell us why something happened, and enable us to link the outcomes of numerous studies and interventions made in a variety of settings. As long as the findings of these efforts remain individual and isolated, they are not particularly valuable to science. Recall that a single observation is viewed with considerable skepticism. Single research findings may be in error, they may be passed over and forgotten, or their broader implications may be missed entirely. Theories enable us to organize these dispersed findings into a larger explanatory scheme. For example, studies of suicide over the years have found, among other things, that married people have lower rates of suicide than single people (Taylor, 1982). Suicide can be considered a form of deviant behavior, and marriage a type of social bond or group connection. Control theory thus provides a link between two disparate and seemingly unrelated sets of behaviors by suggesting that identical social processes involving social bonds and pressures to conform are at work with both "delinquency" and "suicide." Thus theories help us integrate the findings from independent research endeavors.

CONSTRUCTING SOCIAL THEORIES

Theory construction involves the development, elaboration, and change in theories as a consequence of the interchange described above between the abstract, conceptual level, and the concrete, empirical level (see Figure 2.1). Actually, the scientific process gets more complicated because scientists have to make a number of decisions about the nature and types of theories that will be constructed. One choice involves whether theory construction is initiated by beginning with the abstract (deductive theories) or the concrete (inductive theories).

Deductive Approaches

Some scientists argue that the best or preferred theories in science are approached deductively. **Deductive reasoning** involves deducing or inferring a specific conclusion from some general or abstract premises or propositions. A simple form of deductive reasoning is the syllogism, which is an argument composed of two premises and a conclusion. For example:

> *Premise: All cats have four legs.*
> *Premise: Fluffy is a cat.*
> *Conclusion: Therefore, Fluffy has four legs.*

There are important things to note about this syllogism. One is that it is a logical system: If the first two premises are true, then, logically, the conclusion must also be true. If the conclusion is not true, then there must be something wrong with one or both of the premises (such as, "Not all cats have four legs" or "Fluffy is not a cat"). The second thing to note about the syllogism is that it is deductive: It begins with a general statement ("All cats . . .") and ends with a concrete statement about a specific thing or event ("Fluffy has . . .").

Let's look at a more sociological example of this kind of deductive logic. Sociologist Emile Durkheim in the late-nineteenth century conducted an analysis of the social factors that produce suicide. A few decades ago, George Homans (1964) cast Durkheim's theory into a clearly deductive format; part of Durkheim's theory could be cast thusly:

> *Proposition: Suicide increases as individualism in a community increases.*
> *Proposition: Individualism in a community increases as the rate of Protestantism increases.*
> *Conclusion: Suicide increases as the rate of Protestantism increases.*

When we talk about a theory rather than a logical syllogism, we change the word *premise*

FIGURE 2.4   Induction and Deduction in the Process of Research

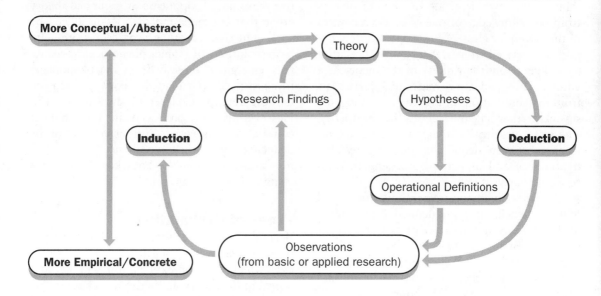

to *proposition* because it is stating a relationship between two variables, such as *suicide* and *individualism*. The logic of the theory is identical with the logic of the syllogism. In both cases, the conclusion is a testable hypothesis. If the first two propositions are true, then (other things being equal) the conclusion should be true. If the conclusion proves false, then the theory may have to be revised to account for that. Also note that the conclusion relates two concepts—"suicide" and "Protestantism"—that had not been related previously in the theory.

In the case of theories as deductive systems, the hypotheses are the conclusions that are derived logically from the propositions that make up the theory. If the propositions are correct, then hypotheses derived logically from them should also be correct. If the hypotheses prove to be false, then the researcher must consider why that occurred. It may be that some of the propositions in the theory are in some way inaccurate and need to be modified. Or it could be that the hypotheses were not derived logically. In either case, the

point is that we could not advance until we can figure out why the hypotheses turned out to be false.

In Figure 2.4, deduction involves beginning with the level of theory and then moving to that of hypotheses, operational definitions, and observations. If the hypothesis derived from a theory proves to be true, it gives us some confidence that the propositions of the theory are true. If we test the hypothesis and determine it to be false, then we need to reassess the validity of the propositions (and the theory from which they were derived).

Theories have this deductive power because they are abstract: The broader and more abstract the propositions and their related concepts are, the more numerous are the specific relationships that can be deduced from them.

Inductive Approaches

In the preceding discussion, I talked about using the conclusions to say something about the premises from which they are derived; this process actually involves *inductive reasoning*.

Inductive reasoning involves inferring something about a whole group or class of objects from our knowledge of one or a few members of that group or class. We test one or a few hypotheses derived from a theory and then infer something about the validity of the theory as a whole. Thus inductive reasoning carries us from the observations in Figure 2.4 to some assessment regarding the validity of the theory. The logic of scientific analysis involves an interplay between deduction (deriving testable hypotheses) and induction (assessing theories based on observations).

At times, inductive research is conducted without benefit of prior deductive reasoning. This can occur in descriptive or exploratory research when no theory exists from which to deduce hypotheses or when our knowledge of a topic is sufficiently scanty that we are unsure about which variables are important. In addition, some proponents of inductive approaches to theory development, such as grounded theory, argue that deducing hypotheses from existing theories can sometimes be limiting, especially in the early stages of theory development when the theory may not include some relevant variables (Strauss & Corbin, 1994). If the variables are not included in the theory, they cannot be part of hypotheses and thus may be ignored. In other words, strict adherence to deductive theory construction might blind researchers to some key phenomena. Instead, it is sometimes useful to begin by making observations. Those who use grounded theory, for example, often make direct observations of behavior in natural settings. Without the restrictions of a preexisting theory, they describe what happens, try to identify relevant variables, and search for explanations of what they observe. Beginning with these concrete observations, they develop more abstract concepts, propositions, and theoretical explanations that would be plausible given those observations. Inductive research of this sort can serve as a foundation for building a theory, and the theory that emerges can later serve as

a source of testable hypotheses through deductive reasoning. Proponents of grounded theory argue that concepts and theories produced by such inductive reasoning provide a more valid representation of some phenomena because they emerge so directly from the phenomena being studied. (Grounded theory is discussed in more detail in Chapter 11 and Chapter 15.) However, theory produced in this manner could also be subject to further verification by deducing and testing hypotheses. Thus induction and deduction are both key links in the chain of scientific reasoning.

Types of Explanation

Beyond deciding whether to develop deductive or inductive theories, scientists also need to decide which type of explanation will be contained in the theory. Earlier in this chapter, theories are defined as involving explanations of some phenomena. An explanation is one way of gaining knowledge of something; it tells why something happens or specifies the conditions under which something occurs. Theories can focus on a number of different types of explanations (R. Miller, 1987; Nagel, 1961).

Nomothetic Explanations

Nomothetic explanations focus on a class of events and attempt to specify the conditions that seem common to all those events. Using social control theory as an illustration, a nomothetic explanation might attempt to prove that all juveniles who shoplift have weak attachment to their parents. The focus of the explanation is on understanding the whole category of youths who shoplift. These explanations do not focus on understanding all of the causes of a phenomenon, such as shoplifting. In fact, control theory would recognize that a complex behavior such as shoplifting probably has many causes other than weak social bonds and that other theories would be necessary to locate and identify those factors. For nomothetic explanations, knowledge results from an understanding of a particular cause in relation to a class of events.

Nomothetic explanations are an attempt at developing knowledge that can be generalized beyond a single study or set of circumstances. In a sense, it is designed to produce the conclusion that weak attachment to parents, in all cases, increases the likelihood that shoplifting will result. This doesn't mean that every person who experiences weak attachment to parents will shoplift. It does, however, mean that those people have a higher probability of engaging in shoplifting. Or, put another way, a randomly selected group of teens with weak parental attachments will have a higher rate of shoplifting than a randomly selected group with strong attachment to parents. The explanation or knowledge that we gain is probabilistic in nature: it tells us something about the probability of events occurring. The knowledge gained is about the aggregate—or the whole group—rather than about specific individuals in that group.

Once you understand what nomothetic explanations consist of, you can begin to see their weaknesses. One weakness is that you can't say for sure what will happen in any particular case or to any particular person. You can't say whether Joe Smith, who has experienced weak parental attachment, will become a shoplifter. A second weakness is that you can't make any claims to knowing the totality of causes that produced some event or phenomenon. So the knowledge, while valuable, is incomplete. There may be, for example, some key factors that must occur in combination with weak parental attachments in order to produce shoplifting.

Idiographic Explanations

Idiographic explanations focus on a single person, event, or situation and attempt to specify all of the conditions that helped produce it. An idiographic explanation of shoplifting might focus on one particular juvenile who shoplifts

and attempt to understand the multiple factors that contributed to bringing about the shoplifting for that person. The focus of the explanation is on this particular unique individual or situation. These explanations do not focus on understanding all instances of shoplifting; in fact, they recognize that other shoplifters may be propelled by a different combination of causes. For idiographic explanations, knowledge results from a thorough understanding of the particular.

Idiographic explanations see causality in terms of a complex pattern of factors fitting together over a period of time to produce an outcome. To truly understand something, you need to comprehend that whole patterned sequence, the whole complex context in which something occurs. When particular variables are isolated for study, as in the nomothetic approach, knowledge is incomplete for two reasons. First is that some factors or variables have not been included in the investigation. Second is that the isolating approach cannot see how the combination of or interaction among the various elements plays a critical role in producing an outcome. It may be, for example, that weak parental bonds produce shoplifting only when they combine or interact with a host of other factors. In fact, it may be that the particular combination of factors that produces shoplifting in one person may be unique and not occur in other cases. It may be that each distinct case of shoplifting is produced by a unique combination of factors. In other words, the explanation or knowledge that we gain is idiosyncratic in nature. While nomothetic explanations are probabilistic in nature, idiographic explanations are deterministic in that the event being studied (such as shoplifting) did actually occur in the case being studied, and the idiographic explanation identifies the causes that determined that outcome.

As with nomothetic explanations, idiographic explanations have weaknesses. One major weakness is their limited generalizability. With such explanations, it is difficult to say

APPLIED SCENARIO **Theory-Based Evaluation Research**

As mentioned in Chapter 1, applied research often focuses on evaluating how effectively different programs or policies achieve their goals. As an example of this, I will look at evaluations of programs established to reduce the rate of repeat pregnancies among teenagers who have already had one child. Normally, these programs would be set up to provide services, such as parenting education or social support, that are designed to help these teens avoid or delay future pregnancies. To state this in this chapter's jargon: The services provided are the independent variables that presumably cause changes in pregnancy rates, which is the dependent variable.

Chapter 2 stresses the central importance of theories. What are the theories in these pregnancy programs? Often the theories that underlie social programs are based on common sense, personal experience, or intuition about what would assist teenagers in avoiding pregnancy. Sometimes, nonscientists like to downplay the importance of theories in favor of getting right down to "action" or "doing something." Yet, theories are an indispensable assist. As two applied researchers (Hendershott & Norland, 1990) put it:

> Lack of theory impedes the work of future program designers as well as program evaluators because there is no clear, theoretical basis for design. The result is that program designers and administrators may know whether a particular program was effective, but may not know why (p. 35).

Recall that one of the key functions of theories is to explain why things happen—to identify underlying, abstract social processes and social phenomena that take particular manifestations, such as in the form of services to teenagers.

When confronted with this situation in an evaluation of teenage pregnancy programs, researchers Hendershott and Norland developed a "theory-based evaluation." They began by identifying the theory that, although unstated, seemed to underlie the pregnancy programs. They concluded that the most useful theory was the life events, stress, and social support theory (Pearlin, Lieberman, Menaghan, & Mullan, 1981). This theory postulates that certain life

whether knowledge can be extended beyond the particular case or situation being studied.

Combining Explanations

Since each type of explanation has its strengths and its limitations, you may have guessed that my conclusion is going to be that neither is inherently better than the other. As alluded to in the beginning of this chapter with the excerpt from *Through the Looking-Glass*, numerous routes to gaining knowledge about the world

exist, and each of these types of explanations provides us with a valuable, though incomplete, route. In later chapters, we will see that some research methodologies, such as surveys or experiments, tend to be used to develop nomothetic explanations while other methodologies, such as field research, in-depth interviewing, and historical–comparative research are often used to develop idiographic explanations. Your task as a student is to understand the logic of each type of explanation and be aware that conclusions supported by research

events produce stress in people lives and that these life stressors can have negative outcomes. So, a stressful life event might be a teenager's first pregnancy, and the negative outcomes could be poor parenting skills and additional pregnancies. Thus far, this is no different from what the original program designers probably had in mind, although this theoretical statement is more explicit. However, Hendershott and Norland point out that the theory says more—it includes other variables and processes. The theory also postulates that some life events are more stressful than others and, particularly, that life events that represent an attack on or diminishment of a person's self are likely to be especially stressful. This suggests that a person's self-concept plays a central part in this whole process. This is not something that the original program designers took into consideration. The basic idea is that the original pregnancy confronts the teenager with the extreme difficulties of caring for an infant, which may in turn make that teenager vulnerable to a loss of self-esteem.

This recognition of the role of self-concept in the process has two important consequences: (1) self-concept must now be measured through operational definition in order to determine its impact on the process, and (2) existing program interventions can be revised or new ones implemented so that they focus more directly on positive supports for the teenagers' self-esteem. Therefore, a theory-based evaluation helps to identify important processes and phenomena that were implicit and largely ignored. The result is a more complete and sophisticated evaluation of a program.

You can practice identifying theories and the roles they play in research by doing the following:

1. Identify other theories that might help in the evaluation and understanding of teenage pregnancy programs. Specify the important propositions, hypotheses, and variables in these theories.

2. Identify other programs in your community that could be evaluated. Specify the implicit or explicit theories upon which they are based and suggest other theories that would be helpful.

using both types of explanation are more complete than when supported by only one type.

Levels of Theory

Beyond induction versus deduction and the type of explanation used, theories also differ from one another in a number of other important ways. For one, they differ in terms of what level of social reality they focus on. **Macrotheories** focus on society as a whole, on large social processes such as industrialization or urbanization, or on social institutions such as religion, politics, or the economy. These theories might focus on such broad issues as how stratification systems change as a consequence of industrialization, how urbanization has impacted on the family, or the social conditions necessary for stable democratic governments to survive. **Microtheories**, on the other hand, focus on the social interaction and social process that occurs among individuals or in small groups. These theories attend to such issues as how group size affects conformity, how teachers' treatment of children in school can affect the children's displays of gendered behavior, and/or how the labeling of adolescents can affect their delinquent behavior.

Another way in which theories differ from one another is in how abstract they are. **Formal theories** are the most abstract and deal with general social processes and mechanisms, without necessarily being tied to a specific social context. Included among formal theories might be a theory of social cohesion, a theory of group dynamics, or a theory of social power. **Substantive theories** are less abstract and focus on specific forms and contexts of social behavior. Substantive theories often involve an application of more formal theories. A formal theory of social cohesion could therefore be transformed into a substantive theory of suicide, or a formal theory of social power could be used to explain how work groups in occupational settings can induce conformity among their members.

The Applied Scenario box describes the important role that theories, concepts, and hypotheses can play in applied research.

THEORIES AND PARADIGMS IN THE SOCIAL SCIENCES

Up to this point, science has been presented as if it were a coherent, unified activity about which all scientists were in agreement. It isn't. Or, more accurately, I should say that some people believe it is a coherent and unified activity whereas others criticize that claim. Over the centuries, philosophers and scientists have debated at length over the issues of what is the nature of reality and how can people know that reality (Couvalis, 1997; R. Miller, 1987). While these have been controversial issues for scientists who study the physical world, they are even more contentious among social scientists who study human beings and their psychological and social reality. Part of the reason for this is the belief that human beings are dif-

ferent from the natural world of physical objects and events. People can emote, remember, speculate, love, hate—they can think about what is happening to them and have feelings about it. People can refuse to behave the way a scientist hypothesizes by doing the unexpected or unpredictable. Atoms, molecules, and chemical compounds do not have these elusive, human properties. This is one of the reasons why natural scientists can often make certain, nonprobabilistic predictions about what will happen: Under a certain set of conditions, all water molecules will freeze when the temperature drops below zero degrees on the centigrade scale. Social scientists have been unable, thus far, to make such statements about social reality.

Another reason why the issue of how we know the world has been controversial among social scientists is that the scientists who study social reality are also people who have personal values, goals, desires, and reactions to what they observe. These personal matters may interfere with their ability to comprehend the world accurately. Going a step further, the scientific endeavor is itself a social process; a part of the social world that social scientists attempt to understand. After all, scientific work can advance one's career, help one earn a living, move one up (or down) in the stratification system. In doing their scientific work, scientists may be influenced by a whole variety of social and psychological factors that routinely influence other human beings in their social endeavors.

What does all this mean? It means that science is a far more complicated—and in many respects, messier—enterprise than many people recognize. It also means that there are a number of competing perspectives on the issue of how society works and what implications this has for how the scientific endeavor works. In fact, historian Thomas Kuhn (1970), in a ground-breaking study of scientific work over many centuries, concluded that scientific activity is shaped by *paradigms*, which are general ways of thinking about how the world works and how we gain knowledge about the world. **Paradigms** are fundamental orientations, perspectives, or worldviews, that are often not questioned or subjected to empirical testing. People may not even be aware that their thinking about the world is shaped by an orientation or worldview. In his study of the history of science, Kuhn discovered that, while paradigms change over time, at any given moment scientific research is shaped by the paradigm that is dominant at that time. Research, which falls outside that paradigm, would be considered inappropriate, irrelevant, oddball, or just plain wrong. In a sense, the world of paradigms falls outside the scientific realm in that issues are not accepted or rejected on the basis of empirical evidence; instead, some things are considered true and others false because it is obvious that that is how things work. Evidence that supports the paradigm will be accepted and competing evidence ignored or rejected. Over time, one paradigm might lose favor and be supplanted by another paradigm—what Kuhn called a "scientific revolution." The rise and fall of paradigms is affected in part by how well the paradigm accords with the available evidence, but this process is also affected by other social forces, such as the prominence or the political and social influence of scientists who support particular paradigms. Influential scientists, for example, may resist rejecting a favored paradigm and adopting a new one long after evidence for the new paradigm has become available. The theories discussed for most of this chapter are subject to empirical test; paradigms are, in a sense, broader and more overarching than theories and are taken for granted, as in "of course that's how things work."

Historically, there has not been a single dominant paradigm in the social sciences. Instead, a number of different perspectives vie for people's attention. These perspectives address fundamental issues regarding reality: What is the nature of social reality? How do we gain knowledge about the world? How

does science work? Three general perspectives on the nature of reality and on human social behavior will be reviewed—the positivist, interpretive, and critical approaches—and what each has to say about science and research methods will be explored (Alford, 1998; Benton, 1977; Smart, 1976). This discussion will be brief and somewhat oversimplified, but will serve as an introduction. Keep in mind that these viewpoints are not necessarily mutually exclusive; people may adopt ideas from more than one of them simultaneously. Also, one could agree with some parts of a paradigm but disagree with other parts of the same paradigm. This issue is addressed early on because it is a debate that arises repeatedly as the different research methodologies are discussed. Your goal should be not to search for a resolution of the debate—great minds continue to clash over it—but to understand the different sides and what their implications are for research methodology.

Positivist Approaches

Positivism (sometimes referred to as logical empiricism) argues that the world exists independent of people's perceptions of it and that science uses objective techniques to discover what exists in the world (Blaikie, 1993; Durkheim, 1938; Halfpenny, 1982). Astronomers, for example, use telescopes to discover stars and galaxies, which exist regardless of whether or not we are aware of them. So, too, human beings can be studied in terms of behaviors that can be observed and recorded using some kind of objective technique. Recording people's gender, age, height, weight, or socioeconomic position are legitimate and objective measurement techniques—the equivalent of the physicist measuring the temperature, volume, or velocity of some liquid or solid. For the positivist, quantifying these measurements—assessing the average age of a group or looking at the percentage of a group that is male—is merely a precise way of describing and summarizing an objective reality.

Such measurement provides a solid and objective foundation for understanding human social behavior. Limiting study to observable behaviors and using objective techniques, positivists argue, is most likely to produce systematic and repeatable research results that are open to refutation by other scientists.

The natural and social world is governed by natural and social rules and regularities that give it pattern, order, and predictability. The goal of research in the natural and social sciences is to discover laws about how the world works and to express those discovered regularities in the deductive theories and propositions that have been discussed in this chapter. As scientists conduct research, they move progressively closer to the truth, which involves uncovering the laws and patterns that underlie objective reality. So science is, at least in its ideal, an objective search for the truth where human values are a hindrance, the impact of which should be limited if not altogether eliminated. Values can only interfere with the objective search for truth. One early sociologist, Emile Durkheim, believed strongly that sociologists could study the social world in much the same way that physical scientists study the physical world. Durkheim believed that there were "social facts" that social scientists could observe and then use those observations to discover the social laws that govern the social world. Once we have discovered these social laws, we will be able both to explain and predict human social behavior.

Of the three paradigms that will be reviewed, positivism is clearly the most widely held view among natural scientists and, although to a lesser degree, among social scientists. In fact, for many natural scientists, the positivist paradigm is, in their view, science—the issues raised by the other two paradigms have nothing to do with how science is conducted. Among social scientists, those who adopt the positivist stance often tend toward using certain kinds of research methodologies, such as quantitative measurement, deductive

and nomothetic explanations, experimental designs, and survey research. However, it is important not to oversimplify the link between a paradigm and the preferred research methodology because positivists at times use qualitative research, inductive or idiographic explanations, and field observations when these are appropriate to a research question. For example, many historical and contemporary anthropologists doing fieldwork among indigenous peoples write qualitative, idiographic field ethnographies despite having a positivist orientation of viewing themselves as discovering what social and cultural life is like among these peoples. However, it is also the case that many positivists among social scientists assume that rigorous quantitative measurement and experimental design are indicative of a more exacting and sophisticated scientific analysis. For these people, qualitative research is conducted early in a research endeavor and serves as the foundation for later quantitative analysis, hypothesis testing, and deductive theory development In this view, social science makes progress by moving toward quantitative, experimental research.

Despite the popularity and dominance of the positivist paradigm, it has been subjected to considerable criticism over the years. Some of this criticism has arisen out of empirical studies by social scientists of exactly how science operates (Galison & Stump, 1996; Lynch & Bogen, 1997; Shapin, 1995). What many of these researchers find is that what scientists actually do looks quite different from what the positivist paradigm says that science should look like. This has led some critics to conclude that the positivist model is an idealized conception of science rather than an accurate description of it. Based on these and other concerns, two other paradigms have emerged.

Interpretive Approaches

Interpretive approaches (also labeled interactionist or *verstehen* approaches) see social reality as having a subjective component to it and as arising out of the creation and exchange of social meanings during the process of social interaction; social science must have ways to understand that subjective reality, and, to an extent, science is a part of that process of meaning creation (Holstein & Gubrium, 1994; J.K. Smith, 1989; Wilson, 1970). Interpretivists argue that the objective, quantitative approaches of positivism miss a very important part of the human experience: the subjective and very personal meanings that people attach to themselves and what they do. Reality is seen as something emergent and in constant flux that arises out of the creation and exchange of social meanings during the process of social interaction. Rather than seeing reality as being apart from human perceptions, interpretive social science sees reality—at least social reality—as created out of human perception and the interpretation of meaning. These kinds of ideas led many nineteenth-century and early-twentieth-century theorists, such as Wilhelm Dilthey, Ernst Troeltsch, and Max Weber, to conclude that social life cannot be understood by the same method used to study the natural world (Barnes, 1948).

Weber, for example, argued that we need to look not only at what people do but also at what they think and feel about what is happening to them (Weber, 1925/1957). This "meaning" or "feeling" or "interpretive" dimension cannot be captured adequately through objective, quantitative measurement techniques. Researchers need to gain what Weber called *verstehen*, or a subjective understanding. They need to view and experience the situation from the perspective of the people themselves. To use a colloquialism, researchers need "to walk a mile in their shoes." They need to talk to the people at length and to immerse themselves in the lives of those people so they can experience the highs, lows, joys, sorrows, triumphs, and tragedies as seen from the perspective of the people being studied. Researchers need to see how individuals experience and give meaning to what is happening to

them. Interpretive research methods provide an understanding through empathy or fellow feeling, whereas positivist methods provide understanding through abstract explanation. Yet the important point is that both methods provide an understanding of the world and both could be considered a part of the scientific enterprise.

Qualitative research methods are an attempt to gain access to that personal, subjective experience; for interpretivists, quantitative research, by its very nature, misses this very important dimension of social reality. Positivists, for their part, do not necessarily deny the existence or importance of subjective experiences, but they do question whether the subjective interpretations of the *verstehen* method have any scientific validity.

According to the interpretivist approach, regularity and pattern in social life results not from objective social laws that exist apart from the human experience and are discovered by scientists. Instead, pattern and predictability arise out of mutually created systems of meaning that emerge from social interaction (Rabinow & Sullivan, 1987; Roscoe, 1995). Regularity and pattern are created and maintained by people, not imposed by external force. Proponents of interpretive approaches argue that qualitative research methods enable the researcher to approximate *verstehen*, an understanding of the subjective experiences of people. Of course, actual access to such experience is impossible; thoughts and feelings, by their very nature, are private. Even when someone tells you how he or she feels, this person has objectified that subjective experience into words and thus changed it. Researchers, however, can gain some insight into those subjective experiences by immersing themselves in the lives and daily experiences of the people they study. By experiencing the same culture, the same values, the same hopes and fears, researchers are in a better position to take on the point of view of these people. However, despite its focus on subjective experiences, such research is still empirical in the sense that it is grounded in observation. Qualitative researchers consider their qualitative observations and conclusions no less systematic or scientific than the more positivistic quantitative research techniques. While positivists would argue that subjective meaning is difficult to quantify and study objectively, interpretive researchers would argue that it is nonetheless a key part of human social reality.

Another important difference between positivists and interpretivists has to do with the role of science: While positivists argue that scientists merely discover what exists in the world, some interpretivists claim that scientists actually help to create social reality through their scientific work (Knorr, 1981). As researchers make observations, gather data, and draw conclusions, these activities contribute to the construction of patterns of meaning. Scientific principles and laws about social behavior become another aspect of reality that can influence people's behavior. Even something as simple as computing an average age for a group creates a new reality: Instead of recognizing that some people in the group are age 22, others are age 34, and still others age 43, we now say that the "average age of the group is 36.7 years." This summary statement gives the impression—creates the reality—that the group members share something in common in terms of age and that we know something very precise about the people's ages. But that sense of commonality or precision comes from the numbers created by the scientist, not from reality. Also, although the average age appears to be precise, it is actually less precise than listing all the ages of the group members.

The interpretive approach focuses more on inductive and idiographic theory construction rather than on the deductive and nomothetic: seeing the theories emerge out of people's experiences rather than viewing them as abstractions developed by scientists. Understanding and truth come from an empathetic grasp of the social meanings of a setting rather than from statistical analysis and abstract generalization to large numbers of cases. Once again,

however, the link between paradigms and research approaches is not mutually exclusive. Interpretive social scientists at times do deductive and nomothetic theory construction and, when appropriate, have even been known to use quantitative methods.

Critical Approaches

The third approach to science and research departs most radically from the conventional, positivist perspective. **Critical social science** views society as consisting of various groups that compete for scarce resources. Scientists are one of those groups, struggling to gain respect and legitimacy, to be accorded authority in society, and to be given esteem, power, and economic rewards. Critical social scientists would neither dispute the positivists' belief that reality exists independent of people's perceptions, nor the interpretivist's assumption that social reality rests on the social construction of meaning. But the critical theorist would argue that both approaches ignore another key aspect of social reality: that groups compete for scarce but valued resources and that groups use whatever advantages they may have in this struggle (Fay, 1987; Harding, 1986; Harvey, 1990). Coercion, domination, and conflict are seen as central and inevitable social mechanisms. Karl Marx (1848/1964) formalized a version of this perspective when he described the historical mechanisms that produced the struggle between the bourgeoisie and the proletariat during the early industrialization in Europe.

The critical perspective views people and groups as having selfish or self-interested motivations—they pursue goals that will bring advantages to them even though this may, to a degree, be detrimental to other individuals and groups. In a sense, then, people and groups are oppressors. But at the same time, people are oppressed: There are always subordinates, underlings, the disadvantaged in this reciprocal power relationship. A position of dominance provides some people with the freedom and resources to explore and develop, whereas the ability of the subordinated to explore and develop is limited by their exploitation. And this brings us to the role of science and research in this perspective: Science is a tool that can be used—by both dominant and subordinate groups—either to promote or resist changes in the patterns of exploitation in society. In most cases, dominant groups resist change because existing patterns tend to work to their benefit whereas subordinate groups promote social change for the very same reason. Critical theorists would strongly dispute the notion that science is an objective endeavor of the discovery of reality. Values and interests suffuse the human condition and, since scientists are human beings, their values and interests inevitably shape their research activities.

For most critical social scientists, such as Marx, scientific theories are either justifications for or criticisms of existing economic, political, and social arrangements. While positivists see theory as value neutral, critical social scientists see theory as value laden, as a promotion and justification for some set of social arrangements. In fact, scientific theories provide the understanding and interpretation that enable people to recognize what is truly occurring in the world. The theories become weapons in the struggle between groups in society. Of course, scientific research could be harnessed to the advantage of the oppressor, but critical social scientists would argue that ethical considerations demand that science be used to reduce or eliminate structures of oppression rather than to maintain or increase them. Ultimately, therefore, science is not a neutral or objective process but one that is value laden and advocates particular outcomes in the social world. Scientists who claim to be neutral are often merely ignoring or hiding the fact that their research helps to buttress structures of domination and oppression.

While this discussion has focused mostly on the differences between the critical and the positivist paradigms, critical social scientists also fault the interpretive paradigm for its shortcomings. In particular, interpretivists are criticized for giving too heavy stress to the subjective nature of social reality, to the point that any perspective is seen as good as any other. This ignores the fact that some aspects of reality are not subjective in nature; in particular, structures of oppression are objective—they exist even if people are not consciously aware of them, and their consequences can be determined objectively. Critical theorists also fault the interpretive paradigm for not taking sides in the struggle to change society, for not recognizing that the scientific process contributes to the promotion or resistance of change even when scientists claim to be objective and "above the battle."

Stance toward the Paradigms

The issues discussed in this section are still highly controversial. Based on this, you should begin to see that science and scientific research are more complicated than you might have originally believed. This does not, however, mean that anything goes or that scientists just do what they want. Each of these paradigms is the product of a long historical and intellectual tradition out of which each has developed sophisticated, rigorous, and well-thought-out standards for what is to be considered acceptable knowledge. That they sometimes disagree about what those standards should be does not mean that they have no standards.

The goal for the student should not be to attempt to resolve the disputes or to choose among the paradigms. Instead, the goal should be to understand the dimensions of the debate, to recognize how the paradigms are similar to or different from one another, and to comprehend the implications of each paradigm for the research process. In addition, the paradigms are not completely exclusionary of one another. All of the paradigms would agree with much of what is covered in this book. For example, all of the paradigms base their search for knowledge in systematic observation. Of course, they may not always agree on what makes observations systematic, but then there is not total agreement within each paradigm

EYE ON DIVERSITY Feminist Research Methods: Do Men and Women Have Different "Voices"?

Positivism assumes that "one size fits all": There is one way to know the world, and the diversity of people in the world is irrelevant to how we gain knowledge. One of the areas in which the debates over the three paradigms discussed in this chapter has probably been most intense is the area known as feminist research. Feminist researchers have argued that one size does not fit all—the diversity in the world also produces diverse ways of gaining knowledge of the world. This has led some feminist researchers to explore the interpretive and critical paradigms for a more complete and complex methodology for understanding human behavior (Olesen, 1994; Reinharz, 1992; Schiebinger, 1999). One profound question raised by feminists is whether ways of knowing are gendered: Are males and females socialized to perceive the world and acquire knowledge in fundamentally different ways? Further, is positivism merely one of those ways of knowing rather than the only way of knowing?

The basic argument is that because of differences between the sexes—due either to biology or to socialization—there is a male model of knowledge development and a female one, and that the two are, to an extent, alien to each other (Nielson, 1989). In developing some of these ideas, Carol Gilligan (1982) used the term "voice" to mean modes of thinking about the world and tried to describe the differences between male and female "voices." Women emphasize the importance of relationships and the danger of being separated from others. Women see connectedness between people and feel obliged to protect and nurture those to whom they feel connected. This leads to a concern about the needs of others, but the needs of self and others exist primarily in the context of their relationships with other people. So women's voices focus on the individual embedded in a social network. This emphasis on connectedness and relationships means that problems and people are inextricably intertwined. Neither problems nor the people they affect can be fully understood if they are separated from one another.

By contrast, men's voices speak of separation and autonomy. They emphasize independence and, to an extent, alienation in the sense that people can be abstracted from their relationships, from their context, and even from their own uniqueness. For men, the separate individual has some meaning and importance, and possibly even more value than a person encumbered by relationships and connections. In this view, these abstracted individuals can all be treated the same, ignoring the unique needs or contexts of each person. To this extent, men's voices are abstract and formal. People and their problems can be separated.

Proponents of the male–female distinction argue that these male and female voices are fundamentally contrasting ways of perceiving and developing knowledge about the world. The male voice tends to be logical, objective, and to avoid or downplay feelings while focusing on accomplishing practical goals. It tends to view the world in terms of separation: The researcher extracts data from the "subject" or "respondent" and any personal relationship (or connectedness) between them is avoided in the interests of objectivity. All contact except that necessary to collect data is avoided. A more quantitative approach is preferred, with standardized measuring instruments and procedures used on the assumption that all people can be treated alike and this will produce the most valid and objective data. A number of procedures commonly used in positivist research can be seen as efforts to strip away the context from the individual: placing them in laboratory settings, using random assignment to conditions, using aggregate responses from large-scale surveys, and isolating variables for study. All these procedures assume that the context in which a person lives is merely interfering "noise," and that more meaningful information can be obtained without it. Through quantification, the male voice minimizes the uniqueness and subjectivity of experiences by providing summary responses of the aggregate; so the fact that Jane Doe became pregnant at age 13 under a certain set of unique and meaningful circumstances gets lost when we conclude that the average teenage pregnancy in a high school occurs at age 14.7.

The female voice, on the other hand, stresses feeling, empathy, and the subjective side of life. It places emphasis on connectedness: Researcher and respondent are tied together in a relationship that influences the data that are produced. "Objectivity," in the positivist sense, is impossible because the meaning and importance of data relate to specific individuals and their relationships with one another. The researcher and subject are seen as partners in the relationship of producing research. Subjects are seen more as collaborators. The feminist researcher also disdains the notion of the objective, impartial researcher in favor of involvement with the topics of their research. In the feminist view of gaining "truth," researcher and subject work together, although in different ways, to develop understanding; truth-seeking is a mutual, cooperative effort. Research questions will be developed in the direction of emphasizing connectedness: keeping people in their social contexts and studying the complexity of the whole. In short, more qualitative approaches are preferred. People can be studied in their homes or where they work and play. Some research methods, such as the field methods discussed in Chapter 11, are better suited to emphasizing this connectedness. Feminist research also emphasizes in-depth interviews, where people have an opportunity to express the fullness and complexity of their lives in their own words, and collaborative forms of research (discussed in Chapter 17).

Obviously, it is possible for men to speak with a female voice and women with a male voice. The point is that there are fundamentally different ways of knowing the world. Feminist researchers do not reject positivism; they recognize that it is different, but not superior. In fact, some critics would argue that researchers in the social sciences, whether male or female, have generally been trained in graduate school with a male voice because that is what has dominated over the years among faculty. Yet, neither voice is inherently superior, and neither is complete or totally objective. Each is a valid way of gaining knowledge about the world. Both involve a perspective that has some limitations. The basic insight that feminist researchers have left us with is that all people, including researchers, are fundamentally gendered beings, and this shapes—whether we admit it or not—how we perceive and think about the world, how we do research, and the research results that are produced.

about that issue either. All of the paradigms would also agree that scientific work should be open and public and that scientists should follow such norms as universalism and honesty as discussed in Chapter 1. Again, however, proponents of the various paradigms might disagree over how easy it is to achieve universalism or honesty or whether these things can be completely achieved.

Another reason why the student need not adopt a preferred paradigm is that many researchers do not choose a particular perspec-

tive to follow exclusively (Alford, 1998). Many researchers find that each of the approaches offers some insights into social life and the scientific process that the others ignore. They move back and forth among the paradigms, using the best that each has to offer in understanding a particular aspect of human social life. In later chapters, disputes that arise among the paradigms over particular issues will be addressed. The Eye on Diversity box explores some issues raised by feminist researchers in regard to these three paradigms.

REVIEW AND CRITICAL THINKING

This chapter covers some of the building blocks of the scientific method: theories, propositions, hypotheses, concepts, variables, and so on. In using these building blocks, scientists try to be very careful when describing and analyzing the world in order to avoid misunderstanding. You can utilize some of these same building blocks in trying to critically analyze information in your everyday life. Consider the following when reading or hearing something.

1. Specify any theories, concepts, or hypotheses that are contained in the information. Remember that in everyday realms, these elements may be implicit in what people say and you may have to make them explicit.

2. Is there any evidence that theories have been verified? If so, how much verification exists? If not, why not?

3. What is the source of the knowledge? Are there claims to objectivity (positivism)? Is there any indication that subjectivity, values, or interests play a part in shaping the knowledge?

Computers and the Internet

I covered some fairly core and controversial issues in social research in this chapter, especially as they relate to the three paradigms discussed toward the end of the chapter. Suppose you work for a research organization and your boss wants you to create an online discussion of these paradigms. You need to locate people, organizations, or resources on the Web that espouse one position or another on these issues. One approach is to use a search engine and enter one or more of the terms used to identify the paradigms or that have some relationship to the paradigms: positivism, logical positivism, subjectivism, relativism, postmodernism, feminism, and so on. These terms alone may generate an excess number of hits, so combine some of the terms or link them with other qualifiers, such as "social science," "natural science," "research," or others. From your search, discuss how different social sciences address issues having to do with these paradigms. Do some social sciences address different issues than other social sciences? Do the same kind of analysis for the natural sciences. Also address whether fields that are nonscientific, such as the humanities, address these issues. Can you put together the online discussion your boss has requested?

Main Points

1. Theories play a key role in the scientific process. The building blocks of theories are concepts and variables. The concepts used by scientists are defined more carefully and precisely than the words used in everyday conversations.

2. Some propositions in a theory are abstract, others more concrete. A proposition that states a relationship that is true for many different situations is called an empirical generalization.

3. Concepts are given both nominal definitions and operational definitions. An operational definition of a variable is called an indicator, and the movement from nominal to operational definition is called the process of measurement.

4. Hypotheses are statements that predict relationships between two or more variables and are tested via research.

5. A key part of the scientific process is theory verification, or demonstrating through research that theories are accurate depictions of reality. Theories in research are more precise, detailed, and explicit than are personal or commonsense theories.

6. Theories perform three major functions: They explain phenomena, guide research, and integrate observations from various research efforts.

7. Theories are developed and elaborated by moving back and forth between the ab-

stract, conceptual level and the concrete, empirical level by using either deductive reasoning or inductive reasoning. Theories differ in the types of explanations they seek, some seeking nomothetic explanations and others idiographic ones. Theories also differ in terms of the level of social reality that is their focus (macrotheories versus microtheories) and their level of abstraction (formal versus substantive theories).

8. Three paradigms, or ways of understanding how we know the world, are predominant in the social sciences today: positivism, interpretivism, and the critical approach. Feminist perspectives on how we understand the world contain elements of the latter two. The goal for the student in this debate should be to understand the dimensions of the debate, to recognize how the paradigms are similar to or different from one another, and to comprehend the implications of each paradigm for the research process.

Important Terms for Review

concepts
critical social science
deductive reasoning
dependent variable
empirical generalizations
facts
formal theory
hypotheses
idiographic explanations
independent variable
indicator
inductive reasoning
interpretive approaches
macrotheories
microtheories
nominal definitions
nomothetic explanations
observations

operational definitions
paradigms
positivism
propositions
substantive theories
theory
variables
verification

For Further Reading

Glaser, Barney G., & Strauss, Anselm L. (1967). *The discovery of grounded theory*. New York: Aldine. An excellent book about the virtues and procedures of developing theoretical propositions from data. This approach emphasizes qualitative research and induction.

Hoover, Kenneth R. (1992). *The elements of social scientific thinking* (5th ed.). New York: St. Martin's. A brief and readable initiation into social science thinking and research. It is intended for those who use the results of research and those just entering the field.

Kaplan, Abraham. (1963). *The conduct of inquiry*. New York: Harper & Row. A very good discussion of the logic of scientific analysis in the behavioral sciences. It covers such topics as concepts, theories, and values.

Merton, Robert K. (1968). *Social theory and social structure* (2nd ed.). New York: Free Press. A classic statement by a sociologist of the relationship between theory and research.

Root, Michael. (1993). *Philosophy of social science: The methods, ideals, and politics of social inquiry*. Oxford, U.K., and Cambridge, MA: Blackwell. This book provides an overview of the philosophical underpinnings of social science research. The author examines the position that social science should be objective and value free but concludes that this is not possible.

Turner, Jonathan H. (Ed.). (1989). *Theory building in sociology: Assessing theoretical accumulation*. Newbury Park, CA: Sage. One of the foremost theoreticians in sociology in the United States addresses in this collection of essays a key assertion of the positivist approach: Does knowledge accumulate through the deductive approach of theory building and hypothesis testing?

Chapter 3

ETHICAL ISSUES IN SOCIAL RESEARCH

The subjects of social research are human beings and, because people have rights and feelings, special considerations apply in social research that do not confront the chemist studying molecules or the physicist investigating gravity (although these natural scientists do address many serious ethical issues). When social scientists conduct research, they often touch people's lives, sometimes changing those lives, and perhaps improving them; but social scientists can also leave people in a worsened condition. So, we need to be very careful about whether and how we conduct research. These ethical issues can be addressed only by referring to some cultural, professional, or personal values that help us decide what is right and proper behavior. There are five expressions of values with which most people in the United States would probably agree:

1. The sanctity and worth of each human being should be respected.

2. People have a right to privacy.

3. People have a right to self-determination.

4. Fairness and equity in social relations should be maintained.

5. One should not do anything to intentionally harm another person.

Ethical dilemmas arise as researchers attempt to conduct research while also trying to uphold these values. Yet, as the saying goes, "The devil is in the details." The difficulty arises in trying to apply these general principles to particular situations.

These are complicated issues. I do not presume that this statement of values is the final word or that their application to particular cases is easy or straightforward. They do serve as a beginning, however, to the discussion of ethical issues in social research. **Ethics** is the study of what is proper and improper behavior, of moral duty and obligation (Reese & Fremouw, 1984). Moral principles can be grounded in philosophy, theology, or both. For social researchers, ethics involves the responsibilities that researchers bear toward those who participate in research, those who sponsor research, and those who are potential beneficiaries of research. It covers many specific issues. For example, is it ever permissible to harm people during the course of a research project? Should people who participate in a research project ever be deceived? Is it appropriate to suppress research findings that cast a sponsor's program in a negative light? Should researchers report to the police crimes they uncover while conducting their research? The ethics of a given action depend on the standards used to assess the action, and those standards are grounded in human values. Because of this, there are few simple or final answers to ethical questions, and there are no scientific "tests" that can show us whether our actions are ethical. In fact, debate continues among scientists about ethical issues because such issues involve matters of judgment and assessment.

My purpose in Chapter 3 is to identify the basic ethical issues in social research and to suggest some strategies for assuring that ethical considerations are attended to in the conduct and use of research. The chapter begins by reviewing ethical issues that involve those who participate as respondents or subjects in research and then looks at efforts to codify ethical standards in recent decades. Finally, the chapter reviews ethical issues involving the sponsors of research and the general community before ending with a discussion of the codes of ethics of professional research organizations.

ETHICAL ISSUES INVOLVING RESEARCH PARTICIPANTS

Some ethical issues in social science research have to do with how we treat the people we are observing—the subjects of our research. Six such issues are examined: informed consent, deception, the right to privacy (anonymity and confidentiality), harm or distress, protecting vulnerable groups, and withholding treatment for research purposes.

Informed Consent

Informed consent refers to telling potential research participants about all aspects of the research that might reasonably influence their decision to participate. Very often people are asked to sign a consent form, which describes the elements of the research that might influence a person's decision to participate (see Figure 3.1). General agreement exists today on the desirability of informed consent in behavioral science research, primarily because, in the United States, cultural values place great emphasis on freedom and self-determination. Whether the issue is whom to marry, what career to pursue, or whether to participate in a research project, we value the right of individuals to assess information and weigh alternatives before making their own judgments. To coerce people into participating or to deceive them to gain their participation is to deny them the ability to determine their own destinies.

Although there is consensus about the general principle of informed consent, the debate regarding exactly how far researchers' obligations extend in this realm is volatile. At one extreme, researchers known as ethical "absolutists" argue that people should be fully informed of all aspects of the research in which they might play a part (Baumrind, 1985; Elms, 1982; Kimmel, 1988). Even when research is based on the public record, such as the documents of organizations and agencies, or on observations of behavior in public, some absolutists argue that people about whom observations have been made should be informed of the research. Otherwise, they do not have the full right to decide whether or not to participate.

Rigid adherence to the absolutist position would, however, make social research far more difficult to conduct. One reason for this difficulty is that such adherence rules out many practices some researchers consider important or essential. Many experiments, for example, rely on some degree of deception, at least to the extent of not telling participants the true research hypotheses. The reason for this is that people might respond differently if they knew these hypotheses in advance. However, all such studies would be unacceptable to the absolutists and could not be conducted. Disguised observation, in which people in public settings are not aware they are being observed, would also be disallowed by absolutists. No longer could researchers engage in such effective strategies as infiltrating organizations—unless, of course, they told all employees what they were doing. The net result would be to make social science research highly conservative and extremely limited. It would become the study of people who volunteer to be studied, and research shows that volunteers differ in many ways from people who do not volunteer. This would have the effect of seriously reducing the generalizability of research findings.

A second reason why the absolutist approach makes research more difficult is that it calls for obtaining written informed consent in all research projects, and research has shown that obtaining written consent can reduce people's willingness to participate in research. One study found that formally requesting informed consent prior to conducting an interview reduced the rate of cooperation by 7%, in comparison to cases in which a formal request was not made (Singer, VonThurn, & Miller, 1995). Because any reduction in response rate reduces the generalizability of the findings (explained in Chapter 9), obtaining informed consent in survey research can have serious negative consequences in terms of the validity of the research.

Written informed consent probably reduces people's willingness to participate because, in the eyes of some respondents, it appears to contradict the researcher's assurance of confidentiality. One minute people are being told that their answers will remain confidential, and

FIGURE 3.1   Sample Consent Form for Adult Human Subjects in Research

I, _____, in return for the opportunity of participating as a subject in a scientific research investigation and for other considerations, hereby authorize the performance upon me of the following procedure:

_____

_____

_____

_____

This consent I give voluntarily as the nature and purpose of the experimental procedure, the known dangers and possible

risks and complications have been fully explained to me by _____.

I understand the potential benefits of the investigation to be _____

_____

_____

as well as the above procedure(s) to be used which may involve the following risks or discomforts: _____

_____

_____

I understand that, as a participant, my rights will not be jeopardized, that my privacy will be maintained and that the data obtained in this study will be used in a manner to maintain confidentiality and personal rights.

I knowingly assume the risks involved, and I am aware that I may withdraw my consent and discontinue participation at any time without penalty to myself.

I also am aware of the fact that in the event of physical injury or illness facilities and professional care which are available will not be provided free of charge and that monetary compensation for such injuries or illness will not be made.

Dated: _____   _____

                                              Signature

Dated: _____   _____

                                              Signature

Source   Northern Michigan University Human Subjects Research Review Committee, *Policies and Procedures,* (December 20, 1982). Reprinted with permission.

the next they are being asked to sign a consent form! Even though signing a consent form doesn't necessarily impede the maintenance of confidentiality, it is not surprising that respondents may not view it this way. Anything that undermines respondents' beliefs in the confidentiality of the answers will reduce response rates. Signing a consent form may also affect the quality of the data obtained because those who do give their consent may be less candid in their responses than they otherwise would have been.

Because of these problems with the absolutist approach to informed consent, many researchers take a less extreme position. They argue that people should be informed of factors that might reasonably be expected to influence their decision to participate, such as any harm that might occur or how much time and effort will be involved. They also use a risk-benefit approach: Questionable research strategies, such as deception, are appropriate if they are essential to conduct the research, if they will bring no harm to participants, and if the outcome of the research is sufficiently important to warrant it. Lively debate continues concerning the use of questionable practices and their routine approval by the Institutional Review Boards (IRBs) that regulate much research (see Baumrind, 1985).

A final issue regarding informed consent has to do with the possibility that a person might feel pressured to agree or might not understand precisely what he or she is agreeing to. After all, asking a person to participate in a study can involve social pressures not unlike those in other settings. We often feel pressured to help others when they ask for our assistance, and some people find it difficult to say no to a face-to-face request. In addition, scientific researchers may be perceived as figures of some authority and status, and people are often disinclined to refuse their requests. In other cases, people may be momentarily confused about what is being asked of them. To resolve these problems, a study involving institutionalized elderly people used a two-step consent procedure (Ratzan, 1981). In the first step, people were told what their participation would involve, what risks were entailed, and that several days later they would be asked whether or not they would be willing to participate. This was intended to reduce any immediate pressures to agree and to enable people to talk with others and clarify any confusing issues. The second step, occurring a few days later, involved obtaining the actual consent. In this study, all those who were asked refused to participate, but this was probably due to the fact that a part of the study involved having a needle inserted in a vein for eight days in order for blood samples to be taken. However, some of the people may have agreed to participate had they been asked to sign a consent form during the first interview. Yet it is questionable whether consent obtained at that time would have been as "informed" and considered as it should be. Remember that the goal of obtaining informed consent is not to pressure people into participating in the research but rather to gain participation that is truly "informed" and protects the self determination of the participant.

Deception

Despite its status as a controversial strategy, deception is still fairly common in some areas of social science research, such as social psychology (Korn, 1997). One reason for this is that, as we have seen in discussing informed consent, some research would be difficult or impossible to conduct without some level of deception. Many experiments, for example, rely on some degree of deception, at least to the extent of not revealing to the participants the true research hypotheses. As another example, field research sometimes uses disguised observation, where people in public settings are observed but are not aware that such observation is occurring. Some would argue that this involves an implicit deception of those being observed. Other types of field research can involve more explicit deception in order to gain the cooperation of those who are being observed. Richard Leo (1996), in his field study of police interrogators, withheld or concealed information on his personal views of crime, punishment, homosexuality, and other issues from the police interrogators he was observing. In fact, he dubbed what he did the "chameleon strategy": "I consciously reinvented my persona to fit the attributes, biases, and worldview of my subjects" (Leo, 1995, p. 120). When in the company of the detectives, he feigned opposition to abortion, support for the death penalty, and antipathy toward gays. He created this persona in order to gain their co-operation and to encourage them to act openly and naturally in front of him. If the police interrogators had known his true views on these issues, Leo claims, they might have distrusted him and not acted openly when he was with them. Since his observations produced ground-breaking sociological research on this topic, he felt that the deception was warranted, given that no harm came to the police officers.

Ethical absolutists would rule out all such deception on the grounds that it is unethical to deliberately deceive other human beings (Erikson, 1967). However, Leo argues that the situation is more complicated: "Fieldwork is a morally ambiguous enterprise that is fraught with moral hazards, contingencies and uncertainties" (1996, p. 125). There are trade-offs and compromises to be taken into account. Further, Leo makes the distinction between "acts of commission" where the researcher intentionally falsifies information, and "acts of omission" where information is withheld. While some would argue that Leo did the former when he "reinvented" his persona, Leo himself claims that his research strategy involved only omission and was thus ethically justifiable. Leo argued that his use of deception passed the test of the three major criteria that are applied: (1) the research is important, (2) no other way to conduct the research is available, and (3) no harm came to those being studied.

When deception is used, it is a sound practice to conclude the period of observations with a *debriefing* during which people are told the true purposes of the research and informed of any deceptions that were utilized. This should be done in a positive and supportive way, so that the participants feel that they were joint partners in a worthwhile enterprise rather than that they were duped by the researchers. This can usually be achieved by explaining the reasons why the deception was necessary and how it would be impossible for anyone to have detected the deceptions before being informed of them. In some research, of course, debriefings are not possible, such as in disguised observations in field settings where the researcher never again has the opportunity to contact the participants after observations are made.

The Right to Privacy: Anonymity and Confidentiality

The right to *privacy* is one of the key values and ethical obligations mentioned at the beginning of this chapter. **Privacy** refers to the ability to control when and under what conditions others will have access to your beliefs, values, or behavior. Intrusions into one's privacy have become endemic in modern society, and with the growth of social research in the twentieth century the danger of even greater intrusion arises. Virtually any attempt to collect data from people raises the issue of privacy and confronts the investigator with the dilemma of whether threats to privacy are warranted by the research. Two well-known research projects illustrate the complexity of this issue.

Two Case Studies

The sociologist Laud Humphreys (1970), in an effort to understand a particular type of sexual behavior, observed men having quick and impersonal sexual encounters with other men in public restrooms. In order to gather his data without arousing suspicion, Humphreys played the role of the "watch queen," one who keeps watch and warns participants of approaching police or "straight" males who might disrupt the activities. None of the men who went to the restrooms to engage in sex was aware that his behavior was being recorded by a researcher. Humphreys also noted the license plates on the cars of these men and was able to locate their home addresses through public motor vehicle records. He then interviewed them in their homes but did not inform them of the real reason for the interviews or that he had earlier observed them in the restrooms.

Humphreys was heavily criticized for using deception, not obtaining informed consent, and for violating the privacy of these men engaging in highly stigmatized actions that, were they made public, might disrupt their family lives or threaten their jobs. Many sociologists believe that research on such sensitive topics is not merely a matter of confidentiality; that is, not letting people's identities become known. Rather, such data should not even be collected because these men were obviously trying to conceal their actions from prying eyes. Social science researchers, it is argued, should respect their privacy.

Humphreys defended his research on the grounds that the confidentiality of his subjects was maintained and the results of the study were of significant scientific value. In fact, no one else has devised a different method to study such sexual behavior. Humphreys discovered that the men who do engage in this type of sexual activity are not unusual or deviant in other aspects of their lives. In fact, they were, for the most part, normal, respectable citizens who had found a rather unusual sexual outlet. Humphreys believes that our greater understanding of what had been considered deviant sexual conduct justifies the threat to privacy these men experienced. In addition, the public setting in which they performed their acts, he argues, reduced their right to claim privacy.

Another field study illustrates further difficulties in protecting people's privacy. When research describes an actual group or community of people, a common practice is to report results giving fictitious names to people and places in order to protect privacy. This procedure works fairly well as long as researchers do not become so detailed in their descriptions of places, events, and people that the protection afforded by the fictitious names is undermined. A now infamous example of this was *Small Town in Mass Society* by Arthur Vidich and Joseph Bensman. First published in 1958,

this observational community study described the power relationships and local governmental operations in a small town the authors called Springdale. The authors had assured all interviewed people that confidentiality would be maintained. Even though no identities were, in fact, directly revealed, it was easy for the residents of the small community to recognize the people and events described in the highly detailed report. Because the study was very critical of some residents of "Springdale," these people, understandably, became outraged. Local newspapers vociferously attacked the researchers for betraying the community. The townspeople even held a Fourth of July parade that featured a full manure spreader carrying mired effigies of the authors (Whyte, 1958). In order to avoid such breaches of confidentiality, one must balance one's enthusiasm for producing a highly detailed account against the ethical obligation to protect fully the identities of those observed.

As these two case studies show, the right to privacy is often a difficult ethical issue to resolve. Researchers have come up with three major ways to deal with the problem of protecting people's privacy: let them edit the data, keep the data anonymous, and keep the data confidential.

Editing the Data

One very effective way to protect privacy is to offer participants the opportunity, after the data has been collected, to destroy any data they wish to remain private. This was done in a study of family interaction that utilized videotape cameras installed in apartments to record all interchanges between family members (Ashcraft & Scheflen, 1976). Even though the families consented to the taping, the investigators, sensitive to the issue of privacy, offered the families the opportunity to review the tapes and edit out anything they wished. The assumption was that, despite agreeing to participate, family members might do something on the spur of the moment that they would prefer not be made public. Surprisingly, not one family exercised the option to edit the tapes. This and other research suggests that people can be more tolerant of invasions of privacy than researchers might expect. Nonetheless, it should always be the research subject's decision; the researcher should never assume that the subjects will be tolerant of invasions or breaches of privacy.

Anonymity

A second—and equally effective—means of ensuring privacy is to accord the participants **anonymity**, which means that *no one*—including the researcher—can link any data to a particular respondent. This can be accomplished by not including any identifying names or numbers with the data collected. True anonymity means that the researcher can never link data to a particular respondent. This method of protecting privacy is ideal because the data are collected in such a way that it is impossible for anyone to determine which data comes from which individual. However, in many research situations, it is not possible or feasible to collect data in this fashion. When this is the case, researchers turn to confidentiality.

Confidentiality

A third way of protecting privacy is through **confidentiality**: ensuring that information about or data collected from those who participate in a study not be made public in a way that can be linked to an individual. Researchers, of course, commonly release their data to the public, but usually only in the aggregate, which means reporting on how a whole group responded rather than how individuals responded. In many studies, the data are not of a sensitive nature and thus confidentiality would seem less important, but it is impossible to predict what bits of data all participants will want kept confidential. Because aggregate and anonymous reporting of results is all that is necessary in most research studies, confidentiality is routinely extended in order to encourage people's participation and honest responses.

It is worth pointing out that anonymity and confidentiality are quite distinct. With confidentiality, the researchers can link responses to particular respondents but do not release this information publicly; with anonymity, not even the researchers can link responses to particular respondents. For example, if a researcher mails questionnaires and they are returned with no names or other identifiers on them, then the respondents have true anonymity. If the same questionnaires are mailed back with names or other identifiers on them, then only confidentiality is possible since the researcher would be capable of linking responses to respondents. Even if the identifiers are removed as soon as the questionnaire is received, it would still be ethical to make promises only of confidentiality since the researcher could link responses to respondents at the point at which the questionnaires are received. Obviously, it is unethical to tell respondents that they will have confidentiality when they won't, but it is also wrong to claim that responses will be anonymous when all that will actually be protected is confidentiality.

In some cases, confidentiality can be breached merely by a person's involvement in a research project becoming known. If it becomes known, for example, that an individual was a part of a study of drug abusers, then knowledge that the individual is a drug abuser has become public even if none of the data collected from that individual has been made public. One aspect of confidentiality is, therefore, to ensure that the identities of those who participate in the research, especially when it is sensitive, are kept hidden.

Confidentiality can also be threatened when third parties, such as the people sponsoring the research or the courts, seek to identify research participants. Intrusion by a sponsor is relatively easy to avoid. When establishing a research agreement with a sponsoring agency or organization, one should make clear in the agreement that identities will not be revealed under any circumstances. If the sponsor objects, researchers should refuse to accept the agreement. More will be said about sponsors and ethics later in this chapter.

Court Challenges to Confidentiality

The courts pose a more complicated threat to confidentiality. Most communication between physicians, lawyers, and clergy and their clients is protected from judicial subpoena. Social researchers, however, do not have legal protection of privileged communication with the people from whom they gather data (Kimmel, 1988; Reece & Siegal, 1986). Thus, courts may subpoena research data that reveal participants' identities, and failure to comply with such a subpoena renders researchers open to contempt of court charges. Actually, social science researchers have been treated somewhat inconsistently by the courts in civil cases. In a civil suit in California, for example, the court refused to force a researcher to reveal the identities of respondents in confidential interviews (Smith, 1981).

In criminal cases, however, the courts have generally held that the right of the public to be protected from criminal activity or the right of suspects to a fair trial supersedes any assurance of confidentiality in research. In one case, for example, the researcher had been making field observations in a restaurant when a suspicious fire erupted (Brajuha & Hallowell, 1986). Police wanted the researcher's field notes to determine whether any evidence of arson could be substantiated. One court quashed a subpoena for the field notes but another upheld it. Eventually, a compromise was reached: The researcher's field notes were considered subject to subpoena, but the researcher was allowed to remove material that would have violated confidentiality. With rulings such as this, the courts seem to recognize that the confidentiality that a legitimate social researcher establishes in a relationship should be protected, if that can be done while still protecting the rights of the public and criminal suspects.

In another case, a sociologist actually spent time in jail because he refused to give evidence in court that he believed violated his promise of confidentiality to his research subjects. Sociologist Rik Scarce (1994) conducted research on activists in the animal liberation movement in the early 1990s. A federal grand jury was investigating break-ins by such activists at university laboratories, and some of the activists in whom the grand jury was interested had been interviewed by Scarce as part of his research. Scarce refused to answer certain questions put to him by the grand jury about these activists because he thought that it would violate the confidentiality he had extended to the people he interviewed. He was jailed for four months on contempt of court charges. Leo, in his study of police interrogators, was compelled to testify in court under threat of contempt of court. As these cases illustrate, the courts have generally held that confidential communication between a researcher and a research participant is not protected in criminal cases; however, some courts also recognize that the confidential relationship is an important and special one and that efforts should be made not to violate it. Because of cases such as these, some researchers have called for federal legislation that would give social science researchers protection, even if limited, from being compelled to violate their confidential relationship (Leo, 1995; Scarce, 1994).

Concern with possible subpoena of their research data has led some researchers to adopt elaborate measures to protect the data. For example, it is common to establish computer files with the data identified by numbers rather than by names. Often, it is unnecessary to retain name identification for research purposes once the data have been collected. In such cases, the names should be destroyed. If name identification is required, say because you want to interview the same people at a later time, the names should be stored in a separate computer file. This procedure reduces the possibility of unauthorized persons linking names with data.

One of the best means of securing confidentiality for sensitive research, such as that dealing with AIDS, substance abuse, or criminal behavior, is to use *certificates of confidentiality*, which are made available by the Public Health Service Act Amendments of 1974 (Melton & Gray, 1988). These certificates guarantee the confidentiality of identifying information associated with a research project, and are intended to protect privacy against a wide array of legal actions, including those in federal, state, and local courts as well as in civil, criminal, legislative, and administrative proceedings. The Department of Health and Human Services awards certificates only for research on sensitive topics, such as illegal drug use or deviant sexual behavior, regardless of whether the research receives federal funding. However, the certificates are discretionary, and given out sparingly, so a researcher must make application to receive one. Some granting organizations, such as the Office of Assistance Programs in the Department of Justice, grant certificates to researchers to whom they are providing funds (Melton, 1990).

Despite the fact that intrusion by the courts is a real danger to research on some topics, the reality is that court involvement is extremely rare. Although more than 200 cases have involved courts seeking journalists' information, in only a dozen or so cases have courts sought research data (Reynolds, 1979; Smith, 1981). Thus the odds of researchers becoming embroiled in such a situation are remote. Notwithstanding, researchers have an obligation to inform potential subjects accurately of any possible threats to confidentiality that might arise, including what might happen if their data were subpoenaed by the courts. The lengths to which investigators go to protect privacy indicate the importance of this ethical issue. The bottom line is that no one should be threatened with harm to themselves or their reputation as a result of participating in a scientific study.

Harm, Distress, and Benefit

Researchers should avoid exposing participants to physical or mental distress or danger. If the potential for such distress exists in a research investigation, the participants should be fully informed, the potential research findings should be of sufficient importance to warrant the risk, and no possibility should exist of achieving the results without the risk. People should never be exposed to situations that might cause serious or lasting harm.

Research in the social sciences rarely involves physical danger, but there are research settings in which psychological distress may be an element. Some studies, for example, have asked people to view such things as pornographic pictures, victims of automobile accidents, and the emaciated inmates of Nazi concentration camps. Certainly, these stimuli can induce powerful emotions—in some cases, emotions that the participants had not expected to experience. A strong emotional reaction, especially an unexpected one, can be very distressful. Other research has asked people to report on their sexual behavior, which could be embarrassing, and interviewing people about their past misdeeds might produce bouts of depression or remorse. In some studies, people have been given false feedback about themselves in order to observe their responses. People have been told, for example, that they failed an examination or that tests show that they have some negative personality characteristics. Any situation in which people might learn something about themselves of which they were unaware can be distressful. Even minor deceptions that are a part of much research can be distressful to people who thought they could not be so easily deceived.

Another area where social science research contains the potential for bringing harm to those being studied is in field research of people engaging in stigmatized, deviant, or illegal behavior. The potential harm comes if the researcher says or does anything that might lead the research subjects to be sanctioned for the deviant or illegal behaviors they displayed in front of the researcher (Klockars, 1979). This is a complex topic that will be explored in more detail in Chapter 11, but the problem can arise because field research often involves an implicit or explicit agreement between the researcher and those being observed. In this agreement, those to be observed consent and allow the researcher to join and observe them, but in return they expect the researcher to support their illegal or deviant behavior, or at least not do anything that would cause repercussions. For example, if police interrogations of suspects are likely to include some harsh or even illegal treatment of the suspects, those police would be unlikely to let a researcher observe the interrogations unless they believed the researcher would support their actions or at least not report them to authorities or testify against them in court. In other words, the police would likely expect the researcher, in return for being permitted to observe them, to act like fellow officers and support them in their actions. In developing a relationship with the police, the researcher may have given the impression that this was understood or the police may have assumed the researcher understood it. This actually creates a moral dilemma for the researcher because either choice is morally compromised. If the researcher supports the police and doesn't report the harsh treatment of the suspect, the suspect is harmed by the researcher's actions; if the researcher supports the suspect and reports the actions, the police are harmed both by having their implicit agreement with the researcher violated and by being sanctioned by authorities for their behavior. This kind of ambiguous dilemma is inherent in some field research on deviant lifestyles and is part of what Leo meant by his comment earlier that fieldwork is "a morally ambiguous enterprise" (Leo, 1996, p. 125).

Assuming that the scientific benefits warrant the risk of distress and that the participants are fully informed, it is then the researcher's obligation to alleviate the impact of whatever distress does actually occur. This is most often accomplished through a *debriefing*, in which people's psychological and emotional reactions to the research are assessed (Smith & Richardson,

1983). Because the distress is usually mild and transitory, it can normally be dissipated quickly and with no permanent impact. If people have stronger reactions, then the researchers have an obligation to assist them through this with whatever resources are necessary. The Applied Scenario box describes a research project that addressed a number of ethical issues discussed in this chapter, including the use of repeated informed consent procedures, harm and distress, and the use of debriefings.

For some social scientists, an ethical standard that rests only on avoiding harm is far too limited. They would also argue that those upon whom we conduct research should also gain some positive benefit from the research. This stance has been put forth most clearly by feminist researchers (discussed in Chapter 2) and collaborative researchers (discussed in Chapter 17) (Nyden, Figert, Shibley, & Burrows, 1997; Reinharz, 1992). Viewing research as a two-way street, feminist and collaborative researchers argue that both researcher and research subject should gain something positive from the research. Researchers, of course, obtain data from the research that enables them to publish books and articles, which advances their careers. But what do the research subjects get? This raises the ethical issue of what the research community owes to research participants. The subjects could be paid for their time and effort, or the research results could be translated into some social policies or practices that benefit the community from which the research subjects come. Or, by participating in the research, the subjects might develop some knowledge or skills that would enable them to make their own lives better. The point is that, from this perspective, researchers have an ethical obligation to leave the research participants and their community better off for having participated in the research.

Protecting Vulnerable Groups

Some groups on whom social scientists conduct research might be considered vulnerable because they have few social, psychological, or

APPLIED SCENARIO   Ethical Issues in Socially Sensitive Research

Applied social researchers often find themselves doing socially sensitive research; that is, research on people who arouse intense and sometimes negative emotions in others. This can include research on murderers, rapists, prostitutes, and/or drug abusers, to name but a few. Such research often involves evaluation research or program evaluations (discussed in Chapter 1 and Chapter 12), and it can be very challenging, especially in terms of maintaining ethical standards.

One example of such research is a program evaluation assessing two methods of treating people convicted of child molestation (Jenkins-Hall & Osborn, 1994). The goal of the research was to provide observational evidence regarding which treatment approach was more effective. In conducting the research, the researchers had to consider the rights of the child-molester participants as well as staff safety. There are potential negative consequences for each group in participating in the research. The researchers identified three general areas of concern: (1) informed consent and voluntary participation, (2) confidentiality of data and privacy, and (3) protection against dangers that a participant may pose to himself or others.

Informed consent was considered so important by the researchers that they incorporated a multiphase consent process that required participants to give consent at each stage of the process. All clients completed a general consent form that covered the basic requirements of the Department of Health and Human Services. Prior to a comprehensive psychological and social assessment—a necessary part of the research—the clients completed an evaluation consent form that spelled out the assessment process. Additional consent forms were completed in conjunction with each subsequent treatment component. Given the complexity of the intervention, this multiphase consent process, while unusual, assured the researchers that the participants understood the program and were freely consenting to participation throughout the project.

Confidentiality and privacy were also considered very important because of the negative consequences that might occur to the child molesters if information about their participation in the program was to be released. One protection of privacy was the use of federal Certificates of Confidentiality. However, additional steps were also taken. The project was located in a public office building suite and was not identified as a sex offender treatment site. Each participant was given a six-digit ID number that appeared on all documents. Names and other identifying information were removed from all correspondence, consents, and records from referral sources. Staff addressed participants by first name only, and information was never released without signed waivers of confidentiality that specified which information could be released and to whom. Because the program involved the delivery of treatment services, each participant

financial resources to protect their own interests. Such groups might include the impoverished, recent immigrants, children in day care settings, participants in job-training programs, or persons with emotional disorders. The members of these groups may find themselves vulnerable to pressures to participate in research projects conducted by the organizations providing them with services. In fact, it is common for social programs to be required by their funding sources to conduct research studies on the effects of their programs. Thus persons who obtain services are likely to be solicited for research. Special obligations fall on researchers to

was provided with a client advocate who was a volunteer clinician and not affiliated with the project. This advocate's sole responsibility was to protect the rights of the clients.

In addressing the issue of harm or distress, the researchers considered not only harm to the participants but also, given that the participants were convicted sex offenders, danger to the staff or members of the local community. Several measures were employed to reduce such risks. One concern for the project was assuring quick detection if a participant was in danger of committing another sex offense. Researchers argued that the project had an ethical and, perhaps, a legal duty to protect potential victims. The project provided each participant with a 24-hour crisis call service as well as therapists to deal with minor crises, such as loss of employment or a breakup with a girlfriend. Over a two-year period, three cases required involuntary commitment of participants due to deterioration in their mental health status. In several cases where the participant was exhibiting negative signs, that participant was temporarily removed from the project and offered alternative treatment. If a participant who was at risk of relapse failed to follow the terms of a crisis-intervention plan, the director of the project was notified and authorities were alerted. Failure to comply would result in termination from the project.

To assure staff safety, the program employed a variety of procedures. Staff were directed to: have unlisted personal phone numbers, never reveal home addresses, never be alone with clients in the suite, and never enter a room that was lockable from the inside with a client. Each office and work area contained a "panic button" that sounded an alarm in the office and at the university police office. Entry to the area where the project was conducted was strictly controlled so that client interaction with other office building occupants was minimal. Participants were routinely debriefed after treatment sessions to assure that they would not leave the center in a distressed, agitated, or aroused state.

A research program such as this clearly presents researchers with more ethical challenges than most social science projects. The procedures described here enabled the researchers to carry out a challenging project on a highly sensitive topic successfully.

Locate in the library or on the Internet other examples of research on socially sensitive behavior.

1. What ethical issues arise in these research projects? What procedures are incorporated to alleviate them?
2. Consider three other groups that applied social scientists often do research on: drug addicts, illegal aliens, and people with AIDS. What ethical issues could arise in such research? What procedures could be incorporated to maintain ethical research?

---

safeguard their research subjects from unreasonable pressures to participate in research.

A crucial issue in this regard is the matter of voluntary and informed consent. Can people in certain vulnerable groups actually give consent freely? This is one of the reasons that much research using prison inmates as subjects has been discontinued: It is debatable how free inmates really are to give consent. If participation in a research project brings significant rewards to the inmates in the form of separate living quarters or greater privacy, these rewards may be considered coercive in the Spartan, degrading, and often dangerous world of the prison inmate. Although participation in the research is, on the face of it, voluntary, inmates may not seriously weigh the disadvantages or dangers of the research when participation is perceived as a means of avoiding assault or rape.

Similar ethical questions arise with research on persons with significant psychological disorders: If the psychopathology involves limited comprehension or impaired insight, then such people may be unable to give truly informed consent. Research on such a population must be approached very carefully: Such research could not be justified if the same research goals could be accomplished with a less vulnerable group. If the decision is made to do research on a group with significant psychopathology, then informed consent should be approached in ways that protect potential participants from even covert coercion; for example, consent could be sought by some party other than the researchers (to avoid the force of authority) or it could be sought in the presence of a relative, friend, or other advocate for the individual (to provide the social support for a refusal). If the psychopathology totally impairs the ability to consent, then most forms of research would be considered unethical unless it can be shown that either the patient or society would benefit significantly and that the research could be done in no other way (Fulford & Howse, 1993).

The problem of voluntary consent may be somewhat more subtle among some recipients of social services. Even if a refusal to participate in research would not result in a termination of benefits, clients may not be sure of this and may be disinclined to take the risk of finding out. This concern is not limited to recipients of social services; it might be an issue in any subordinate group. Sociologist Ivy Goduka (1990) conducted research on black schoolchildren in South Africa during the period of apartheid, when black South Africans were rigidly oppressed; had few social, political, or economic rights; and could expect little

or no protection from the power of the state. As Goduka put it, "Because they lack basic resources, blacks are not only illiterate and poverty-stricken, but also ignorant, vulnerable, and powerless to make decisions and choices directly affecting their own lives" (pp. 329–330). Blacks could be told where to live, moved arbitrarily, and had little say in their own lives. In terms of research, these native South Africans could not read or understand a consent form. Established institutions, such as schools and governments, were so authoritarian that black South Africans felt they had no choice about signing a consent form or allowing their children to participate in research if the request came from these institutions or from well-educated South Africans such as Goduka. They therefore consented because they wished to avoid trouble, or hoped that acquiescence would bring some positive improvements to their lives.

It is not suggested here that research should never be conducted on such populations. Rather, researchers need to exercise additional caution—and, in some cases, possibly forgo valuable research projects—in the interests of ensuring truly voluntary informed consent.

Withholding Treatment for Research Purposes

Another important ethical issue arises when some presumed benefit or treatment is withheld from a group to see if that benefit or treatment actually works. This became an issue, for example, in a study of the effectiveness of needle-exchange programs among intravenous drug abusers (Leary, 1996). The goal of the needle-exchange program was to reduce the abusers' risk of being infected with disease viruses, such as HIV. One group of participants would get free clean needles in exchange for old ones; another group would not. This would allow a comparison of the two groups to see which had the lowest risk of infection and thus whether giving clean needles helped to reduce the spread of infection. The study was criticized on the grounds that the control group would not have the same access to the needle exchange even though many do believe that such needle exchanges reduce infection rates. And herein lies the ethical dilemma: The most effective way to test whether the needle exchange is effective is to withhold it from a control group. The control group serves as a comparison group. If the group receiving clean needles shows lower infection than does the control group over a span of time, then we can feel justified in claiming that the needle exchange brought about the difference. Control groups are thus very important to our ability to say whether an independent variable causes change in a dependent variable (see Chapter 2 and Chapter 8). To test the new treatment, we have to deprive people in the control group of the possibility of improvement.

This is a serious and difficult dilemma to resolve, and there are a number of issues that must be considered. First, we might ask whether it is ethical to proffer untested treatments. Services provided by social programs are often expensive, time consuming, and divert resources away from other programs; they also raise people's expectations for improvement. Is it ethical to do this when there is no solid scientific evidence showing that the treatment will be beneficial? That was the argument of the researchers conducting the needle-exchange study: There were no controlled studies showing whether needle exchanges reduce infection rates. We would not think of allowing the marketing of new medications without a thorough test of their effects, both positive and negative. We should expect no less from social programs offered to people. Opponents of the needle-exchange study, while recognizing that there were no randomized studies with control groups on this topic, argued that the overwhelming preponderance of other research suggested that needle exchanges do reduce infection rates; in fact, they felt the evidence on this was so overwhelming that it was

unethical to withhold clean needles from a control group.

There may be, in some cases, alternatives to withholding treatment. One alternative is to offer the control group a known effective treatment and then see whether the new treatment provides more or less improvement. In this way, all groups are receiving a treatment that is either known to be, or suspected of being, efficacious. In a study of the effectiveness of suicide intervention programs, for example, one would certainly be reluctant to evaluate a new intervention by using a control group that received no intervention at all. One recent study of interventions with suicidal young adults used a randomized experiment in which subjects were assigned to either a new experimental treatment or a control group that received a treatment that had been commonly used for these interventions. Although both groups showed similar improvements in a number of areas, the experimental treatment was more effective at keeping the highest risk participants in the program where they could continue to receive assistance (Rudd, et al., 1996). A variant on this was used in the needle-exchange study mentioned above: The people in the control group, although not given new needles in the study, were told where they could purchase new syringes.

Another alternative to withholding treatment would be to delay giving the treatment to the control group for a period of time and make the comparisons between those receiving treatment and the control group over this time period. For example, in a study of two different approaches to controlling people's smoking, all participants in the study were told that, if they were placed randomly in the no treatment group, they would receive treatment when the study was completed and would receive the treatment the study demonstrated to be most effective (Coelho, 1983). This resolves the ethical problem by ensuring that all participants will receive treatment at some point.

EYE ON DIVERSITY Protecting Minorities from Unethical Research

Scientific research does not exist in a vacuum but rather within the context of the distribution of power in a particular society. When lopsided power relationships exist between groups, it raises the potential that one group might exploit another and that scientific research could become an instrument in that exploitation. Two episodes involving the exploitation of minorities occurred in the twentieth century—one abroad and the other in the United States—that starkly documented how this could become a reality and served as major catalysts for efforts to codify a set of ethical standards for research.

The first episode occurred in Europe during World War II: the heinous series of experiments conducted by the Nazis on Jews and others in concentration camps (Beauchamp, Faden, Wallace, & Walters, 1982). Prior to this, there had been no codification of scientific ethics, and researchers were left largely to their own devices and consciences in deciding how to conduct their studies. The revelation of the German atrocities, exposed during the Nuremberg trials in 1945 and 1946, shocked the sensibilities of the world and left an indelible imprint on research ethics. The cruelty of the experiments is almost inconceivable. Healthy people were intentionally infected with such serious diseases as spotted fever or malaria. Others were used as subjects to test the effects of various poisons. Some had parts of their bodies frozen to test new treatments. Still others were purposefully wounded to study the effects of new antibiotics and other treatments. Perhaps most evil of all were the excruciating decompression studies designed to test reactions to high-altitude flight (Katz, 1972). These gruesome experiments forcefully brought home exactly how far some people would go in using research to further their own ends. They also brought home the vulnerability of minorities to exploitation in research, especially when the resources and authority of powerful groups support the research by persons with few or no ethical standards. Public outrage after the war led to heightened concern for establishing codified standards for the ethical conduct of medical research on human subjects.

The second episode that influenced the development of a codified set of ethical standards was an infamous study of syphilis conducted by the U.S. Public Health Service (PHS). The study began with African American males from Tuskegee, Alabama, in 1932: 425 with syphilis

---

In situations where such alternatives cannot be devised, however, researchers must use the risk–benefit approach: Does the benefit to be gained from the research outweigh the risks of withholding treatment from the people in the control group? If we were studying people at high risk of suicide, we would be cautious about withholding treatment. In the treatment of nonassertiveness, on the other hand, we might decide that a delay of a few weeks in treatment would not be highly detrimental, and whatever damage occurred could be undone

once treatment was initiated. As with so many other ethical issues, this is ultimately a judgmental one over which there will be disagreement.

EFFORTS TO REGULATE ETHICAL BEHAVIOR

The preceding discussion focuses mostly on researchers' efforts to "do the right thing," but such individual consideration is no guarantee of ethical action. The immoral treatment of

and 200 without syphilis. All were poor and semiliterate; many were deceived about the true nature of their ailment and led to believe that the tests being conducted by the researchers were actually treatments for the disease. The intent of the researchers was to keep the men in the study and prevent them from receiving treatment in order to observe how the disease progressed over a span of years. When the study began, there was no known cure for syphilis. Fifteen years later, penicillin was discovered to be an effective cure for this dreaded affliction. Despite this discovery, the PHS continued the syphilis study an additional 25 years, into the 1970s and after the government had established the ethical guidelines discussed in this chapter. The researchers even went to great lengths to prevent the men from receiving treatment when they sought it. The justification for deceiving the men was that, because they were poor and semiliterate, they would in all likelihood not seek treatment even if they knew of it. And, unbelievably, the men were allowed to spread the disease to their sexual partners!

When syphilis is untreated, it can cause paralysis, insanity, blindness, and heart disease. Ultimately, it can be fatal. Many of the afflicted participants in the Tuskegee study suffered serious physical disorders or died as a result of not receiving treatment for the disease (Jones, 1992; Kimmel, 1988).

Most Americans today would agree that the actions of the PHS were repugnant, racist, and unethical in the extreme. When the study began, however—and even when effective treatment for syphilis became available—the subordinate position of African Americans and impoverished people in the United States resulted in their receiving fewer of the political, economic, and even medical rights that Anglos and the more affluent receive. It is unlikely that well-to-do Anglos would have been treated in the same arrogant fashion, and one would hope that no such debacle would at all be contemplated today. Yet, we should keep in mind that the PHS study continued into the 1970s—an era we like to consider a more enlightened one regarding human rights—and ended only when it received public notoriety.

These two episodes, then, document the extent to which minorities and other groups that lack adequate protections from the powerful are at risk of dangerous and inhumane treatment by scientific researchers who are not governed by clear and enforceable ethical standards.

some minorities in research during the twentieth century, described in the Eye on Diversity box, served as the catalyst for the development of the enforceable ethical standards that govern research today. To avoid such ethical debacles and to protect research subjects, researchers must have guidelines and regulations backed up by effective sanctions.

The first effort along these lines was the Nuremberg Code, developed in 1946 in direct response to the atrocities committed during World War II. The Nuremberg Code was lim-

ited to issues of ethics in medical research. In 1966, the United States Public Health Service (PHS) established ethical regulations for medical research that emphasized that: (1) full disclosure of relevant information should be made to the participants; (2) the decision to participate must be completely voluntary; and (3) researchers must obtain documented, informed consent from participants (Gray, 1982; Reynolds, 1979). In 1974, the Department of Health, Education, and Welfare (DHEW), now the Department of Health and

The Northern Michigan University Human Subjects Research Review Committee approves/disapproves all requests to conduct research involving human subjects. In completing the following application, be advised that the persons reviewing it may be entirely unfamiliar with the field of study involved. Present the request in nontechnical terms understandable to the Committee. It is the investigators' responsibility to give information about research procedure that is most likely to entail risk but not to express judgement about the risk. Please submit a copy of your complete proposal and attach a curriculum vitae or biographical sketch if project is to be submitted to an outside funding source.

Principal Investigator   Department   Telephone   Date of Request
(Please include mailing address for students submitting an application)

Type: New [ ]   Renewal [ ]   Continuation [ ]   If renewal or continuation, has procedure changed?   Yes [ ]   No [ ]

Project Title

Agency to which Proposal has been Submitted   Date   Location of Project

Purpose of Research

| Data Obtained by | [ ] Mail | [ ] Telephone | [ ] Interview | [ ] Observation |
| | [ ] Experiment | [ ] Secondary Source | [ ] Questionnaire | [ ] Other |

Provide a detailed description, in nontechnical terms, of the procedures that will allow the Committee to evaluate the risks to the subjects. Assurance from the investigator, no matter how strong, will not substitute for a description of the transaction between the investigator and subject. If a questionnaire is used, attach a copy.

FIGURE 3.2 (*continued*)

Give ages, gender, source and total number of subjects.

_____

How will subjects be selected, enlisted or recruited?

_____

How will subjects be informed of the procedures, intent of study and potential risks?

_____

What steps will be taken to allow subjects to withdraw at any time without prejudice?

_____

How will the subjects' privacy be maintained?

_____

<u>The following documents must be attached:</u> 1) Complete research proposal, 2) principal investigator's curriculum vitae, 3) questionnaire, 4) informed consent form(s), and 5) supporting documentation (e.g. letters from agency directors.)

_____

In making this application, I certify that I have read and understand the "rules governing the Participation of Human Beings as Subjects in Research," (available in the Office of Continuing Education) and that I will comply with the letter and spirit of University policy. Proposed changes in the protocol will be submitted to the Dean of Graduate Studies and Research for written approval prior to those changes being put into practice. Records will be kept of informed consent of the subject for at least two years after the subject's participation.

_____   _____
Principal Investigator's Signature   Date

APPROVALS:

_____   _____
Department Head   Date

_____   _____
College Dean   Date

A. This application has been reviewed and [   ] approved; [   ] disapproved by the NMU Human Subjects Research Review Committee or [   ] deferred until adequate application has been made.

_____   _____
Chair of University Committee (or representative)   Date

B. Exemption: [   ] Approved   [   ] Disapproved

_____   _____
Dean of Graduate Studies and Research   Date

Source   Northern Michigan University Human Subjects Review Committee, (1999). Reprinted with permission.

Human Services (DHHS), decreed that the PHS guidelines would apply to social science research as well. Further, DHEW recognized that codes of conduct alone would not ensure ethical research without some oversight procedures. To this end, DHEW required research institutions, such as universities, to establish institutional review boards (IRBs) to review research proposals to ensure that the guidelines were followed. These DHHS regulations have been broadened so that they now apply not only to those projects directly funded by that agency but also to any research carried on in organizations that obtain DHHS funding.

Reaction to the imposition of DHHS guidelines and the IRBs among social scientists has been somewhat hostile. A common feeling is that the risks to participants in most social science research are minimal and differ from those risks that may occur in medical research (Murray, Donovan, Kail, & Medvene, 1980). Researchers see the major risk in social science research being a breach of confidentiality, not in direct harm caused during the research process. Researchers also feel that the regulations obstruct important research, hinder response rates, and reduce the validity and generalizability of findings.

In response to these concerns, the DHHS, under a policy issued in 1981, exempted the following research methodologies from IRB review: evaluation of teaching procedures or courses, educational testing, many surveys (especially when anonymous), observation of public behavior, and documentary research (Huber, 1981). The only exceptions are research dealing with "sensitive" behavior, such as drug and alcohol abuse, illegal behavior, or sexual conduct. In assessing social science research not exempt from IRB review, the IRBs have tended to subscribe to the risk-benefit doctrine (Smith, 1981). This means that questionable practices, such as deception or disguised observation, may be utilized if the purpose of the study is viewed as sufficiently important to justify them.

Although the DHHS regulations exempt much social science research from IRB review,

many universities and other research institutions require that all research projects be reviewed in order to determine whether they are legitimately exempt. Further, it has become increasingly commonplace for private foundations to require IRB review as a prerequisite for funding (Ceci, Peters, & Plotkin, 1985). When planning research, an application to conduct the research should be routinely submitted to the appropriate IRB in order to determine whether the research is exempt and, if not exempt, gain permission to conduct the research. Figure 3.2 illustrates one such IRB application form. On that form you can see that applicants are required to describe how they will address many of the ethical issues discussed in this chapter. Even if a project is exempt from IRB review, researchers are not exempt from ethical concerns. Whether one's research will or will not be reviewed, one still has the responsibility of considering the impact of the research on those who participate and on those who might benefit from it. And ethical guidelines do not eliminate the controversy that surrounds ethical issues—because guidelines need to be interpreted and applied to specific contexts. Much debate surrounds such interpretation and application.

ETHICAL ISSUES INVOLVING SPONSORS AND THE COMMUNITY

Some ethical dilemmas in research have to do not with the persons who agree to be the subjects in the research but rather with those who directly sponsor the research or with the community at large, or possibly society as a whole, that is in some way affected by the research.

Sponsored Research

Because much social research is conducted under the auspices of a third-party sponsor, certain ethical considerations arise from that relationship. When research is sponsored, some type of research agreement, essentially a contract, is developed. Researchers and sponsors alike should exercise great care in drafting this agreement. The potential for ethical problems to arise later is reduced when the research agreement clearly specifies the rights and obligations of the parties involved. Three areas are of particular concern in sponsored research.

First, it is common for sponsors to want to retain control over the release of the collected data. The precise conditions of release should be specified in the research agreement to avoid conflicts. One limitation that should not be tolerated, however, is the conditional publication of results; that is, agreeing to publish results only if they turn out a certain way (usually so they support the preconceived notions of the sponsor). Such conditional publication violates the integrity of the research process and the autonomy of the researcher (Wolfgang, 1981). If the researcher agrees to some other type of limitation on release, however, it must be honored. To do otherwise would be a breach of the agreement and therefore unethical.

The second major concern in sponsored research is the nature of the research project itself. The precise purposes and procedures of the study should be specified in the agreement. Ethical questions arise if the researcher modifies the study to cover matters not in the agreement. Sponsors will often allow researchers to use the data gathered for scientific purposes beyond the needs of the sponsor, but to agree to do a study that a sponsor wants and then change it for personal reasons so that it no longer meets the expectations of the sponsor is unethical.

A third area of ethical concern in sponsored research is the issue of informed consent; namely, revealing the sponsor's identity to participants. Although controversy exists in this regard, some researchers take the stance that potential participants can only give truly informed consent if they know who is sponsoring the study and for what purpose. Some people might object, for example, to providing data for a research project that would help a company better market a product, a political party present a candidate, or the government disseminate propaganda. In fact, studies show that people are less likely to participate in research they know to be sponsored by commercial organizations, which suggests that information about sponsorship can influence the decision of whether or not to participate (Fox, Crask, & Kim, 1988). So, although there might be exceptions, the general rule is to inform potential participants of sponsorship as a part of the informed consent process. At a minimum, to deceive people regarding the sponsorship of a study in order to gain their participation is certainly unethical.

Scientific Misconduct and Fraud

When a research project reaches the final stage—dissemination of results—the primary consideration regarding ethical conduct shifts from avoiding harming the participants in research to making sure the consumers of the research are not adversely affected. Results of a study are typically communicated in the form of a report to a sponsoring organization, a publication in a professional journal, or, possibly, a news release to the media. The preeminent ethical obligation in this regard is not to disclose inaccurate, deceptive, or fraudulent research results. Inaccurate results create at least two problems. First, they mislead scientists who depend on previous research to guide their work. Ethical violations concerning disclosure of results undermine the very nature of the scientific process that, as we have seen in Chapter 1 and Chapter 2, depends on building future knowledge on the foundation of existing knowledge. If we cannot depend on the accuracy of existing knowledge, then the scientific endeavor is threatened. In that case, the credibility of all research is damaged by such violations. Second, inaccurate results create difficulties for those who utilize the results of social science research in developing social policies or other applied activities. Deceptive or fraudulent disclosures of research results can cause these people to design useless—or in some cases even dangerous—programs, policies, or interventions based on the faulty studies.

Many ethical violations can occur in the process of reporting research (Morrison, 1990). **Fraud** is the deliberate falsification, misrepresentation, or plagiarizing of data, findings, or the ideas of others. This includes such things as falsifying data, embellishing research reports, reporting research that has not been conducted, or manipulating data in a deceptive way. **Misconduct** is a broader concept that includes not only fraud but also carelessness or bias in recording and reporting data, mishandling data, and incomplete reporting of results. Other questionable practices are irresponsible claims of authorship (listing co-authors who made no significant contributions to the research) and premature public release of results without undergoing the peer review process.

Estimates of the actual amount of scientific fraud and misconduct suggest that the problem is relatively minor. One investigation found only 26 cases between 1980 and 1987 in which research fraud was either admitted by the researcher or proved by an investigating body. Yet it does happen; at times, researchers have even published articles describing experiments that were never conducted (Engler, Covell, Friedman, Kitcher, & Peters, 1987). As for scientific misconduct, between 1988 and 1991, the National Science Foundation received only 99 allegations of such misconduct (Teich & Frankel, 1992). Of course, underreporting could hide a more serious problem. But even if the number of cases is small, any misconduct in research can still damage the credibility of all research and place at risk the users of that research.

The issue of who has responsibility for detecting scientific misconduct is still very controversial (Teich & Frankel, 1992). For many years, governmental agencies that fund research pushed this responsibility onto the shoulders of the universities and other institutions where research was actually conducted. By the late 1980s, however, the National Institutes of Health and the Department of Health and Human Services had both established their own offices to watch for misconduct. Most scientists, however, still consider protection against misconduct to be primarily a responsibility of the scientific community, which uses a number of mechanisms to control or detect misconduct. One is the *peer review* of article submissions (discussed in Chapter 1). Yet peer review, while a critically important mechanism, cannot detect plausible, internally consistent fabrications. In addition, the sheer number of research articles submitted to publications for review requires a large number of qualified reviewers who understand both the methodology and the content of the article and, in some fields, sufficient numbers of reviewers with the appropriate qualifications are not available.

A second mechanism for detecting misconduct is *replication*. Inaccurate or fraudulent results will be thrown into question when future researchers cannot repeat the results. However, the effectiveness of replication has been severely hobbled by the modern research system because research funds are not often appropriated for replication. Funding agencies prefer to fund research efforts that delve into new areas. In addition, many journals are disinclined to publish replication studies. Further, replication, where it does occur, tends to be reserved for those projects that have produced unusual results. Fraudulent studies that enhance the prestige of researchers but that do not run counter to accepted findings in a field are not likely to arouse enough attention to warrant replication (Engler et al., 1987).

Beyond peer review and replication, there are several other ways of reducing the chances of fraud or detecting it if does occur. One is to *supervise novice researchers* until they demonstrate good practices and ethical conduct as well as technical competency (in other words, socialize them into the scientific community). Second, journals can reduce fraud by *requiring more complete data to be submitted to reviewers,* even if they cannot be included in the published article. Third, professionals in the field can also help reduce fraud by *considering it their ethical obligation to report suspected cases* to appropriate authorities.

Even though not as unethical as purposeful deception, careless errors in research have the same effect of creating misinformation. Social researchers owe the scientific community carefully conducted research that is as free of error as possible. As with frauds, errors are discovered and corrected through peer review of research, reanalysis of data, and replication. For example, a study done in the 1980s claimed to document substantial negative economic consequences of divorce for women and their children (Weitzman, 1985). The results of the study served as an important foundation for developing social policies, and was cited by numerous state appellate and supreme courts and even the U.S. Supreme Court to justify rulings on divorce law reform. In part because the findings in this study were somewhat at variance with the results of other studies, a reanalysis of the data was undertaken (Peterson, 1996). The reanalysis showed that the original conclusions were due in part to errors in the original analysis. The reanalysis showed that women suffer economically after divorce but not nearly as severely as the original analysis had suggested. Had the original research conclusions not been based on an inaccurate analysis, the subsequent policy development might have been different. Scientists are human, and human beings make mistakes. But, as researchers, we must maximize safeguards to keep mistakes to the absolute minimum.

Beyond the problems of fraud and carelessness, researchers also have an obligation to report their results thoroughly. Researchers must take care to ensure that what is reported does not give a distorted picture of the overall results. Additionally, researchers should point out any limitations that might qualify the findings.

Some social science research is highly newsworthy, and this can also create special ethical dilemmas. For example, a study of crime, poverty, and other problems in an urban public housing project in St. Louis in the 1960s found itself under intense scrutiny by the press (Rainwater & Pittman, 1967). Journalists exerted considerable pressure on the researchers, before their study was complete, to release their findings about the housing project and its problems. Fearing that premature disclosure of results might be inadvertently inaccurate or deceptive, the researchers chose to say absolutely nothing: "Ethically, we did not want to be in a position of asserting findings before we were really sure of what we knew" (Rainwater & Pittman, 1967, p. 359).

Scientific Advocacy

Scientific knowledge rarely remains the exclusive domain of the scientific community. It typically finds its way into public life in the form of inventions, technological developments, or social policy. This raises potential ethical dilemmas in terms of the role of researchers as advocates of particular uses of their research results. What responsibility, if any, do scientists have for overseeing the use to which their results are put? To what extent should scientists become advocates for applying knowledge in a particular way? Quite naturally, disagreements exist on how to resolve these issues, and the disagreements partially reflect the different paradigms explored in Chapter 2. The classic approach to these issues derives from the positivist paradigm and the exhortations of the sociologist Max Weber (1946) that science should be "value free." Social scientists, according to Weber, should create knowledge, not apply it. They therefore have no special responsibility for the ultimate use to which that knowledge is put. Further, according to Weber, they are under no obligation to advocate particular uses of scientific knowledge. Indeed, advocacy is frowned upon as possibly threatening objectivity, which, as we have seen, is a central concern of those who adopt the positivist paradigm. So while remaining value free is difficult, many argue, abandoning the effort would be disastrous in that it would prevent us from acquiring an accurate body of knowledge about human social behavior and might threaten the researcher's credibility as a disinterested expert (Gibbs, 1983; Gordon, 1988; Halfpenny, 1982).

The opposite stance in this controversy was originally developed by Karl Marx (1848/1964) and reflects the critical paradigm. Marx championed the cause of the poor and downtrodden, and believed that social researchers should bring strong moral commitments to their work. They should strive to change unfair or immoral conditions. Following Marx, some researchers today believe that social research should be guided by personal and political values and should be directed toward alleviating social ills (Fay, 1987). Scientists should also advocate for how their research should be used by others who would help accomplish those personal goals.

A compromise position on the value-free controversy was proposed by sociologist Alvin Gouldner (1976). He pointed to the obvious; namely, that scientists have values just as do other human beings. He noted further that those values can influence research in so many subtle ways that their effects can never be totally eliminated. Instead of denying or ignoring the existence or impact of these personal values, scientists need to be acutely aware of them and up-front about them in their research reports. Being thus forewarned, other scientists and consumers of their research are better able to assess whether the findings have been influenced by personal bias. In addition, Gouldner argued that social scientists have not only the right but also the duty to promote the constructive use of scientific knowledge (Becker, 1967; Gouldner, 1976). Because decisions will be made by someone concerning the use of sci-

entific knowledge, scientists themselves are best equipped to make those judgments. As Gouldner states, technical competence would seem to provide a person with some warrant for making value judgments. Without necessarily adopting the critical paradigm, people who take Gouldner's position view the value-free stance of many positivists as a potentially dangerous dereliction of their responsibility as those most knowledgeable about the research. That responsibility accrues to scientists by virtue of their role in developing new knowledge and their expertise.

Nothing is wrong with researchers who openly press for the application of scientific knowledge in ways they see as desirable, so long as their advocacy does not prevent them from producing honest and accurate scientific results. The danger of advocacy is that scientists can come to feel so strongly about issues they promote that it might hamper the objective collection and analysis of data.

CODES OF ETHICS

A point that has been emphasized in this chapter is that ethical judgments are difficult and often controversial because they involve interpretation and assessment. Most professional organizations establish written codes of ethics to serve as guides for their members to follow. Though these codes by no means settle all debate, they do stand as a foundation from which professionals can begin to formulate ethical decisions. These codes of ethics can be found at the Web sites of the major organizations of professionals who conduct social science research. See, for example, the Web pages of the American Sociological Association (**www.asanet.org/ecoderev.htm**) the Society for Applied Sociology (**www.appliedsoc.org/ethics.htm**) and the American Association for Public Opinion Research (**www.aapor.org/ethics/**).

REVIEW AND CRITICAL THINKING

People in their everyday lives grapple with some of the same issues that have been described as research ethics issues in this chapter. Just as researchers operate in a context of moral principles and cultural values, all people have moral, philosophical, or theological beliefs that tell them what is proper behavior. Everyone has to take these beliefs or standards and apply them to particular situations when deciding how to behave. Applying ethics is no easier in our everyday lives than it is when doing research. You can incorporate some lessons from this chapter in terms of how to critically analyze ethical issues in your everyday life.

1. In your everyday relationships with other people, how do things like informed consent, deception, confidentiality, privacy, and the other ethical issues discussed in this chapter come into play, especially as these relate to assessing information? Give an example of each from your everyday life.

2. How do you regulate ethical behavior in your everyday life?

3. What "codes of ethics" do you rely on in order to do this? How are these codes of ethics similar to or different from the codes of ethics of the professional organizations discussed in this chapter?

Psychological Society or the American Criminological Society, or they could be practitioner organizations, such as the National Association of Social Workers, the American Medical Association, or the American Nurses Association. Explore these Web sites for topics related to ethics. Also use the search engines to look for a variety of key words, such as *ethics, science, research fraud, IRB*, or some combination.

The National Institute of Health has an extremely useful Web site from which ethics materials can be downloaded: **http://grants.nih.gov/80** The guidebooks at this site cover practically all the issues discussed in this chapter. Also, universities often publish their guidelines for research on their Web sites, so you can learn how their IRBs operate and how they protect human subjects. If you go to a university's Web page, you may find a link such as "Guidelines for Researchers" or "Research Policies and Procedures." This is where you will find issues related to the ethical conduct of research. Visit several sites and compare them to your own university's policies.

A final Web site to mention is called Ethics in Science: **http://www.chem.vt.edu/ethics/ethics.html.** This Web site has links to many resources related to ethical issues in research. Many of the links relate to the physical and natural sciences, but some apply to the social sciences, and many ethical issues span disciplines.

Computers and the Internet

Use Internet search engines to locate professional organizations' Web pages, other than those mentioned in the text. They could be research organizations, such as the American

Main Points

1. What is ethical in research is based on human values and varies as those values change.

2. Some ethical issues have to do with how researchers treat the subjects of their research. Informed consent refers to telling potential research participants about all aspects of the pending research before they agree to participate. Rigid adherence to the doctrine of informed consent can limit social research by eliminating some useful research methods and forcing the study of only those persons who volunteer.

3. Some level of deception is sometimes used in research, but it has been highly controversial whether, or under what conditions, deception is acceptable.

4. Research subjects have a right to privacy, and this can be protected by letting them edit data about themselves from a dataset, keeping the data anonymous, or keeping the data confidential. Confidentiality and anonymity are not the same thing. Intrusion by third parties, such as courts of law, can occasionally be a threat to the guarantee of confidentiality.

5. Exposing subjects to physical or mental distress should be kept to a minimum and should never be done without fully informed consent. Subjects should be thoroughly debriefed at the conclusion of the research. Research subjects should also gain some benefit from participating in the research.

6. Research subjects who occupy a disadvantaged status may be vulnerable to coercion to participate in research projects and, therefore, require special protection.

7. Withholding treatment from control groups raises an ethical dilemma, but it is often justified when testing unproved approaches.

8. The mistreatment of minority peoples in research has been a major impetus to the development of ethical standards for research in the United States and abroad.

9. Some ethical issues deal with how sponsors or the community at large is affected by the research. When conducting research for a sponsor, many ethical difficulties can be avoided by a detailed research agreement that covers such things as the purpose and nature of the research, rights of publication, and revealing to participants the sponsor of the research.

10. Researchers have an obligation to report their results fully and honestly and to avoid any kind of scientific misconduct or fraud. Replication is a major tool of science for correcting research errors and fraudulent reports.

11. It is the researcher's own decision regarding the degree to which he or she will become an advocate; however, caution is required so that objectivity is not undermined.

Important Terms for Review

anonymity
confidentiality
ethics
fraud (scientific)
informed consent
misconduct (scientific)
privacy

For Further Reading

Beauchamp, Tom L., Faden, Ruth R., Wallace Jr., R. Jay, & Walters, LeRoy (Eds.). (1982). *Ethical issues in social science research*. Baltimore, MD: Johns Hopkins University Press. A volume of readings that covers the range of ethical issues one is likely to face in social research.

Caplan, A. L. (Ed.). (1992). *When medicine went mad*. Totowa, NJ: Humana Press. This volume contains articles relating to the research conducted by the Nazis during World War II. Although it deals with an extreme situation that many hope will never occur again, it does provide an illustration of what can happen when science "goes mad."

Crossen, Cynthia. (1994). *Tainted truth: The manipulation of fact in America*. New York: Simon & Schuster. An enlightened book by a journalist about the many frauds that can be perpetrated by scientists. The author shows how frauds can be artfully crafted and how they can adversely affect us all.

Homan, Roger. (1991). *The ethics of social research*. London: Longman, 1991. This book provides a general overview of ethical issues in social science research and some ways of resolving ethical dilemmas.

Kimmel, Allan J. (1988). *Ethics and values in applied social research*. Beverly Hills, CA: Sage. An excellent overview of the ethical dilemmas that confront applied researchers.

Lee, Raymond. (1993). *Doing research on sensitive topics*. Newbury Park, CA: Sage. Ethical issues become especially important and complicated when doing research on sensitive topics, and this author suggests ethical guidelines for navigating through such treacherous waters.

Miller, Arthur G. (1986). *The obedience experiment: A case study of controversy in social science*. New York: Praeger. An exhaustive discussion of the controversial Milgram experiments on obedience to authority and their ethical implications, especially in terms of the psychological harm done to the participants. Other experiments involving deception and stressful conditions are also presented.

Penslar, Robin Levin. (Ed.). (1995). *Research ethics: Cases and materials*. Bloomington and Indianapolis: Indiana University Press. This book covers ethical dilemmas in psychology, history, and biology, but the issues are relevant to all the social sciences. Issues are explored by using a case-study method where cases are analyzed in detail.

Sieber, Joan E. (1992*). Planning ethically responsible research: A guide for students and internal review boards*. Newbury Park, CA: Sage. This short volume offers a good deal of practical information on how to translate ethical principles and federal regulations into valid applied research methods.

Chapter 4

PROBLEM FORMULATION AND RESEARCH DESIGN DEVELOPMENT

Suppose you were required, as are many students in courses on social research, to design and conduct a research project. My experience teaching research courses in the social sciences is that some students respond to this assignment by drawing a total blank. Other students grasp eagerly onto a topic, such as "the cause of drug addiction" and rush off with total confidence that they are about to solve this enduring problem. In each case, the student is having difficulty adequately formulating a research problem. In the first case, the difficulty is in locating a problem to investigate, whereas in the second, it is in formulating a problem sufficiently specific that it is amenable to scientific research. This problem is not unique to students. Every researcher must grapple with the issue of problem formulation. Because it is the initial step and provides the basis for the complete research project, problem formulation is of crucial importance. Many potentially serious difficulties can be avoided—or at least minimized—by careful problem formulation.

Chapter 4 provides an overview of the entire research process so that you can see the "big picture." It begins with a discussion of how researchers initially locate research topics. Then the major issues and processes involved in shaping and refining the problem into one that is clearly researchable are discussed. Much of this is accomplished during the first two steps of the research process described in Figure 1.1. This is followed by an overview of the remaining steps in actually conducting research as laid out in Figure 1.1. Finally, factors relating to the feasibility of research are discussed.

SELECTING A RESEARCH PROBLEM

The first hurdle confronting a researcher is to select an appropriate topic for scientific investigation. Actually, this is not as difficult as it may first appear because the social world around us is teeming with unanswered questions. Selecting a problem calls for some creativity and imagination, but there are also a number of places to which you can turn for inspiration.

Personal Interest

Research topics are often selected because a researcher has an interest in some aspect of human behavior, possibly owing to some personal experience. One social scientist, for example, conducted research on battered women and women's shelters in part because of her own earlier experience of being abused by her husband; another researcher, who had grown up in the only African American family in a small rural town, later researched prejudice, discrimination, and the experience of minorities (Higgins & Johnson, 1988). Research sometimes focuses on behavior that is unique or bizarre and thus compelling to some. Examples of such research abound, including studies of nudist colonies (Weinberg, 1968), pool hustlers (Polsky, 1967), juvenile gangs (Horowitz, 1987), striptease dancers (Dressel & Petersen, 1982), serial murderers (Holmes & DeBurger, 1988), and homeless heroin addicts (Bourgois, Lettiere, & Quesada, 1997).

Researchers who select topics out of personal interest must be careful to demonstrate the scientific worth of their projects. The purpose of research is to advance knowledge, not merely to satisfy personal curiosity. In his study of strippers, for example, sociologist James Skipper (1979) was interested in learning about how people adapt to a job that many consider deviant. Such a focus placed his research firmly in an established area of study and amplified its scientific contribution. A researcher who chooses a topic based on personal interest—especially if it deals with some unusual or bizarre aspect of human behavior—should be prepared for the possibility that others will fail to appreciate its worth. Even though Skipper, as noted, established a scientific rationale for his study of strippers, he was subjected to much abuse by those who failed to appreciate its scientific value.

Social Problems

In selecting a topic for research, you often need look no further than the daily newspaper, where you can read about the many social problems faced by modern societies. Problems such as crime, delinquency, poverty, pollution, overpopulation, drug abuse, alcoholism, mental illness, discrimination, and political oppression have all been popular sources of topics for social research. The Society for the Study of Social Problems—a professional organization to which many social scientists belong—publishes a journal titled *Social Problems* the sole purpose of which is to publish the results of scientific investigations into current social issues.

Within each of these general categories of social problems, a range of issues can be studied. Many studies, for example, focus on the sources of a problem. Others are concerned with the consequences these problems have for individuals or for society as a whole. Still others focus on assessing the effectiveness of social policies and social programs that are intended to ameliorate these problems. So there are a variety of different directions from which one can conduct research on any social problem.

Testing Theory

Some social scientists are primarily theoreticians—their professional work focuses on developing and verifying some abstract theory. For these researchers, research problems are selected primarily on the basis of their utility in testing and verifying a particular theory rather than because of interest in a particular substantive problem. Chapter 2 notes that theoretical concerns should be at issue to some degree in all research. At a minimum, nearly all research has some implications for existing theory. Certain research topics, however, are selected specifically for the purpose of testing

some aspect of a particular theory. Many theories in the social sciences have not been thoroughly tested. In some cases this means we do not know how valid the theories are, whereas in other cases it means that we do not know how wide the range of human behavior is to which the theory can be applied.

Prior Research

One of the most fruitful sources of research problems is prior research because the findings of all research projects have limitations. Even though some questions are answered, others always remain. In addition, new questions may be raised by the findings. It is, in fact, common for investigators to conclude research reports with a discussion of the weaknesses and limitations of their research, including suggestions for future research that follow from the findings that have been presented. Focusing on these unanswered questions or expanding on previous research is a good way to find problems to research.

Prior research can also lead to new research problems if we have reason to question the findings or validity of the prior research. Chapter 3 discusses a study done in the 1980s, which found that women experienced severe declines in their economic status after divorce, far more severe than men. Because the findings of this study were widely at variance with the existing research up to that time, it raised a question in some people's minds as to whether the study was done properly or whether some errors had crept into the analysis. This led to a reanalysis of the data, which changed the magnitude of the findings, although not the general direction (Peterson, 1996). As emphasized in Chapter 2, it is imperative that we not complacently accept research findings, especially when conclusions are based on a single study or are at variance with the results of other studies because there are numerous opportunities for error or bias to influence results.

Program Evaluation

Program evaluation focuses on assessing the effectiveness or efficiency of some program or practice. As noted in Chapter 1, evaluation has become an increasingly important activity of social science researchers. Agencies or organizations that fund social programs today typically demand that evaluation research be conducted if funding is to be granted or continued. Such research, developed for practical reasons, can take many forms. A social agency, for example, may require some needs assessment research to gather information about its clients if it is to deliver services to them efficiently. Prison officials need to know, before granting a parole, which criminal offenders pose the greatest risk to society. Governmental agencies that fund delinquency intervention programs need to know if their programs actually reduce delinquency among the teenagers they serve. In all these cases, the practical information required by an agency or organization determines the focus of the research effort.

The Politics of Research

From the preceding discussion, one could get the impression that the problem selection process is largely a matter of personal preference: Social scientists select problems they find interesting or important. But problem selection, like most other human endeavors, cannot be explained solely in such individual terms. In fact, if you were to ask students in research courses why they chose the term paper topics they did, you might find that, in addition to personal interest, theoretical orientation, or social problems, their choices were also governed by such factors as: "My instructor had a dataset available on this problem"; "I got financial aid to work as a research assistant"; or "I knew my prof was interested in this topic, so I hoped studying it might help me get a better grade." In other words, issues of political efficacy can influence problem selection.

The world of professional research is not unlike that of the student's world. However, the stakes are much higher, and the consequences much greater. Although the number of problems to be studied may be infinite, the resources society can allocate to research them are not. Research is a major societal enterprise

in which universities, governmental organizations, private research corporations, and independent researchers compete with each other for limited resources. In this competition, there are forces working in society to assure that the concerns of the powerful, the affluent, or vested interest groups receive attention from the research community. Thus, problem selection is, at least at times, very much a political issue involving the exercise of power. In this competition, the problems that affect minorities or other groups with little or no clout may not receive the research attention they deserve. The Eye on Diversity box provides some examples of this and explores the consequences.

The factors that influence the allocation of research funds include the following (Lally, 1977; Strickland, 1972):

- the existence of a powerful, articulate, and effective interest group that can push for research on a particular problem;

- support for research by influentials at the national policy-making levels;

- definition of a condition as a social problem by national influentials;

- public awareness of and concern about the problem;

- the severity, extent, and economic costs of the problem;

- the amount of publicity about the problem; and

- the amount of support for research on the problem in the major funding agencies.

On this last point, it is important to recognize that major agencies of the government, such as the Department of Health and Human Services (DHHS) and the National Science Foundation (NSF), dispense millions of dollars for research each year. Support from congressional leaders and key personnel in these major departments is essential for problem areas to be deemed worthy of financial backing for research. Typically, funding sources publish "requests for proposals" (RFPs), which outline

EYE ON DIVERSITY   Gender, Sexual Orientation, and the Politics of Problem Selection

Over the years, both gender and sexual orientation have been powerful influences on which problems researchers select as worthy of investigation. Consider the example of spouse abuse. Spouse abuse existed long before there was social research. Prior to the 1960s, however, one would have been hard pressed to find much research on the topic. Today, social science literature is replete with studies on the topic. What explains this change? Certainly there is no single answer, but a major factor has been the women's movement. Before the 1960s, women as a group had considerably less political power than they do today, and the special problems of this minority often received little research attention. Spouse abuse is now an important issue to the women's movement, and this politically powerful group has been able to translate its concerns into public policies that support research on this topic. Partly because of its pressure, the federal government now allocates substantial sums of money for domestic violence research. In addition, with the changing roles of women in society, more women in the past four decades have chosen to pursue careers as researchers in the social sciences. One consequence of more women in research positions is that, given the role of personal interest in the selection of research topics, they are more likely to identify spouse abuse as a problem warranting research investigation. As the topic gained more prominence in the social sciences, editors of journals became more receptive to publishing research articles on the topic. And all these factors, over time, had a snowball effect. Researchers seeking problems to study were attracted to the area by the availability of funds, the potential for publication, and the desire to contribute knowledge to an area other researchers recognized as important.

Another case where minority status influenced research funding was with Acquired Immune Deficiency Syndrome (AIDS) research. AIDS, of course, has become a huge health problem, with over 200,000 people in the United States having AIDS and close to 1 million being infected with the HIV virus (Stine, 1998). This will translate into massive health-care costs as these people develop full-blown AIDS. In the early years of the spread of AIDS in the United States, a substantial proportion of the people contracting the virus were men who had sex with other men, or intravenous drug abusers. During these years, AIDS was associated in many

---

the organization's funding priorities and requirements (see Chapter 16). Researchers are invited to submit proposals for competitive consideration with other researchers. Proposals may be for hundreds of thousands of dollars, and the competition for funding is as intense and high pressured as any big business deal.

The Eye on Ethics box addresses the ethical dimensions of these issues of problem selection. Selecting a research topic is often influenced by more than one of the factors discussed above.

Finding a research topic, however, is only the first step in problem formulation. The next step is to shape it into a problem that can be researched.

SHAPING AND REFINING THE PROBLEM

As mentioned, one frustrating trap into which novice researchers often become enveloped is

people's minds with two groups of people—gay men and IV drug abusers—who were viewed as marginal and highly stigmatized, especially by those in positions of power. In fact, AIDS was widely defined for a number of years as a "gay disease" and, thus, something that most people in the United States need not worry about. As long as AIDS was defined as a disease of "those deviants," funds for research on the cause and prevention of AIDS were slow in coming.

There were other reasons for the delayed response to the AIDS crisis, beyond the marginality and stigma suffered by its early victims (Shilts, 1987). One reason was the policy of the Reagan administration, which entered office in 1981, to emphasize smaller government and austerity in social and health programs. The first AIDS victims in the United States appeared in 1980. The competition for government funds in the early 1980s was fierce, and AIDS researchers typically lost the battle, partly because those suffering from AIDS had less clout in the policy process than did other groups competing for dwindling funds. Yet another reason the battle against AIDS was slow to start had to do with urban politics. New York City had the largest number of AIDS cases in the early 1980s, yet New York Mayor Edward Koch refused to do anything about it for a number of years, apparently because of his and his constituents' belief that support for gay causes would link him with the gay rights movement and hurt his chances for reelection.

In short, society's response to the minority status of those afflicted with AIDS, along with other political factors, contributed to a significant delay in attacking the problem. At the time, gays did not have clout in the domains where decisions were made about the funding of research. When AIDS began to threaten the blood supply for blood transfusions, and the fear rose that AIDS was entering the heterosexual population, then considerable research support was forthcoming. Even though researchers cannot dismantle this political process, they need to be aware of its influence and search for worthy research topics even though those topics may not draw the attention of influential groups. While politics will always influence problem selection, researchers must work to ensure that the other factors described here also play a part in which problems are the focus of research.

choosing a topic that is so broad and encompassing that, by itself, it offers little guidance in terms of how to proceed. Finding the "causes of juvenile delinquency" or the "weaknesses of the modern family" sounds intriguing, but these topics provide little direction concerning specifically where to begin to look. The next step in the research process, then, is to begin translating a general topical interest into a precise, researchable problem. The scope of the problem needs to be narrowed to manageable proportions. One investigation is unlikely to uncover "the causes of juvenile delinquency," but it might provide some insight regarding the influence of particular variables on the emergence of particular forms of delinquency. Refining, narrowing, and focusing a research problem do not occur quickly or necessarily simultaneously; in fact, they can form a continuous process that may evolve over much of the duration of the research project itself. A discussion of the major

EYE ON ETHICS **Power Structures and the Selection of Research Problems**

One of the statements of ethical principles presented in the beginning of Chapter 3 has to do with maintaining fairness and equity in social relationships: Social science research impacts on people's lives. Sometimes it brings them advantages, at other times the consequences for people are negative. An enduring, if not inevitable, feature of societies is an inequitable distribution of power and resources: Some groups have more social and economic resources, and these resources can at times mean that these groups are better equipped to sponsor research that may benefit them. Does this mean that social science research merely becomes the tool of those groups who can afford to sponsor the research? Nothing, of course, is quite that clear cut. But it does mean that disadvantaged groups find an inordinately small share of research resources being devoted to their concerns and benefits. Sociologist Edna Bonacich describes what can occur:

> Many people become sociologists because they want to make society more just. Yet they find themselves either doing research on behalf of existing power structures because that is where funding is obtained; or doing research for their own professional career advancement (Bonacich, 1990, p. 7).

Bonacich argues that this is an ethical dilemma for all the social sciences in that it deals with the equitable distribution of the benefits derived from social research. One way to deal with this problem is for researchers to seek out research problems that are not on the agenda of more powerful groups. This is essentially what happened, over a period of years, with the topic of spouse abuse. A second way to deal with this problem is for researchers to listen to the voices of all the stakeholders in the research, irrespective of who is funding the research. A

procedures involved in this shaping process follows.

Conceptual Development

Chapter 2 discussed the role of theories and hypotheses in the research process, pointing out that concepts are one of the central components of theories. In refining a research problem, one of the key steps is *conceptual development*: identifying and properly defining the concepts that will be the focus of the study. In an exploratory study, a researcher may enter an area where there is little conceptual development, and a major purpose of the research itself may be to identify and define concepts. In cases where there is existing theory and re-

search to rely on, however, some conceptual development still occurs as a part of formulating a research problem. One part of this process (discussed in Chapter 2) is to clearly define the meaning of concepts. Another part of the process is to narrow the focus of the concepts so that they encompass something that is feasible to research in a single study. In a study of juvenile delinquency, for example, researchers might ask: "Are we interested in all forms of delinquent behavior or only in some types?" In reality, the concept "delinquency" is an extremely broad category that includes all actions by juveniles that violate criminal or juvenile codes. There is no reason to assume that all types of delinquency can be explained on

stakeholder is any person or group who stands to gain or lose from the research. The powerful stakeholders, especially those who are funding the research, will have their voices heard. It is the less powerful stakeholders that the researcher needs to seek out and listen to. For example, if a state department of corrections funds a study on some aspect of prison life, then the corrections administrators will certainly have an influence on what is researched and how it is done. But other stakeholders to this research include the prisoners themselves and the correctional officers who are in daily contact with them. What kind of research might benefit these groups or work to their disadvantage? Social scientists striving to make research more equitable would make efforts to ensure that the voices of the inmates and correctional officers are heard.

Bonacich proposes an even more radical way of dealing with this problem: Circumvent the power structure altogether by having social science disciplines establish a program of "social research aid," similar to legal aid. Just as legal aid lawyers provide their services to the poor and disadvantaged at low or no cost, social scientists could provide free or low cost research services to the same groups who otherwise could not afford to pay for such services. In this way, the interests of disadvantaged groups would be addressed to some degree and serve as a counterbalance to the fact that advantaged groups fund and use research for their own benefit. This would devote some research expertise to the problems of the impoverished and others whose problems would not normally draw attention in the political competition for research funds. Certainly everyone agrees that researchers need to earn a living, but it is also important to recognize the ethical obligation to ensure that science and scientific research are broadly beneficial to the whole of society and not just to limited groups.

the basis of a single cause. The focus of the research, therefore, might be narrowed so that it includes only certain behaviors, such as theft or truancy. The goal of this specification process, then, is to make clear exactly what the focus of the research effort is to be.

Once key concepts have been clearly defined, the next consideration is their measurability. Only concepts that are in some way measurable can be used in the research process. Eventually, of course, concepts will have to be operationalized (as pointed out and explained in Chapter 2), so any that are not readily measurable will be eliminated. Measuring concepts can sometimes be difficult (discussed in more detail in Chapter 5 and Chapter 6). In

fact, theories at times include concepts that are difficult to operationalize. Theorists are sometimes criticized for this practice, although the criticism is misdirected. Theorists, of necessity, must be free to create and utilize whichever concepts are deemed necessary without regard to their immediate measurability. To do otherwise would limit theoretical development to those concepts that we currently have the skill to measure (Denzin, 1989b; Shearing, 1973). Theorists' use of concepts that are not immediately measurable allows for theoretical advances, but it also presents researchers with the task of creating ways to measure the concepts. However, if concepts in a proposed study cannot be measured, then some modification in the project—and possibly in the theory—will be necessary. This process of refining and developing concepts as a part of the research process illustrates a point made in Chapter 2 regarding the interplay between theory and research: Theories provide concepts and hypotheses for research, whereas research modifies theories through conceptual development.

Review of the Literature

An important assist in problem formulation in general, and conceptual development in particular, is to conduct a thorough review of previous research that relates to the current research problem. This review of the literature is a necessary and important part of the research process (Locke, Silverman, & Spirduso, 1998). Researchers do it in order to familiarize themselves with the current state of knowledge regarding a research problem and to learn how others have delineated similar problems. Unless researchers are planning a replication study, it is unlikely that they will formulate a problem precisely like any of those previous studies. Rather, researchers are likely to glean ideas from several previous studies that can be integrated to improve the proposed study. Through reviewing the relevant literature, a researcher can ensure that his or her proposed study does not unnecessarily duplicate what others have already done. Researchers will undoubtedly find that pitfalls can be avoided by learning from others' experiences. It may be, for example, that several studies using specific approaches have failed to find significant results or strong relationships. Unless there is good reason to believe that another attempt will be more successful, there is little point in reusing the identical approach.

A thorough literature review calls for familiarity with basic information retrieval resources, including how to locate and evaluate books, professional journals, public documents, and online sources of information. To help with this important aspect of doing research, see Appendix A, which discusses information retrieval strategies and resources, including how to use the library and the Internet. This appendix shows you how to find the books, journals, government documents, and other sources in which the reports of research are found. Even those with some experience using the library will likely find some helpful new information in this appendix. Additionally, the Computers and the Internet section of each chapter provides some guidance on doing literature reviews on the Internet.

In a literature review, a systematic search is conducted of each research report for certain kinds of information. First, the reviewer pays attention to *theoretical* and *conceptual issues*: What concepts and theories are used? How well developed are they? Have they been subjected to empirical test before? If the theories and concepts are well developed, they can serve as an important guide in designing the planned research and explaining the relationships between variables. If they are not well developed, one will have to rely more on personal insight and creativity. In this case, researchers sometimes consider doing exploratory research, which may involve loosely structured interviews and less quantitative measuring devices, as a way of advancing conceptual and theoretical development.

A second component of a literature review is the *research hypotheses*, including identification of the *independent* and *dependent variables*. Are the hypotheses clearly stated and testable? Are they related to the variables and hypotheses being considered in the planned study? Existing research can provide some fairly specific direction in terms of already tested relationships between independent and dependent variables.

A third focus of a literature review should be the *measurement* and *operational definitions* that have been used in previous research. As noted, successful operationalization of concepts is often difficult. Previous work in this area is invaluable in finding workable measures for concepts. Measures used in the past may, of course, require modification to meet current needs, but making these modifications is likely to be easier than developing completely new measures, which is a difficult and time-consuming process.

The literature review will also provide information about a fourth important element of research: the most appropriate *research* or *observational technique* for a particular research problem. Successful approaches by others should be noted, and unsuccessful ones should be avoided. It is of the utmost importance that the problem determine the research technique used, not the other way around. A variety of data-gathering techniques exists because no one method is always best. As will be noted in subsequent chapters, each technique has its strengths and weaknesses; each is suitable for answering some questions but not others.

A fifth element obtained from a literature review is the *sampling strategy*. Previous research can be useful in determining the sampling strategy that should be used and in avoiding sampling problems encountered by others. Suppose, for example, that the study we propose calls for the use of mailed questionnaires. An ever-present problem with mailed questionnaires is being sure that a sufficient number of people complete and return them. It would be useful for us to know what the experience of other investigators has been with people like those we plan to survey. If the group we are proposing to sample has exhibited notoriously low return rates in previous studies with mailed questionnaires, we have to plan accordingly. We would likely increase the number of questionnaires mailed and would certainly use all available means of obtaining the highest response rate possible. Or, we may want to search for another group to study, choose another method of data collection, or even consider whether this particular project is feasible.

A sixth element of a literature review is the *statistical techniques* used. In Chapters 13, 14,

and 15, issues relating to appropriate use of statistics are discussed. In a literature review, one must be aware of whether the appropriate procedures were used; if any inappropriate techniques were used; and what constraints the concepts, variables, and hypotheses placed on the kind of statistics that would be appropriate.

Finally, a literature review would note the *findings* and *conclusions* of the study: Which hypotheses were confirmed? What guidelines for future research were presented? One aspect of the findings to watch for is the *effect size*, which refers to how great an effect an independent variable has on a dependent variable. You need to assess whether a dependent variable is affected only in a small, albeit measurable, way or whether the impact is dramatic.

A thorough literature review would involve evaluating and comparing many research reports, identifying where they used similar procedures and reached similar outcomes, and where there were discrepancies between studies. This can be a complicated process, especially when hundreds of studies may be involved. It is sometimes helpful to produce a summary table to make comparisons. An example of such a table, comparing studies on the relationship between self-esteem and teenage pregnancies, is presented in Table 4.1. Studies that show an association between the two variables are in the top half of the table; those showing self-esteem as not related to pregnancy are in the bottom half. Note that this table cites each separate study in the left-hand column, including a notation as to whether it was a longitudinal study. The second column provides information about the sample used; the third column indicates which scale was used to measure self-esteem. The right-hand column indicates, for those studies showing an association, the direction of the relationship. In this case, the relationships are all negative, indicating that teenagers with high self-esteem are less likely to become pregnant.

A systematic literature review of this sort provides the most useful information from previous studies. One can see at a glance how

TABLE 4.1 A Summary Table of a Literature Review of the Association between Self-Esteem and Teenage Pregnancy

	Sample Used	Self-Esteem Scale	Association[a]
Self-Esteem Related to Teenage Pregnancy			
Barth, Schinke, and Maxwell, 1983	117 nonwhite females, 68 white females; 11–21 years old	Rosenberg Self-Esteem Scale	Negative
Kaplan, Smith, and Pokorny, 1979 (longitudinal)	410 females	Self-Derogation Scale	Negative
Robbins, Kaplan, and Martin, 1985 (longitudinal)	2,158 males and females; white, black, and Hispanic	Self-Derogation Scale	Negative, for females only
Werner and Smith, 1977 (longitudinal)	614 males and females; 18 years old	California Personality Inventory	Negative
Self-Esteem Not Related to Teenage Pregnancy			
Brunswick, 1971	196 black females; 12–17 years old	Rosenberg Self-Esteem Scale (subset)	
Streetman, 1987	93 nonwhite females; 14–19 years old	Coopersmith Self-Esteem Inventory; Rosenberg Self-Esteem Scale	
Vernon, Green, and Frothingham, 1983 (longitudinal)	745 black females, 22 other nonwhite females; 91 white females; 13–19 years old	Coopersmith Self-Esteem Inventory	

[a] A negative association indicates that low self-esteem is associated with a greater risk of becoming pregnant during adolescence.

Source Adapted from Susan B. Crockenberg and Barbara A. Soby, "Self-esteem and teenage pregnancy," in *The social importance of self-esteem,* edited by Andrew M. Mecca, Neil J. Smelser, and John Vasconcellos (Berkeley: University of California Press, 1989), p. 148.

many studies came to similar conclusions and how certain measuring devices were commonly used. The ability to compile and summarize succinctly the features of studies in this fashion is essential to formulating a research problem and refining it into a research question that can be investigated empirically.

Levels and Units of Analysis

Chapter 2 discusses different levels at which theories can be focused and introduces the notions of microtheories and macrotheories. Reflecting these different theoretical levels, re-searchers also have to identify the level at which the analysis in their research will occur. Generally, two levels of analysis can be identified. **Microlevel research** focuses on the face-to-face social interaction and social process that occurs among individuals. **Macrolevel research** focuses on large-scale social structures and the social processes that occur among them. These different levels of research reflect the fact that there are many specific units of analysis that can be investigated in a research project. **Units of analysis** are the specific objects or elements whose characteristics we wish to describe or explain and about which data

TABLE 4.2   Possible Units of Analysis in Research on Juvenile Delinquency

Unit of Analysis	Example	Appropriate Variables	Research Problem
Individuals	Adolescents arrested for larceny	Age, sex, prior arrests	Do males receive different penalties from females for similar offenses?
Groups	Delinquent gangs	Size, norms on drug usage	Are gangs involved in drug trafficking more violent than other gangs?
Organizations	Adolescent treatment agencies	Size, auspices, funding level	Do private agencies serve fewer minority and lower-class delinquents than public agencies?
Programs	Delinquency prevention programs	Theoretical model, type of host setting	What services are most frequently included in prevention programs?
Social artifacts	Transcripts of adjudication hearings	Number of references to victim injury	To what extent does the level of violence in the offense affect the kind of penalty imposed?

will be collected. Although there are many units of analysis, five used commonly in social science research are individuals, groups, organizations, programs, and social artifacts (see Table 4.2). The first would be the unit of analysis in most micro-level research; the others would be the units of analysis appropriate to most macrolevel research. (There are different units of analysis used in studying documents, and these are discussed in Chapter 10.)

Much social research focuses on the *individual* as the unit of analysis. The typical survey, for example, obtains information from individuals about their attitudes or behaviors. Anytime we define our population of inquiry with reference to some personal status, we are operating at the individual level of analysis. For example, unwed mothers, welfare recipients, hockey players, Nobel prize winners, and similar categories all identify individuals with reference to a status they occupy.

If we identify our unit of analysis as individuals, it is important to recognize that the entire analysis will remain at that level. For the sake of describing large numbers of individuals, it is necessary to utilize summarizing statistics such as averages. We might, for example, as a part of a study of teenage mothers, note that, when giving birth, their average age was 16.8. Aggregating data in this fashion in no way changes the unit of analysis. Our data are still being collected about individuals.

Social scientists sometimes focus on social *groups* as their unit of analysis and collect data on some group characteristic or behavior. Some groups are made up of individuals who share some social relationship with the other group members. For example, in families, peer groups, occupational groups, or juvenile gangs, the members have some sense of membership or belonging to the group. If we study families in terms of whether they are intact or not, we are investigating the characteristics of a group—the family—not the individuals. Other groups of interest to social scientists are merely aggregates of individuals with no necessary sense of membership, such as census tracts, cities, states, or members of a particular social class. For example, we might study the relationship between poverty and delinquency by comparing rates of delinquency in census tracts with low income and in census tracts with high income. In this case, we have collected data regarding the characteristics of census tracts rather than individuals.

Social scientists also deal with *organizations* as the unit of analysis. Formal organizations are deliberately constructed groups organized for the achievement of some specific goals. Examples of formal organizations include corporations, schools, prisons, unions, governmental bureaus, and social service agencies. For example, a social scientist might be interested in the social conditions under which these organizations are likely to change. A theory might propose that larger or older organizations are, for a variety of reasons, less likely to exhibit organizational change under conditions that are conducive to change (Barnett & Carroll, 1995). If the researcher looked at rates of change in organizations of differing sizes and ages, then the study would be utilizing as the unit of analysis organizations, not individuals or groups. *Rate of change, organizational size,* and *organizational age*—the variables being studied—are characteristics of organizations, at least as defined here.

Research can also focus on *programs* as the basic unit of analysis. The program may provide services for individuals, and exist as part of an organization, but it is still a separate unit of analysis about which data can be collected. Like organizations, programs can have success rates or be assessed in terms of overall costs. For example, one research project investigated 25 programs providing services for pregnant and parenting teenagers (Fernandez & Ruch-Ross, 1998). Each program was assessed by its overall "success rate": a successful program was one in which the clients were more likely to remain in school or stay employed and less likely to become pregnant than were clients in the other programs. Note that a program can have a success *rate* (in other words, a certain proportion of its clients succeeding) but an in-

dividual can only succeed or fail (rather than showing a rate of success). The researchers in this study compared the programs to determine the characteristics of successful and unsuccessful programs. Programs might cut across a number of different organizations, such as social service agencies, in which case the unit being observed is the effectiveness of the combination of services provided by these organizations.

Finally, the unit of analysis may be *social artifacts*, which are simply any material products produced by people. Examples are virtually endless: newspapers, buildings, movies, books, magazines, automobiles, songs, graffiti, and so on. As reflections of people and the society that produces them, analysis of social artifacts can be very useful. Books and magazines, for example, can be used as artifacts in the assessment of sex-role stereotyping. Children's books have been attacked for allegedly reinforcing traditional sex roles through their presentations of men and women (Peterson & Lach, 1990; Purcell & Stewart, 1990). Another creative approach to social artifacts is the study of urban graffiti (Ferrell, 1995). Social scientists have studied hip-hop graffiti on buildings and other locations in terms of their social and cultural significance, seeing it as a form of youthful communication and protest. Such social artifacts provide social scientists with another window on social life.

Clearly specifying the unit of analysis in research is highly important in order to avoid a serious problem: an illegitimate shift in the analysis from one unit to another. Careless jumping from one level to another can result in drawing erroneous conclusions. One way in which this can happen is called the **ecological fallacy**: inferring something about individuals based on data collected at higher units of analysis, such as groups: There is a mismatch between the unit of analysis about which data are collected and that about which conclusions are drawn. Suppose, for example, a study found that census tracts with high rates of teenage drug abuse also had a large percentage of single-parent families. We might be tempted to conclude that single-parent families are a factor promoting teenage drug abuse. Such a conclusion, however, represents an illegitimate shift in the unit of analysis. The data have been collected about census tracts, which are at the group level. The conclusion drawn, however, is at the individual level, namely, that teenage drug abusers live in single-parent families. The data do not show this. They only show the association of two rates—substance abuse and single parenthood—in census tracts. It is, of course, possible that relationships found at the group level will hold up at the individual level, but they may not. It is always an empirical question whether relationships found at one level of analysis will hold up at other levels. In our hypothetical study, it may be that some other characteristic of census tracts leads to both high rates of drug abuse and single-parent families.

Fallacious reasoning can occur in the opposite direction and is called the **reductionist fallacy**: inferring something about groups, or other macrolevels of analysis, based on data collected from individuals. Suppose we collected data from individual teenagers about their drug abuse and family environments and found an association; namely, that teens from single-parent families are more likely to abuse drugs. Could we then conclude that communities with high rates of single-parent families would also have high rates of teenage drug abuse? The answer once again is that we could not draw that conclusion about the macrolevel (communities) with any certainty because the data we have is about social process at the microlevel (what happens in the lives of individual teens). It well may be that the social process that produces high rates of drug abuse in communities differs from the social process that leads individuals to abuse drugs. When data is collected at one level of analysis, it is always an empirical question as to whether conclusions can be drawn from that data about

other levels of analysis. A clear awareness of the unit of analysis with which we are dealing can help ensure that we do not make such illegitimate shifts.

A final point needs to be made about the unit of analysis in contrast to the source of data. The unit of analysis refers to the element *about which* data are collected and inferences made, but it is not necessarily the source *from which* data are collected. A common example is the U.S. Census, which reports data on *households*. We speak of "household size" and "household income," but households don't fill out questionnaires; people do. In this case, individuals, such as the heads of households, are the *source* of the data, but the household is the unit of analysis *about which* data are collected. When the unit of analysis is something other than the individual, attention must be paid to the source of the data because this might introduce bias into the data analysis. For example, when the household is the unit of analysis, data are often collected from one member of the household. In single-parent families, which are primarily headed by women, we would be gathering data mostly from women. In two-parent families we would be obtaining data both from men and from women since either could be the head of the household and, in some cases, men might represent the majority from whom data are collected. If men tend to give different answers to some questions than women, there could be a sex bias in the results, even though our unit of analysis was not linked to sex. A difference that we attribute to single-parent as compared to two-parent families may be due to the fact that the former involves mostly women answering questions, whereas the latter involves more men.

Reactivity

Another consideration in refining a research problem is the issue of *reactivity*. **Reactivity** refers to the fact that people can react to being studied and may behave differently from when they don't think they are being studied. In other words, the data collected from people who know they are the object of study might differ from data collected from the same people if they did not know. A reactive research technique therefore changes the very thing that is being studied. Suppose, for example, that you are a parent. A researcher enters your home and sets up videotaping equipment to observe your interactions with your children. Would you behave in the same way that you would if the observer were not present? You might, but most people would feel some pressure to be "on their toes" and present themselves as "good" parents. You might be more forgiving of your child, for instance, or give fewer negative sanctions. If you do this, then your behavior is being changed by the fact that you are being observed.

Reactivity in research can take many forms and is a problem for virtually all the sciences. However, it is especially acute in social research because human beings are so self-conscious and aware of what is happening to them. Refining a research problem and choosing a research design are done with an eye toward reducing as much as possible the extensiveness of reactivity. This will be considered in assessing the various research strategies in later chapters.

Qualitative and Quantitative Research

Another aspect of refining a research problem is to decide whether qualitative or quantitative research, or a combination of the two, is most appropriate to a research problem. This issue has already been discussed at some length in Chapter 1 and Chapter 2. Recall that **qualitative research** basically involves data in the form of words, pictures, descriptions, or narratives; whereas **quantitative research** uses numbers, counts, and measures of things. The decision as to which general orientation to follow in a given research project depends primarily on two factors: the state of our knowledge on

a particular research topic, and the researcher's assessment regarding the nature of the phenomenon being studied. When knowledge is sketchy or when there is little theoretical understanding of a phenomenon, it may be impossible to develop precise hypotheses or operational definitions. In such cases, researchers often turn to qualitative research because it can be more exploratory in nature. The research can be very descriptive, possibly resulting in the formulation rather than the verification of hypotheses. When there is enough previous research on a topic, it may be more feasible to state concepts, variables, and hypotheses precisely. It also may be possible to develop quantifiable operational definitions of what you are interested in, which then allows research to take on a more quantitative nature.

The second consideration in choosing between qualitative and quantitative research stems from a more fundamental controversy over the nature of social phenomena, which is explored in Chapter 2 in the context of presenting the various theoretical paradigms in the social sciences. It is a complicated issue but basically involves debate over whether the human experience can be reduced to numbers and measures meaningfully. Those who adopt the interpretivist paradigm argue that the human experience has a subjective dimension to it—the very personal meanings and feelings that people have about themselves and what they do. These meanings or feelings cannot be well captured through numbers or measures. They are better captured through narrative descriptions of people going about their daily routines or through lengthy and broad-ranging interviews. Interview techniques are better able to capture the very critical subjective meanings that are an essential element of understanding human behavior. For researchers who adopt this approach, knowledge and truth arise from gaining insight into the social meanings that pervade and inform people's lives. Positivists, on the other hand, argue that science is attempting to discover an objective reality and that quantitative research can tell us much about that objective social reality. In addition, they argue, quantification provides us with far more objective and precise statements about human behavior than does qualitative research.

The line between qualitative and quantitative approaches is not always completely clear, and the choice between the two can be difficult. Many research projects incorporate both approaches in order to gain the most benefit. In later chapters, additional considerations relevant to this choice between quantitative and qualitative research are discussed.

Cross-Sectional versus Longitudinal Research

Refining a research problem also requires a decision about the time dimension. The basic issue involved is whether you want a single "snapshot" in time of some phenomenon or an ongoing series of photographs over time. The former is called **cross-sectional research**, and focuses on a cross section of a population at one point in time. Many surveys, for example, are cross-sectional in nature.

Although all the data in cross-sectional research are collected at one time, such studies can nonetheless be used to investigate the development of some phenomenon over time. For example, to study the developmental problems of children of alcoholic parents, one could select groups of children of varying ages, say, one group at age 5, another at age 10, and a third group at age 15. By observing differences in developmental problems among these groups, we may infer that a single youngster would experience changes as he or she grew up similar to the differences observed among these three groups. Yet, one of the major weaknesses of such cross-sectional studies is that we have not actually observed the changes an individual goes through; rather, we have observed three different groups of individuals at one point in time. Differences among these groups

may reflect something other than the developmental changes that individuals experience. Because of this disadvantage, researchers with hypotheses about change over time often prefer the other way of handling the time issue: *longitudinal* studies.

Longitudinal research involves gathering data over an extended period, which might span months, years, or, in a few cases, decades. One type of longitudinal approach is the **panel study,** in which the same people are studied at different times. This allows us to observe the actual changes that these individuals go through over time. For example, a study of the social, psychological, and familial characteristics that influence whether drug addicts can successfully remain free of drugs followed the same 354 narcotics addicts for 24 + years, collecting data on their family experiences, employment records, and a host of other factors (Bailey, Hser, Hsieh, & Anglin, 1994). A second longitudinal approach is the **trend study,** in which different people are observed at different times. Public opinion polling and research on political attitudes are often trend studies. Some people would argue that trend studies are not really longitudinal because they don't observe the same people over time. However, they do study different groups over time, which makes them an improvement over cross-sectional research for studying change over time.

The decision of whether to use a longitudinal or cross-sectional approach is typically determined both by the nature of the research problem and by practical considerations. Longitudinal studies, especially panel studies, have the advantage of providing the most accurate information regarding changes over time. A research question regarding such changes, then, would probably benefit from this approach. A disadvantage of panel studies is that they can be reactive: People's responses or behavior at one time may be influenced by the fact that they have been observed earlier. For example, a person who stated opposition to abortion in one survey may be inclined to respond the same way six months later so as not to appear inconsistent or vacillating, even if those attitudes had changed in the interim. Another disadvantage of panel studies is that persons who participated early in a panel study may not want to, or may be unable to, participate later. People die, move away, become disinterested, or in other ways become unavailable as panel studies progress. This loss of participants can affect the validity of the research findings adversely. The disadvantages of all longitudinal studies are that they can be difficult and expensive to conduct, especially if they span a long period of time.

Cross-sectional research is cheaper and faster to conduct, and one need not worry about the loss of participants. However, cross-sectional research may not provide the most useful data for some research questions involving change over time. Thus, the decision on the issue of the time dimension should be based on considerations of both the nature of the research problem and practical issues. There are times, of course, when practical feasibility can play a large part in the decision.

ASSESSING RELATIONSHIPS BETWEEN VARIABLES

We saw in Chapter 2 that theories are developed by looking at hypotheses, variables, and the relationships between variables. In formulating a research problem, one aspect of the relationship between variables that must be considered is whether any of the variables are causally linked. While not all research involves stating causal relationships, many research projects involve tests of whether variation on some independent variable is the cause of variation on a dependent variable. If causal relationships are being assessed, then this may influence how a problem is formulated and a research design developed.

Causality and Determinism

For some scientists, one of the more important yet difficult tasks in scientific research is the search for causes—the reasons why particular forms of behavior occur. By **causality**, we mean that some independent variable (X) is the factor, or one of several factors, whose change produces variation in a dependent variable (Y). Why do some juveniles become delinquent whereas others present no behavioral problems? Why do some societies develop stable democratic institutions while others do not? Discovering causal relationships is a difficult task because causality cannot be observed directly. Rather, it must be inferred from the observation of other factors. Because of this, the philosopher John Kemeny has labeled causality "the mysterious force" (1959, p. 49). We cannot see, feel, or hear it, but we often assume it is there and many scientists search for causality with hopefulness and diligence.

This search for causes rests on an assumption of **determinism**: the belief that there is order in the universe, that there are reasons why everything happens, and that scientists, using the procedures of science, can discover what those reasons are. This assumption of determinism contradicts some commonsense beliefs about human behavior: that people are unpredictable or that their behavior is a product of free will or voluntary choice. This is a complicated philosophical issue. Certainly, there may be aspects of people's behavior that we will never be able to predict, and people do at times make choices. Yet we don't have to look very far to see how predictable people are. When I drive into the university later this morning, I can predict that, on two-way streets, all the drivers going in my direction will be on my right-hand side of the road while the drivers going in the opposite direction will be on the other side of the road. Amazing predictability!

As for the issue of making choices, making choices does not necessarily mean that social influences are not causing, at least in part, people's behavior. For example, students in my classes presumably "choose" each day whether to attend class. Yet, I can have a profound influence on their "choice" by announcing in class today that all students who attend class tomorrow will receive five additional points toward their final grade in the course. In all likelihood, most students will "choose" to attend class tomorrow, including some students who, prior to my announcement, may have decided not to attend class tomorrow. Is their behavior a product of free will, or is it constrained (caused) by the external force of the reward structure that I had the authority to create in the situation?

Clearly, the situation is complicated, but just as clearly people's behavior is predictable to a degree and determined to some extent by social conditions. To assume that human behavior is not caused by something is to say that there is no order or pattern to social life, and that contradicts what we seem to see around us and what we can see throughout human history. This is not to say that the search is easy or that some things may not always be beyond our powers to explain through the methods of science. Some instances of human behavior may be truly unpredictable, yet science can still understand and explain those parts that are predictable.

One final point must be made about issues of determinism and predictability in the social sciences. Social science predictions are mostly *probabilistic* in nature rather than absolute. We make predictions about the probability that certain events will occur among a whole group or aggregate of people; we don't, and can't, for the most part, make predictions about how a particular individual will behave. So, I can predict that, when I offer extra credit for attendance in my class, the percentage of students who attend will increase, but I can't say which particular students will attend. One reason for this is that many factors shape the behavior of individuals—some pushing them toward attendance (interest in the subject matter, enjoyment of the professor's humor, and the like) and others deflecting them from attendance (a sick child at home, working extra hours to obtain income, and so on). This complex of factors may produce, say, a 75% attendance rate on any given day. When I add to

that mix of causal factors the five points toward the final grade, this will push some students who might have missed class to attend that day, increasing the aggregate attendance rate that day to, say, 90%. But I could not necessarily predict whether a particular student would attend.

Criteria for Inferring Causality

To infer the existence of a causal relationship, one must demonstrate the following.

1. A statistical association between the independent and dependent variables must exist.

2. The independent variable must occur prior in time to the dependent variable.

3. The relationship between independent and dependent variables must not be spurious; that is, the relationship must not disappear when the effects of other variables are taken into account.

An Association

Each requirement of causal inference will be considered in the context of an issue that is much in the news today: the campaign to eliminate cigarette smoking. Over the years, there have been reports in the media about the negative impact of cigarette smoking on people's health. Some have argued that making these reports public as a part of a health campaign can motivate people to quit smoking. Table 4.3 presents hypothetical data seeming to show a

TABLE 4.3   Effectiveness of Reading Media Reports on Smoking Cessation

		Person Reads Report	
		Yes	**No**
Person Quits Smoking	Yes	200 (50%)	135 (27%)
	No	200 (50%)	365 (73%)
	Totals	400 (100%)	500 (100%)

link between *reading reports* about smoking (the independent variable) and actually *quitting smoking* (the dependent variable): Fifty percent of those who read the reports quit smoking, compared to only 27% of those who did not read the reports. Finding such a statistical relationship satisfies the first criterion for establishing a causal relationship.

The Proper Time Sequence

The second requirement—that the independent variable occur prior in time to the dependent variable—is not as easy to establish. A major factor in this is the nature of the study. Some research techniques, such as the experiment or participant observation, are inherently *longitudinal*, which means that the researcher is in a position to trace the development of behavior as it unfolds over time. In these cases, establishing the time sequence of events is generally simple. Questions of temporal order are more difficult to resolve when dealing with *cross-sectional data*, such as surveys, in which measurements of the independent and dependent variables occur simultaneously. This is especially true if the question of temporal sequence is not addressed until after the data have been collected. It is sometimes possible to sort out the time sequence of variables in survey data by asking additional questions. However, if the necessary information is not gathered at the time of the survey, establishing the appropriate time order of the variables may be impossible—ergo the emphasis on the importance of carefully considering issues of data analysis when originally developing a research design.

The data in our illustration may suffer from this problem. One interpretation of the data is that *reading reports* is the independent variable that has an influence on whether people *quit smoking*, the dependent variable. For this interpretation to be correct, the reports would have to have been publicized and read before the people quit smoking. If the respondents were not asked when they quit smoking, it

would be impossible to say whether they quit smoking before or after reading the reports. If they quit smoking before reading the reports, then such health campaigns could not have caused their quitting. In our example, without knowing the temporal sequence, one could argue logically for either event being the cause of the other. Obviously, the health campaign could cause people to quit smoking if they became frightened by learning the dire consequences of their habit. However, it could also be that those who quit smoking are happy with and proud of their victory and enjoy reading reports on what could have happened to them had they not quit smoking. In this second scenario, *quitting smoking* would be the independent variable that increases the likelihood that people will *read reports* about the health threat of smoking, the dependent variable.

A Nonspurious Relationship

The final criterion necessary for inferring causality is that the relationship between the independent and dependent variables not be *spurious*. A spurious relationship is like a false reading—it is a relationship between two variables that appears only because each of these variables is related to some third variable. This spurious relationship disappears when the effect of the third variable is taken into account. The logic of causal and spurious relationships is compared in Figure 4.1. If the relationship between X and Y is spurious, then it will disappear when controlling for a third variable, Z.

Considerable effort is expended during the problem formulation and design stage of research to control as many potentially troublesome extraneous variables as possible. Experiments, for example, are particularly good for avoiding spurious relationships, owing to the high degree of control the experimental situation affords the researcher. Surveys, on the other hand, provide far less control, such that several variables capable of producing spuriousness will typically have to be considered

FIGURE 4.1   Causal and Spurious
Relationships

Causal Relationship Between *X* and *Y*

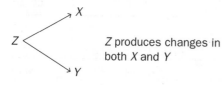

**Spurious Relationship Between *X* and *Y* When
Controlling for *Z***

FIGURE 4.2   Causal and Spurious
Relationships between Reading a
Report and Quitting Smoking

Causal Relationship

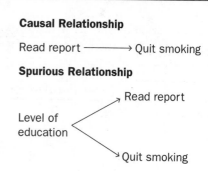

Spurious Relationship

during data analysis. Several statistical techniques exist to control extraneous variables when the data are analyzed.

Returning to our example of smoking cessation, suppose we had solved the time sequence problem and had thus satisfied the first two requirements for establishing a causal relationship. We would now begin to consider variables that might render the relationship spurious. One variable that might do this is the *level of education* of the people studied. The logic of this is outlined in Figure 4.2. Considerable research links *education* with *health behavior*: Generally, people with higher levels of education engage in more health-promoting activities such as quitting smoking or exercising. How do we determine whether the link between *report reading* and *smoking cessation* is spurious? We introduce *level of education* as a control variable, which is illustrated with the hypothetical data in Table 4.4. In this table, the respondents from Table 4.3 are divided into those with at least a high school education and those with less than a high school education. First, we can see by examining the row totals in each table that *level of education* is related to *health behavior*: Sixty percent of the better-educated group have quit smoking compared to only 19% of the less-educated group. However, we are really interested in what happens to the link between *report reading* and

smoking cessation. Careful inspection of Table 4.4 shows that the relationship largely disappears: Within each educational group, the same percentage of people quit smoking among those who read the report as among those who did not. So it is *educational level*, not whether one has *read the report*, that influences a person's likelihood of quitting smoking. Further, in our hypothetical example, *educational level* also influences whether one reads the report: 300 out of 400 (75%) of those with a high school education did so compared with only 100 of 500 (20%) of those with less than high school education. In our example, the link between *reading the report* and *quitting smoking* is therefore spurious; it occurs only because each of those two variables is affected by the same third variable—*level of education*.

If we had found the link between *report reading* and *smoking cessation* to be nonspurious when we controlled for *level of education*, could we conclude that the relationship was causal? The answer is no—at least not yet. All that would have been shown was that the relationship remained when one alternative explanation was ruled out. Other variables could render the relationship spurious and would also have to be investigated. This makes spuriousness the most difficult of the three criteria for causality to satisfy. Only after conducting a number of tests for spuriousness would we be

TABLE 4.4 Effectiveness of Reading Media Reports on Smoking Cessation, Controlling for Education

Less Than High School Education

		Person Reads Report		
		Yes	No	Totals
Person Quits Smoking	Yes	20 (20%)	75 (19%)	95 (19%)
	No	80 (80%)	325 (81%)	405 (81%)
	Totals	100 (100%)	400 (100%)	500

High School Education or More

		Person Reads Report		
		Yes	No	Totals
Person Quits Smoking	Yes	180 (60%)	60 (60%)	240 (60%)
	No	120 (40%)	40 (40%)	160 (40%)
	Totals	300 (100%)	100 (100%)	400

able to argue with any confidence that a relationship is causal. (More of the intricacies of this sort of analysis are addressed in Chapter 13 and Chapter 14.)

Correlational versus Experimental Research

This discussion of causality leads us to another decision to be made while shaping and refining a research problem: Will the research be correlational or experimental in nature? **Correlational research** is research in which the statistical analysis shows that two variables are correlated, or vary together in some systematic way. Correlational research can satisfy only the first, and possibly the third, criteria for assessing causality. It can demonstrate an association and possibly show that the relationship is not spurious. However, it cannot usually satisfy the second criterion; namely, that the time sequence of the changes in the variables is appropriate to inferring causality. **Experimental**

research, on the other hand, is designed such that changes in an independent variable are known to occur before changes are observed in a dependent variable, establishing the existence of the appropriate time sequence to infer that changes in the former caused changes in the latter (if the other two criteria are also satisfied, which experimental research is capable of determining).

From what has been discussed so far, it might seem that experimental research is always superior and so should always be preferred. And, at some level, that is true. But conducting research in the real world is a complicated enterprise, and the limitations that researchers confront often make experimental research impractical or impossible. In Chapter 8, experiments and their limitations will be described, but one example is supplied here. The best way to study the relationship between *smoking* and *lung cancer* would be experimental research in which a randomly selected group of people is prevented from smoking while another randomly selected group is forced to smoke every day. No such experimental research has been done because researchers would consider it highly unethical: It is highly likely to bring significant and permanent harm to one group, and this makes it an offensive intrusion into these people's lives. What if we were studying how religion affects people's lives? Could we take a random group of 10-year-olds and force them to attend church and religious education (let's not even get into the question of which religion we would force upon them!) while we prevented another group from receiving such religious contact? Again, most researchers would consider it highly unethical to manipulate people's lives in such intrusive ways. Many considerations are addressed when making the decision about whether to conduct correlational or experimental research, and sometimes correlational research is more practical or feasible.

IMPLEMENTING THE RESEARCH

Considerable time has been spent describing how to develop and refine a research problem. It would be helpful now to have a brief overview of what is done during the remainder of the research project—in other words, to see how the research is actually implemented. This will provide an overview that will be fleshed out in detail in the remaining chapters.

Choosing a Research Design

Once a problem has been adequately formulated and refined, the researcher must devise a way to make observations that will achieve the research goals. There are a number of elements to this process, but it generally involves figuring out how to measure variables, how to choose a sample, and which research technique to use to make observations.

Measurement

Measurement refers to the procedures that are used to determine whether something exists or how much of something exists. Whether we are studying fairly concrete things like sex or age, or rather abstract phenomena such as self-esteem or social cohesion, we need to be able to determine, for each case or person that we study, what the sex, age, or self-esteem is. If the focus of attention is groups or organizations, we might need to know the level of social cohesion in the group. We often measure variables by asking people questions or having them respond to verbal statements. Sex and age are often measured in this way: We ask people their sex and age on questionnaires, and our question and their response is the operational definition of the variables. With a more abstract variable such as *self-esteem*, we might ask people to respond to a series of questions or statements and then develop a composite score from their responses that constitutes their *level of self-esteem*. These multiple-item measuring devices are called *indexes* or *scales*. This type of quantitative measurement and the use of indexes and scales are explored in Chapter 5 and Chapter 6.

Measurement can also be based on information found in a variety of documents, such as people's diaries or the records of an organization (see Chapter 10). In these cases, measurement is done by observing people's behavior directly, and this measurement might also be quantitative in nature. You could record the number of times that a person makes hostile remarks to others (you would also need to operationally define what is meant by *hostile*). However, in some observational settings, measurement would be more qualitative than quantitative. But qualitative researchers also need to address issues of measurement to justify any claims they make about what occurs in a setting (see Chapter 11 and Chapter 15).

Sampling

Sampling deals with deciding exactly who to observe. Oftentimes it is neither feasible nor necessary to observe all the people, groups, or organizations of interest. Instead, a subset is selected on which to make observations. As will be discussed in Chapter 7, this selection must be done carefully in order to ensure the greatest confidence that what we learn about the sample can be generalized to the whole group. In some cases, we can use an understanding of probability theory to select a sample that has the highest likelihood of being generalizable. In other cases, where probability theory will not help, we make a careful and considered, although somewhat more judgmental, decision about whom to select for the sample.

Observational Technique

Researchers have to decide on exactly which technique they will use to make observations. This decision flows out of the nature of the research problem, the manner in which variables will be operationalized, and the resources available to the researcher. It is a complex decision, with a considerable amount of weighing of pros and cons of various choices. Much of the remainder of this book—but especially chapters 8 through 12—analyze how this decision is made. One option is to perform an experiment, which is really a logical way of organizing observations rather than a type of observational technique. In experiments, observations are organized such that we can determine whether changes in an independent variable are the factors that produce changes in a dependent variable. The actual observations in an experiment might be made by survey, direct observation, or available data.

Another option would be to conduct a survey in which persons are asked questions about their attitudes, feelings, or behaviors. Their answers to these questions are the observations, or data, the researcher uses. Yet another option would be to use the direct observation of behavior by the researcher as the data for the research. Sometimes direct observation involves going into the field, or the settings where people live out their daily lives.

Field research enables one to see people behaving naturally, as they do in their normal lives when researchers are not observing them. Generally, direct observation is preferred over survey data because the former is a more accurate measurement of behavior than what people say about their own behavior. Many factors can, however, mitigate against using direct observation: It can be expensive and time consuming, and might place the researcher in personal danger if the behavior of interest were some form of criminal behavior or crowd rioting.

In other cases, researchers might find it possible to use data that was collected by someone else for some other purpose. For example, organizations such as hospitals or social service agencies, keep voluminous records about their employees, clients, and organizational activities. This is called *available data*, and can be an extremely useful source of data. Although available data has its problems (discussed in Chapter 10), it often contains information that researchers would never be able to collect themselves, and is often a very inexpensive way of collecting data (because the organization has expended its resources to do most of the data collection).

You can begin to see that the choice of a research technique is complex, and researchers might in some cases disagree about a particular choice for a particular research project. In the following chapters, you will learn the advantages and disadvantages of each technique and the kinds of criteria that researchers use in choosing among them.

Collecting and Analyzing the Data

Once all the preceding tasks have been accomplished, the researcher begins the most critical part of the research process: the actual collection of the data that are at the core of the scientific endeavor. Great care and attention must be devoted to this because any sloppiness or errors introduced at this point will jeopardize the entire endeavor. Normally, a researcher has only one opportunity to collect data and, if this is not done well, the weaknesses introduced cannot be rectified. What can happen at this stage? Many things: Poorly trained interviewers may forget to ask some questions or neglect to get a consent form signed; a research assistant might make errors while transferring data from an organizational record to a coding sheet; or field observers might miss some critical observations because they hadn't adequately prepared for making observations in the field. It is often not possible to rectify errors introduced at the stage of data collection.

Once the data are collected, they must be analyzed, which means to extract some meaning from them. Some data analysis in social science research is quantitative in nature, sometimes based on highly sophisticated statistical procedures (see Chapter 13 and Chapter 14). Researchers who focus on positivist approaches often do this. However, data analysis can also be qualitative in nature, where the meaning extracted from the data takes the form of narrative descriptions, interpretations, or subjective analyses of what was observed at a scene. While students sometimes believe that doing statistical analysis is one of the most complicated and difficult parts of research, qualitative data analysis is equally difficult to do well—some would argue it is more difficult. As you will see in Chapter 15, qualitative data analysis requires sophisticated thinking and the application of rigorous criteria in order to produce valid knowledge that will stand up to critical assessment.

Drawing Conclusions and Disseminating the Results

Once the data have been collected and analyzed, the researcher "sums up." An assessment is made of whether hypotheses were confirmed, theories were verified and, in general, of what was learned from the research. If the research was applied in nature, then the researcher makes recommendations regarding what the research results say about changing an organization, agency, or program. The researcher also makes recommendations regarding future research, suggesting weaknesses in the current project that could be rectified, or directions for advancing knowledge beyond the current project.

Within the scientific community, there are standard vehicles for disseminating the results of research. Generally, research results are disseminated to other researchers in the field, to the lay public who may be interested in the research, and to clients and sponsors of the research who will use the results for some preordained purpose. To communicate with other researchers, the results might be presented to audiences at the meetings of professional organizations, such as the American Sociological Association or the American Association for Public Opinion Research. These and many other regional and national professional organizations in the social sciences meet at least once a year. Beyond these professional meetings, the research could be submitted for publication in these organizations' professional journals, such as the *American Sociological Review* or *Public Opinion Quarterly*. There are hundreds of these journals, which are read by other professional social scientists as well as by some clients and sponsors of research (and, of course, by students doing term papers in college courses in the social sciences). For clients and sponsors, researchers also submit special reports that provide sponsors with complete information about the research and its implications. Finally, researchers will sometimes communicate with the lay public through popular books or magazine and newspaper articles. Sometimes journalists interview researchers and write articles that summarize their research.

Many ways to disseminate research results exist. No matter how the dissemination occurs, however, researchers are obliged to give clear, honest, and accurate reports of their research. As we saw in Chapter 3, fraud or misconduct in reporting research is never acceptable.

FEASIBILITY OF A RESEARCH PROJECT

During this entire process of problem formulation and research design development, consideration must be given to practical issues involving the feasibility of a project. Such considerations can force researchers, sometimes painfully, to change or even reduce the scale of a project. In making a feasibility assessment, one should keep in mind a couple of axioms that apply to research projects: "Anything that can go wrong will," and "Everything will take longer than possibly imagined." The practical aspects of a project's feasibility center primarily on two related concerns: time and money (Kelly & McGrath, 1988).

Time Constraints

In developing a research project, one of the major considerations is whether there will be sufficient time to complete adequately what you hope to accomplish. In later chapters, as specific research techniques are considered, we will see how different techniques vary in terms of how much time they take. Here, some of the major factors related to time considerations are discussed. One factor concerns the population that is the focus of the research. If that population has characteristics that are fairly widespread, then a sufficient number of people will be readily available from which to collect data. If you were studying the attitudes of the public toward work-release programs for prison inmates, for example, you could select a sample of men and women from whatever city or state you happened to be in. If, however, your study focuses on persons with special characteristics that are somewhat rare, problems may arise. In general, the smaller the number of people who possess the characteristics needed for inclusion in a study, the more difficult and time consuming it will be to contact a sufficiently large number necessary to draw scientifically

valid conclusions. For example, a study of incestuous fathers, even in a large city, may encounter problems obtaining enough cases because relatively few such people will be openly known.

A second problem relating to time constraints involves the proper development of measuring devices. All techniques used for gathering data should be tested before the actual study is conducted, and this can be very time consuming. A pretest, as we saw in Chapter 1, refers to the preliminary application of the data-gathering techniques for purposes of assessing their adequacy. A pilot study is a small-scale trial run of all the procedures planned for use in the main study. In some studies, several pretests may be needed as data-collection devices are modified based on the results of previous pretests. All in all, a great deal of time can be consumed in such refining of data-gathering procedures.

A third major factor related to time considerations is the amount of time required for actual data collection, which can range from a few hours for a questionnaire administered to a group of "captive" students to the years that are necessary in many longitudinal studies. Because the amount of time required for data collection varies greatly, it should be given close scrutiny when addressing the question of the feasibility of a particular research design.

A fourth consideration related to the time issue is the amount of time necessary to complete the analysis of the data. In general, the less structured the data, the more time will be required for analysis. The field notes that serve as the data for some observational studies, for example, can be very time consuming to analyze (see Chapter 11 and Chapter 15). Likewise, the videotapes collected during an experiment may take many viewings before they are understood adequately (see Chapter 8). However, highly structured data in quantified form can also be time consuming to analyze (see Chapter 13 and Chapter 14). Although computers can manipulate data rapidly, it takes considerable time to prepare the data for entry into the computer, and the amount of time needed increases as the number of cases increases. So, as with the time required for data collection, the time needed for analysis should be considered carefully, owing to wide variation in the amount that may be necessary.

Financial Considerations

The financial expenditures associated with a research project are another constraint on feasibility. Good research is not always expensive. In many instances, researchers are able to get by with only modest costs because data are easy to obtain, analysis procedures are simple, and the labor is inexpensive. But even small projects are likely to require money for long-distance telephone calls, typing and duplicating questionnaires, and other services that can quickly stress the tight resources of a small research organization. At the other extreme, it is not unusual for the price tag of major research projects to run into six figures. In 1996, for example, the National Institute of Justice (1999) awarded numerous research grants to study crime control and prevention, under the Crime Act of 1995–1996, that ranged from $100,000 to $1 million. Table 4.5 shows the budget for a modest survey research project and indicates some of the general expenditure categories that should be considered in assessing the feasibility of any project.

The salary paid to those who conduct the study is potentially the most expensive item, especially for studies that require large interviewer staffs. Not only must the interviewers' wages be paid, but transportation costs and living expenses, which can be sizable, must also be covered. Interviewers may also require hours of training time before they can begin to collect data. If respondents are unavailable, callbacks may be necessary, which further increases the cost of each interview. To get the work done in a timely fashion and to ensure reliability, investigators may need to hire individuals or contract with an organization to code the raw data into an analyzable format.

TABLE 4.5 A Budget for a Study Involving Face-to-Face Interviews of 520 People

	Total Cost (dollars)	Percent of Total [a]
Prepare for survey		
Purchase map for area frame	200	1.0%
Print interviewer manuals	29	<1.0
Print questionnaires (690)	379	1.9
Train interviewers (20-hour training session)	1,134	5.7
Miscellaneous	25	<1.0
Conduct the survey		
Locate residences; contact respondents; conduct interviews; field edit questionnaires; 3.5 completed interviews per 8-hour day	9,655	48.8
Travel cost ($8.50 per completed interview; interviewers use own car)	4,420	22.3
Office edit and general clerical		
(6 completed questionnaires per hour)	728	3.8
Total, excluding professional time	**16,570**	
Professional time (160 hrs @ $35,000 annual salary plus 20% fringe benefits)	3,231	16.3
Total, including professional time	**19,801**	**100.0**

[a] Totals to more than 100% because of rounding.
Source Source Adapted from Priscilla Salant and Don A. Dillman, *How to conduct your own survey*, pp. 46–49. Copyright © 1994 by John Wiley & Sons. Reprinted by permission of John Wiley & Sons, Inc.

Computer expenses can also be formidable. These expenses may include the purchase of computer hardware and software that will enable you to perform the data analysis. Formerly, data analysis comprised the bulk of computer costs, but today computers are also used for questionnaire design, project management, data collection, literature review searches, and report preparation. A significant cost here is for the software packages that actually perform the procedures in addition to the cost of the computer itself and peripherals like printers, scanners, and modems.

Another major cost consideration is expenditures for office supplies and equipment. Under this category are such items as paper, envelopes, postage, tape, printing, and the like. Paper products might appear inexpensive at first glance, but given the large samples used for some surveys, the cost can be substantial. Different kinds of studies present different cost issues. For example, studies based on direct observation of behavior may necessitate high-cost equipment items such as video cameras, recorders, and videotape. Unless these are already available, substantial outlays will be needed.

Providing incentives to ensure cooperation of persons in the study may also be a cost factor. This may range from giving stickers or balloons to schoolchildren for completing a questionnaire to paying respondents in recognition of the large time commitment required for a longitudinal study. A study evaluating the effectiveness of advocacy services to women who were leaving their abusive partners illustrates participant incentive costs. Women were interviewed before the program and at 5, 10, and 20 weeks following the program. Not only was it difficult to maintain contact with this highly mobile population, but completing a survey form was very likely a low priority for them, given the stress and disruption in their lives. In order to encourage the women to participate, the 46 participants were paid $10, $20, $30, and $40, respectively, for the four interviews for a total cost of $4,600 (Sullivan, 1991).

Dissemination of research findings also generates costs. Besides the additional printing and office supplies for preparing reports, this may include travel to professional meetings to present papers. In the case of program evalua-tion studies, it may also entail hosting workshops or conferences with sponsors and other interested parties to ensure that the findings are incorporated into the policy and intervention planning process.

Finally, some costs associated with a research project are paid by the organization, such as a university or research center, under whose auspices the research is being conducted. For example, the organization will spend money to heat and light the offices where the research project is housed, but this cost is difficult to assess precisely. The organization may also have an extensive research library, sophisticated computer facilities, or laboratory equipment that will be used in the research. The organization maintains these facilities for general use, not just for a particular research project, and it is difficult to ascertain how much of the overall cost of supporting the facilities should be assigned to a given project. Consequently, the organization is paid what are called "indirect costs," typically a percentage of the total grant request, to cover these real but hard-to-specify costs. Each organization negotiates its own rate for indirect costs with the funding organization on the basis of the facilities and equipment that the organization has for research. The amount charged to indirect costs can be a sizable percentage of the overall research budget.

Anticipating and Avoiding Problems

Problems related to time and financial considerations arise during virtually all research projects, but their impact on the outcome of the research can be minimized if they can be anticipated as much as possible, especially during the planning stage when the details of the project are easier to change. A number of things can be done to anticipate problems. First, learn as much as possible from the experiences of others. One way to do this is through the studies consulted during the literature review. I mentioned earlier that noting problems these other researchers encountered is one focus of this review. Personal advice from experienced researchers who might be available for consultation can also be solicited. A knowledgeable researcher may be able to identify potential trouble spots in your plans and suggest modifications to avoid them.

Second, obtain whatever permissions or consents may be needed early in the planning stages of the project. Depending on the people you wish to study, it may be necessary to obtain official permission to collect data. For example, some studies are aimed at school-age children and seek to gather data while the children are in school. School administrators, to protect students from undue harassment and themselves from parental complaints, are frequently cool to allowing researchers into their schools. It may take considerable time to persuade whatever authorities are involved to grant the permissions you need—if they are granted at all. It is certainly wise to obtain any needed permissions before expending effort on other phases of the project, which might be wasted if permissions cannot be obtained.

The final—and perhaps most important—suggestion for avoiding problems is to conduct a pilot study. As noted in Chapter 1, a pilot study is a preliminary, small-scale, run through of all the procedures planned for the main study. For surveys, a small part of the sample, say 20 participants, should be contacted and interviewed. The data should be analyzed just as they will be in the complete project. In experiments, the researcher should run a few groups through all procedures, looking for any unexpected reactions from participants. Observational researchers should visit the observation sites and make observations as planned for the larger study. Any problems that surface during the pilot study can then be dealt with through modifications in the research plans before the main project is launched.

Given all the pitfalls that a project might encounter, it is quite possible that at some point you may conclude that the project is not feasible as planned. Before throwing up your hands and calling it quits, however, you should give careful consideration to possible modifications in the project that would enhance its feasibility. If inadequate time or money is the problem, perhaps the project can be scaled down. It might be possible to reduce the sample size or the number of hypotheses tested to make the project manageable. If interviewing was originally planned, consider a mailed questionnaire or even a telephone survey as cost-cutting and time-reducing measures. The point is that a project should not be abandoned until all efforts to make it feasible have been investigated.

REVIEW AND CRITICAL THINKING

Chapter 4 gives an overview of the entire research process; the critical thinking lessons focus only on the problem formulation and development issues. Issues related to measurement, sampling, and selection of observational techniques will be covered in later chapters. Problem formulation and development is something we can experience in any part of our lives. It doesn't have to be a scientific problem, which is the focus of this book; it could be a problem confronting an individual, a family, or an organization. Careful and systematic thinking about a problem can pay off in a clearer view of the problem:

1. Why is this considered to be a problem at all? Is there some personal interest or political issues that lead some people to see it as a problem?

2. Are the nature and terms of the problem clear and precise? Might some conceptual development produce a clearer and sharper focus on *exactly* what the nature of the problem is?

3. What have others said about this problem? Might a more thorough review of the literature uncover other ideas, perspectives, or data about the problem and its solution?

4. Are any causal inferences being made? Does the evidence presented satisfy the criteria that warrant a causal inference?

Computers and the Internet

The Internet can be a valuable tool throughout the problem formulation process. When selecting a problem based on personal interest, a social problem, or testing a theory, it can be worthwhile to begin by using a search engine on the Web to locate information on that general topic. For example, if you were interested in conducting research on such topics as self-esteem, drug abuse, or mental illness, you could use a search engine to locate a wide array of sites devoted to these topics. An initial perusal of such sites can help identify important issues for developing the research problem, such as: How is the concept defined? Does there appear to be consensus on what the concept is? What controversies or issues are being debated about the topic? Exploring such questions can help you focus on potentially interesting research problems. You may also find organizations and resources that may be useful in later steps in the research process, such as measurement tools.

The Internet can prove highly useful in discovering prior research. Many Web sites consist of bibliographies or include bibliographies as a special section. One example is the National Criminal Justice Reference Service. Access this site at **http://www.ncjrs.org/homepage.htm** and select "victims" from the home page. This will lead to a diverse listing of sources on victims of crime. Select one of the victim categories, such as "domestic violence," and peruse some of the available sources. Note that Web sites such as this also contain links to other sites that may be specifically devoted to your topic of interest. Other sites provide searchable databases. For example, the National Clearinghouse for Alcohol and Drug Information (**http://www.health.org/DBarea/index.htm**) provides immediate access to a wealth of articles related to substance abuse. The material at this site tends to focus very much on applied research, with studies on drugs, alcohol, tobacco, and various treatment programs. By beginning with a basic search engine, such as Lycos, and using a general term, such as "crime," "health," or "education," you can locate a major center or clearinghouse for that topic, as illustrated for domestic violence and alcohol. These sites will then lead you to the specific articles and bibliographies.

For those seeking to develop research questions around program evaluation, the Internet can be an excellent source. Sites associated with organizations that sponsor program evaluations often include descriptions of previous evaluations, projects that are currently underway, and information on future funding opportunities. As an example, explore "Research and Evaluation" from the Justice Information Center home page (**http://www.ncjrs.org/homepage.htm**).

Main Points

1. Suitable topics for research may be obtained from a variety of sources, including personal interest, social problems, theory testing, prior research, and program evaluation.

2. Problem selection is also influenced by political factors, which means that powerful interest groups encourage the expenditure of research resources on issues that are of interest to them.

3. After a general topic for research is selected, it must be narrowed and focused into a precise, researchable problem. An important part of refining a research problem is conceptual development: identifying and defining the concepts that will be the focus of the study.

4. Reviewing previous research related to the selected topic is a crucial step in problem development and preparing to conduct a research project. The nature of the research problem should dictate the choice of the specific research technique employed; do not attempt to make the problem fit a particular method of data gathering.

5. The levels and units of analysis must be clearly specified and may be either individuals, groups, organizations, programs, or social artifacts. Continual awareness of the operative unit of analysis is necessary to avoid errors such as the ecological fallacy and the reductionist fallacy.

6. Reactivity refers to the fact that people may behave differently when being observed than when they are not being watched, and its effects must be considered when shaping a research problem.

7. Consideration must be given to whether quantitative or qualitative research is most appropriate and whether the research question calls for longitudinal or cross-sectional data.

8. Research often assesses whether the relationship between two variables is causal in nature. The search for causality rests on the assumption of determinism: that there are reasons things happen and that we can discover those reasons. To demonstrate a causal relationship one must establish a statistical association between two variables, show that the independent variable occurs temporally first, and demonstrate that the relationship is not spurious. Experimental research is more appropriate to assessing causal relationships than is correlational research.

9. Once a research problem has been refined, it must be implemented. This involves issues of measurement, sampling, and choosing an observational technique. The data are then collected and analyzed, conclusions are drawn, and the results are disseminated to an appropriate audience.

10. Once a proposed research problem is fully refined, the practical feasibility of the project should be assessed realistically.

Important Terms for Review

causality
correlational research
cross-sectional research
determinism
ecological fallacy
experimental research
longitudinal research
macro-level research
micro-level research
panel study
qualitative research
quantitative research
reactivity
reductionist fallacy
trend study
units of analysis

For Further Reading

Berg, Bruce L. (1998). *Qualitative research methods for the social sciences* (3d ed.). Boston: Allyn & Bacon. Although this book focuses primarily on how to conduct good qualitative research, it also contains a good comparison of qualitative and quantitative research and assesses when each is the most appropriate design.

Bransford, John D., & Stein, Barry S. (1995). *The ideal problem solver: A guide for improving thinking, learning, and creativity* (2nd ed.). New York: Freeman. Sound thinking combined with creativity is clearly important to formulating research problems. This guide assists you in improving thought processes, drawing logical deductions, enhancing creativity, and even improving communication skills.

Gross, Ronald. (1993). *The independent scholar's handbook*. Berkeley, CA: Ten Speed Press. This book contains many examples of how successful scholars developed personal hunches and notions into serious research inquiries. It is also filled with practical advice about such things as obtaining resources and communicating with other researchers who share similar research interests.

Higgins, P. C., & Johnson, J. M.. (1988). *Personal sociology*. New York: Praeger. This book includes many illustrations of how personal life events and experiences shaped the research interests of several sociologists.

Locke, Lawrence F., Silverman, Stephen J., & Spirduso, Waneen Wyrick. (1998). *Reading and understanding research*. Thousand Oaks, CA: Sage. This is an excellent and thorough overview of how to do a good literature review and extract the appropriate information from it in an organized fashion.

Menard, Scott. (1991). *Longitudinal research*. Newbury Park, CA: Sage. This book provides a readable overview of both longitudinal and cross-sectional research. It discusses when each is an appropriate design and some of the problems confronted in doing good longitudinal research.

Reinharz, Shulamit. (1992). *Feminist methods in social research*. New York: Oxford University Press. This is a massive compilation of examples of all types of research conducted by researchers identified as feminists. Anyone interested in conducting research from this perspective is well advised to consult this impressive work.

Part II

MEASUREMENT AND SAMPLING

CHAPTER 5:
The Process of Measurement

CHAPTER 6:
Constructing Questions, Indexes, and Scales

CHAPTER 7:
Sampling

Part I of this book described how researchers begin with a general idea of a research topic and then use a series of techniques to hone and refine the problem into something that can be researched to produce useful and meaningful findings that add to the body of scientific knowledge. Part II describes the next critical steps in the process. Chapter 5 and Chapter 6 focus on the measurement process: the procedures social scientists use to determine whether something exists or how much of it exists. These are the fundamental observations upon which scientific knowledge rests; scientists are therefore very careful in how they measure. Chapter 5 deals at a general level with what scientists use to measure and how they evaluate their measurements. Chapter 6 goes into more detail on one common type of measurement: The use of verbal questions and statements to measure social phenomena. The chapter addresses how to construct questions and statements and how to combine them into measurement instruments called *indexes* and *scales*. Chapter 7 focuses on the topic of sampling: selecting elements from a population that will be the focus of observation.

Chapter 5

THE PROCESS OF MEASUREMENT

Reality is complicated, and in order to know that something is the case, we search for signs or indications of it. For example, a fire watcher in a tower interprets a plume of smoke observed on the horizon as an indication of a forest fire; a physician in an emergency room interprets a bruise on a child's arm as an indication of child abuse; a police officer sees a young person duck into an alley and takes this as an indication that a crime has been committed. In all these cases, the observer doesn't see the thing itself (the fire, the abuse, the crime), at least not immediately; instead, they see something that they assume is an indication of the thing. In some cases, the indication is very reliable: It is almost a certainty that fire will be found where there is smoke. However, in other cases, the indicators are not quite as certain: The bruise could be due to something other than abuse, and a law-abiding citizen might have reason to appear to "duck" into an alley.

Researchers also use various observations as indicators of the concepts of interest in a research project. However, as you will see, researchers take far greater care in selecting such indicators and determining how accurate they are than do people in their everyday lives. **Measurement** refers to the process of describing abstract concepts in terms of specific indicators by assigning numbers or other symbols to these indicants in accordance with rules. At the very minimum, we must have some means of determining whether something is either present or absent. In many cases, however, measurement is more complex and involves assessing how much of or to what degree some phenomenon is present.

Measurement is a part of the process of moving from the abstract or theoretical level to the concrete. Recall from Chapter 2 that scientific concepts have two types of definitions: nominal and operational. Before research can proceed, nominal definitions have to be translated into operational definitions. The operational definitions indicate the exact procedures—or operations—that will be used to measure the concepts. Measurement is essentially the process of operationalizing concepts. Figure 5.1 illustrates the place of measurement in the research process. Measurement will be a topic of interest throughout this book.

FIGURE 5.1   THE MEASUREMENT PROCESS

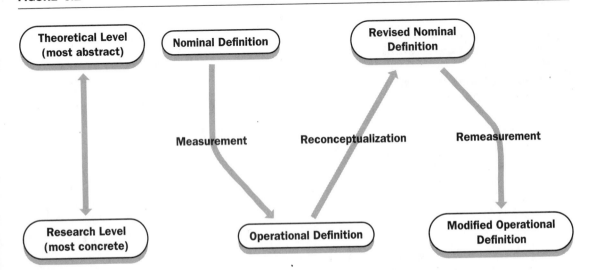

Discussed in this chapter are the general issues that relate to all measurement, beginning with some of the different ways in which measurements can be made. Then how measurements can be made at different levels is analyzed, and their effects on the mathematical operations that can be performed on them are examined. Finally, ways of evaluating and choosing among measures and determining the errors that can occur in the measurement process are presented.

MEASUREMENT AND OPERATIONALIZATION

From Concepts to Indicators

The concepts and variables that are the focus of research cannot normally be directly observed. For example, such things as "social structure," "poverty," and "social class" are abstract in nature and cannot be seen directly but only inferred from something else. Take something as seemingly obvious as child abuse. Can you directly observe child abuse? Not really. What you directly observe is a bruise on a child's back; an infant's broken leg; a father slapping his daughter. Even the slap may not relate to child abuse because parents sometimes slap their children without its being a case of child abuse. Keep in mind that child abuse is not merely a behavior, such as a "slap"; it is a behavior or series of behaviors that have a certain meaning or are interpreted or judged in a particular way by people. In other words, "child abuse" involves social meanings, and these are much less tangible than a slap. However, all these things—the bruise, the broken leg, the slap—may be used as *indicators* of child abuse. In research, an **indicator** is an observation that is assumed to be evidence of the attributes or properties of some phenomenon. What we observe are the indicators of a variable, not the actual properties of the variable itself. Emergency room personnel

may take a child's broken leg as an indicator of child abuse even though they have not observed the actual abuse.

Child abuse represents a good illustration of the difficulties of moving from nominal to operational definitions with variables involving social and psychological events. At the nominal level, we might define *child abuse* as an occurrence in which a parent or caretaker injures a child, not by accident but in anger or with deliberate intent (Gelles, 1987). But what indicators would we use to operationalize this definition? Some things would obviously seem to indicate child abuse, such as a cigarette burn on a child's buttock. What about a bruise on the arm? There are subcultures in the United States that view hitting children, even to the point of bruising, as an appropriate way to train or discipline a child. Further, some people argue that a serious psychological disorder suffered by a child is an indicator of child abuse because it shows the parents did not provide the proper love and affection for stable development. In short, one of the problems in operationalizing *child abuse*, as is true with many other variables in social science research, is that its definition is culture-bound and involves subjective judgments. This illustrates the importance of good conceptual development and precise nominal definitions for research. It also shows how the theoretical and research levels can mutually influence one another: As nominal definitions are shaped into operational ones, difficulties that arise often lead to a reconceptualization, or change, in the nominal definition at the theoretical level (see Figure 5.1).

The example of child abuse also illustrates another point about measurement; namely, that there can be more than one indicator of a variable. The term **item** is used to refer to a single indicator of a variable. Items can take numerous forms, such as an answer to a question or an observation of some behavior or characteristic. Asking a person her age or noting her sex, for example, would both produce "items" of measurement. In many cases, however, operationalizing variables involves combining a number of items into a composite score called an **index** or **scale**. (Although scales involve more rigor in their construction than do indexes, for our purposes we can use the terms interchangeably at this point; they will be discussed further in Chapter 6 and the distinction between them explained.) Attitude scales, for example, commonly involve asking people a series of questions, or items, and then summarizing their responses into a single score that represents their attitude on an issue. A major reason for measuring with scales or indexes rather than single items is that these enable us to measure variables in a more precise and usually more accurate fashion. To illustrate the value of scales over a single item, consider your grade in this course. In all likelihood, your final grade will be an index, or composite score, of your answers to many questions on many tests throughout the semester. Would you prefer that your final grade be determined by a one-item measure, such as one multiple-choice or essay question? Probably not because it would not measure the full range of what you had learned in the course. Further, an error on that one question would indicate that you had not learned much in the course even though you could miss one question and still have learned quite a bit; multiple-item tests are a more accurate measure of what you have learned. For these reasons, then, multiple-item measures are usually preferred over single-item indicators.

This discussion began by noting that many variables are abstract and cannot normally be directly observed. Actually, variables differ in their degree of abstraction, and this affects the ease with which they can be measured. In general, the more abstract the variable, the more difficult it is to measure. For example, a study of child abuse might include the variable *number of children in family*, on the theoretical presumption that large families create more stress for parents and are therefore more likely

to precipitate abusive attacks on children. This is a rather easy variable to measure because the concepts "children" and "family" have readily identifiable empirical referents; they are relatively easy and unambiguous to observe and count. Suppose, however, that the child abuse study also included as a dependent variable *positiveness of a child's self-concept*. *Self-concept* is a difficult notion to measure because it is abstract and can take many different forms. Although we have narrowed it to the positive-negative dimension, it is still more difficult to measure than *number of children in family* because a wide array of questions could be asked to explore how positively people feel about themselves. In addition, *self-concept* can be measured not only by asking people how they feel but also by observing certain behaviors since people who feel positively about themselves may behave differently from those who do not. The point is that with highly abstract concepts there is usually no single empirical indicator that is clearly and obviously preferable to others as a measure of the concept.

Ways of Measuring

You will be taught specific techniques for measuring variables in other chapters in this book, but it will help you understand the issues surrounding measurement to read an overview of these techniques at this point. Measurement techniques in the social sciences vary widely because the concepts we measure are so diverse, but these measurement techniques mostly fall into one of three categories (see Figure 5.2).

Verbal Reports

This is undoubtedly the most common measurement technique in the social sciences and involves people answering questions, being interviewed, or responding to verbal statements (see especially Chapter 6 and Chapter 9). For example, research on people's attitudes typically uses this technique by asking individuals how they feel about commercial products,

FIGURE 5.2   The Major Strategies Used by Social Scientists to Measure Variables

Verbal Reports

people answering questions, being interviewed, or responding to verbal statements

Measuring School Performance:

ask students to tell us what their grades are or how much they know about a particular subject

Researcher

Observation

directly watch people at school, work, or other setting and make note of what they say and do.

Measuring School Performance:

directly observe their behavior in the classroom, noting how often they answer questions posed by teachers, how often their answers are correct, and how they get along with teachers and students

Researcher

Archival Reports

review available recorded information

Measuring School Performance:

use the school records to locate students' grades, performance on exams, attendance records, and disciplinary problems

Researcher

political candidates, or social policies. In a study of school performance, to mention another example, we could measure how well students do in school by asking them to tell us what their grades are or asking them to answer questions about how much they know about a particular subject.

Observation

Social scientists also measure concepts by making direct observations of some phenomena (see Chapter 10 and Chapter 11). We watch people at school or at work and note what

they say and/or do. We may even make an audio or video recording as a way of preserving the observations. In a study of school performance, we could measure how well students do in school by directly observing their behavior in the classroom, noting how often they answer questions posed by teachers, how often their answers are correct, and how they get along with teachers and students.

Archival Records

Researchers also use available recorded information to measure variables (see Chapter 10).

TABLE 5.1   Elements in the Process of Measurement

X Observation	=	T True Phenomenon	+	E Error
Reading on a weight scale	=	Your actual weight	+	Clothing you are wearing Heavy object in your pocket
Grade on an examination in social research class	=	Actual knowledge you acquired in social research class	+	Heat and humidity in test room Distraction due to fight with partner
Score on a scale measuring self esteem	=	Your actual level of self esteem	+	Recent compliment Questions on self esteem scale that are difficult to understand

This information might take the form of statistical records, governmental or organizational documents, personal letters and diaries, newspapers and magazines, or movies and musical lyrics. All of these archival records are the products of human social behavior and can serve as indicators of concepts in the social sciences. In the study of school performance, for example, a researcher could use the school records to locate students' grades, performance on exams, attendance records, and disciplinary problems as measures of how well they are doing in school.

Whichever way concepts are measured, the researcher must specify exactly what aspects of the verbal reports, observations, or available data will serve as indicators of the concepts being measured. There are also some key criteria that researchers use to help them decide whether a particular indicator is a "good" measure of some concept. These criteria will be discussed later in this chapter.

Positivist and Nonpositivist Views of Measurement

Much of the foundation for measurement and operationalization in the social sciences derives from the work of statisticians, mathematicians, philosophers, and scientists in a field called "classical test theory" or "measurement theory" (Bohrnstedt, 1983; Stevens, 1951). It provides the logical foundation for issues discussed in this chapter and derives largely from the positivist view of science discussed in Chapter 2. The logic of measurement can be described by the following formula:

$$X = T + E$$

In the formula, X represents our observation or measurement of some phenomenon; it is our indicator. For example, X might be your grade in an examination in a social research class or your responses to a self esteem scale (see Table 5.1). T represents the true, actual phenomenon that we are attempting to measure with X. T would be what you have actually learned in your social research class, or what your true self-esteem is. The third symbol in the formula, E, represents any measurement error that occurs, or anything that influences X other than T. E might be the heat and humidity in the classroom on the day of the social research exam that made it difficult for you to concentrate during the exam, or it might be the fact that someone gave you a huge compliment that temporarily boosted your self-esteem just prior to filling out the self esteem scale.

What the formula says, in short, is very simple but also very profound and important: Our measurement of any phenomenon is a product of the characteristics or qualities of the phenomenon itself and any errors that occur in the measurement process. What we strive for is measurement with no error:

$$E = 0$$

and therefore

$$X = T.$$

This would be the ideal to strive for, where our measurement of the phenomenon is determined only by the true state of the phenomenon itself. However, scientists recognize that this ideal state is normally not achievable in its entirety. In reality, we attempt to reduce E as much as possible. Later in the chapter this measurement formula becomes more complex, but for now it can stand as a shorthand way to understand the process of measurement.

Before going more into the process of measurement, however, it is important to consider the nonpositivists' critique of classical measurement theory. The interpretivists argue that there is a huge assumption in all this that hasn't been examined and that may render it all somewhat problematic. The assumption is that the phenomenon being measured (T) exists objectively in the world and that our measuring device is merely discovering it and its properties. Some things do exist in the world independent of our perceptions of and judgments about them. The computer monitor upon which I am writing these words has a screen that is 9 inches tall; I just measured it with a ruler. My measurement of it was a discovery of its properties; the measurement process did not create or change those properties. However, think about a social science concept such as "self-esteem." We measure it by asking you to agree or disagree with a series of 10 statements; we score a "strongly agree" response as 4 and a "strongly disagree" response as 1; then we sum up your responses to all the separate items in the scale and give you a self-esteem score that ranges from 10 to 40. But what is the objective reality behind this measurement? If you receive a score of 32 on our measuring device, what does that 32 correspond to in

your subjective world, or mind, or consciousness? If 32 is the X in our measurement formula, what is the T that it corresponds to? Is the link between my measurement of the computer screen and its actual height as direct as the link between the 32 score on the self-esteem measure and your actual subjective experience of yourself?

Nonpositivists would say no. They argue that many social science concepts do not have such clear and objective referents in the world. Social science concepts are based on an intuitive and theoretical understanding of what parts of the world are like. In other words, we are constructing the world, not just discovering it. We believe that something like self-esteem exists, but it is our construction of it that we measure with the self-esteem scale, not the thing itself (if the thing itself even exists). This doesn't make measurement theory useless, but it does suggest that the entire process is more complicated and not nearly as objective as positivists would suggest. Nonetheless, many nonpositivists agree that some social science measurement can follow the model of measurement theory; social phenomena, such as age and gender, do have some objective existence in the world. Your age has something to do with how many times the Earth has circled the sun since you were born, and your gender has something to do with your physical genitalia. The social significance of these characteristics is of course another matter, but the measurement of your age and gender in many cases can follow classical measurement theory. The Applied Scenario box further explores these issues by analyzing a research project where the issue of how to measure variables became complicated because of this debate between the positivists and nonpositivists.

A major problem in most measurement has to do with which indicators are to be used in a particular research project. This depends in part, of course, on theoretical concerns, but there are other matters that influence the decision. One such concern has to do with what

Applied researchers can become embroiled in intense controversies that have important implications for advocacy and social policy. Sometimes the controversies have to do with how to measure phenomena and tell us something about the complexities and ambiguities of the workings of science. Sociologist Peter Rossi and his colleagues confronted this in a study they did (Rossi, 1987; Rossi, Wright, Fisher, & Willis, 1987). Rossi and colleagues were asked to perform an assessment of the extent of homelessness in Chicago. Their research was to provide an estimate of the number of homeless persons and their characteristics. The researchers roamed Chicago in the middle of the night (accompanied by off-duty Chicago police officers) and interviewed individuals they encountered in shelters, in all-night businesses, in abandoned buildings, and on the street. Rossi and colleagues concluded that there were at most 7,000 homeless people in Chicago on any given night. This enraged the advocates for the homeless who believed that at least 25,000 people were homeless.

The advocates' reaction to Rossi's conclusion was to attack the validity of his operational definition. This controversy arose, in part, because there is no inherently or objectively "best" operational definition of some concepts, and this is true of the concept of "homelessness." Even when using quantitative scientific measurement, some judgment still needs to be applied in deciding just who exactly is homeless. As Rossi and colleagues (1987, p. 1336) point out:

> [T]here is a continuum running from the obviously domiciled to the obviously homeless, with many ambiguous cases to be encountered along the continuum. Any effort to draw a line across that continuum demarcating the homed from the homeless, is of necessity somewhat arbitrary and therefore potentially contentious.

Rossi and colleagues limited the definition of *homelessness* to persons with no regular access to a conventional dwelling or residence. This excluded those who had found temporary shelter in a single-room-occupancy hotel or in the home of a family member, or who had a bed in a dormitory-type shelter. Advocates for the homeless believed that all these people should be included under the concept "homeless," which would certainly increase the estimate of the number of homeless substantially over the estimate of Rossi and colleagues.

As the interpretivist and critical perspectives discussed in Chapter 2 suggest, values and interests can play a part in the process of operationalizing the concept of "homelessness." As

mathematical operations can be performed on a particular measure, and this issue is addressed next.

LEVELS OF MEASUREMENT

We have seen that there are numerous ways of measuring phenomena, such as by asking questions or by directly observing behavior. Measures also differ from one another in terms of what is called their **level of measurement**, or the rules that define permissible mathematical operations that can be performed on a set of numbers produced by a measure. There are four levels of measurement: nominal, ordinal, interval, and ratio. Keeping in mind that variables are things that can take on different values, measurement basically involves assessing

Rossi and colleagues point out, "A definition of homelessness is . . . a statement as to what should constitute the floor of housing adequacy below which no member of society should be permitted to fall" (1987, p. 1336). Is it acceptable for people to live crowded in with an uncle or brother? If you believe it is, then you will not see these people as homeless and will tend to restrict the definition of *homelessness*. A different political ideology might find such living accommodations unacceptable in an affluent nation such as the United States. People who believe this will be inclined to expand the definition of *homelessness* to include these people under the rubric "homeless." Stretching the definition in this direction will result in a larger estimate of the number of homeless. A larger estimate will presumably place more pressure on politicians and the public to devote resources toward alleviating the problem.

The point is that where you draw the line across the continuum of homelessness described by Rossi is a function, in part, of personal ideology, values, and interests. Despite what positivists might say, there is no purely scientific way to decide this issue. By drawing the line, you simplify a more complex reality and go beyond discovering what is in the world to instead engage in creating and defining what that world is like. And this is true of many operational definitions. However, things can be done to help clarify situations such as this. One possible clarification is to engage in further conceptual development. Perhaps reality is too complex in this realm to be divided into two categories, the homeless and the homed. Possibly we need further categories to cover the ambiguous cases that fall in between these two categories (this is the reconceptualization process presented in Figure 5.1). A second clarification would be to carefully and critically assess the impact of values on decisions, even if the impact of those values cannot be eliminated. In this way, you can assess how much bias, and in what direction, is created by using a given operational definition. It is impossible, for example, to make a full assessment of Rossi and colleagues' findings without knowing how *homelessness* was operationally defined.

Look in the library or on the Internet for other applied research projects.

1. Can you find any research projects where measurement issues became complicated and seemed to be influenced, at least potentially, by personal values or interests?

2. What about domestic violence? Locate some ways in which this has been measured. To what extent might values and interests influence these operational definitions?

the value, or category, into which a particular entity falls. Measuring age, for example, is the process of placing each person into a particular age category.

Nominal Measures

Nominal measures classify observations into mutually exclusive and exhaustive categories. They represent nominal variables at the theoretical level. Variables such as *sex, ethnicity, religion*, or *political party preference* are examples. Thus we might classify people according to their religious affiliation by placing them into one of five categories: Protestant, Catholic, Jewish, other, or no religious affiliation. These are mutually exclusive categories because membership in one precludes membership in another. For purposes of data analysis, we might assign numbers to represent each of the categories. We could label Protestant as 1, Catholic as 2, Jewish as 3, other as 4, and none as 5. It is important to recognize, however, that the assignment of numbers is purely arbitrary; the numbers comprising a nominal measure have none of the ranking, ordering, and magnitude properties that we usually associate with numbers. None of the usual arithmetic operations, such as adding, subtracting, multiplying, or dividing, can legitimately be performed on numbers comprising a nominal scale. The reason for this is that the numbers in a nominal scale are merely symbols or labels used to identify a category of the nominal variable. Protestant could have been labeled 2 as easily as 1.

Ordinal Measures

Ordinal measures have the same characteristics as nominal measures (mutually exclusive and exhaustive categories), but in addition there is an inherent, fixed order to the categories of the variable. Ordinal measures are considered to be at a higher level than nominal because ordinal measures have more characteristics, which contain more information and precision than nominal measures. *Socioeconomic status*, for example, constitutes an ordinal variable, and measures of *socioeconomic status* are ordinal scales. Table 5.2 illustrates how *socioeconomic status* might be divided into ordinal categories. With ordinal measurement, we can speak of a given category as ranking higher or lower than some other category; lower-upper class, for example, is higher than middle class but not as high as upper-upper class. It is important to recognize that ordinal measurement does not assume that the categories are equally spaced. For example, the distance between lower-upper and upper-upper is not necessarily the same as that between lower-middle and middle even though in both cases the classes are one rank apart. This lack of equal spacing means that the numbers assigned to ordinal categories do not possess the numerical properties necessary for arithmetic operations. Like nominal scales,

TABLE 5.2   Ordinal Ranking of Socioeconomic Status

Category	Rank
Upper-upper	7
Lower-upper	6
Upper-middle	5
Middle	4
Lower-middle	3
Upper-lower	2
Lower-lower	1

ordinal scales cannot be added, subtracted, multiplied, or divided. The only characteristic they have that nominal scales do not is the fixed order of the categories.

Interval Measures

The next highest level of measurement is called *interval*. **Interval measures** share the characteristics of ordinal scales—mutually exclusive categories and an inherent order to the categories. But interval measures have equal spacing between the categories. Equal spacing comes about because some specific unit of measurement, such as the degrees of a temperature scale, is a part of the measure. Each of these units is considered to possess a certain value, which creates the equal spacing characteristic of an interval scale. We would have an interval scale if the difference between, say, scores of 30 and 40 was the same as the difference between scores of 70 and 80. A 10-point difference is a 10-point difference regardless of where on the scale it occurs.

Common temperature scales—Fahrenheit and Celsius—are true interval scales. Both of these temperature scales have, as units of measurement, degrees and the equal spacing characteristic of interval scales. A difference of 10 degrees is always the same, no matter where it occurs on the scale. These temperature scales illustrate another characteristic of true interval scales: The point on the scale labeled zero is arbitrarily selected. Neither 0°C nor 0°F is absolute zero, the complete absence of heat. Because the zero point is arbitrary in true interval scales, we cannot make statements concerning ratios. That is, we cannot say that a given score is twice or three times as high as some other score. For example, a temperature of 80°F is not twice as hot as a temperature of 40°F. Despite not having this ratio characteristic, interval scales contain numbers possessing all the other arithmetic properties. If we have achieved the interval level of measurement, we can legitimately perform all the common arithmetic operations on the numbers.

Considerable controversy exists over which measures used in social science research are true interval measures, with only a few measures clearly of interval level. One, for example, is intelligence, as measured by IQ tests. With IQ tests, there are specific units of measurement—points on the IQ scale—and each point on the scale is mutually exclusive. Further, the distance between IQs of 80 and 90 is equivalent to the distance between IQs of 110 and 120, in that both intervals involve 10 measurement units. However, there is no absolute zero point on an IQ scale, so we cannot say that a person with an IQ of 150 is twice as intelligent as a person with an IQ of 75. As with temperature scales, the IQ scale is in part an arbitrary construction that allows us to make only some comparisons, not all. Beyond a few measures such as intelligence, however, the debate ensues. Some researchers argue, for example, that some attitude scales can be treated as interval scales (Kenny, 1986). Attitudes are often measured by having people respond to verbal statements by choosing one of four responses: strongly agree, agree, disagree, or strongly disagree. The argument is that people see the difference between "strongly agree" and "agree" as roughly equivalent to the distance between "disagree" and "strongly disagree." This perceived equidistance, some argue, makes

it possible to treat these as interval level scales. Other researchers argue that there is no logical or empirical reason to assume that such perceived equidistance exists and, therefore, that attitude scales should always be considered ordinal measures rather than interval measures.

I will not presume to settle this debate here. Rather, I raise the issue because level of measurement influences which statistical procedures can be used at the data analysis stage of research, which is discussed further in Chapter 13 and Chapter 14. You should be sensitive to the issue of levels of measurement when utilizing research. The results of research in which a statistical procedure inappropriate for a given level of measurement has been used should be viewed with caution.

Ratio Measures

The highest level of measurement is *ratio*. **Ratio measures** have all the characteristics of interval measures, but the zero point is absolute and meaningful rather than arbitrary. As the name implies, with ratio measures we can make statements to the effect that some score is a given ratio of another score. For example, one ratio variable commonly used in the social sciences is *income*. With *income*, we have the dollar (or penny, depending on how precise you wish to be) as the unit of measurement. Also, as many are all too well aware, there is such a thing as no income at all, so the zero point is meaningful and not arbitrary. And because it is meaningful, it is perfectly legitimate to make the kinds of statements about *income* that are commonly made: An income of $20,000 is twice as much as $10,000, but only one-third as much as $60,000. (Recognize, of course, that income is a ratio measure only as an indicator of the amount of money available to a person; if income is used as a measure of a person's social status, for example, then a difference between $110,000 and $120,000 does not necessarily represent a shift in status equivalent to that between $10,000 and $20,000.)

TABLE 5.3 The Characteristics of the Four Levels of Measurement

Level of Measurement	Mutually Exclusive	Possesses a Fixed Order	Equal Spacing Between Ranks[a]	A True Zero Point[a,b]
Nominal	*y*			
Ordinal	*y*	*y*		
Interval	*y*	*y*	*y*	
Ratio	*y*	*y*	*y*	*y*

y = possesses that characteristic
[a] Permits standard mathematical operations of addition, subtraction, multiplication, and division.
[b] Permits statements about proportions and ratios.

Given that ratio scales have all the characteristics of interval scales, we can, of course, perform all arithmetic operations on them.

Selecting a Level of Measurement

The characteristics of the four levels of measurement are summarized in Table 5.3. It is important to understand that the primary determinant of the level of measurement of a variable is the nature of the variable being measured. The major concern is to have an accurate measurement of a variable (a topic discussed at length in the next section). *Religious preference*, for example, is a nominal variable because that is the nature of the theoretical concept "religious preference." There is no way to treat *religious preference* as anything other than merely a nominal classification.

However, although researchers have no control over the nature of a variable in reality, they do have some control over how they will define variables, at both the nominal and operational levels, and it is sometimes possible to change the level of measurement of a variable by redefining the variable at the nominal or operational level. For example, if, instead of *religious preference*, we decided to change our focus to *religiosity*, or the *strength of religious beliefs*, we would have a variable that could be conceptualized and measured as ordinal and

perhaps even interval or ratio. Figure 5.3 illustrates this process of reconceptualization. If we asked people how strongly they held their religious preference, then people could be ranked into ordered categories of greater or lesser religiosity. Or we could ask people how often they attend religious services. One way this is categorized in Figure 5.3 achieves only ordinal level of measurement because there is no equidistance between the categories. On the other hand, if we asked people to state the exact number of times they attend religious services, then it would be a ratio level variable with a meaningful zero point (although some would question how accurate this indicator would be). However, when we measure *religiosity* instead of *religious preference*, we are measuring a different variable, and this should be done only if religiosity is an appropriate conceptualization given our theoretical concerns. We would never go for a higher level of measurement if the resulting variable was inappropriate for our theoretical concerns.

It should be clear from this example that the theoretical nature of the variable plays the most critical part in determining the level of measurement. Within the constraints created by this concern, however, researchers generally strive for the highest level of measurement possible and will sometimes change the way a

FIGURE 5.3 An Illustration of Measuring a Variable at Differing Levels of Measurement

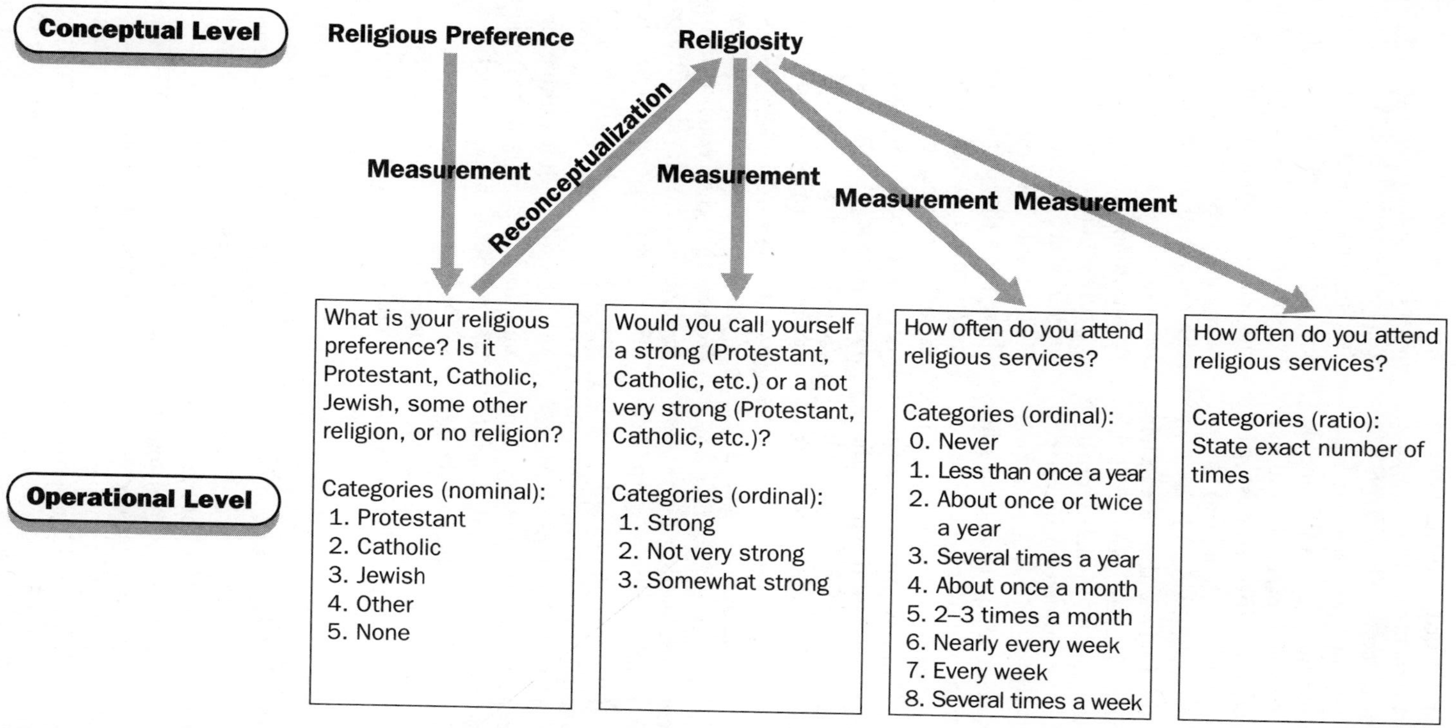

Conceptual Level

Religious Preference → Religiosity

Reconceptualization

Operational Level

Measurement | Measurement | Measurement | Measurement

What is your religious preference? Is it Protestant, Catholic, Jewish, some other religion, or no religion?

Categories (nominal):
1. Protestant
2. Catholic
3. Jewish
4. Other
5. None

Would you call yourself a strong (Protestant, Catholic, etc.) or a not very strong (Protestant, Catholic, etc.)?

Categories (ordinal):
1. Strong
2. Not very strong
3. Somewhat strong

How often do you attend religious services?

Categories (ordinal):
0. Never
1. Less than once a year
2. About once or twice a year
3. Several times a year
4. About once a month
5. 2–3 times a month
6. Nearly every week
7. Every week
8. Several times a week

How often do you attend religious services?

Categories (ratio):
State exact number of times

Source Questions are taken from the General Social Survey conducted by the National Opinion Research Center, University of Chicago, and distributed by the Roper Center, University of Connecticut (**http://www.icpsr.umich.edu/GSS/**).

variable is measured in order to achieve a higher level of measurement if that does not change their theoretical focus. They strive for the highest measurement possible because higher levels of measurement generally enable us to measure variables more precisely and use more powerful statistical procedures (see Chapter 13 and Chapter 14). It is also desirable to measure at the highest possible level of measurement because it gives the researcher the most options: Once data are collected, the level of measurement can be reduced during the data analysis, but it cannot be increased. Thus, measuring variables at too low a level introduces a permanent limitation into the data analysis. Therefore, if a number of indicators to measure a variable are available, one rule in deciding which one is preferred is this: Everything else being equal, choose the indicator that is at the highest level of measurement.

Despite the benefits of higher levels of measurement, there is nothing inherently undesirable about nominal variables. It would be quite wrong to get the impression that variables capable of being measured at higher levels are always better than nominal variables. The first consideration should be to select variables on theoretical grounds and not on the basis of their possible level of measurement. Thus, if a research study is really concerned with *religious preference* and not *religiosity*, the nominal measure is the correct one to use, and not a measure of *religiosity* even though it is an ordinal or possibly interval measure.

Discrete versus Continuous Variables

In addition to considering the level of measurement of a variable, researchers also distinguish between variables that are *discrete* or *continuous*. **Discrete variables** are those variables possessing a finite number of distinct and separate values or categories. Variables such as *sex*, *race, household size, number of days absent,* or *number of arrests* are all examples of discrete variables. *Household size* is a discrete variable because households can be measured only in a discrete set of units, such as having one member, two members, and so on. No meaningful measurement values lie between these distinct and separate values. **Continuous variables** are those variables that, at least theoretically, can take on an infinite number of values. *Age* is a continuous variable because it can be measured by an infinite array of values. We normally measure *age* in terms of years, but theoretically we could measure it in terms of months, weeks, days, minutes, seconds, or even nanoseconds! There is no theoretical limit to how precise the measurement of *age* might be. For most social science purposes, the measurement of *age* in years is quite satisfactory, but in some cases more precise measurement may be called for, such as in the study of infant development where time periods of days or weeks may reveal significant changes in development.

Nominal variables are, by definition, discrete in that they consist of mutually exclusive or discrete categories. Ordinal variables are also mostly discrete. The mutually exclusive categories of an ordinal variable may be ranked from low to high, but there can be no partial rank. For example, in a study of the military, *rank* might be ordered 1 = private, 2 = corporal, and so on, but it would be nonsensical to speak of a *rank* of 1.3. In some cases, interval and ratio variables may be discrete. For example, *family size* or *number of arrests* can only be whole numbers, or discrete intervals; you can be arrested once or twice, but not 1.7 times. We can, however, summarize discrete interval and ratio data for a whole group by saying that the average family size is 1.8 people, but this is a summary statistic of a group, not a measurement of a particular case or household. Many variables at the interval and ratio level are continuous, at least at the theoretical level. A researcher may settle for discrete indicators because the study does not demand greater precision or because there are no existing tools that can measure the continuous variable with sufficient reliability. In some cases, there is debate among researchers over whether a particular variable is discrete or continuous in nature. For example, I used *social class* as an illustration of an ordinal variable, suggesting that there are several distinct classes. Others argue that *social class* is inherently a continuous interval variable and that we only treat it as ordinal because of the lack of instruments that would permit researchers to reliably measure it as a true continuous, interval variable (Borgatta & Bohrnstedt, 1981).

It is important to recognize that a variable is continuous or discrete by its very nature and the researcher cannot change that. It is possible to measure a continuous variable by specifying a number of discrete categories, as is typically done with *age*, but this does not change the nature of the variable itself. Whether variables are discrete or continuous may influence how they are used in data analysis. For example, in setting up categories for a frequency distribution on *household size*, which is a discrete variable, intervals of 1, 2, 3, and so on would be used instead of .4–1.7, 1.8–2.3, or 2.4–3.8. Sometimes discrete data are treated as continuous in order to use statistical models, but care must be taken to assure that the results will be meaningful. Knowing the level of measurement and whether variables are discrete or continuous has implications for selecting the best procedures for analyzing the data.

EVALUATING MEASURES

As just discussed, when choosing a measuring device, one important consideration is the level of measurement. Another important consideration is whether the device provides an accurate measure of the phenomenon of interest. Researchers evaluate the accuracy of measuring devices by looking at their *validity* and *reliability*.

Validity

Validity refers to the accuracy of a measure: Does it accurately measure the variable that it is intended to measure? If we were developing a measure of "self-esteem," for example, a major concern would be whether our measuring device measures the concept as it is theoretically defined. The validity of many measures is difficult to demonstrate with any finality. However, several procedures are used by researchers to assess the validity of measures, and good research always evaluates any measures used by applying at least some of these procedures (Zeller & Carmines, 1980).

Face and Content Validity

Face validity involves assessing whether a logical relationship exists between the variable and the proposed measure. Essentially, it amounts to a rather commonsense comparison of what comprises the measure and the theoretical definition of the variable: Does it seem logical to use this measure to reflect that variable? We might measure *child abuse* in terms of the reports made by physicians or emergency room personnel of injuries suffered by children. Although this is not a perfect measure because health personnel might be wrong, it does seem logical that an injury reported by such people might reflect actual abuse.

No matter how carefully done, face validity is clearly subjective in nature. All we have is logic and common sense as arguments for the validity of a measure. This serves to make face validity the weakest demonstration of validity, and should usually be considered as no more than a starting point. All measures must pass the test of face validity. If they do, we should attempt one of the more stringent methods of assessing validity.

An extension of face validity is called **content validity** or **sampling validity**. It has to do with whether a measuring device covers the full range of meanings or forms that would be included in a variable that is being measured.

In other words, a valid measuring device would provide an adequate, or representative, sample of all content, or elements, or instances of the phenomenon being measured. For example, in measuring general *self-esteem*, it would be important to recognize that *self-esteem* can relate to many realms of people's lives, such as at work, at school, or in the family. Self-esteem might be expressed or come into play in all those settings. A valid measure of *self-esteem* would take that variability into account. If a measure of *self-esteem* consisted of a series of statements to which people expressed degrees of agreement, then a valid measure would include statements that relate to those many settings in which self-esteem might be expressed. If all the statements in the measuring device had to do, say, with school, then it would be judged a less valid measure of general *self-esteem*.

Content validity is a more extensive assessment of validity than is face validity because it involves a detailed analysis of the breadth of the concept being measured and whether that breadth is reflected in the measuring device. Content validity involves two distinct steps: (1) determining the full range or domain of the content of a variable, and (2) determining whether all those domains are represented among the items that constitute the measuring device. It is still, however, a subjective assessment in that someone has to judge what the full domain of the variable is and whether a particular aspect of a concept has been represented adequately in the measuring device. There are no agreed-upon criteria that can be applied to determine whether a measure has content validity. It is ultimately a judgment, albeit a more systematic and carefully considered judgment than with face validity.

One way to strengthen confidence in face or content validity is to gather the opinions of other investigators, especially those knowledgeable about the variables involved, regarding whether particular operational definitions seem to be logical measures of the variables. This extension of face or content validity, sometimes referred to as *jury opinion*, is still subjective, of course. However, because there are more people to serve as a check on bias or misinterpretation, jury opinion is superior to using individual tests of face or content validity.

Criterion Validity

Criterion validity refers to establishing validity by showing a correlation between a measuring device and some other criterion or standard that we know or believe accurately measures the variable under consideration. Or we might correlate the results of the measuring device with some properties or characteristics of the variable the measuring device is intended to measure. For example, a scale intended to measure *risk of suicide* should correlate with the occurrence of *self-destructive behavior* if it is to be considered valid. The key to criterion validity is to find a criterion against which to compare the results of our measuring device.

Criterion validity moves away from the subjective assessments of face and content validity and provides more objective evidence of validity. One type of criterion validity is **concurrent validity**, in which the instrument being evaluated is compared to some already existing criterion, such as the results of another measuring device. If I had developed a new measure of *self-esteem*, for example, I could compare its results to the results from existing measures of *self-esteem*. Any measuring devices used as the criterion in this assessment should have already been tested for validity. Existing measures to be used to establish the concurrent validity of a newly developed measure can be located during the literature review. The following are some compilations of such measures available in the social sciences: Bonjean, Hill, and McLemore, 1967; Brodsky & Smitherman, 1983; Fischer & Corcoran, 1994; Lake, Miles, & Earle, 1973; McDowell & Newell, 1996; Miller, 1991; Mueller, 1986; Robinson, Shaver, & Wrightsman, 1991; Schutte & Malouff, 1995; Straus, 1969. Additional measures can be found in research articles in professional journals. This is an important reason why, as pointed out in Chapter 4, a thorough review of existing literature should be undertaken: It can help you locate criterion measures. Should a suitable measure for comparison be found, it is a matter of applying both measures to the same sample and comparing the results. If a substantial correlation is found between the measures, we have reason to believe that our measure has concurrent validity. As a matter of convention, a correlation of $r = .50$ is considered the minimum required for establishing concurrent validity, but most researchers look for considerably higher correlations than that in order to have confidence in the validity of a measure.

The inherent weakness of concurrent validity is the validity of the existing measure that is used for comparison. All we can conclude is that our measure is as valid as the other. If the measure we select to use for comparison is not valid, the fact that ours correlates with it hardly makes our measure valid. For this reason, only those measures established by research to be valid should be used for comparison purposes in concurrent validity.

A second form of criterion validity is **predictive validity**, in which an instrument is used to predict some future state of affairs. In this case, the criterion used to assess the instrument is certain future events. The Scholastic Aptitude Test (SAT), for example, can be subjected to predictive validity by comparing performance on the test to how people perform in college. If people who score high on the SAT do better in college than do low scorers, then the SAT is presumably a valid measure of *scholastic aptitude*. Some measures are created for the specific purpose of predicting a given behavior, and these measures are obvious candidates for assessment by predictive validity. For example, attempts have been made to develop a measure that can predict which convicted criminals are likely to revert to high involvement with crime when released from prison (Chaiken & Chaiken, 1984). To make these predictions, the measure gathers information about the number and types of crimes

people commit, the age at which they commit their first crimes, and involvement with hard drugs. Ultimately, a measure such as this is validated by its ability to make accurate predictions about who actually experiences high crime involvement after release.

Sometimes the future state of affairs used to validate a measure is too far in the future and an earlier assessment of validity is desirable. If so, a variation on predictive validity, called the *known groups approach*, can be used. If it is known that certain groups are likely to differ substantially on a given variable, a measure's ability to discriminate between these groups can be used as an indicator of its ability to predict who will be in these groups in the future. For example, with the already discussed measure to predict high involvement in crime, the scales can be initially assessed for validity on their ability to differentiate between high and low crime involvement among current criminals. It is expected that if a measure can make this differentiation, it can also predict future involvement in crime. As another example of using known groups, suppose we were developing a measure of *prejudice*. We might apply the measure to a group of Protestant ministers, who we would expect to be low in *prejudice*, and to people involved in white supremacist groups who we would expect to be high in *prejudice*. If these groups differed significantly in how they responded to the instrument, then we would have reason to believe that the measure is valid. If it failed to show a substantial difference, we would certainly have doubt about its validity.

Despite the utility of the known groups approach, it does have its limitations. One is that there may be no groups known to differ on the variable we are attempting to measure. In fact, the purpose of developing a measure is often to allow the identification of groups who do differ on some variable. The upshot of this is that the known groups technique cannot always be used. A second limitation of the known groups technique is that it cannot tell us whether a measure can make finer distinctions between less extreme groups than those used in the validation. It may be, for example, that the measure of *prejudice* just described shows the members of white supremacist groups to be high in *prejudice* and the ministers low. With a broader sample, though, the measure may show that only the white supremacists score high and everyone else—not just the ministers—score low. Since we would expect a range of variation in prejudice among broad groups of people—from low to moderate to high levels of prejudice—a measure that produced this outcome is not very sensitive and can distinguish between groups only in a very crude fashion.

Construct Validity

Construct validity, the most complex of the types of validity discussed here, involves relating a measuring instrument to an overall theoretical framework in order to determine whether the instrument confirms a series of hypotheses derived from an existing and at least partially verified theory (Cronbach & Meehl, 1955; Zeller & Carmines, 1980). In this case, instruments are not assessed in terms of how they relate to any criterion but rather to measures of concepts that are derived from a broader theory. While construct validity can be based on one such comparison, more confidence that a measure has construct validity is adduced with numerous comparisons with a variety of concepts derived from the theory.

For example, Murray Straus and his colleagues (1996) developed a Conflict Tactics Scale (CTS) to measure how partners resolve conflicts in relationships. It is partly a measure of the use of psychological and physical aggression, but it also measures forms of conflict resolution in general. The CTS consists of a number of subscales, and Straus and colleagues assessed the construct validity of the subscales. One of their validity assessments was based on previous theory and research that suggested

that physical assaults of partners by men produce more serious injuries than do such assaults by women. If this is the case, they reasoned, then the correlation between the CTS's physical assault subscale and injury subscale should be strong among men but not among women. In other words, if these two scales measure what they claim to measure (namely, *occurrence of assault in relationships* and *extent of injury in relationships*), then previous theory and research suggests that *assault* and *injury level* will go together with men but be unrelated among women. And this was exactly what they found, providing some evidence of the construct validity of their subscales. Straus and colleagues conducted a series of assessments of this sort based on a variety of different theoretical predictions. The difference between criterion validity and construct validity is that the latter is based on the predictions of an existing and established theoretical framework and involves numerous comparisons. If a construct validity comparison is based on only one comparison, it comes closer to being criterion validity. After conducting a program of construct validity assessment, if some or all of the predicted relationships are not found, then we may question the validity of the new measuring instrument. Of course, it may be that the theories themselves are flawed, and this possibility must always be considered in assessing the results of construct validity.

There are some complex forms of construct validity. One is the **multitrait-multimethod approach** (Campbell & Fiske, 1959). This is based on two ideas. First is that two instruments that are valid measures of the same concept should correlate rather highly with each other even though they are different instruments. Second, two instruments, although similar to each other, should not correlate highly if they measure different concepts. You can readily imagine that this approach to validity involves the simultaneous assessment of numerous instruments (multimethod) and numerous concepts (multitrait) through the computation of intercorrelations. This approach has been used to assess the validity of instruments that are basically someone's subjective judgments about a phenomenon. For example, it has been used to assess the validity of people's ratings of how closely nations approximate a political democracy and the validity of committees' assessment of the relative value of particular occupations (Bollen & Paxton, 1998). The point is that assessing construct validity can become highly complex, but the very nature of this complexity offers greater evidence of the validity of the measures.

Another part of establishing construct validity is to demonstrate what is called **discriminant validity**: a measure should not show a correlation with variables or measures that are irrelevant to it or with which theoretical considerations suggest that it should not be correlated (Campbell & Fiske, 1959). Straus and colleagues (1996) did this as a part of their development of the CTS. The theory on which the scale is based suggests that the negotiation subscale should not correlate with the *sexual coercion* subscale or the *injury* subscale. And, in fact, they did not find any such correlations, which, along with their other assessments of validity, gave them more confidence that the CTS had construct validity.

The types of validity we have discussed—face, content, criterion, and construct—involve a progression in which each builds on the previous one to provide a move sophisticated assesment of validity. Each requires more information than prior ones but also provides a better assessment of validity. Unfortunately, many studies limit their assessment to face or content validity, with their heavy reliance on the subjective judgments of individuals or juries. Although sometimes this is necessary, measures that have been subjected only to face or content validity should be used with caution.

Reliability

In addition to validity, measures are also evaluated in terms of their **reliability**, which refers to a measure's ability to yield consistent results each time it is applied. In other words, reliable measures do not fluctuate from time to time unless the thing being measured has changed. In general, a valid measure is reliable. If we were certain of the validity of a measure, we would not therefore need to concern ourselves with its reliability. However, evidence of validity is almost always less than perfect, and that is why we turn to other ways of evaluating measures, including reliability. Reliability provides more evidence for validity since a reliable measure may be a valid one.

Fortunately, reliability can be demonstrated in a more straightforward manner than validity. Many specific techniques exist for estimating the reliability of a measure, but they are all based on one of two principles: *stability* or *equivalence*. *Stability* is the idea that a reliable measure should not change from one application to the next, assuming the concept being measured has not changed. *Equivalence* is the idea that all items that make up a measuring instrument should be measuring the same thing and thus be consistent with one another. The first technique for estimating reliability, test-retest reliability, is based on the stability approach. The others discussed use the equivalence principle.

Test-Retest

The first and most generally applicable assessment of reliability is called "test-retest." As the name implies, this technique involves applying a measure to a sample of individuals and then, somewhat later, applying the same measure to the same individuals again. After the retest, we have two scores on the same measure for each person, as illustrated in Table 5.4. These two sets of scores are then correlated by using an appropriate statistical measure of association (see Chapter 14). Because the association in test-retest reliability involves scores obtained from two identical questionnaires, we fully expect a high degree

TABLE 5.4 Hypothetical Test-Retest Data

Subjects	Initial Test	Retest
1	12	15
2	15	20
3	22	30
4	38	35
5	40	35
6	40	38
7	40	41
8	60	55
9	70	65
10	75	77

$r = .98$

of association. As a matter of convention, a correlation coefficient of .80 or better is normally necessary for a measure to be considered reliable. In Table 5.4, the r means that the particular statistic used was Pearson's correlation coefficient, and the value of .98 indicates that the measurement instrument is highly reliable according to the test-retest method. If a reliability coefficient does not achieve the conventional level but is close to it, the researcher must make a judgment about whether to assume the instrument is reliable (and that the low coefficient is due to factors other than the unreliability of the instrument) or to rework the instrument in order to obtain higher levels of association.

In actual practice, the test-retest method sometimes cannot be used quite as simply as suggested because exposing people to the same measure twice creates a problem known as "multiple-testing effects" (Campbell & Stanley, 1963). A group of people may not react to a measure the second time the same way as they do the first. They may, for example, recall their previous answers, and this could influence their second response. People might respond as they

recall doing the first time to maintain consistency or purposefully change responses for the sake of variety. Either case can have a confounding effect on testing reliability. If people strive for consistency, their efforts can mask actual unreliability in the instrument. If they purposefully change responses, they can make a reliable measure appear unreliable. This is especially likely to occur with measures that people are likely to remember how they had responded or when the two administrations are only a short time apart.

A solution to this dilemma is to divide the test group into two groups randomly: an experimental group that is tested twice and a control group that is tested only once. Table 5.5 illustrates the design for such an experiment. Ideally, the measure will yield consistent results in all three testing sessions, and, if it does, we have solid reason to believe the measure is reliable. On the other hand, substantial differences among the groups may indicate unreliability. If, for example, the experimental group shows consistency in both responses to the measuring instrument but the control group differs, the measure may be unreliable, and the consistency of the experimental group might derive from the multiple-testing effects. Another outcome would be for the experimental group to yield inconsistent results but the control group to show results similar to the experimental group's initial test; this outcome could also be due to multiple-testing effects and results from the experimental group's purposefully changing their answers during the retest. Despite the inconsistency in the experimental group, the measure still might be reliable if this outcome is observed. Finally, the results of all three testing sessions may be inconsistent. Such an outcome would suggest that the measure is not reliable. If either of the outcomes leaves the reliability of the measure in doubt, a second test-retest experiment should be conducted with the hope of obtaining clearer results. If the same result occurs, the instrument needs to be redesigned.

TABLE 5.5   Design for Test–Retest

	Initial Test	Retest
Experimental Group	Yes	Yes
Control Group	No	Yes

The test-retest method of assessing reliability has advantages and disadvantages. Its major advantage is that it can be used with many measures, which is not true of the other ways of assessing reliability. Its disadvantage is that it is slow and cumbersome to use, with its required two testing sessions and the desirability of having a control group. Also, as we have seen, the outcome may not be clear, leading to the necessity of repeating the whole procedure. Finally, the test-retest method cannot be used on measures of variables whose value might have changed in the interval between tests. For example, people's attitudes can change for reasons that have nothing to do with the testing, and a measure of attitudes might appear unreliable when indeed it is not.

Multiple Forms

If our measuring device is a multiple-item scale, as is often the case, we can approach the question of reliability through the technique of *multiple forms*. When developing the scale, we create two separate but equivalent versions made up of different items, such as different questions. These two forms are administered to the same individuals at a single testing session. The results from each form are correlated with each other, as was done in test–retest, using an appropriate statistical measure of association, with the same convention of $r = .80$ required for establishing reliability. If the correlation between the two forms is sufficiently high, we can assume that each scale is reliable.

The advantage of multiple forms is that only one testing session is required and no control group is needed. This may be a significant advantage if either multiple-testing sessions or using a control group is impractical. In addition, one need not worry about changes in a variable over time because both forms are administered during the same testing session.

The multiple forms technique relies on the two forms appearing to the respondents as though they were only one long measure so that the respondents will not realize that they are really taking the same test twice. This necessity of deluding people points to one of the disadvantages of multiple forms. To maintain the equivalence of the forms, the items in the two forms will likely be quite similar—so similar that people may realize that they are responding to essentially the same items twice. If this occurs, it raises the specter of multiple-testing effects and casts doubt on the accuracy of the reliability test. Another disadvantage of multiple forms is the difficulty of developing two measures with different items that are really equivalent. If we obtain inconsistent results from the two forms, it may be due to differences in the forms rather than the unreliability of either one. In a way, it is questionable whether multiple forms really test reliability and not just our ability to create equivalent versions of the measure.

Split-Half Approach

In the split-half approach to reliability, the test group responds to the complete measuring instrument. The items that make up the instrument are then randomly divided into two halves. Each half is then treated as though it were a separate scale, and the two halves are correlated by using an appropriate measure of association. Once again, a coefficient of $r = .80$ is needed to demonstrate reliability.

One complication in using the split-half reliability test is that the correlation coefficient may understate the reliability of the measure because, other things being equal, a longer measuring scale is more reliable than a shorter one. Because the split-half approach divides the scale in two, each half is shorter than the whole scale and will appear less reliable than the whole scale. To correct for this, the correlation coefficient is adjusted by applying the Spearman-Brown formula:

$$r = \frac{2r_i}{1 + r_i}$$

where

r_i = uncorrected correlation coefficient, and

r = corrected correlation coefficient
(reliability coefficient).

To illustrate the effect of the Spearman-Brown formula, suppose we have a 20-item scale with a correlation between the two halves of $r_i = .70$, which is smaller than the minimum needed to demonstrate reliability. The Spearman-Brown formula corrects as follows:

$$r = \frac{(2)(.70)}{1 + .70} = \frac{1.40}{1.70} = .82$$

It can be seen that the Spearman-Brown formula has a substantial effect, increasing the uncorrected coefficient from well below .80 to just over it. If we had obtained these results with an actual scale, we would conclude that its reliability was now adequate.

Using the split-half technique requires that two preconditions be met that can limit its applicability. First, all the items in the scale must be measuring the same variable. If the scale in question is a jumble of items measuring several different variables, it would be meaningless to divide it and compare the halves. Second, the scale must contain a sufficient number of items so that when it is divided the halves do not become too short to be considered scales in themselves. A suggested minimum is 8 to 10 items per half (Goode & Hatt, 1952, p. 236). Because many measures are shorter than these minimums, it may not be

possible to assess their reliability with the split-half technique.

Randomly dividing the items in a scale into halves could result in many different arrangements of items. For a 20-item scale, for example, there are 184,756 possible arrangements of the items into two halves. Thus, there are conceivably 184,756 different tests of reliability that could be made when using the split-half technique, and each would yield a slightly different correlation between the halves. Which should be used to assess reliability? The most popular approach to this problem is to use *Cronbach's alpha,* which may be thought of as the average of all possible split-half correlations. Theoretically, the scale is divided into all possible configurations of two halves; a correlation is computed for each possibility, and the average of those correlations is computed to derive alpha (Cronbach, 1951). The computation of alpha is complicated and beyond the scope of this book, but many common statistical software packages compute Cronbach's alpha as well as other complex reliability tests.

The split-half reliability test has several advantages. It requires only one testing session, and no control group is required. It also gives the clearest indication of reliability. For these reasons, it is the preferred method of assessing reliability when it can be used. The only disadvantage, as noted, is that it cannot always be used. In addition, it is preferable to use more than one test, if possible, in assessing both reliability and validity. The issues are sufficiently important that the expenditure of time is justified.

The Eye on Diversity box discusses some issues to be considered regarding measurement—including validity and reliability—when doing research on diverse populations.

Errors in Measurement

The range of precision in measurement is quite broad: from the cook who measures in terms of pinches, dashes, and smidgens to the physicist who measures in tiny angstrom units (0.003937 millionths of an inch). No matter whether measurement is crude or precise, it is important to recognize that all measurement involves some component of error. There is no such thing as an exact measurement. Although some measuring devices in the social sciences are fairly precise, others contain substantial error components. The major reason for this error is that most social science measures deal with abstract or shifting phenomena, which are difficult to measure with any high degree of precision. The large error component in many social science measurements means that researchers must be concerned with the different types and sources of error so they can be kept to a minimum. In measurement, researchers confront two basic types of error: *random* and *systematic*. In fact, we can modify the formula from measurement theory that was introduced earlier in this chapter with the recognition that the error term in that formula, *E*, is actually made up of two components:

$$E = R + S,$$

where *R* represents random error, and *S* represents systematic error. Now our measurement formula looks like this:

$$X = T + R + S.$$

Our measurement or observation of a phenomenon is a function of the true nature of the phenomenon along with any random and systematic error that occurs in the measuring process.

Random Errors

Random error is neither consistent nor patterned; the error is as likely to be in one direction as another. Random errors are essentially chance errors that in the long run tend to cancel themselves out. In fact, in measurement theory, mathematicians often assume that, in the long run, $R = 0$. For example, a respondent may misread or mismark an item on a questionnaire; an interviewer may misunderstand and thus record incorrectly something said during an interview; or a data entry assistant may enter data incorrectly into a computerized database. All these examples are random sources of error and can occur at virtually every point in a research project. Cognizant of the numerous sources of random error, researchers take steps to minimize them. Careful wording of questions, convenient response formats, and "cleaning" and double checking of computerized data are some of the things that can be done to keep random error to a minimum. Despite all efforts, however, the final data may contain some component of random error.

Because of their unpatterned nature, it is assumed that random errors tend to cancel out each other. For example, the data entry assistant mentioned earlier would be just as likely to incorrectly enter a score lower than the actual one as to enter a higher score. The net effect is that random errors, at least in part, offset each other. However, random error will weaken the precision with which a researcher can measure variables, and this can reduce the ability to detect a relationship between variables when in fact one is present. For example, consider a hypothetical study on the relationship between *level of education* and *satisfaction with one's life*. Let's assume that higher levels of *life satisfaction* actually are associated with higher levels of *education*. In such a study, measurements would be taken of the *level of education* of each individual and of their *satisfaction with life*. If some random error occurs, then some cases will be scored too high and some too low on one or both variables. Even with such random errors, the mean *education* and *satisfaction* levels of all cases can be expected to be quite close to the true averages because the random errors are canceling out one another in the whole group. However, in terms of individual cases, random error will produce some *satisfaction* scores that are erroneously low for their associated *education* levels. Conversely, random error will produce

EYE ON DIVERSITY   Culturally Sensitive Measurement

When conducting research on minority populations, considerable opportunity for bias exists if concepts and operational definitions are not developed carefully. One area where poorly constructed operational definitions have produced misleading conclusions is in the area of spouse abuse (Lockhart, 1991). Many of the earlier studies found rates of spouse abuse to be considerably higher among African Americans than among Anglos. Typically, these studies used one of the following as an operational definition of the occurrence of abuse: a homicide involving a domestic killing, a battered woman seeking care in an emergency room or social service setting, a wife abuse claim handled by a domestic court, or a domestic dispute call to a police department. Note, however, that this measurement of *abuse* depends on an episode becoming known to some authorities before abuse is considered to have occurred. It is well known that African Americans are overrepresented among people who come to the attention of the police, emergency room personnel, or social workers. Because they are generally overrepresented among these populations, they appear to have higher rates of abuse than Anglos when abuse is operationalized in this fashion. These problems can be reduced by selecting a sample of subjects from a broad-based community and having these people answer questions about the amount of conflict and violence that occurs in their family. This avoids the biasing effect of looking only at certain locales (Straus & Gelles, 1988).

The Committee on the Status of Women in Sociology (1986) has indicated another area in which operational definitions have led to misleading results: studies of work and social contribution. Work is often operationalized in terms of paid employment, but this excludes many types of work from consideration, such as community service or home-based work. With such an operational definition, if an employee of a carpet-cleaning firm shampoos the carpets in a home for a fee, that would be counted as "work," but if a woman does the same activity on her own time in her own home, it would not be classified as "work." Such an operationalization of *work* tends to underestimate the extent of productive activity engaged in by women because women are less likely than men to be paid for those activities.

Another problem that arises in measuring diverse populations is that measuring devices that have been tested for validity and reliability on one cultural group are often erroneously assumed to be valid and reliable measures for other cultural groups. Validity and reliability is often first assessed by applying them to Anglo, non-Hispanic respondents because such people are often the most accessible to researchers. However, it should almost never be assumed that such assessments can be generalized to minority populations (Manson, 1986). The unique cultural characteristics and attitudes of minorities are, typically, not considered in the development of such instruments. For some minorities, such as Asians and Hispanics, language differences mean that an English-language interview would have some respondents answering in a second language. It cannot be assumed that such a respondent understands words and phrases as well, or in the same way, as a person for whom English is his or her first language. Additionally, idiomatically, there may be concepts in English that do not have a precise equivalent or mean something completely different in another language.

Measuring instruments, therefore, usually need to be refined if they are to be valid and reliable measures among minorities. A study of mental health among Native Americans, for example, had to drop the word *blue* as a descriptor of depression because that word had no equivalent meaning among the Native Americans (Manson, 1986). Researchers also had to add a category of "traditional healer" to a list of professionals to whom one might turn for help. A study of Eskimos found that, because of the cultural context, the same question could be used but had to be interpreted differently. Because Eskimo culture emphasizes tolerance and endurance, Eskimos are less likely than Anglo Americans to give in to pain and not go to work. A positive response from an Eskimo to a question like "Does sickness often keep you from doing your work?" is thus considered a far more potent indicator of distress than is the same answer given by an Anglo American.

These illustrations should make clear that measurement in social research must be culturally sensitive. When conducting research on a group with a culture that differs from that of the researchers, there are a number of things that researchers can do to produce more valid and reliable measurement instruments (Marin & Marin, 1991).

1. In conceptual development and operationalization, care must be taken to assess whether measurement instruments might lead to a distorted view of minorities. In some cases, this calls for careful consideration of what a concept is intended to mean. Using the earlier example, is the focus of the research on "paid employment" or is it on "social contribution"? In other cases it calls for careful assessment of whether a definition will lead to an inaccurate, over- or underrepresentation of minorities.

2. Researchers can immerse themselves in the culture of the group to be studied, experiencing the daily activities of life and the cultural products as the "natives" do.

3. Researchers should use key informants—people who participate routinely in the culture of the group to be studied—to help develop and assess the measurement instruments.

4. When translating an instrument from English into another language, researchers should use the most effective translation methods, usually "double translation" (translate from English into the target language and then translate by an independent person back into English to check for errors or inconsistencies).

5. After developing or translating measuring instruments for use with minority populations, the instruments should be tested for validity and reliability on that population.

some *education* levels that are high for their associated *satisfaction* scores. Thus, the random error will tend to mask the true correlation between *education* and *satisfaction*. Despite the fact that *education* and *satisfaction* really are correlated, if this type of random measurement error is too high, it will result in concluding that a relationship between the variables does not exist or is smaller than it actually is.

Fortunately, researchers can combat random error with a variety of strategies. One is to increase sample size; generally, random error has less impact on a larger sample. A second strategy is to increase the number of items on the measurement scales or in other ways refine the tools to be more precise. Although such strategies can reduce the impact of random error, the same cannot be said for systematic error.

Systematic Errors

Systematic error is consistent and patterned. Unlike random errors, systematic errors may not cancel themselves out. If there is a consistent over- or understatement of the value of a given variable, then the errors will accumulate. For example, it is well known that systematic error occurs when measuring crime with official reports of crimes known to the police; the Uniform Crime Reports (UCR) of the Federal Bureau of Investigation counts only crimes that are reported to the police. The Department of Justice supplements these statistics with the National Crime Survey (NCS), which measures the number of people who claim to be the victims of crime. Comparisons of these two measures consistently reveal a substantial amount of "hidden" crime—crimes reported by victims but never brought to the attention of the police. For example, NCS and other crime data indicate that only about 45% of all violent crimes and only one-third of other crimes are reported to the police (Rennison, 1999). So there is a very large systematic error when measuring the amount of crime the way the

UCR does because of the underreporting of most crimes.

Systematic errors are more troublesome to researchers than random errors because they are more likely to lead to false conclusions. For example, official juvenile delinquency statistics consistently show higher rates of delinquency among children of lower socioeconomic status families. Self-report studies of delinquency involvement suggest, however, that the official data systematically overstate the relationship between delinquency and socioeconomic status (Binder, Geis, & Bruce, 1988). It should be easy to see how the systematic error in delinquency data could lead to erroneous conclusions as to possible causes of delinquency as well as to inappropriate prevention or treatment strategies.

Improving Validity and Reliability

When a measuring device does not achieve acceptable levels of validity and reliability—when there is too much error—researchers often attempt to redesign the device so that it is more valid and reliable. How to develop valid and reliable measuring devices is discussed at length in the following chapters where you are shown how to design good measuring tools. However, a few techniques can be mentioned here as a preview of what is done when a measuring device does not yield adequate validity and reliability.

More extensive conceptual development

Often validity and reliability are compromised because the researcher has not been sufficiently clear and precise about the nature of the concepts being measured and possible indicators of them. Rethinking the concepts often helps in revising the measuring instrument to make it more valid.

Better training of those who will be applying the measuring devices

This is especially useful when a measuring device is based on someone's subjective assessment of an attitude or state. The people applying the device can be shown how their judgments can be biased or produce error and how they can guard against it in their assessments.

Interview the subjects of the research about the measuring devices

Those being studied may have some insight into why the verbal reports, observations, or archival reports are not producing accurate measures of their behavior. They may, for example, be able to tell you that the wording of questions is ambiguous or that some words are interpreted differently by members of their subculture than the researcher intended.

Higher level of measurement

This does not guarantee greater validity and reliability, but a higher level of measurement can in some cases produce a more reliable measuring device. So, when the researcher has some options in terms of how to measure a variable, it is worth considering a higher level of measurement.

Use more indicators of a variable

This also does not guarantee enhanced reliability and validity, but a multiple-item measuring device can in some cases produce a more valid measure than does a measuring device with fewer items.

Conduct an item-by-item assessment of multiple-item measures

If the measuring device consists of a number of questions or items, it may be that only one or a few items are the problem: It is the invalid items that are reducing the validity and relia-

bility of the instrument. Revising or deleting them may improve validity and reliability.

When a measuring device is revised based on these ideas, the revised version would, of course, have to be subjected to tests of validity and reliability.

CHOOSING A MEASURING DEVICE

We have seen that there are normally a number of indicators, sometimes a large number and at different levels of measurement, that can be used to measure a variable. For a particular study, how do we choose the best of these measuring devices? Or if we are developing a new measuring device, how do we decide whether it is or is not good? This can be a complicated and sometimes difficult decision for researchers, but a number of factors, discussed in this or earlier chapters, can serve as guidelines in making this decision.

1. Indicators should be chosen that measure the variables in ways that are theoretically important in the research (discussed in Chapter 2).

2. Indicators should be chosen based on their proven validity and reliability.

3. If two measuring instruments are equivalent in all other ways except for level of measurement, then the indicator at the higher level of measurement should be chosen.

4. Indicators that produce the least amount of systematic and random error should be chosen.

5. Indicators should be chosen with matters of feasibility in mind (discussed in Chapter 4).

REVIEW AND CRITICAL THINKING

Measurement is all about being careful and precise. It is about linking the abstract world of concepts and ideas to the concrete world of observation. Much can be learned from how scientists do these things, and translated into tips for critically analyzing any information or situations that you might confront.

1. When something is being discussed, can you identify any explicit or implicit indicators (or operational definitions) that are being used? In other words, how do people know if that thing exists or how much of it exists?

2. Would any particular group's values or interests be promoted by such operational definitions? If so, does this produce any bias?

3. Is there any effort to evaluate these operational definitions (that is, assessment of validity and reliability)?

4. Could error in measurement be affecting what is being said or observed?

Computers and the Internet

Searching the Internet provides many opportunities for locating information on issues of measurement in science. You can use a search engine to locate many of the terms used in this chapter, such as "measurement," "error," "validity," "reliability," and so on. If you combine some of the terms, it will focus the search and reduce the number of less useful sites that are located. Many of these Web sites will not pertain to the social sciences, but you will see that many of the measurement issues are the same in any fields that use a scientific approach. You should use this opportunity to identify the commonalities across scientific disciplines. You will also notice in this search that issues of qualitative versus quantitative research arise frequently.

One specific area that you might find valuable to search through is psychometrics, which is the study of personality and mental states and attributes. This is a field primarily of interest to psychologists, and psychometric measurement is often used for diagnostic and clinical, rather than research, purposes. However, it is generally a highly quantitative field and thus a good area in which to explore measurement issues. Identify Web sites related to this topic and list the measurement issues explored. What issues related to qualitative research do the Web sites address?

Main Points

1. Measurement is the process or act of operationalizing concepts, or describing abstract concepts in terms of specific indicators by assigning numbers or other symbols to them. An indicator is an observation that is assumed to be evidence of the attributes or properties of some phenomenon.

2. Measurement might be accomplished with a single item or a multiple-item index or scale. Social scientists measure most variables through either verbal reports, observation, or archival records. Positivists and nonpositivists disagree about the nature of measurement.

3. The four levels of measurement are nominal, ordinal, interval, and ratio. The level of measurement achieved with a given variable is determined by the nature of the variable itself and by the way it is measured.

4. Discrete variables have a limited number of distinct and separate values. Continuous variables theoretically have an infinite number of possible values.

5. Validity refers to how accurately a measuring device measures the variable it is intended to measure. Face validity, content or sampling validity, jury opinion, criterion validity, and construct validity are techniques of assessing the validity of measures.

6. Reliability refers to a measure's ability to yield consistent results each time it is applied. Test-retest, multiple forms, and split-half are techniques for assessing the reliability of measures.

7. Random errors are those that are neither consistent nor patterned and can reduce the precision with which variables are measured. Systematic errors are consistent and patterned and can potentially lead to erroneous conclusions. There are a number of steps that researchers can take to improve the validity and reliability of measuring devices.

8. Measuring devices are chosen on the basis of theoretical considerations, their validity and reliability, their level of measurement, the amount of systematic and random errors, and feasibility.

Important Terms for Review

concurrent validity
construct validity
content validity
continuous variables
criterion validity
discrete variables
discriminant validity
face validity
index
indicator
interval measures

item
level of measurement
measurement
multitrait–multimethod approach to validity
nominal measures
ordinal measures
predictive validity
random error
ratio measures
reliability
sampling validity
scale
systematic error
validity

For Further Reading

Burgess, Robert G. (Ed.). (1986). *Key variables in social investigation*. London: Routledge & Kegan Paul. This collection of essays is unusual in that it focuses on 10 of the most commonly used social science variables, analyzing their underlying concepts and how they have been operationalized.

Hindelang, Michael J., Hirschi, Travis, & Weis, Joseph G. (1980). *Measuring delinquency*. Beverly Hills, CA: Sage. A good description of the development of a measuring device. The volume covers all the issues related to problems of measurement.

Kirk, Jerome, & Miller, Marc L.. (1986). *Reliability and validity in qualitative research*. Beverly Hills, CA: Sage. This work presents the measurement issues of reliability and validity as they apply to qualitative research such as field research (see Chapter 11). Unfortunately, reliability and validity are often only presented in the context of quantitative research.

Martin, Lawrence L., & Kettner, Peter M. (1996). *Measuring the performance of human service programs*. Thousand Oaks, CA: Sage. This short book explains in detail how to measure and assess human service programs, especially with outcome measures. It includes such measures as levels of functioning scales and client satisfaction.

Miller, Delbert C. (1991). *Handbook of research design and social measurement* (5th ed.). Newbury Park, CA: Sage. A good resource work for scales and indexes focusing on specific social science concerns.

Price, James L. (1986). *Handbook of organizational measurement*. Marshfield, MA: Pitman. Using concepts that are a part of organizational theory, this book illustrates how to link concepts with indicators.

Chapter 6

CONSTRUCTING QUESTIONS, INDEXES, AND SCALES

easurement in social science research often involves obtaining people's answers to a question or responses to a verbal statement. These questions and statements are the operational definitions of the variables being studied. This is how variables are measured in survey research (see Chapter 9), but such questions and statements are also sometimes used to measure variables in field research, experiments, and other types of research designs in the social sciences. These questions and statements are the social scientist's equivalent of the chemist's microscope or the astronomer's telescope. Just as a microscope or a telescope is constructed with great care, social scientists take great care in constructing the linguistic questions or statements that serve as their measuring devices.

As Chapter 5 discussed, measurement is sometimes fairly straightforward and involves the use of only a single item, or indicator, of a variable. We can, for example, measure an individual's age with one question that asks how old the person is. Many other variables, however, are far more difficult and complex to measure. In some cases, there may be more than one indicator of a variable. In other cases, the variable may involve a number of dimensions that call for multiple indicators. In still other cases, we may be concerned with the degree to which a variable is present. In all these cases, a single-item measuring instrument may be inadequate, and an *index* or a *scale* would be used—a number of items that are combined to form a composite score on a variable. To measure people's attitudes toward having children, for example, we could ask how much they agree with a series of questions, which are the items that make up the scale. Their composite score on the scale would indicate their overall attitude toward having children.

Chapter 5 discussed procedures for evaluating measuring devices once they have been developed. The topic of Chapter 6 is the actual development of good questions, statements, indexes, and scales and begins with a discussion of how to design good questions or statements. Then I distinguish between indexes and scales and describe the kinds of indexes and scales used and how they are developed.

DESIGNING QUESTIONS

When social scientists measure variables by having people respond to questions or statements, this can be done in one of two ways: with *questionnaires* or with *interviews*. A **questionnaire** contains written questions that people respond to directly on the questionnaire form itself, without the assistance of an interviewer. An **interview** involves an interviewer reading questions to respondents and recording their answers. It is important to keep in mind that when social scientists talk about their "measuring instrument," they are referring to the whole questionnaire or interview, not just to the individual questions. After all, people are affected by and react to the whole thing, including such aspects as the physical appearance and the order in which questions or statements are presented. Chapter 9 explores how to develop whole questionnaires and interviews, including such aspects as physical appearance and question ordering. Chapter 6 focuses on a more limited issue: how to design the individual questions, statements, indexes, and scales that make up questionnaires and interviews. Designing valid questions and statements is a complex and challenging process, and I will review some of the major guidelines that can be used in doing it well (Sheatsley, 1983).

Conceptual Relevance

To design good questions, a researcher needs to be clear about the nature and dimensions of the concepts to be measured and take care that the questions link directly with the concepts. For example, if the concepts relate to people's behavior, then the questions should ask about behaviors—not attitudes, thoughts, or emotions. Likewise, concepts that refer to emotions or attitudes should be tapped by questions that address emotions or attitudes. As another example, a concept that relates to poverty is not necessarily adequately measured by questions having to do with welfare. Although poverty and welfare are related, they are not identical, and researchers need to continually relate the questions being developed to the abstract concepts they are intended to measure. This again illustrates the critical importance of conceptual development discussed in Chapter 2 and Chapter 4 and the continual interplay between the abstract conceptual level and the concrete level of operational definitions.

Directions

One of the simplest but most important parts of a measuring device is the directions that guide the respondents' answering of the questions. Good directions go a long way toward improving the quality of data generated by questionnaires and interviews. If you want respondents to put an X in a box corresponding to their answer, tell them precisely that. Questionnaires often contain items requiring different ways of answering. At each place in the questionnaire where the format changes, additional directions should be included. As the development of questions, indexes, and scales are described in this chapter, examples of how to design good directions will be provided. Clear directions are especially important in questionnaires since there may be no one present to clarify any ambiguities, as when a questionnaire is distributed through the mail or over the Internet. However, directions are also important in interviews where the interviewer may read directions that guide the respondent's answers.

Closed-Ended versus Open-Ended Questions

Two basic types of questions can be used: *closed ended* or *open ended* (Sudman & Bradburn, 1982). **Closed-ended questions** are those that provide respondents with a fixed set of al-

ternatives from which to choose. The response formats of multiple-item scales, for example, are all closed ended, as are multiple-choice examination questions, with which you are undoubtedly familiar. **Open-ended questions** are questions to which the respondents write their own responses, much as you do for an essay-type examination question.

The proper use of open- and closed-ended questions is important for the quality of data generated. Theoretical considerations play an important part in the decision about which type of question to use. In general, closed-ended questions should be used when all the possible, theoretically relevant responses to a question can be determined in advance and the number of possible responses is limited. For example, a question relating to marital status would almost certainly be treated as a closed-ended question. A known and limited number of answers are possible: married, single, divorced, separated, or widowed. In current research, people may be offered an additional alternative—"living together" or "cohabitating"—in order to reflect more accurately the living arrangements that people choose today. Another obvious closed-ended question is about sexual status. To leave such questions open ended runs the risk that some respondent will either purposefully or inadvertently answer in a way that provides meaningless data. Putting "sex" with a blank after it, for example, is an open invitation for some character to write "yes" rather than give the information needed.

Open-ended questions, on the other hand, are used in qualitative research when the researcher wants individuals to describe their feelings or discover the meanings that are important to people. For such data, open-ended questions allow people to provide in their own words as complete an accounting as possible of some phenomenon. Data of this sort will be discussed in more detail in Chapter 11 and Chapter 15. Quantitative research also sometimes uses open-ended questions, but in this case the ultimate goal will probably be to quantify people's responses. Open-ended questions might therefore be appropriate in an exploratory study in which the lack of theoretical development makes it difficult to know how to categorize people's responses before seeing the responses. In addition, when researchers cannot predict all the possible answers to a question or when too many possible answers exist to list them all practically, then open-ended questions are appropriate. Suppose we wanted to know the reasons people moved to their current residence. So many possible reasons exist that such a question would probably be treated as open ended. If interested in the county and state in which respondents reside, we could generate a complete list of all the possibilities and thus create a closed-ended question. But the list would consume so much space on the questionnaire that it would be excessively cumbersome, especially considering that respondents should be able to answer this question correctly in its open-ended form.

Some questions lend themselves to a combination of both formats. Religious affiliation is a question usually handled in this way. Although there are a great many religions, there are some to which only a few respondents will belong. Thus, religions with large memberships can be listed in closed-ended fashion with a category "Other," where a person can fill in the name of a religion not found on the list (see question 4, Table 6.1). Any question with a similar pattern of responses—numerous possibilities, but a few popular ones—can be handled efficiently in this way. The combined format maintains the convenience of closed-ended questions for most of the respondents but also allows those with less common responses to express them. When the option of "Other" is used in a closed-ended question, it is a good idea to prompt respondents to write in their response by indicating "please specify." These answers can then be coded into whatever response categories seem appropriate for data analysis.

Another issue in choosing between open- and closed-ended questions is the ease with which each can be handled at the data-analysis stage. Closed-ended questions can be quickly entered into a computer file and prepared for data analysis. Open-ended questions, if they are to be handled quantitatively, must first be coded, which means creating a category system into which everybody's answers can be placed. The process of coding is discussed in chapters 11, 13, and 15, but the point here is that it is a more time-consuming form of data analysis. If the nature of the variables warrants using closed-ended questions, it will be quicker and more efficient at the data-analysis stage.

As you can see from this discussion, the decision about whether to use open- or closed-ended questions is complex and can have substantial effects on the type and quality of the data that are collected, as was illustrated in a survey of attitudes about social problems confronting the United States. The Institute for Social Research at the University of Michigan asked a sample of respondents open- and closed-ended versions of essentially the same questions (Schuman & Presser, 1979). The two versions elicited quite different responses. For example, with the closed-ended version, 35% of the respondents indicated that crime and violence were important social problems, compared to only 15.7% in the open-ended version. The same pattern occurred with a number of other issues: People were more likely to identify an issue as a problem if it was on the closed-ended list. One reason that the type of question has such an effect is that the list of alternatives in the closed-ended questions tends to serve as a "reminder" to the respondent of issues that might be problems. Without the stimulus of the list, some respondents might not even think of those issues. A second reason is that people tend to choose from the list provided in closed-ended questions rather than writing in their own answers even when provided with an "Other" category.

It is possible, in some cases, to gain the benefits of both open- and closed-ended questions by using an open-ended format in a pretest or pilot study and then, based on the

TABLE 6.1   Formatting Questions for a Questionnaire

Please indicate your response to the following questions by placing a X in the appropriate box.

1. Which of the following best describes where you live?

   ☒ In a large city (100,000 population or more).

   ☐ In a suburb near a large city

   ☐ In a middle-sized city or small town (under 100,000 population) but not a suburb of a large city.

   ☐ Open country (but not on a farm)

   ☐ On a farm

2. Have you ever shoplifted an item with a value of $10 or more?

   ☐ Yes

   ☐ No

   If Yes: How many times have you taken such items?

   ☐ Once

   ☐ 2 to 5 times

   ☐ 6 to 10 times

   ☐ More than 10 times

Filter Questions

Contingency Question

3. Have you purchased a new automobile between 1995 and the present?

   ☐ Yes

   ☐ No (*If No*, please skip to Section C, question 1.)

4. Please indicate the religion to which you belong:

   ☐ Protestant

   ☐ Catholic

   ☐ Jewish

   ☐ Other. Please specify. _____

results, designing closed-ended questions for the actual survey.

Wording of Questions and Statements

Because the questions or statements that are used to measure variables are the basic data-gathering devices, they need to be worded with great care. This is true especially for questionnaires that allow no opportunity to clarify questions for the respondent. With these questionnaires, ambiguity in questions can be a source of substantial measurement error. I will

review some of the major issues in developing good questions (Gorden, 1992; Sudman & Bradburn, 1982).

Pretesting

The wording of questions should always be subjected to empirical assessment. In other words, whenever possible, a researcher should attempt to determine whether a particular wording might lead to unnoticed bias. Words, after all, have connotative meanings (that is, emotional or evaluative associations) that the researcher may not be aware of but that may influence respondents' answers to questions. In a study of people's attitudes about social welfare policy in the United States, for example, survey respondents were asked whether they believed we should spend more or less money on welfare (T. W. Smith, 1987). However, they were asked the question in three slightly different ways: Group one was asked whether we were spending too much or too little on "welfare"; group two was asked about spending on "assistance for the poor"; and group three was asked about money for "caring for the poor." At first glance, all three questions would seem to have much the same meaning; yet people's responses to them suggested something quite different. Basically, people responded far more negatively to the question containing the word *welfare*, indicating substantially less willingness to spend more money on "welfare" than they were to "assist the poor." This seems to be a minor semantical difference in wording, yet the impact was dramatic. Although the study didn't investigate why these differing responses occurred, it seems plausible that, for many people, the word *welfare* has connotative meanings that involve images of laziness, waste, fraud, bureaucracy, or the poor as disreputable. "Assisting the poor," on the other hand, is more likely to be associated with giving and Judeo-Christian charity. These connotations lead to very different responses. In many cases, the only way to assess such differences is to compare people's responses to different versions of the same question during a pretest or to conduct assessments of validity and reliability, as discussed in Chapter 5.

So, once questions and statements are developed, they should be pretested to see if they are clearly and properly understood and are unbiased. Pretesting can be done by having people respond to the questionnaire or interview and then reviewing their responses with them to find any problems. The way in which a group responds to the questions themselves can also point to trouble. For example, if many respondents leave a particular answer blank,

then there may be a problem with the question. Once the instrument is pretested, modifications should be made where called for, and it should be pretested again. Any change in the questionnaire requires more pretesting. Once it is pretested with no changes called for, it is ready to be used in research.

Tense

In general, questions should be stated in the present tense. An exception would be specialized questions that focus on past experiences or expectations for the future. In these situations, the appropriate tense would be used. Of major importance is that tenses not be mixed carelessly. Failure to maintain consistent tense of questions can lead to an understandable confusion on the part of respondents and therefore more measurement error.

Simple, Direct, and Clear

Questions should be simple, direct, and express only one idea. Complex statements expressing more than one idea should be avoided. For example, the first statement illustrated in Table 6.2 is not acceptable for this reason and needs to be changed as indicated. Another practice to avoid is reference to things that are not clearly definable or that depend on the respondent's interpretation. This is a problem with the last two questions in Table 6.2, and a revision is suggested as a solution.

Always be aware that statements that seem crystal clear to a researcher may prove unclear to many respondents. When writing questions, it is preferable to err on the side of designing questions that may be too simple for the intended audience than to err in the opposite direction. This concern is especially relevant when respondents are suspected of having low levels of education or poor reading skills. Accordingly, the researcher should avoid the use of technical jargon. For example, it would not be advisable to include a statement that reads, "The current stratification system in the United States is too rigid." *Stratification* is a technical

TABLE 6.2   Common Errors in Writing Questions and Statements

Original Question	Problem	Solution
The city needs more housing for the elderly and property taxes should be raised to finance it	**Two questions in one:** Some respondents might agree with the first part but disagree with the second.	Questions should be broken up into two separate statements, each expressing a single idea.
In order to build more stealth bombers, the government should raise taxes.	**False premise:** What if a person doesn't want more bombers built? How do they answer?	First ask their opinion on whether the bomber should be built; then, for those who respond "Yes," ask the question about taxes.
Are you generally satisfied with your job, or are there some things about it that you don't like?	**Overlapping alternatives:** A person might want to answer "Yes" to the first part (i.e., they are generally satisfied) but "No" to the second part (i.e., there are also some things they don't like).	Divide this into two questions: one measures their level of satisfaction while the other assesses whether there are things they don't like.
How satisfied are you with the number and fairness of the tests in this course?	**Double-barreled question:** It asks about both the "number" and the "fairness," and a person might feel differently about each.	Divide this into two questions.
What is your income?	**Vague and ambiguous words:** Does "income" refer to before-tax or after-tax income? To hourly, weekly, monthly, or yearly income?	Clarify: What was your total annual income, before taxes, for the year 2000?
Children who get into trouble typically have had a bad home life.	**Vague and ambiguous words:** The words *trouble* and *bad home life* are unclear. Is it trouble with the law, trouble at school, trouble with parents, or what? What constitutes a *bad home life* depends on the respondent's interpretation.	Clarify: Specify what you mean by the words: *trouble* means "having been arrested" and *bad home life* means "an alcoholic parent."

word used in the social sciences that many people outside the field may not understand in the same sense that social scientists do.

Length

Questions and statements should be kept as short as possible while still providing the respondent with the complete and accurate intended meaning. One guideline that has been offered is that questions not be more than 25 words in length (Sheatsley, 1983). While this is not a hard and fast rule, it would certainly be wise to look critically at any question approaching 25 words to determine whether it is potentially confusing or actually contains more than one thought and should be broken down into two or more shorter questions.

Insider Language and Slang

For most questions, especially those designed for the general public, slang words should not be used. Slang usage tends to arise in the context of particular groups and subcultures. Slang words may have a precise meaning within those groups but confuse people outside those groups. Occasionally, however, the target population for a survey will be more specialized than the general population, and the use of their insider language or "in-group" jargon may be appropriate. It would demonstrate to the respondents that the researcher cared enough to "learn their language" and could increase rapport, resulting in better responses. In interviewing prostitutes, for example, it might be preferable to refer to prostitutes as "girls" or "working girls" (if this is the terminology used by the prostitutes themselves); referring to them as "prostitutes" may create social distance and make it more difficult to gather valid responses from them. Having decided to use slang, however, the burden is on the researcher to be certain the slang is used correctly.

Some additional problems that can arise in writing questions and statements are presented in Table 6.2.

Providing a Context

To elicit valid responses, it is important to be sure that people possess adequate information to give accurate answers. In part, this means asking specific questions, but it also means placing the question in a clear context for the respondent. One way to provide a clear context is to place the question and answer in a specific *time perspective*. For example, if you were interested in individuals' perceptions of changes in their financial circumstances, you might ask "Do you think that your financial circumstances are getting better or worse?" Yet, two people whose circumstances are identical might give very different responses if one interpreted the question to mean "over the past few months" while the other was looking "over the past few years." Based on whatever time period seems appropriate given conceptual considerations, you could revise the question to read, "Considering just the past two-year period, do you think that your financial circumstances are getting better or worse?"

Another way to provide a context for a question is to provide a *spatial* or *geographic perspective*. A quality of life study, for example, might be interested in whether people feel safe in their community. You might ask, "Do you feel safe walking the streets at night alone?" but this could elicit different responses if the person is thinking of his or her own neighborhood as opposed to any neighborhood in a city. Again, based on conceptual considerations, you could contextualize the response by asking, "Do you feel safe walking the streets *in your own neighborhood* at night alone?" Or, you could be even more specific: "Do you feel safe walking the streets *within five blocks of your home* at night alone?"

A third way to provide a context for a question is to provide an *interpretive perspective*. When the issues being asked about are sensitive, people may be unwilling to admit them to an interviewer. With spouse abuse, for example, respondents may be reluctant to admit being either a perpetrator or a victim. The scale presented in Table 6.3 begins with instructions that provide an interpretive statement that essentially says "some people do these things" and "if you have done these things, you are not alone." This provides a way of gradually moving into some issues that may be very sensitive, and eases the feeling people may have that they will be judged negatively based on their responses. It makes it more socially acceptable to admit to a behavior that could carry some stigma.

A fourth way to provide a context is to include any *definitions of terms* that may be misunderstood or any *facts* that are essential to giving accurate answers. If it were necessary to use the words *cohabitation* or *partner* (meaning one's cohabitant) in a question, for example, it might be worth preceding the question with a brief and clear statement of what you mean by (your definition of) those words.

Sensitive Questions

Questions on sensitive or potentially embarrassing topics can be a challenge to design. Sensitive topics are not limited to things such as sexual behavior or drug abuse; some people may also consider telling someone their income

TABLE 6.3 Two Subscales (Psychological Aggression and Physical Assault) of the Revised Conflict Tactics Scale (CTS2)*

Instructions:

No matter how well a couple gets along, there are times when they disagree, get annoyed with the other person, want different things from each other, or just have spats or fights because they are in a bad mood, are tired, or for some other reason. Couples also have many different ways of trying to settle their differences. This is list of things that might happen when you have differences. Please circle how many times you did each of these things in the past year, and how many times your partner did them in the past year. If you or your partner did not do one of these things in the past year, but it happened before that, circle "7."

How often did this happen?

1 = Once in the past year

2 = Twice in the past year

3 = 3–5 times in the past year

4 = 6–10 times in the past year

5 = 11–20 times in the past year

6 = More than 20 times in the past year

7 = Not in the past year, but it did happen before

0 = This has never happened

5. I insulted or swore at my partner.	1 2 3 4 5 6 7 0
25. I called my partner fat or ugly.	1 2 3 4 5 6 7 0
29. I destroyed something belonging to my partner.	1 2 3 4 5 6 7 0
35. I shouted or yelled at my partner.	1 2 3 4 5 6 7 0
49. I stomped out of the room or house or yard during a disagreement.	1 2 3 4 5 6 7 0
65. I accused my partner of being a lousy lover.	1 2 3 4 5 6 7 0
67. I did something to spite my partner.	1 2 3 4 5 6 7 0
69. I threatened to hit or throw something at my partner.	1 2 3 4 5 6 7 0
7. I threw something at my partner that could hurt.	1 2 3 4 5 6 7 0
9. I twisted my partner's arm or hair	1 2 3 4 5 6 7 0
17. I pushed or shoved my partner.	1 2 3 4 5 6 7 0
21. I used a knife or a gun on my partner.	1 2 3 4 5 6 7 0
27. I punched or hit my partner with something that could hurt.	1 2 3 4 5 6 7 0
33. I choked my partner.	1 2 3 4 5 6 7 0
37. I slammed my partner against a wall.	1 2 3 4 5 6 7 0
43. I beat up my partner.	1 2 3 4 5 6 7 0
45. I grabbed my partner.	1 2 3 4 5 6 7 0
53. I slapped my partner.	1 2 3 4 5 6 7 0
61. I burned or scalded my partner on purpose.	1 2 3 4 5 6 7 0
73. I kicked my partner.	1 2 3 4 5 6 7 0

*Note   (1) The first eight items are the psychological aggression subscale. The numbers indicate the order in which the items would appear on the scale, but items of other subscales would be interspersed among them. Also, this table does not show the items where the respondent indicates that their partner did this to them (e.g. for Item 5, the next item would be: "My partner insulted or swore at me.") (2) Permission to use this test must be obtained from the copyright owners (Straus et al.).

Source   Adapted from Murray A. Straus, Sherry L. Hamby, Sue Boney-McCoy, and David B. Sugarman, "The Revised Conflict Tactics Scale (CTS2): Development and Preliminary Psychometric Data," *Journal of Family Issues*, 17 (May 1996), 310–312.

or other aspects of their lifestyle to be private and sensitive. The problem is that the respondent may prefer to keep some things private, may be too embarrassed to respond completely, or may not know exactly how to respond, in terms of using language that will be acceptable to the interviewer.

One way of handling this problem is to provide respondents with acceptable wording for their answers in the question itself. In a question on sexual practices, for example, the response alternatives could be provided in the question itself: "We want to ask you about sexual behaviors that people sometimes engage in. People are known to engage in vaginal intercourse, oral intercourse, and anal intercourse . . ." In this way, the topic is defused to an extent because the terminology is out in the open, and the appropriate vocabulary to use for various activities is suggested. Another way to do this is to have the response alternatives listed by letters on a card; hand that card to the respondent, and ask: "Tell me the letter of the activities that you have engaged in." An illustration of this with the variable *income* is presented in Chapter 2, Table 2.2 (p. 35). This procedure is sometimes used with a variable such as *income* because people are less reluctant to tell an interviewer a letter than they are to tell the interviewer their exact income.

Response Formats

All efforts at careful wording of questions will be for naught unless the questions are presented in a manner that facilitates response. The goal is to make responding to the questions as straightforward and convenient as possible and to reduce the amount of data lost because of uninterpretable responses.

When presenting response alternatives for closed-ended questions, best results are obtained by having respondents indicate their selection by placing an X in a box corresponding to that alternative, as illustrated in question 1 of Table 6.1. Most word processing programs permit you to format a bulleted list of alternatives where the bullet can be designed in various ways, including an empty box. If a word processing program won't permit such boxes or a typewriter is being used, then the response area can be delimited with open and close brackets [ ], or parentheses ( ), with space between. These formats are preferable to open blanks and check marks (✓) because it is too easy for respondents to get sloppy and place check marks between alternatives, rendering their responses unclear and therefore useless as data. Boxes increase the likelihood that respondents will give unambiguous responses. As an alternative to X marks in boxes, it generally works well to number each response alternative and have the respondents circle the number of their choice.

When all or part of a question is closed ended, good questions also depend on providing the respondent with appropriate response alternatives that reflect the complete range of responses that might be chosen in answer to the question. In some cases, the response alternatives are appropriately either "Yes" or "No," as in question 3 in Table 6.1. Oftentimes, however, variables involve a range of opinions or attitudes. If we ask people if they favor campaign finance reform, we could offer alternatives that range from "Favor it very much" to "Favor it not at all." But does this range provide all the positions people might hold on this issue? For example, if someone feels "Strongly opposed" to campaign finance reform, does the alternative "Favor it not at all" really encapsulate that position? A more appropriate wording of a range of alternatives might be: "Favor strongly," "Favor moderately," "Unsure," "Oppose moderately," "Oppose strongly." In the first way of designing the response alternatives, all the choices include the positive word (*favor*) but that may not communicate opposition, which is a legitimate position someone could take. In Table 6.4, a number of illustrations of response formats for survey questions are presented to give you an

TABLE 6.4   A Variety of Response Formats for Survey Statements

Measuring a Student's Evaluation of a College Instructor

Overall, I rate the quality of this professor as:

☐ Excellent   ☐ Good   ☐ Fair   ☐ Poor

Measuring Interpersonal Conflict

How many times in the past 12 months have you insulted or swore at him or her?

☐ Once   ☐ Twice   ☐ 3–5 times   ☐ 6–10 times   ☐ 11–20 times   ☐ More than 20 times

Measuring Job Anxiety

How likely do you think it is that you will lose your job or be laid off?

☐ Very likely   ☐ Fairly likely   ☐ Not too likely   ☐ Not at all likely

Measuring Job Satisfaction

On the whole, how satisfied are you with the work that you do?

☐ Very satisfied   ☐ Moderately satisfied   ☐ Moderately dissatisfied   ☐ Very dissatisfied

Measuring Attitude Toward Military Service

For most young women, do you think military service is a good experience or not?

☐ Definitely good   ☐ Probably good   ☐ Probably not good   ☐ Definitely not good

Measuring Characteristics of Jobs

How important do you personally consider job security as a characteristic of a job?

Unimportant   **Important**

1   2   3   4   5   6   7

Measuring Delinquency

Have you ever skipped school without a legitimate excuse?

☐ Never   ☐ Once or Twice   ☐ Several times   ☐ Often   ☐ Very often

idea of some of the variations possible. (As a test of what you have learned in this section of the chapter, you might consider whether any of the statements in Table 6.4 are poorly worded and see if you can improve on their wording.)

When considering the number of response alternatives to provide, research suggests that questions or statements with more response alternatives are more reliable and valid measures than those with fewer alternatives (Alwin, 1997). Although there is considerable variation in how many alternatives should be used, five

alternatives is a common number, three alternatives is probably too few, and seven or more may be better.

Real Attitudes versus Nonattitudes

When asking questions that measure people's subjective state or their attitudes, researchers often want to provide an opportunity for people to state that they don't have an attitude on an issue or that they are uncertain about how they feel. This is based on the reasonable theoretical assumption that some persons may not have an attitude about a particular issue or that they have an opinion but it is confused or uncertain. A common way of handling this is to provide a "Don't know" or "Uncertain" response alternative among those available to the respondent. If the response alternatives involve an intensity level, such as from "Strongly agree" to "Strongly disagree," then the "Uncertain" alternative is sometimes placed at the middle of the intensity range or is placed away from the other alternatives. Placing the "Uncertain" alternative in the middle implies that it is a middle position on the range of alternatives when a person's choice of "Uncertain" may not reflect that. An individual, for example, could feel strongly about an issue but may also possess a high degree of uncertainty. Setting the "Uncertain" alternative off to the side allows this expression of attitude to be clearer.

The "Don't know" and "Uncertain" response alternatives raise another problem: whether you are measuring real attitudes or nonattitudes (Converse, 1970; Gilljam & Granberg, 1993). A real attitude is, of course, one that a person actually holds and is measured by your question. A nonattitude is an attitude that is expressed but not really held by a person; it is what is called a *false positive*. It may be expressed, for example, because the design of a question does not permit the person to avoid expressing it. If there is no "Don't know" or "Uncertain" alternative, then the person is forced to express an attitude (or to not answer the question). This would suggest the importance of including such response alternatives, but there is an opposite problem: *False negatives* are attitudes that a person actually holds but are not expressed. People who do possess an attitude may choose the "Don't know" alternative because it is easier and quicker to do so.

What we do know about this problem is that both false positives and false negatives occur when assessing people's attitudes. What we

know much less about is how to rectify the problem. One of the difficulties is that designing questions to reduce one problem often increases the other. Therefore, a first step is to decide which type of error is more important to avoid. For a variety of reasons, researchers are often more concerned about false positives, and these can be reduced by including a "Don't know" or "Uncertain" alternative or by asking a *filter question*. A filter question's answer determines which question the respondent goes to next. In Table 6.1, questions 2 and 3 are both filter questions. In question 2, the part of the question asking "How many items have you taken" is called a *contingency question* because whether a person answers it depends on (is contingent upon) their answer to the filter question. Note the two ways filter questions can be designed. With question 2, the person answering "Yes" is directed to their next question by the arrow and the question is clearly set off by a box; also in the box, the phrase "If Yes" is included to be sure the person realizes that this question is only for those who answered "Yes" to the previous question. With question 3, the answer "No" is followed by a statement telling the person which question they should answer next. Either format is acceptable, but again the point is to provide clear directions for the respondent.

When measuring attitudes, we sometimes use filter questions that ask persons if they have an attitude on the issue of interest as a way of reducing false positives. If the respondents state that they have an opinion, then they are presented with the question that measures their position on that issue. If false negatives are of more concern, then these options will not be included, which forces respondents to express an opinion but increases the likelihood of false positives.

Whichever of these alternatives we adopt will, of course, produce some error, and for this reason some survey researchers recommend great caution in interpreting the meaning of responses to questions. A more complex way to deal with this problem is to ask a series of questions, with the earlier ones focusing on false positives and the later ones on false negatives. A study of attitudes toward nuclear power in Sweden did this by first asking a filter question that has an "easy out" alternative (Gilljam & Granberg, 1993):

> There are various views regarding nuclear power as an energy source. What is your view? Are you generally for or against the use of nuclear power as an energy source in Sweden, or don't you have any particular opinion on this question?

A second question did not include a filter or "Don't know" alternative but did include a clear neutral point on the scale of opinion; this forces an opinion but also allows an "uncertain" type of response.

> I want to ask your position on nuclear power. Where would you place yourself on this scale?
>
> $-5 \ -4 \ -3 \ -2 \ -1 \ \ 0 \ +1 \ +2 \ +3 \ +4 \ +5$
>
> Very negative toward nuclear power   Very positive toward nuclear power

A third question was asked that offered five alternatives on how to use nuclear power (from "Shut down all plants now" to "Build more plants"), but with no neutral position. This question forces a statement of opinion with none of the easy outs. Each of these three questions is slightly different, and respondents were asked all three. This enabled the researchers, in the data-analysis phase, to assess how much response is real opinion and how much is false positives or false negatives. A final alternative to alleviate this problem is to follow up a question with an open-ended question or an in-depth interview where a respondent's attitude could be probed at some length to assess his or her real attitude.

So, special care must be taken in designing response alternatives; consideration must be given to the underlying concept being measured. Clarifying the concept can often help one design a valid set of response alternatives. In addition, pretests and tests of validity and reliability may help discover any difficulties with the alternatives. A number of additional issues relative to designing response alternatives are discussed in the remainder of this chapter while discussing index and scale construction.

MULTIPLE-ITEM INDEXES AND SCALES

Some social phenomena are too abstract or complex to be measured accurately by an individual's response to a single question or statement. When this is so, social scientists turn to multiple-item measuring devices that produce a quantitative score that is a composite of the subject's responses to a number of separate items.

Indexes versus Scales

Social scientists use two different kinds of multiple-item measuring devices: *indexes* and *scales* (DeVellis, 1991; Zeller & Carmines, 1980). However, be forewarned at the outset that some confusion surrounds the use of these two terms. Some people use them interchangeably, whereas others distinguish between them but then disagree over whether a particular measuring device is an index or a scale. In addition, some measuring devices may have properties of both. It is worth exploring these terms because they are widely used, and understanding the distinction between them will help you understand more about multiple-item measuring devices and how they are constructed. With these qualifications, let's explore what they are and examples of each.

An **index** is a composite measure in which separate indicators of a phenomenon are combined to create a single measure. In some cases,

TABLE 6.5 An example of the Bogardus Social Distance (SD) Scale

Check the "Yes" box to all of the following statements with which you agree.

I would be willing to have Norwegians as my close kin by marriage.

☐ Yes

I would be willing to have Norwegians in my club as personal friends.

☐ Yes

I would be willing to have Norwegians on my street as neighbors.

☐ Yes

I would be willing to have Norwegians working alongside me on my job.

☐ Yes

I would be willing to have Norwegians as citizens in my country.

☐ Yes

I would be willing to have Norwegians as visitors to my country.

☐ Yes

scores on each individual indicator are summed to give an overall score on the composite phenomenon. Suppose, for example, that we wanted to construct an index of involvement in delinquent activities. We could create a number of separate items that reflect particular instances of delinquent behavior, such as: Has the person been truant from school in the past year? Has the person shoplifted items from a store in the past year? Has the person vandalized any property in the past year? If each item had only two alternatives—Yes/No—we could sum up the number of Yes responses as our index of the level of delinquent activity. If there were 10 items, then a low score of 0 would represent *no delinquent activity* on our index, whereas a high score of 10 would mean that a person chose a "Yes" option to all 10 items in the index. Indexes achieve ordinal, and in some cases interval-ratio, level of measurement, as does this one. This multiple-item measure enables us to measure something more abstract (*extent of involvement in delinquency*) than the specific behaviors identified in the separate items (truancy, shoplifting, and vandalism).

A **scale** is a multiple-item measuring device in which there is a built-in intensity structure, potency, or natural levels of feeling to the items that make up the scale. A scale is made up of separate items or indicators, as is an index, but in a scale the variation in intensity among the items means that there tends to be more pattern to people's responses to the various items. Table 6.5 contains a scale that illustrates these properties. It is a version of what is called the "Bogardus Social Distance Scale," and measures people's willingness to interact with others who are different from them in terms of race, ethnicity, or culture (Converse, 1987). Like an index, a scale measures an abstract phenomenon (in this case, social distance) by combining people's responses to much more concrete and specific items (such as, "have a person as a neighbor"). In addition, if you review the items, you can see that, with some items, many people would probably agree ("willing to have Norwegians as citizens in my country"), whereas with other items, fewer people might agree. Also, an inherent order seems to exist among the items, in the sense that a "Yes" response to some items

TABLE 6.6   Rosenberg Self-Esteem Scale

	1 Strongly Agree	2 Agree	3 Disagree	4 Strongly Disagree
(1) On the whole, I am satisfied with myself.	SA[4]	A[3]	D[2]	SD[1]
(2) At times, I think I am no good at all.	SA[1]	A[2]	D[3]	SD[4]
(3) I feel that I have a number of good qualities.	SA[4]	A[3]	D[2]	SD[1]
(4) I am able to do things as well as most other people.	SA[4]	A[3]	D[2]	SD[1]
(5) I feel I do not have much to be proud of.	SA[1]	A[2]	D[3]	SD[4]
(6) I certainly feel useless at times.	SA[1]	A[2]	D[3]	SD[4]
(7) I feel that I'm a person of worth, at least on an equal plane with others	SA[4]	A[3]	D[2]	SD[1]
(8) I wish I could have more respect for myself.	SA[1]	A[2]	D[3]	SD[4]
(9) All in all, I am inclined to feel that I am a failure.	SA[1]	A[2]	D[3]	SD[4]
(10) I take a positive attitude toward myself.	SA[4]	A[3]	D[2]	SD[1]

Source   Morris Rosenberg, *Conceiving the Self* (rev. ed.) Malabar, FL: Krieger, Publishing (1986).

probably means that the person responded "Yes" to other items. For example, if you would admit someone as your neighbor, you would probably also admit that person to be a citizen in your country. This is the intensity structure to the items of a scale that distinguish it from an index where such an intensity structure doesn't exist. With scales, people's composite score tends to represent more of a pattern in the responses. Indexes, without the intensity structure, tend not to show such patterns; with the delinquency index, if you had answered "Yes" to shoplifting, that doesn't mean that you had also responded "Yes" to truancy or vandalism.

As we look at particular indexes and scales in this and other chapters, consider whether each has the properties of an index, a scale, or some combination of the two. Don't be misled by what others *call* a multiple-item measuring device they are using: Just because someone labels something a scale or index doesn't mean that it necessarily is. As I said, the two terms are not always used consistently.

Advantages of Multiple-Item Measures

Indexes and scales have four major advantages over single-item measures.

Improved Validity

When measuring abstract or complex variables, a multiple-item measure is generally more valid than a single-item measure. Consider the variable *self-esteem*. The Rosenberg Self-esteem Scale, developed to measure *self-esteem*, contains 10 statements (see Table 6.6). It does so because no single question or statement could possibly measure something as complex, multifaceted, and constantly changing as a person's self-esteem. What single question could be asked that might encompass all the feelings that you have about yourself? Clearly, self-esteem involves many aspects of a person's life situation—family, occupation, financial and social status, to name a few. Multiple-item scales provide more valid measures of such complex phenomena.

Improved Reliability

In general, as shown in Chapter 5, the more items contained in a measure, the more reliable it will be. This is so because the statements comprising a scale are actually just a sample of the entire universe of statements that could have been used. A single-item measure is a sample of one, and is less likely to be representative of the universe of statements than multiple items would be. Multiple-item scales are larger samples from this universe and are therefore more likely to be representative. Being more representative, they are more reliable than single-item measures.

Reduced Measurement Error

In addition to providing a more representative sampling of the items that measure a variable, multiple-item measures also reduce the impact of measurement error, especially random error. The reason for this is that any single indicator of a variable will probably measure somewhat high or low of the actual state of the variable. As more items are added to the measuring device, the likelihood that some measures will err in the opposite direction of the original items is increased. Eventually, a point is reached where sufficient items are obtained that approximately 50% measure high and 50% low, thus canceling out the random errors—one of the goals of measurement discussed in Chapter 5.

Increased Level of Measurement

Single-item measures often produce data that are nominal or, at the very best, partially ordered. The term *partially ordered data* refers to data with a few ordered categories but with many cases falling into each category, or having the same value for the variable. Although superior to nominal, these data are less desirable than fully ordered data in which every— or nearly every—case has its own rank (see Chapter 14). Multiple-item scales are capable of producing data that are closer to fully ordered and, in some cases, interval-level data.

As we saw in Chapter 5, a higher level of measurement is generally preferred because it involves more precision and increased flexibility in data analysis.

When the concept to be measured is complex, multiple-item measures offer substantial advantages for the researcher—advantages that often outweigh the difficulty of their construction.

DEVELOPING INDEXES AND SCALES

Once it is decided to use a multiple-item measuring device, an appropriate index or scale needs to be found or developed. In many cases, indexes and scales consist of questions to which individuals respond, or statements to which they indicate their level of agreement. In other cases, especially for indexes, the measuring device may consist of items of data collected elsewhere. Earlier in this chapter, guidelines were discussed for wording questions or statements in questionnaires or interviews, and those same general rules apply to the development or selection of index and scale items. In many cases, it is possible to use a complete index or scale developed by someone else if it is a valid and reliable measure of the variables under investigation. Or you might be able to use some items, or revise some items, from previously developed scales. (When using all or part of an existing measuring device, be sure to check whether you need to get permission to use it or pay copyright fees.) A major advantage of using existing indexes or scales is that their validity and reliability have usually already been established. A second advantage is that comparisons of research findings are more direct when different research projects use the same operational definitions of variables. If two projects measure a variable in different ways and reach different conclusions, it may be because they were actually measuring *different variables*. Finally, keep in mind that if you revise or use only part of an existing measuring device, it must be subjected to validity and reliability checks. A few of the many compilations of measurement scales are listed in the "For Further Reading" section of this chapter. In addition, indexes and scales are reported in the many research journals in the behavioral sciences.

If no existing index or scale will do the job, then a new scale should be developed (DeVellis, 1991; Zeller & Carmines, 1980). While there are some differences in developing indexes and scales, a certain logic is common to both. Developing multiple-item measures generally involves the following steps:

1. developing or locating many potential items, far more than will appear in the final measure;

2. eliminating items that are irrelevant, redundant, ambiguous, or for some other reason inappropriate for measuring the variable;

3. pretesting the remaining items for validity, reliability, or other measurement checks to be described shortly;

4. eliminating items that do not pass the tests of step 3; and

5. repeating steps 3 and 4 as often as necessary to reduce the index or scale to the number of items required.

Sources of Index and Scale Items

In developing or locating items to constitute an index or scale, the researcher begins with theoretical considerations: What kind of direction or guidance is provided by the theory of which the abstract concept is a part. For example, in developing the Conflict Tactics Scale (CTS), sociologist Murray Straus and his colleagues (1996) point out that "the theoretical basis of the CTS is conflict theory. . . . This theory assumes that conflict is an inevitable part of all human association, whereas violence as a tactic to deal with conflict is not" (Straus et al., 1996, p. 284). This theoretical insight led them

to the conclusion that the items for the CTS had to cover not only violent tactics but also nonviolent ones. The CTS therefore includes items measuring psychological aggression, negotiation, sexual coercion, and other nonviolent modes of conflict as well as items measuring physical assault (see Table 6.3). Conflict theory also led these researchers to realize that interpersonal conflict can be a two-way street: Both partners can engage in conflictual actions. So the CTS contains items that measure the conflict tactics by both partners in a relationship. So, you can see that careful theoretical analysis gives direction to at least the *kind* of items that might be appropriate for an index or scale.

As for finding the specific items themselves, one of the most accessible sources of items for any multiple-item measure is a researcher's own imagination. Once a concept has been developed and refined, the researcher has a pretty good idea of what is to be measured. The researcher can then generate a range of statements that seem to satisfy the criteria to be discussed. At this early stage in index and scale construction, one need not be overly concerned with honing and polishing the statements to perfection because much pretesting remains before any statement is ever seen by an actual respondent.

A second source of items is people, sometimes called "judges," who are considered to be especially knowledgeable in a particular area. If one is seeking items for a delinquency scale, for instance, it would seem reasonable to discuss the issue with juvenile probation officers and others having daily contact with delinquents. Two social psychologists used this approach to find items for a scale to measure people's tendency to manipulate others for their own personal gain (Christie & Geis, 1970). They turned to the writings of Niccolò Machiavelli, a sixteenth-century adviser to the prince of Florence, Italy. In his classic book *The Prince*, Machiavelli propounded essentially a con artist's view of the world and politics: People, according to Machiavelli, are to be manipulated for one's own benefit, in a cool and unemotional fashion. In the writings of this Florentine of four centuries ago, these social psychologists found such statements as: "It is safer to be feared than to be loved," and "Humility not only is of no service but is actually harmful." They constructed a scale made up of Machiavelli's statements, somewhat revised, and asked individuals whether they agreed with each statement. The scale is now known as the "Machiavellianism Scale" and has been used widely in social science research. This illustrates a particularly creative use of "judges" in the development of multiple-item measures.

A third source of index and scale items is the persons who are the focus of the research project. Claudia Coulton (1979), for example, was interested in person-environment fit among consumers of hospital social services. In developing her scale, she obtained a large number of verbatim statements from hospital patients and then began to form these into a scale. In like manner, if one were interested in attitudes among teenagers toward unwanted pregnancies, an excellent beginning would be to discuss the topic with teenagers themselves and gather from them as many statements as possible regarding the issue. When items are garnered from individuals in this fashion, only rarely would statements be usable without editing. Many statements would ultimately be rejected, and most would have to be considerably rewritten. Such persons, however, are likely to provide a range of statements that have meaning from the perspective of the group under investigation.

Characteristics of Index and Scale Items

Once a large number of items has been found, the best have to be selected for the final measuring device. Good items have the following characteristics.

Validity

A primary concern in item selection is the validity of the statements (see Chapter 5). Each statement considered for inclusion should be assessed for face validity and content validity. For example, if we were creating a self-report delinquency scale, each statement would be assessed as to how it relates to measuring delinquent activity. Statements concerning a person's participation in various delinquent acts would be reasonable as valid measures of how delinquent that person is. On the other hand, an item relating to how well the respondent gets along with his or her siblings would probably not be a valid indicator of delinquency.

Range of Variation

Variables that are measured with multiple-item measures are normally considered to consist of a number of possible values or positions that a person could take. If we wanted to measure *attitudes toward growing old*, for example, people's positions on that variable could be extremely positive, extremely negative, or anywhere in between. In selecting items for measuring devices, we should ensure that the items cover the actual range of possible variation on the variable being measured. Failure to do so will result in a poor measuring device. When selecting items on the basis of variability, the researcher needs to exercise care to avoid defining the range either too narrowly or too broadly. Failure to include a sufficiently wide range of items will result in respondents "piling up" (clustering) at one or both ends of the scale's range. If many respondents tie with either the lowest or highest possible score, the range in the scale is inadequate. This piling up effect reduces the precision of the measurement because we are unable to differentiate among the respondents with tied scores.

Going to extremes with items to define the range is also not desirable. If we include items that are too extreme, they will apply to few, if any, respondents. In the case of a delinquency scale, for example, an item pertaining to engaging in cannibalism would be such an extreme item as to warrant exclusion. The act is so rare that it is unlikely that any juvenile has done it, and it thus contributes nothing of benefit to the scale. The goal is to select items with enough range of variation to cover the actual range of alternatives that individuals are likely to choose, without including items that are so extreme that they do not apply to anyone. In fact, one guideline is to provide a range of alternatives such that, in the long run, 50% of the respondents will choose alternatives on the positive side of the range and 50% on the negative side. However, this is only a guideline and is sometimes not necessary or possible to achieve in actual practice.

Unidimensionality

In the construction of multiple-item measures, especially scales, often the goal is to measure one specific variable. We do not want the results confounded by items on the scale that actually measure a different, although possibly related, variable. The items of a **unidimensional scale** measure only one variable. If a scale actually measures more than one variable, it is called *multidimensional*. In creating our delinquency scale, we might be tempted to include an item about school performance on the grounds that delinquents seem to perform poorly in school. Although an empirical relationship may exist between *delinquency* and *school performance*, they are separate variables and should be treated and measured as such.

Relationships between Items

To gain systematic evidence of the unidimensionality of a scale, the researcher can intercorrelate each item on the scale with every other scale item. This is often done during a pretest. If some items do not correlate with the others, it is possible that these items do not measure

the same variable or they measure separate aspects of the variable that vary independently from one another. If we suspect that these items measure a different variable, they should be eliminated from the scale. If we find a few items that have nearly perfect correlations, we only need to use one of them in the scale. Two items to which individuals respond identically are merely redundant, and using both adds nothing to the measurement abilities of the scale. Occasionally, however, highly correlated items are included to detect response inconsistency or random answering. That exception notwithstanding, the final scale will be composed of statements that correlate fairly highly, but not perfectly, with one another.

A knowledge of the characteristics and sources of items provides an important and necessary foundation for the development of indexes and scales. There are many different types of indexes and scales, and the remainder of the chapter discusses them and how they are constructed.

INDEX CONSTRUCTION

One commonly used and easily understood index is the Federal Bureau of Investigation's Crime Index, which is published each year in the FBI's *Uniform Crime Reports*. The items that make up the Crime Index are the seven crimes that the FBI classifies as Part I, or more serious, crimes: homicide, rape, robbery, aggravated assault, burglary, larceny-theft, and motor vehicle theft. (Arson is also classified as a Part I offense but is not included in the Crime Index.) The Crime Index consists of the total of these seven offenses that are known to the police for every 100,000 persons (see Table 6.7).

There are several things to note about the crime index. First, each of the seven items that make it up is a part of the broader phenomenon being studied, namely, the extent of serious crimes. These items are chosen based on theoretical and conceptual considerations

TABLE 6.7   Construction of a Crime Index

Crime rates per 100,000 People, 1995		Crime Index, 1995
Homicide	8.2	
Rape	37.1	
Robbery	221	
Assault	418	5,278
Burglary	988	
Larceny-theft	3,045	
Motor vehicle theft	561	

Source   Federal Bureau of Investigation, *Uniform Crime Reports.*

regarding what constitutes a serious crime; for the criminologists who developed the index, serious crimes are those that pose the greatest and most direct threat to personal safety, property, and public order. Such crimes as gambling, prostitution, and commercialized vice are not included in the index because they were judged to pose a considerably weaker threat to the public order.

A second thing to note is that the seven items—the seven index crimes—constitute the universe of elements that make up the more abstract concept of "serious crimes." In fact, the concept of "serious crimes" is defined by its operational definition: A serious crime is one of the Part I offenses. This is often the case with indexes: The items used in their construction are all the elements or constructs that make up the broader phenomenon.

A third thing to note about the Crime Index is the relationship between the index and the items that measure it: the values of each item determine the level of the variable measured by the index, rather than the other way around. When the incidence of rape or robbery rises, this causes the level of the index to rise. However, causality does not flow in the other direction: Changes in the index cannot occur first and account for changes in the individual

items. For example, a rise in serious crimes does not cause changes in the amounts of robbery that occur; in fact, the index (serious crimes) could rise while the robbery rate falls, as long as some of the other crimes in the index rise. With an index, changes in the items produce changes in the index, rather than the other way around. This is a characteristic of many, although not all, indexes. When we discuss scales, we will see that they involve something different (DeVellis, 1991). The Applied Scenario box discusses the development and use of an index for an applied research problem.

In developing indexes, it is often the case that the separate items are not given equal weighting as was the case with the crime index. Theoretical considerations may suggest that some indicators of a variable are more important in determining the state of the variable. This is the case with some indexes that have been developed to measure socioeconomic status (SES). Most measures of SES use a combination of two or more indicators, such as occupation, income, and education. A number of years ago, sociologists Seymour Parker and Robert Kleiner (1966), building on earlier work, developed an SES measure, described in Table 6.8, that combined all three. Since education, occupation, and income are measured in different units, they need to be transformed into common measurement units. As you can see, this was done by dividing the variables in a way such that each had seven values. Parker and Kleiner did this based on theoretical considerations of the variables as well as on the empirical distribution of persons into the various categories. For example, the occupational categories were created in part based on the relative prestige of various occupations, trying to place occupations with similar prestige ratings into the same or close categories. Then Parker and Kleiner were confronted with the decision of how much to weight each indicator of SES. Some SES measures give them equal weight. Based on theoretical considerations regarding socioeconomic status, Parker and Kleiner decided that education

TABLE 6.8 An Index of Socioeconomic Status (SES)

Education (Years Completed)	Item Value	Annual Income	Item Value	Occupation	Item Value
0–4 years	1	$0–1,000	1	Unskilled workers	1
5–8 years	2	$1,001–2,000	2	Sales personnel, semi-skilled workers	2
9–11 years	3	$2,001–3,000	3	Skilled craftsmen and clerical, minor government workers	3
12 years (high school graduate)	4	$3,001–4,000	4	Minor administrative, supervisors, office managers	4
13–15 years	5	$4,001–5,000	5	Minor professionals, medical technicians, teachers	5
16 years (college graduate)	6	$5,001–6,000	6	Major administrative, managers	6
17 years or more	7	$6,001 and over	7	Major professionals (doctors, university professors)	7

Source Reprinted and adapted with the permission of The Free Press, a Division of Simon & Schuster, Inc. from *Mental Illness in the Urban Negro Community* by Seymour Parker and Robert J. Kleiner. Copyright © 1966 by The Free Press.

was by far the most important determinant of a person's SES and occupation least important. After lengthy analysis, they settled on the following weights: 4.4 for education, 2.5 for income, and 1.0 for occupation. Then, each individual's SES level is determined by the following formula:

$$\frac{(education\ value \times 4.4) + (income\ value \times 2.5) + (occupation\ value \times 1.0)}{3}$$

So, a person who had received item values of 6 for education, 4 for income, and 5 for occupation, would have an SES index score of 13.8. In this case, the index of socioeconomic status is a weighted average of the values of the three separate indicators. Weighting of indexes can be done in ways that differ from this; for example, it might be a weighted total, rather than an average. But this illustration gives you an example of index construction using weights.

Evaluating Indexes

Indexes are subjected to the various kinds of measurement assessments (discussed in Chapter 5). In particular, indexes would need to pass the various assessments of validity and reliability that all measuring devices are subjected to. In addition, however, there are comparisons that can be made among the items themselves in order to assess particular items and an overall index. Especially important is something called *item analysis*. With some indexes, an assessment can be made by looking at the correlations among the various items that make up the index. The basic principle is that there should be a fairly steady and strong, although not perfect, relationship between the items of the index if all the items are good and the index is valid. If the items are at the interval-ratio level of measurement, then a correlation between each pair of items can be calculated; if the items are at the ordinal level, then a contingency table could be utilized to see if there is a relationship (see Chapter 13 and

APPLIED SCENARIO   Measuring the Success of Teenage Pregnancy Prevention Programs

Teenage pregnancies are a serious problem in the United States because most teenagers are ill equipped to be effective parents and the demands of parenting often limit the educational and occupational opportunities available to teenage parents. Programs to prevent teenage pregnancies and to help teenage parents are common across the country, and applied social scientists are often enlisted to provide systematic assessments of how well these programs work. Social scientists Marilyn Fernandez and Holly Ruch-Ross (1998) evaluated a number of such programs in Illinois, and their research illustrates ways in which simple multiple-item indexes can be used to measure variables in applied research.

Teenage pregnancy programs typically pursue a number of goals. Virtually all such programs, for example, strive to prevent, or at least postpone, future pregnancies, and they also typically try to enhance the self-sufficiency of teenage parents by supporting them in school or assisting them in getting and keeping a job. These goals are the dependent variables that applied researchers attempt to measure. Fernandez and Ruch-Ross used these goals to develop a composite index of success, which they called a Result Score. The unit of analysis (see Chapter 4) in this research was the organization, in particular the social service or other agencies that were administering the pregnancy prevention program. The researchers wanted to distinguish the successful from the unsuccessful agencies. The Result Score was the agency's measure of success and was a simple index composed of only two items. One item was whether the rate of repeat pregnancy among clients of an agency was below the average for all the agencies. The second item was whether the clients of an agency had a school attendance rate or employment rate that was higher than the average for all agencies. Success was indicated by an agency's having a below average repeat pregnancy rate or an above average school attendance or employment rate. An agency received a score of 1 for each success; so, the index could take on three values: 2 (if an agency was successful by both measures), 1 (if successful by one measure but not the

Chapter 14 on contingency tables). We could also look at the relationship between more than two items at a time by conducting an appropriate multivariate statistical analysis.

When an index is used to measure some subjective state or attitude, then another type of assessment can be made: comparing an individual's response to each item to the results of the overall index. When people's responses to individual index items is presumed to be caused by the underlying variable, then each item should correlate with the results of the overall

scale. For example, persons who score high on a self-esteem index should tend to choose the high-esteem alternative of an item that makes up that index. Any item that shows no relationship or a negative relationship would have to be assessed very carefully in terms of whether it is a good item for that index.

However, this kind of item analysis is appropriate only for some indexes; namely, those in which the values of the items that make up the index are caused by the value of the underlying phenomenon being measured. This would

other), and 0 (if successful by neither measure). Like other indexes, the Result Score measured an abstract concept, "success," by combining measures of two more concrete phenomena: repeat pregnancy rates and education/employment rates.

Fernandez and Ruch-Ross (1998) found some expected and some unexpected outcomes in their study. Not surprisingly, they found that better funded agencies had better success rates. However, unexpectedly, they found that agencies that devoted more hours of service to their clients had lower success rates! This anomalous finding will certainly motivate researchers and agencies to try to figure out what is going on, and this is exactly the point of doing such evaluations—to improve how agencies provide services.

Most of the indexes and scales discussed in Chapter 6 are based on survey questions asked of individuals. The Fernandez and Ruch-Ross index is different in that it uses the organization as the unit of analysis and the agency records as the source of data in constructing the index, rather than survey questions. (Agency records are a form of what is called "available data" and is discussed in greater detail in Chapter 10.) It is important to recognize that indexes and scales can be developed from virtually any sort of data, not just data based on survey questions or statements. The key to indexes and scales as measuring devices is not the source of the data but rather that one is constructing a composite score by using multiple indicators of a phenomenon.

Identify some organization or agency with which you have contact (your university, your place of employment, some social service agency) and consider ways that you could develop indexes and scales from the data they collect.

1. Identify what data these organizations collect and which variables they measure.
2. Develop some ways that these data could serve as the items for a multiple-item index or scale. Which abstract concepts do these indexes and scales measure?

be true of a self-esteem index where a person's underlying self-esteem level is what causes his or her responses to the items of the index. In other indexes, such as the index of SES, this is not the case, and we would not necessarily expect to see all items being correlated with the overall index.

More sophisticated statistical techniques exist for evaluating the items that make up indexes and scales. Going by such names as *factor analysis* and *Q-sort methodology*, these techniques use complex statistical procedures

for deciding which items on an index or scale seem to be measuring a single dimension of the variable being studied. Basically, it involves correlating each item with the overall index and with each dimension or factor that emerges in the statistical analysis. The researcher can determine which items seem to correlate highly or cluster together and thus represent a single factor. These procedures can be used to decide whether a variable is unidimensional or whether it contains a number of distinct dimensions. It can also be used to assign weights to the items of an index or scale and to eliminate items that don't contribute much to measuring a variable. These procedures are fairly complex and require a good grounding in basic statistics to fully understand them.

SCALE CONSTRUCTION

A scale is made up of separate items or indicators, as is an index, but in a scale there is an intensity structure to the items. In addition, in scales, people tend to respond to the items in more of a pattern; people with a similar scale score show a more similar response pattern. Scaling can utilize a number of formats, and each format calls for some unique elements in its design.

Likert Scales

One of the most popular approaches to multiple-item measures is that developed by Rensis Likert (1932). A **Likert scale** consists of a series of statements, each followed by a series of response alternatives for the respondent to express himself or herself about the statement. An illustration of a Likert scale is presented in Table 6.6, with response alternatives ranging from "Strongly agree" to "Strongly disagree." Some Likert scales have the intensity structure built into the items in the scale. With the Conflict Tactics Scale, for example, some items are clearly a stronger or more intense expression of the variable being measured (see Table 6.3). So, "used a knife or gun on my partner" is a stronger or more intense form of conflict resolution than is "insulted or swore at my partner." Other Likert scales, such as the self-esteem scale in Table 6.6, have the intensity built into the response format, with the "Strongly agree" to "Strongly disagree" providing the intensity structure to each item (Anderson, Basilevsky, & Hum, 1983; Nunnally, 1978).

Likert originally developed his scale with the agree-disagree format for his alternatives, and some social scientists still maintain that a true Likert scale should contain those response alternatives. However, other response alternatives are often used today, such as strongly approve–strongly disapprove or very satisfied–very dissatisfied (see Table 6.4), and most researchers still consider them Likert scales (Alwin, 1997). In a Likert scale the most common number of alternatives is five because it offers respondents a sufficient range of choices without requiring unnecessarily minute distinctions in attitudes. More or fewer than five alternatives are sometimes used, but recall the research mentioned earlier that concludes that more rather than fewer alternatives make for more valid and reliable measures. The exact wording of the response alternatives in a Likert scale must be grammatically consistent with the wording in the statements (see Table 6.4).

Note in Table 6.6 the numbers ranging from 1 to 4 in brackets next to each response alternative. These numbers are included on the scale here for purposes of illustration only; they would not be printed on a scale for actual use because their presence might influence respondents' answers. The numbers are used when scoring the scale. The numbers associated with each response are totaled to provide the overall score for each respondent. In this case—a 10-item scale—individual scores can range from a low of 10 (if alternative 1 were chosen every time) to a high of 40 (if alternative 4 were chosen every time). Remember, as discussed in Chapter 5, each item in a Likert scale is an ordinal measure, ranging from a low of "Strongly disagree" to a high of "Strongly agree." Because the total score of a Likert scale is the sum of individual ordinal items, some researchers contend that a Likert scale is therefore ordinal in nature. However, other researchers maintain that the composite score produced by a Likert scale is actually at the interval level, and these researchers use interval-ratio statistics to analyze data produced by Likert scales.

The Likert scale is one example of scales known as **summated rating scales**, in which a person's score is determined by summing the

TABLE 6.9   Calculation of Discriminatory Power (DP) Score for One Item on a Scale

Quartile	N	1	2	3	4	5	Weighted Total	Weighted Mean	DP Score
				Response Value					
Upper	10	0	1	2	4	3	39	3.90	
Lower	10	2	8	0	0	0	18	1.80	2.10
								2.10	

number of questions answered in a particular way. We could, for example, ask respondents to agree or disagree with statements and then assign a 1 for each statement they agree with and a 0 for each disagreement. Their scale score is then the sum of their responses. Summated rating scales can take a number of different forms, although the Likert format is the most common.

Constructing a Likert scale, as with all scales, requires considerable time and effort. One begins by developing a series of statements relating to the variable being measured using the general criteria for statements outlined previously in the chapter. A common rule of thumb is to begin with three times the number of statements desired for the final scale since many of the statements will prove unacceptable for one reason or another and be deleted.

In deciding which items will ultimately be used in a Likert scale, an important criterion is whether the scale items *discriminate* among people. That is, we want people's responses to an item to range over the four or five alternatives rather than cluster on one or two choices. Imagine a scale with an item that reads: "Persons convicted of shoplifting should have their hands amputated." If such an item were submitted to a group of college students, it is likely that most would respond with "Strongly disagree" and maybe a few "Disagrees." It is highly unlikely that any would agree. Of what use is this item to us? We cannot *compare* people—assess who is more likely to agree or disagree—because they all disagree. We cannot correlate responses to this item with the social or psychological

characteristics of the students because there is little or no variation in responses to the item.

For our scale, then, we want to eliminate nondiscriminating items from consideration. *Nondiscriminating items* are those that are responded to in a similar fashion both by people who score high and by people who score low on the overall scale. Nondiscriminating items on a scale can be detected on the basis of results from a pretest in which people respond to all the preliminary items of the scale. One way of identifying nondiscriminating items is by computing a **discriminatory power score** (DP score) for each item. The DP score essentially tells us the degree to which each item differentiates between respondents with high scores and respondents with low scores on the overall scale. The first step in obtaining DP scores is to calculate the total scores of each respondent and rank the scores from highest to lowest. We then identify the upper and lower quartiles of the distribution of total scores. The *upper quartile* (Q_3) is the cutoff point in a distribution above which the highest 25% of the scores are located; the *lower quartile* (Q_1) is the cutoff point below which the lowest 25% of the scores are located. With the quartiles based on *total* scores identified, we compare the pattern of responses to *each* scale item for respondents whose scores fall above the upper quartile with the pattern for respondents whose scores fall below the lower quartile. Table 6.9 illustrates the computation of DP scores for one item on a scale to which 40 persons responded. Ten respondents are above the upper quartile, and 10 are below the lower quartile. It can be seen that the high scorers tended to agree with this item because most had scores of 4 or 5. Low scorers tended to disagree because they are totally concentrated in the 1 and 2 score range. The next step is to compute a weighted total on this item for the two groups. This is done by multiplying each score by the number of respondents with that score. For example, for those above the upper quartile, the weighted total is:

$$(1 \times 0) + (2 \times 1) + (3 \times 2) + (4 \times 4) \\ + (5 \times 3) = 0 + 2 + 6 + 16 + 15 = 39.$$

Next, the weighted mean (average) is computed by dividing the weighted total by the number of cases in the quartile. For the upper quartile, we have

$$39/10 = 3.9.$$

The DP score for this item is then obtained by subtracting the mean of those below the lower quartile from the mean of those above the upper quartile. In this example, we have: 3.9 — 1.8 = 2.1. This process is repeated for every item in the preliminary scale so that each item has a calculated DP score.

Once we have DP scores for all the preliminary items, final selection can begin. The best items are those with the *highest* DP scores because this shows that people in the upper and lower quartiles responded to the items very differently. As a rule of thumb, as many items as possible should have DP scores of 1.00 or greater, and few if any should drop below 0.50. Applying this rule to the item in Table 6.9, we could conclude that it is a very good item and would include it in the final scale. Under no circumstances should an item with a negative DP score be included because this means that high scorers on the overall scale scored lower on this item than did the low scorers. If the size of the

negative DP score is small, it is probably an ambiguous statement that is being variously interpreted by respondents. If the negative DP score is large, however, it is possible that the item was accidentally misscored; that is, a negative item was scored as if it were positive or vice versa.

The Likert format is one of the most popular multiple-item formats because of the many advantages it possesses. First, it offers respondents a range of choices rather than the limited Yes-No alternatives used in some other scales. This makes Likert measures valuable if our theoretical assessment of the manifestations of a variable is that they range along a continuum rather than being either present or absent. Second, data produced by Likert-type scales are at least ordinal level and many consider them interval level, which enables us to use more powerful statistical procedures than with nominal-level data. Third, Likert measures are fairly straightforward to construct.

Likert scales have the same disadvantages as many other scales. In particular, one must be careful in interpreting a single score based on a Likert scale because it is a summary of so much information (separate responses to a number of items). Whenever data are summarized, some information is lost. (Your grade in this course is a summary measure of your performance, and in calculating it your instructor loses information regarding those high—or low—scores you received on individual exams.) The summary score might hide information about patterns of variation in responses or about possible multidimensionality of the scale.

Thurstone Scales

Another approach to scaling was developed by L. L. Thurstone and E. J. Chave (1929). **Thurstone scales** are constructed so that they use *equal-appearing intervals*—that is, it is assumed that the distance between any two adjacent points on the scale is the same. This feature, it is argued, provides some justification for treating the data as interval-level and using

FIGURE 6.1   Equal-Appearing Intervals as Used in Thurstone Scale Construction

Unfavorable				Neutral				Favorable		
1	2	3	4	5	6	7	8	9	10	11

all the powerful statistical procedures that require interval-level data.

Construction of a Thurstone scale begins much the same way as for Likert scales: with the selection of many statements that relate to the variable being measured. Once a sufficient number of statements is at hand, the next step is to provide a value between 1 and 11 for each statement. As illustrated in Figure 6.1, Thurstone scales utilize an 11-point scale ranging from 1 (the least favorable statement regarding an object, event, or issue) to 11 (the most favorable). Point 6 on the scale is labeled "neutral" and is used for statements that are neither favorable nor unfavorable. For example, the statement "Teenage girls who get pregnant are immoral" would be considered highly unfavorable toward teenage pregnancies.

The task of rating each statement as to how favorable or unfavorable it is with regard to the measured variable is assigned to a group of people known as "judges." With each of the preliminary statements printed on a separate card, the judges rate the items by placing them in piles corresponding to points on the 11-point scale. The judges place in each pile statements that they assess to be roughly equivalent in terms of their favorability. This use of judges affords some confidence that a Thurstone scale has the intensity structure among the items necessary to be considered a scale rather than an index.

Once the scale values are computed for all the preliminary items, the next step is to determine which items are the least ambiguous and therefore best for inclusion in the final scale.

If the judges differed widely in their ratings of an item, it is likely something is unclear about the statement itself that leads to varying interpretations. Therefore the degree of agreement among judges about the rating of an item is used as one indicator of ambiguity.

Scales should include the items with the most agreement among judges, and there should be a roughly equal number of items for each of the 11 scale values ranging from unfavorable to favorable, moving upward in half-point increments. This would mean that a minimum of 21 items is required, although some argue that if reliability of .90 or better is desired, as many as 50 statements may be needed (Seiler & Hough, 1970). Regardless of the number actually used, the last step in Thurstone scale construction is to order the items randomly for presentation to respondents.

Table 6.10 presents the first 13 statements contained in the original 45-item Thurstone scale developed by Thurstone and Chave, with the scale value of each item indicated in parentheses. This particular scale is designed so that items with high scale values are "Unfavorable" toward the church, and items with low scale values are "Favorable." The scale values would not, of course, be included on a working version of the scale and are presented here for illustration only. Note that respondents are required only to check the statements with which they agree, making the Thurstone format particularly easy for respondents.

Scoring a Thurstone scale differs from the simple summation procedure used with Likert scales. Because the respondents will agree to differing numbers of statements with different values, the simple sum of the item values is worthless; two people could both agree with four statements, but these may be different statements at different levels of intensity, which would indicate quite different attitudes. Rather, a respondent's score is either the mean or median of the scale values of the items that the person agrees with. For example, if a person agreed with statements 2, 4, 8, and 12 (a

TABLE 6.10   Attitude Toward Church Scale

Check (✓) every statement below that expresses your sentiment toward the church. Interpret the statements in accordance with your own experience with churches.

(8.3)*   1. I think the teaching of the church is altogether too superficial to have much social significance.

(1.7)   2. I feel the church services give me inspiration and help me to live up to my best during the following week.

(2.6)   3. I think the church keeps business and politics up to a higher standard than they would otherwise tend to maintain.

(2.3)   4. I find the services of the church both restful and inspiring.

(4.0)   5. When I go to church, I enjoy a fine ritual service with good music.

(4.5)   6. I believe in what the church teaches but with mental reservations.

(5.7)   7. I do not receive any benefit from attending church services, but I think it helps some people.

(5.4)   8. I believe in religion, but I seldom go to church.

(4.7)   9. I am careless about religion and church relationships, but I would not like to see my attitude become general.

(10.5)   10. I regard the church as a static, crystallized institution, and as such it is unwholesome and detrimental to society and the individual.

(1.5)   11. I believe church membership is almost essential to living at its best.

(3.1)   12. I do not understand the dogmas or creeds of the church, but I find that the church helps me to be more honest and creditable.

(8.2)   13. The paternal and benevolent attitude of the church is quite distasteful to me.

---

* Scale value.
Source   L. L. Thurstone and E. J. Chave, *The Measurement of Attitude.* Chicago: University of Chicago Press (1929). Used with permission of the University of Chicago Press.

total of four statements) in Table 6.10, that person's Thurstone scale score would be 3.13. Another person, agreeing with 1, 7, 10, and 13 (still four statements), would have a score of 8.18. This scoring procedure distributes respondents along the original 11-point scale.

Thurstone and Likert scaling techniques are essentially interchangeable methods of measuring attitudes. A major advantage of the Thurstone technique is that it provides interval-level data. However, if you claim that Likert scales are also interval level or if the interval-data properties are not needed, the Likert technique is probably preferable owing to its higher reliability with fewer items and its greater ease of construction. A second advantage of Thurstone scales is that people can respond to the items more quickly than with a Likert scale because they need only indicate whether they agree with an item and need not ponder to what degree they agree or disagree. However, because reliability calls for Thurstone scales to be longer, this advantage may be minimal. In fact, this can even become a disadvantage if the longer scale leads people to be overly quick or careless in responding to statements. Another major disadvantage of Thurstone scales is that they are costly and difficult to construct.

Semantic Differential Scales

Another scaling format, which has proved quite popular, is the semantic differential (SD) scale developed by Osgood, Suci, and Tannenbaum (1957). The **semantic differential** format presents the respondent with a stimulus, such as a person or event, that is to be rated on a

TABLE 6.11   Semantic Differential Scale Assessing Attitudes Toward the Elderly

				Scale				
Active	7	6	5	4	3	2	1	Passive
Competent	—	—	—	—	—	—	—	Incompetent
High IQ	—	—	—	—	—	—	—	Low IQ
Powerful	—	—	—	—	—	—	—	Weak
Healthy	—	—	—	—	—	—	—	Sickly
Secure	—	—	—	—	—	—	—	Insecure
Creative	—	—	—	—	—	—	—	Uncreative
Fast	—	—	—	—	—	—	—	Slow
Attractive	—	—	—	—	—	—	—	Ugly
Pleasant	—	—	—	—	—	—	—	Unpleasant
Reliable	—	—	—	—	—	—	—	Unreliable
Energetic	—	—	—	—	—	—	—	Lazy
Calm	—	—	—	—	—	—	—	Irritable
Flexible	—	—	—	—	—	—	—	Rigid
Educated	—	—	—	—	—	—	—	Uneducated
Generous	—	—	—	—	—	—	—	Selfish
Wealthy	—	—	—	—	—	—	—	Poor
Good memory	—	—	—	—	—	—	—	Poor memory
Involved	—	—	—	—	—	—	—	Socially isolated

Source   William C. Levin, "Age Stereotyping: College Student Evaluations," *Research on Aging*, Vol 10 (March 1988), pp. 134–148. Copyright © 1988 by Sage Publications, Inc. Reprinted by permission of Sage Publications, Inc.

scale between a series of polar opposite adjectives. Normally, the scale has 7 points, but scales can have fewer or more points if theoretical or methodological considerations call for it. Table 6.11 illustrates an SD designed to measure people's attitudes toward the elderly. In this study, college students were shown pictures of people of varying ages and then asked to describe the characteristics of each person by placing an X on the line between each adjective pair that best represented their assessment of the person. So, on the first line, placing an X over the 6 means that you view the person as quite active, whereas placing an X over the 1 is an assessment of very passive. In this example, all the positive adjectives are on the left and the negative adjectives are on the right. Sometimes the positive responses to some adjectives are put on the right in order to discourage disinterested respondents from placing all their responses in the same column.

Semantic differential responses are analyzed somewhat differently from Likert or Thurstone scales. First, the responses to the adjectives are investigated to determine if they reflect some underlying, more abstract, dimension or factor. The adjectives that make up each factor are presumed to be indicators of one underlying attitudinal dimension. Identification of these attitudinal dimensions, or factors, can be accomplished with a rather complex statistical procedure called *factor analysis*, which basically correlates responses to each adjective pair with responses to every other adjective pair. For example, recent analyses of stereotypes toward elderly people using SDs suggest that the adjective pairs yield four factors (Intrieri, von Eye, & Kelly, 1995): acceptability (socially at ease and pleasing to others), instrumentality (vitality and active pursuit of goals), autonomy (self-sufficiency and active participation in life), and integrity (personal satisfaction and peacefulness with oneself). In Table 6.11, for example, the pairs Active–Passive, Powerful–Weak, and Energetic–Lazy would be some of the indicators of the instrumentality factor. One of the challenges of analyzing SDs is figuring out the nature of the abstract dimension that is reflected in a particular grouping of adjective pairs.

Once the factors being tapped by an SD have been determined, then the SD can be scored. One way to do this is to treat the response to each adjective pair separately. This approach is appropriate when the attitude dimension being tapped is validly measured by one item. Thus, if we were specifically interested in whether or not college students viewed the elderly as socially involved, the last item in Table 6.11 could be used as a measure of the variable. We could compare whether the average social involvement score given to the elderly differs from that given to other adults. Often, however, we are interested in one or more of the abstract dimensions which are more validly measured by a number of adjective pairs. For this to be accomplished, responses on the adjective pairs that constitute each dimension can be summed to provide an overall score on each of the dimensions measured—another variant of the summated ratings scale. For example, the college students who were given the SD in Table 6.11 consistently stereotyped the elderly as less instrumental than young adults.

Semantic differentials have several advantages when compared both to Likert and to Thurstone formats (Nunnally, 1978). Unlike the other scaling techniques that require 20 or more items for adequate reliability, SDs require only four to eight adjective pairs for each dimension to reach reliabilities of .80 or better. This brevity means that an SD can be filled out quickly (Heise, 1970; Miller, 1991). Another advantage is that SDs are much easier and less time-consuming to construct than either Likert or Thurstone scales. Adjective pairs are easier to develop than are unambiguous and unbiased statements about an issue. In addition, adjective pairs from prior studies are more readily adaptable to new studies because of their general and nonspecific nature. This is particularly important if a measuring scale is needed quickly. If, for example, we wanted people's reactions to some unanticipated event while it is still fresh in their minds, time would be of the essence. Only an SD-type scale could be readied in time.

About the only disadvantage of an SD is that identifying the abstract dimensions tapped by the adjective pairs is somewhat subjective and judgmental. The validity of the conclusions drawn is only as good as the judgment of those who identify the dimensions.

Guttman Scales

At the outset of this chapter, I noted that efforts are made to create scales that are unidimensional; that is, they measure a single variable or a single aspect of a variable. With a **Guttman scale**, the procedures used in construction give us the greatest confidence that the resulting scale is unidimensional (Guttman, 1944).

Researchers using Guttman scales achieve unidimensionality by developing the items in such a way that, in a perfect Guttman scale, there is only one pattern of responses that will yield any given score on the scale. (In fact, some argue that this is a characteristic of a true scale.) For example, if an individual's score is 5, we would expect that he or she had agreed with the first five items on the scale. This can be contrasted to other scaling techniques that allow obtaining the same score by agreeing or disagreeing with any number of items and having completely different response patterns.

TABLE 6.12   Response Patterns in a Guttman Scale

Response Alternatives		Guttman Scale Patterns							Error Pattern
		0	1	2	3	4	5	6	
Harder Items	Have person as close kin by marriage	No	No	No	No	No	No	Yes	No
	Have person in my club as personal friend	No	No	No	No	No	Yes	Yes	Yes
	Have person on my street as neighbors	No	No	No	No	Yes	Yes	Yes	No
	Have person working alongside me on my job	No	No	No	Yes	Yes	Yes	Yes	Yes
	Have person as citizen in my country	No	No	Yes	Yes	Yes	Yes	Yes	Yes
Easier Items	Have person as visitor to my country	No	Yes	Yes	Yes	Yes	Yes	Yes	Yes

Guttman scaling is able to do this because the items in the scale have an inherently progressive nature relating to the intensity of the variable being measured. In the parlance of Guttman scaling, the least intense items are referred to as "easy" because more people are likely to agree with them; the most intense items are considered "hard" because fewer are expected to agree with them. If a person agrees with a certain item, we would expect him or her also to agree with all the less intense items; conversely, if a person disagrees with a particular item, we would also expect that person to disagree with all the more intense items.

The Bogardus Social Distance Scale in Table 6.5 can be considered a Guttman scale. The items are arranged with the "easiest" at the bottom to the "hardest" at the top. Often, only two response categories are provided with Guttman scales, either Agree/Disagree or Yes/No. Some Guttman scales make use of the Likert-type response categories, but the categories are collapsed to a dichotomy in the data analysis.

The fact that the items in a Guttman scale are progressive and cumulative leads to the basic means of assessing whether a set of items constitutes a Guttman scale. This criterion is called *reproducibility*, which is the ability of each individual's composite score to predict exactly which items he or she had agreed and disagreed with. For example, in a true Guttman scale, all persons with scores of 2 will agree with the two easiest items and disagree with the rest; persons with scores of 3 will agree with the three easiest items and disagree with the rest; and so on. In a perfect Guttman scale, each respondent's score will reproduce one of these patterns, as is illustrated in Table 6.12. There is always one more perfect response pattern in a Guttman scale than there are items in the scale because one pattern will involve disagreeing with all the items; therefore, the six-item scale in Table 6.12 would have seven possible response patterns. You can see in each of the response patterns that once the "No" response changes to a "Yes," the person then answers "Yes" to all the easier questions. In actual practice, perfect Guttman scales are virtually nonexistent. Usually, some respondents will deviate from the expected pattern. Nevertheless, Guttman scales with very high levels of reproducibility have been developed.

Constructing a Guttman scale is difficult and to an extent risky because we will not know whether the scale we have devised will have sufficient reproducibility to qualify as a Guttman scale until after we have applied it to a sample of respondents. As with the other scaling techniques, a basic first step is creating and selecting items for inclusion in the scale. In Guttman scaling, this task is further complicated by the need for the items eventually selected to have the characteristic of progression. The procedure for selecting items for a Guttman scale is known as the *scale discrimination technique* (Edwards & Kilpatrick, 1948). As was done with both Likert and Thurstone scaling techniques, we begin by writing a large number of statements relating to the variable to be measured. These statements are rated by a group of judges along the 11-point Thurstone equal-appearing interval scale. The items on which judges are in the greatest agreement are given a Likert-type response format and presented to a pretest group. The pretest results are used to calculate discriminatory power (DP) scores as described under Likert scaling. Items for inclusion in the final Guttman scale are selected so that they cover the full Thurstone scale range and have the highest DP scores. Despite the effort involved in this approach, all it accomplishes is to increase the likelihood that the selected items will have sufficient reproducibility to constitute a Guttman scale; it does not guarantee reproducibility.

The only way to determine if we have succeeded in developing a true Guttman scale is to administer it to another pretest group and see if it has adequate reproducibility. This is done by determining how many errors occur in the response patterns. In Guttman scaling, an *error* refers to any response pattern by an individual that does not follow one of the expected patterns presented in Table 6.12. Table 6.12 presents one possible error pattern, where an individual responded "Yes" to an item that was harder than another item to which that person had responded "No."

The total number of these errors is calculated for all respondents and is used in the following formula to calculate the *coefficient of reproducibility* (R_c):

$$R_c = 1 - \frac{number\ of\ errors}{(no.\ of\ items) \times (no.\ of\ subjects)}$$

Guttman (1950) suggested that a coefficient of reproducibility of .90 is the minimum acceptable for a scale to qualify as a Guttman scale. In general, the more items in a Guttman scale, the more difficult it is to achieve a high level of reproducibility. For a very short scale, .90 would certainly be the minimum acceptable; with a longer scale, a slightly lower coefficient of reproducibility would be acceptable.

Suppose that we developed a scale and found its reproducibility too low. It is perfectly legitimate to then rearrange the order of the items or delete items in an effort to achieve the necessary reproducibility. We might, for example, delete one or two of the items containing the most error responses to see if the remaining items would produce adequate reproducibility to qualify as a Guttman scale.

The data generated by Guttman scaling is ordinal level. Given the relatively few items characteristic of these scales and the common Agree-Disagree format used, there are few possible scores for respondents to achieve. This means that large numbers of respondents will have tied scores on the scale, so many statisticians believe it is better to consider these numbers as ranks (ordinal) rather than interval- or ratio-level data (see Chapter 5). Guttman scales are unique, however, for the characteristics of unidimensionality and reproducibility. If these attributes are desired, they are apt to more than outweigh the presence of all the tied scores.

Table 6.13 provides a summary of the key features, advantages, and disadvantages of the indexes and scales I have discussed, as well as the tools that are used in their development and evaluation. Scales are most commonly used in research problems in which the unit of analysis is the individual (see Chapter 4), and

TABLE 6.13   A Comparison of Various Indexes and Scales

Measuring Device	Key Feature	Assessment Tools	Advantages	Disadvantage
Indexes	Separate indicators combine to create a single measure	Validity, reliability, item analysis	More valid and reliable than single-item measure	Not unidimensional
Likert scale	Evaluate statements with 4–7 response alternatives	Validity, reliability, item analysis, discriminatory power scores, factor analysis	Range of response alternatives, easy to respond to, easy to construct	Hard to interpret a single summary score
Thurstone scale	Equal-appearing intervals	Validity, reliability item analysis	Easy and quick to respond to, interval-level data	Difficult and costly to construct
Semantic Differential (SD) Scales	Choose points between polar-opposite adjectives	Validity, reliability, factor analysis	Easy to construct, can achieve reliability with few items, easy and quick to respond to	May not be unidimensional (depending on choice of adjectives)
Guttman Scale	Reproducibility of items	Validity, reliability, item analysis, coefficient of reproducibility	Unidimensionality, true scale with intensity structure to items	Difficult to construct

this emphasis is reflected in the preceding discussion. However, scales can also be developed to measure other units of analysis, such as the characteristics of organizations or political entities.

AVOIDING RESPONSE BIAS

As we saw in Chapter 5, a key issue in measurement is whether people's answers to questions are accurate reflections of their actual feelings, beliefs, or behaviors. In other words, our measure of some phenomenon should be determined by the nature of the phenomenon itself and not by systematic or random errors (review the measurement formula on p. 139). One source of such error in people's responses to questions or statements is called **response bias**: the tendency for an individual's answers to questions to be influenced by things other than their true feelings, beliefs, and behaviors. It can result in a patterned overestimation or underestimation of variables (Bradburn, 1983).

Sources of Response Bias

One source of response bias is called **response set**: Some people tend to be either yea-sayers or nay-sayers, tending either to agree or disagree with statements regardless of their content. This is sometimes called the *acquiescence response set* because it more often takes the form of people being predisposed to agree with statements. To illustrate this, look again at the self-esteem scale in Table 6.6. If the scale was constructed so that "Strongly agree" always indicated high self-esteem, then people who tend to agree with statements would score higher on self-esteem than they actually should because they tend to agree with statements irrespective of content. This would throw into question the validity of the scale because it would produce the systematic error discussed in Chapter 5.

Another source of response bias is **response pattern anxiety**: Some people become anxious if they have to repeat the same response all the time and change their responses to avoid doing so. If this occurs, then their reactions to statements do not reflect their actual attitudes but rather their reaction to a certain response pattern, and the validity of the scale is reduced. (As students, you have probably had this experience when taking a multiple-choice exam. If several consecutive questions all have the same answer, you become concerned and may doubt answers that you are fairly sure of just because the pattern of responses differs from the more random pattern you expect.)

Another source of response bias is the **social desirability effect**: people's tendency to give socially acceptable, popular answers in order to present themselves in a good light. It is very socially unacceptable, for example, to admit using a knife or gun on your spouse, and this may affect how people respond to the Conflict Tactics Scale (Table 6.3). People may deny using a knife or gun, even if they have done so, in order to avoid appearing socially unacceptable to an interviewer. The Eye on Diversity box discusses some ways in which cultural diversity has to be taken into account in assessing response bias.

Reducing Response Bias

Researchers use a number of strategies in an attempt to reduce response bias. Response set and response pattern anxiety can be avoided by designing statements so that positive statements are not always an expression of the same attitude. Likert scales are routinely designed like this. You will note in the items in Table 6.6 that choosing "Strongly agree" on items 1, 3, 4, 7, and 10 would be an expression of high self-esteem; choosing "Strongly agree" on items 2, 5, 6, 8, and 9, on the other hand, would be an expression of low self-esteem. If "Strongly agree" were an expression of high self-esteem for all items, then some respondents would have to choose the same alternative on every item in order to express their opinion. Mixing the response pattern of items is taken into account in scoring Likert scales. The alternatives that indicate an expression of the same opinion or feeling are given the same numerical score. In our example, for instance, all high-esteem alternatives (whether they be "Strongly agree" or "Strongly disagree") are given a score of 4. Then each person's responses to all items can be summed for a total scale score.

Another technique for avoiding response bias is to present sensitive issues in a neutral and nonjudgmental context. In developing the CTS (see Table 6.3), Straus and his colleagues presented questions about violent acts in the context of disagreements and conflicts, which would presumably appear more socially acceptable to people than abuse and violence.

A third way to reduce response bias has to do with the ordering of questions: Questions can be asked in a hierarchical order, beginning with the less sensitive and gradually moving on to the more sensitive issues. The CTS does this by beginning with a few items that are positive ways of resolving conflict ("I explained my side of a disagreement to my partner") before moving onto questions about psychological and physical abuse. Questions about the use of violence don't appear until well into the instrument. The rationale for this design is that people feel less reticent about divulging acts of violence if they have been given the chance to show that such acts were "the last straw" after attempting other means of conflict resolution.

A fourth strategy for reducing response bias is to use an interspersed pattern for the items, where socially acceptable items are interspersed with the less acceptable items. In the Conflict Tactics scale, positive items, such as "I said I was sure we could work out a problem," are followed by such items as "My partner needed to see a doctor because of a fight with me." The reason for this pattern is that a straight hierarchical ordering may open the door to a form of response set: A respondent may blindly answer "Never" to every item once items begin referring to violent acts. Interspersing sensitive with

Whether response bias occurs depends in part on cultural values, norms, and modes of interaction. After all, answering questions or responding to verbal statements is a mode of interaction, and how people do it will be shaped by the ways they have learned to interact with others. As an example, some studies of mental health have found that women report higher incidence of psychiatric symptoms than do men (Gallagher, 1987). From this it might be concluded that women are less mentally healthy than men and, at a more general level, that one's gender has an impact on one's mental health. Critics respond that women may report more symptoms because it is more socially acceptable for women in our society to "complain," show weakness, and the like. The relationship between an independent variable (*gender*) and a dependent variable (*mental health status*) may in part be an artifact due to response bias: People are answering questions in terms of the social acceptability of their responses rather than in terms of their actual feelings.

In addition to gender, ethnic and cultural background can also influence what people learn and how they relate. Research suggests, for example, that Hispanics as a group tend to give more extreme responses to questions, tend more toward acquiescent responses, and are more inclined to give socially acceptable responses than do non–Hispanics (Marin & Marin, 1991). This is truer of Hispanics who have come to the United States recently, and who have less education and lower socioeconomic status. This response bias may be in part an interactional strategy of the less powerful: a deferential and submissive demeanor that is a way for the less powerful to ingratiate themselves. It may also be a learned response in a culture that encourages politeness and respect and discourages confrontation when interacting.

A survey in Kazakhstan also found that culture affected acquiescence response set which in turn influenced the attitudes expressed in surveys. The two major ethnic groups in Kazakhstan are the Kazakhs and the Russians (Javeline, 1999). While the acquiescence tendency was found in both groups, it was a much stronger effect with the Kazakhs than with the Russians. Kazakh culture places a great deal of importance on hospitality, deference, and respect in social relations. These norms and customs undoubtedly play out in interview situations as the Kazakhs are more inclined than the Russians to agree with questions that are asked of them.

Research being conducted among diverse groups, then, needs to address the issue of whether response bias of one sort or another might be affecting measurement. If all groups in the research have the same tendencies toward acquiescence, then it may not change what is discovered about between-group differences, although it does reduce measurement precision. However, when the groups compared have differing acquiescence tendencies, as with the Kazakhs and Russians, then we may discover between-group differences that are due to response bias rather than to some independent variable.

positive items encourages participants to think more carefully about each item before responding. So the CTS actually uses a combination of hierarchical and interspersed ordering of sensitive items. For any given scale, whether a hierarchical pattern or an interspersed ordering produces the least bias is an empirical question to be settled through research on the scale itself.

A fifth technique that helps to reduce response bias is called "funneling." A researcher might ask respondents first about conflict in their city, then about conflict in their local community and among neighbors, and finally about conflict in their own families. As another example of this, Moser and Kalton (1972) suggest phrasing questions so that respondents can answer in the third person. For example, "Many men have hit their wives at one time or another. I wonder if you know under what circumstances it happens?" This can be followed with a direct question asking if the respondent has done it.

REVIEW AND CRITICAL THINKING

As an extension of Chapter 5, this chapter is all about measurement, or being careful and precise. Chapter 6 provides some of the nuts and bolts of how researchers design valid questions, indexes, and scales used to measure much social phenomena. Much can be learned from how scientists do these things and translated into tips for critically analyzing any information or situations that you might confront.

1. If questions or statements are used as the basis for collecting information, how well designed are they? Is there anything in their design (wording, context, etc.) that might lead to misunderstanding, misinterpretation, or bias in the information that results?

2. Is the information based on some kind of multiple-item measuring device? Is the phenomenon appropriate to the use of a multiple-item measure?

3. If a multiple-item measuring device is used, is it properly constructed to support the conclusions drawn? Can you identify ways in which its design could lead to misinterpretation or bias in the conclusions?

Computers and the Internet

Most major survey research centers maintain Web sites, some of which are extremely useful for learning more about designing questions, indexes, and scales. Some sites provide the opportunity to examine questions used in actual surveys. It is also possible in some cases to download and read entire questionnaires and survey instruments. By reviewing these surveys, you can explore how the instrument is structured, the design and ordering of questions, and other features that enhance the quality of the research instrument.

Four of the best sites that I have found are:

- The Odum Institute for Research in Social Science : **http://www.irss.unc.edu/**.

- The General Social Survey: **http://www.icpsr.umich.edu/gss/**.

- The Centre for Applied Social Surveys in England: **http://www.scpr.ac.uk/cass/**.

- The Survey Research Center at Princeton University: **http://www.princeton.edu/~abelson/index.html**. (This site has some of the best links to organizations involved in survey research throughout the world that I have seen.)

Use the search engine on your Web browser to look for other survey research centers like these. You can also access search engines not associated with your particular Web browser or Internet service, by typing in its name in the address field. Most Internet providers accept addresses without the assignment of http or other delimiters. For example, if Yahoo was not included by your service provider, keyboarding in yahoo (all lowercase letters) will automatically bring you to Yahoo's Web site, which you could use as a guest.

Many survey center Web sites have a search feature that enables you to search that Web site for questions on almost any topic imaginable. Try the search features at the above sites by entering a topic of interest to you, such as domestic violence, health care, or poverty. Determine if the questions that result are open- or closed-ended. Examine how the questions are worded and consider how you might improve or adapt them to a project of interest to you. Depending on the source, you may be able to use or adapt questions from these surveys for use in a survey instrument that you are designing.

Main Points

1. Questions or statements, either singly or in combination, are often what social scientists use to measure variables, and these questions and statements must be designed very carefully. The questions must be related to the concepts being studied, and good directions and instructions for answering questions must be provided.

2. A decision must be made as to whether closed-ended questions or open-ended questions are most appropriate. Questions must be worded very carefully and pretested to ensure that they are appropriately understood. The present tense should generally be used, and questions should be simple, direct, and clear.

3. Insider language should be used carefully and when useful, and questions sometimes need a context in order to be properly understood. Care also needs to be taken in asking sensitive questions, and the response formats should be designed so that people know clearly how to respond.

4. Multiple-item measuring devices can take the form of indexes or scales. Multiple-item measures are particularly valuable for measuring complex variables because they enhance validity and reliability, reduce the impact of measurement error, and increase the level of measurement.

5. The five basic steps common to developing most indexes and scales are: develop many preliminary items, delete obviously poor items, pretest the remaining items, delete poor items on the basis of pretest results, and select items for the final scale. Sources of items can be already existing measures developed by someone else, one's imagination, judges and experts, and the people who are the focus of the research. Good index and scale items should have validity, a range of variation, and unidimensionality.

6. Likert scales are very popular; an important consideration in selecting items for inclusion in a Likert scale is their discriminatory power. Thurstone scaling uses judges to assign a value from 1 to 11 to each item on the scale. Items for inclusion on a Thurstone scale have scale scores that cover the full 1 to 11 range and were most agreed upon by the judges.

7. The semantic differential scaling format presents respondents with a concept to be rated and a series of opposite adjective pairs separated by a 7-point scale that is used to evaluate the concept. Guttman scales have

the unique characteristic of reproducibility, meaning that a given total score reflects one and only one pattern of responses to the items on the scale.

8. Response bias can be produced by, among other things, response set, response pattern anxiety, and the social desirability effect. Researchers use a variety of techniques in designing questions to avoid response bias.

Important Terms for Review

closed-ended questions
discriminatory power score
Guttman scale
index
interview
Likert scale
open-ended questions
questionnaire
response bias
response pattern anxiety
response set
scale
semantic differential
social desirability effect
summated rating scales
Thurstone scale
unidimensional scale

For Further Reading

Brodsky, Stanley L., & H. O'Neal Smitherman. (1983). *Handbook of scales for research on crime and delinquency*. New York and London: Plenum. This volume contains many indexes and scales on crime and delinquency, along with information on validity, reliability, and other criteria for assessing the instruments.

De Vellis, Robert F. (1991). *Scale development: Theories and applications*. Newbury Park, CA: Sage.

This is an understandable guide to all the various stages in developing good scales. It includes discussions of how to generate items, how long scales should be, and other useful topics.

McDowell, Ian, & Claire Newell. (1996). *Measuring health: A guide to rating scales and questionnaires* (2nd ed.). New York: Oxford University Press. This volume discusses the theoretical and technical aspects of constructing and evaluating scales relating to such health issues as social health, psychological well-being, and depression. It presents and evaluates many actual scales.

Mueller, Daniel J. (1986). *Measuring social attitudes: A handbook for researchers and practitioners*. New York: Teachers College Press. A step-by-step guide to developing any of the major scaling formats (Likert, Thurstone, semantic differential, etc.). It also contains many useful examples of existing scales.

Robinson, J. P., P. R. Shaver, & L. S. Wrightsman (Eds.). (1991). *Measures of personality and social psychological attitudes*. San Diego: Academic Press. Yet another compendium of indexes and scales.

Schuman, Howard, & Stanley Presser. (1996). *Questions and answers in attitude surveys: Experiments on question form, wording, and content*. Thousand Oaks, CA: Sage. This is a comprehensive handbook of the rules, problems, and pitfalls of designing questions for surveys. It goes far beyond what this chapter is able to cover on this important topic.

Straus, M. A. (1969). *Family measurement techniques*. Minneapolis: University of Minnesota Press. This book is an excellent source for indexes and scales relating to a wide variety of aspects of the family.

Sudman, S., N. Bradburn, & N. Schwarz. (1996). *Thinking about Answers: The application of cognitive processes to survey methodology*. San Francisco: Jossey-Bass. An exploration of what answers mean in relation to how people understand the world around them and communicate with one another. The authors present the survey as a social conversation and investigate how to determine the meanings of the answers respondents give in surveys.

Chapter 7

SAMPLING

The issues at the center of this chapter are those of *sampling*, or selecting a few cases out of some larger grouping for study. All of us have had experience with sampling. Cautiously tasting a spoonful of soup to see if it is hot is a process of sampling; or taking a bite of a new brand of pizza is sampling to see if we like it. Scientists also do sampling. A study of political attitudes, for example, might be based on a nationwide sample of 1,200 people, whereas a study of recidivism might focus on the inmates of one prison. All sampling involves attempting to make a judgment about a whole something—a bowl of soup, a brand of pizza, the political attitudes of all citizens, the recidivism of all prison inmates—based on an analysis of a part of the whole. Scientific sampling, however, is considerably more careful and systematic than casual, everyday sampling. In this chapter, I discuss the fundamentals of sampling along with the benefits and disadvantages of various sampling techniques.

THE PURPOSE OF SAMPLING

A **sample** consists of one or more elements or cases selected from some larger grouping or population. When the subject of sampling is first encountered, it is not uncommon to ask, Why bother? Why not just study the whole group? One major reason for studying samples rather than the whole group is that the whole group is sometimes so large it is not feasible to study it. For example, social scientists might be interested in learning about welfare recipients, the mentally ill, prison inmates, or some other large group of people. It might be difficult and expensive—and often impossible—to study all members of these groups. Sampling allows us to study a workable number of cases from the large group to derive findings that are relevant for all members of the group.

A second reason for sampling is that, as surprising as it may seem, information based on carefully drawn samples can actually be better than information from an entire group. This is especially true when the group being studied is extremely large. For example, a census of all residents of the United States is taken at the beginning of each decade. Despite the vast resources the federal government expends on the census, substantial undercounts and other errors occur. In fact, after each census in recent decades, numerous cities filed lawsuits complaining of alleged undercounts. Between the decennial censuses, the Census Bureau conducts *sample* surveys to update population statistics and collect data on other matters. The quality of the data gathered by these sample surveys is actually superior to that of the census itself. The reason for this is that, for a total population survey to be accurate, it must contact everyone, or almost everyone, in the population. This is extremely difficult to accomplish with large populations, and if those not contacted possess some special characteristics, such as being poor or non-Anglo, then the resulting population survey will be biased. In doing a sample survey, on the other hand, only a few thousand people must be contacted, and the task is more manageable: Better-trained interviewers can be used, greater control can be exercised over the interviewers, and more resources can be devoted to contacting hard-to-find respondents. In fact, the Bureau of the Census conducts a sample survey after each population census as a check on the accuracy of that census. Indeed, were it not a constitutional requirement, the complete census might well be dropped and replaced by sample surveys.

Much research, then, is based on samples of people. Samples make possible a glimpse at the behavior and attitudes of whole groups of people, and the validity and accuracy of research results depend heavily on how samples are drawn. An improperly drawn sample renders the data collected virtually useless. An important consideration regarding samples is how *representative* they are of the population from which they are drawn. A **representative sample** is one that accurately reflects the distribution of relevant variables in the target population. In a sense, the sample should be considered a small reproduction of the population. Imagine, for example, that you were interested in the success of unmarried teenage mothers in raising their children in your community. Your sample should reflect the relevant characteristics of unmarried teenage mothers in your community. Such characteristics might include age, years of education, and socioeconomic status. To be representative, the sample would have to contain the same proportion of unmarried teenage mothers at each age level, educational level, and socioeconomic status that exists among unmarried teenage mothers in the community as a whole. In short, a representative sample should have all the same characteristics as the population. The representative character of samples allows the conclusions based on them to be legitimately generalized to the populations from which they are drawn. As we will see later in this chapter, nonrepresentative samples can be useful for some research purposes, but researchers must always assess the representativeness of their samples in order to make accurate conclusions. Before discussing specific techniques for drawing samples, some of the major terms used in the field of sampling will be defined.

Populations and Samples

A sample is drawn from a **population**, which refers to all possible cases of what we are interested in studying. In the social sciences, the target population is often people who have some particular characteristic in common, such as all citizens of the United States, all eligible voters, all school-age children, and so on. A population need not, however, be composed of people. Recall from Chapter 4 that the unit of analysis can be something other than individuals, such as groups or organizations. Then the target population will be all possible cases of whatever is our unit of analysis. A sample consists of one or more elements selected from a population. The manner in which the elements are selected for the sample has enormous implications for the scientific utility of the research based on that sample. To select a good sample, you need to define clearly the population from which the sample is to be drawn. Failure to define the population clearly can make generalizing from the sample observations highly ambiguous and result in drawing inaccurate conclusions.

The definition of a population should specify four things: content, unit, extent, and time (Kish, 1965, p. 7). These can be illustrated by the sample used by James Greenley and Richard Schoenherr (1981) to study the effects of the characteristics of social service agencies on the delivery of services to clients. First, the *content* of the population refers to the particular characteristic that the members of the population have in common. For Greenley and Schoenherr, the characteristic held in common by the members of their population was that they were health or social service agencies. Second, the *unit* indicates the unit of analysis, which in our illustration is organizations rather than individuals or groups. Although Greenley and Schoenherr collected data from practitioners and clients in the organizations, their focus was on comparing the performance of agencies. Third, the *extent* of the population refers to its spatial or geographic coverage. For practical reasons, Greenley and Schoenherr limited the extent of their population to health and social service agencies serving one county in Wisconsin. It would not have been financially feasible for them to define the extent of their population as all agencies in Wisconsin or the United States. Finally, the *time* factor refers to the temporal period during which a unit would have to possess the appropriate characteristic in order to qualify for the sample. Greenley and Schoenherr conducted a cross-sectional study, and only agencies that were in operation at the time they collected their data qualified. A longitudinal study might have included agencies that came into existence during the course of the study.

With these four points clearly defined, a population will normally be adequately delimited, and what is called a *sampling frame* can be constructed.

Sampling Frames

A **sampling frame** is a listing of *all* elements in a population. In many studies, the actual sample is drawn from this listing. The adequacy of the sampling frame is crucial in determining the quality of the sample drawn from it. Of major importance is the degree to which the sampling frame includes *all* members of the population. Although there is an endless number of possible sampling frames, depending on the research problem, a few illustrations follow that describe some of the intricacies of developing good sampling frames.

In social science research, some of the most adequate sampling frames consist of lists of members of organizations. If we wanted, for example, to study recidivism among inmates being released from prison, we could draw a sample of inmates in one prison. The sampling frame would be quite straightforward, consisting of all inmates currently listed as incarcerated in that institution. Given the care with which correctional facilities maintain accurate records of inmates, the sampling frame would undoubtedly be complete and accurate. Other examples of sampling frames based on organizational affiliation would be the membership rosters of professional groups, such as the National Association of Social Workers, the American Medical Association, or the American Society of Criminology. These lists would not be quite as accurate as the inmate roster because people who had recently joined the

organization might not appear on the official list; also, clerical errors might lead to names being missed. These few errors, however, would have little effect on the adequacy of the sampling frame.

When using organizational lists as a sampling frame, caution must be exercised regarding what is defined as the population and about whom generalizations are made. The population consists of the sampling frame, and legitimate generalizations can be made only about the sampling frame. Many physicians, for example, do not belong to the American Medical Association (AMA). Thus a sample drawn from the AMA membership roster would represent only AMA members—not all physicians. In the use of organizational lists as sampling frames, then, it is important to assess carefully who the list includes and who it excludes. Sometimes research focuses on a theoretical concept that is operationalized in terms of an organizational list that does not include all actual instances of what is intended by the concept. For example, a study of poverty could operationalize the concept "poor" as those receiving welfare payments. Yet, many people with little or no income do not receive welfare. In this case, the sampling frame would not completely reflect the population intended by the theoretical concept.

Some research focuses on populations that are quite large, such as residents of a city or state. To develop sampling frames for these populations, three listings are commonly used: telephone numbers, utility subscribers, or city directories (Lavrakas, 1987). A listing of telephone numbers in an area can of course be found in telephone books, but telephone books have a number of problems when used as sampling frames. Even today some people do not have telephone service, although the number of households that lack telephone service has diminished over the past few decades. As recently as 1970, 13% of households in the United States were without telephone service; today, probably no more than 6% lack such service (U.S. Bureau of the Census, 1998). Those without telephones, however, are concentrated among the poor, those living in rural areas, and transient groups, such as the young. For a research project in which these groups are significant, sampling based on telephone books would therefore be very unrepresentative.

Another problem with using telephone books as sampling frames is unlisted numbers. The extent of unlisted numbers varies from one locale to another, but it can be greater than 50% in some areas. Because of these problems with telephone book listings, they are not typically used, at least not by themselves, for drawing samples. Instead, there are other techniques, such as *random digit dialing* (RDD), that can assure that every household with telephone service has an equal chance of appearing in the sample. With RDD, telephone numbers are selected for the sample by using a table of random numbers or having a computer generate random telephone numbers (see Appendix B). (If the researcher knows the telephone prefixes of the areas to be sampled, then only the last four digits need to be chosen randomly.) Because the phone numbers are determined randomly, RDD gives all telephone numbers an equal chance of being selected, regardless of whether they are listed in the directory, and therefore removes the problem of unlisted numbers. Random digit dialing, of course, does nothing about noncoverage due to the lack of telephone service in some households.

Another population listing that can be used for sampling is a list of customers from the local electric utility company. Even though some households do not have telephone service, relatively few lack electricity, and the problem of noncoverage is therefore less significant. Utility listings do, however, have their own problems that must be handled in order to draw a satisfactory sample. The major problem comes from multiple-family dwellings, which often have utilities listed only in the name of the owner rather than all the individual residents. Multiple-family dwellings are more likely to be inhabited by the young, the elderly, the poor, and the unmarried. Unless the utility listings are supplemented, samples will systematically underrepresent people in these groups. This problem can be overcome by visiting the dwellings and adding the residents to the list of utility subscribers, but this is a very time-consuming task. Beyond the problem of multiple dwellings, the elderly, the poor, and those living in rural areas are more likely to be without utilities and thus not appear in the sampling frame.

As a source of population listings, city directories are quite useful. City directories can be found in most libraries and are generally divided into four sections. The first is a listing of commercial firms and is analogous to the yellow pages of the telephone book. The second section is an alphabetical listing of residents together with their addresses, phone numbers, and the head of household's occupation. This section of the directory is useful if the research problem calls for a sample of people or households with particular occupational characteristics. Next comes an alphabetical listing of streets and addresses with residents' names. For sampling purposes, this is very useful for locating housing units, and the address listing will exclude only new construction. However, residents' names can become outdated quickly as people move. The last section is a listing of telephone numbers in numerical order, together with the name of the person to whom the number is assigned. The accuracy of city directories is quite high, certainly as good a sampling frame as a researcher could compile starting from scratch (Sudman, 1976). In addition, city directories are the least likely to exclude people with low income.

Probability Theory and Sampling Distributions

A sound understanding of sampling issues requires that we delve into a topic known as *probability theory*, which focuses on determining the likelihood or probability that certain events will occur. This is studied in greater detail in Chapter 14, but a basic understanding is essential here. To begin with a familiar example: Everyone recognizes that a coin tossed in the air will land on a flat surface, on average, 50% of the time with the head showing and 50% with the tail showing. When we say "on average," we mean that if we flipped the coin many times, we would get this result. In fact, when we say we flip it "many" times, we really mean an infinite number of times. The infinite number of times that our coin could be tossed is the population of all coin tosses. If I were to toss the coin 10 times, then those 10 tosses would be one sample of all the coin tosses in the population. I just took a quarter from my pocket and flipped it onto a flat surface 10 times. The result: 7 tails and 3 heads. I just flipped the coin another 10 times (a second sample): 4 tails and 6 heads. So we have two samples here, and neither is identical with what we would expect to happen in the long run. However, if I were so inclined, I could continue tossing the coin and recording the combination of heads and tails in each set of 10 flips. If I continued for 1,000 such sets of 10 flips each, that would constitute 1,000 samples from the infinite number of flips that I could possibly make (of course I could flip the coin an infinite number of times only if I were immortal or could convince others to carry on the flipping tradition after my demise). With those 1,000 sets of 10 flips, the most common occurrence would be 5 heads and 5 tails; less commonly we would obtain 6 heads and 4 tails; much less commonly, we would get 9 heads and 1 tail or 10 heads and no tails.

The 1,000 sets of 10 flips each is called a **sampling distribution**, a distribution of sample statistics (in this case, a distribution of proportions of heads and tails). In a sampling distribution, the most likely events tend to occur most frequently and the least likely events least frequently. The same thing happens when we

select good random samples from populations in the social sciences: The most likely outcome is that the sample will be identical to, or at least very similar to, the population; and the least likely outcome is that the sample will be very different from the population.

Let's demonstrate this. Assume there is a set of surveys that were completed by 1,495 people. Consider this group of 1,495 people to be the population we are interested in. The mean age in this population is 46.2 years. I can instruct my computer program to select a random sample of any size from this population. So I tell the computer to select a sample size of 50 people out of the total of 1,495 and calculate the mean age in that sample. Additionally, I tell the computer to select 49 more separate and independent samples of size 50 from that population and calculate the mean age for each sample so that we can see the results. We end up with 50 sample means, and Figure 7.1A shows the sampling distribution that results. (In the figure, each average age has been rounded to the nearest whole number for ease of presentation.) Those 50 sample means are what are called *sample statistics*, or statistics computed on samples, and the distribution of those 50 means is a *sampling distribution*. The first thing to note about that sampling distribution in Figure 7.1A is that most of the 50 samples produced a mean that was not the same as the mean in the population. The second thing to note is that all 50 sample statistics are quite close to the population value of 46.2 years. Even though the ages in this population range from 18 to 89, the 50 samples produce means that range over only 11 years, and the population parameter falls almost exactly in the center of that range. The third thing to note, although it is not obvious from the distribution in Figure 7.1A, is that, if I continue selecting samples of size 50, I will eventually obtain samples with means very different from the population parameter, say 34.7 years or 59.1 years. However, to reiterate: with a sampling distribution of random samples, the most

FIGURE 7.1   The Impact of Sample Size on Sampling Distributions, with 50 Random Samples

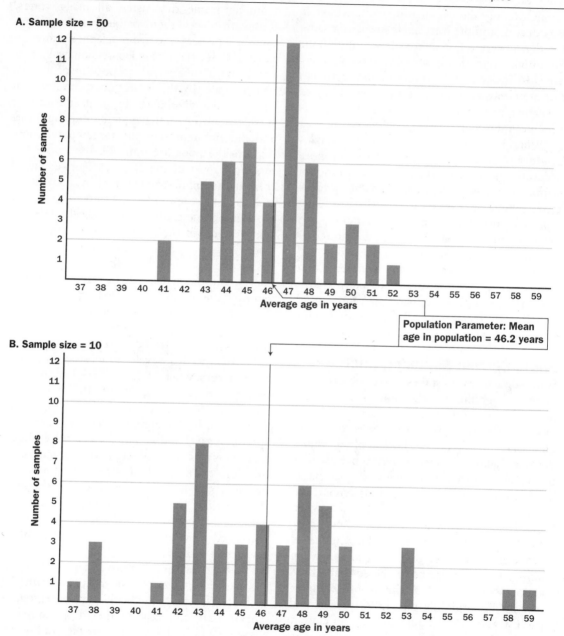

likely event is a sample that looks like (is representative of) the population from which it was drawn, and the least likely event is a sample that is very different from the population. So, in our sampling distribution, most samples are close to the population mean and relatively few are far away from it. You can see in Figure 7.1A that, as you move away from the population mean, fewer and fewer samples occur.

When I stated earlier that random samples produce representative samples, I should more accurately have said: When we draw a random sample, the most likely outcome is a representative sample or one that is very close to being representative. However, good sampling can produce, on occasion and by chance, unrepresentative samples. The advantage of what will be defined below as *probability samples* is that (1) in the long run, they are more likely to produce representative samples than are other sampling strategies, and (2) we can determine the probability of being in error.

Some Classic Sampling Mistakes

Some disastrous mistakes have occurred in sampling that illustrate the importance of some of the issues discussed in this section. These mistakes occurred, in part, because of inadequate sampling frames: The sampling frames did not include all elements in the populations to which generalizations were to be made. These mistakes resulted in serious embarrassment, and sometimes financial loss, when the investigators made a precise—and easily refutable—prediction based on the sample.

A classic example of this occurred when *Literary Digest* magazine predicted that Alfred Landon would beat Franklin Roosevelt by a substantial margin in the 1936 presidential race (Cahalan, 1989; Squire, 1988). Of course, Roosevelt won the election. Why the error in prediction? The erroneous prediction resulted from two factors, each serious by itself but deadly in combination. The first problem was a flawed sampling frame. In predicting elections, the target population is all likely voters.

The *Literary Digest*, however, did not utilize a sampling frame that listed all likely voters. Rather, they drew their sample from lists of automobile owners and from telephone directories. In 1936, the Great Depression was at its peak, and a substantial proportion of eligible voters, especially the poorer ones, did not own cars or have telephones. In short, the sample was drawn from an inadequate sampling frame and did not represent the target population. Because the poor are more likely to vote Democratic, most of the eligible voters excluded from the sampling frame—and thus having no chance to be counted in the sample—voted for the Democratic candidate, Roosevelt. The second problem in the *Literary Digest* poll was a poor response rate. Although employing a massive sample size, the pollsters used a mailed survey, and the percentage of respondents who returned the surveys was very low, about 23%. An independent follow-up investigation in a city where 50% of the voters voted for Roosevelt and 50% for Landon found that only 15% of the Roosevelt supporters returned their surveys, whereas 33% of the Landon supporters did (Cahalan, 1989). If there was a bias in the sampling frame against Roosevelt supporters, it was compounded by the nonresponse bias: Landon supporters were much more likely to return their surveys to the *Literary Digest*. The result was the magazine's embarrassingly inaccurate prediction. Although the *Literary Digest* was a popular and respected magazine before the election, it never recovered from its prediction and went out of business a short time later.

George Gallup and most other public opinion pollsters, despite using flawed sampling techniques, had better luck than the *Literary Digest* and correctly predicted a Roosevelt victory in 1936. However, their luck ran out in 1948 when they wrongly predicted that Thomas Dewey would defeat Harry Truman in the presidential contest (Katz, 1949). A number of mistakes led to this erroneous prediction. For example, the last poll in 1948 was conducted many weeks before the election, and there was a substantial shift in support toward Truman during that period. However, a part of the problem was that these pollsters used a then-popular sampling technique called "quota sampling." This technique will be discussed later in the chapter, but it basically involves giving interviewers quotas of characteristics of people to be interviewed. Each interviewer might be required to interview 10 males between age 20 and 30 with at least a high school education, and eight females with similar age and educational characteristics. Quotas would also be set for characteristics such as race, ethnicity, and geographic location. Exactly which proportions of people with different characteristics are included in the sample depends on the distribution of those characteristics in the population. The idea is to get a sample that is representative of the population along those characteristics. However, exactly which individuals are interviewed is up to the interviewer to determine, as long as the people have the appropriate characteristics. Therein lies one of the flaws in the technique: Apparently the interviewers in 1948, while remaining faithful to selecting people with the appropriate characteristics, had a tendency to choose Dewey supporters more than Truman supporters. So, quota samples are not constructed from a complete sampling frame; they depend more on the sometimes arbitrary and sometimes biased choices of the interviewers. Another flaw was that the pollsters based the quotas for various characteristics on census data that was probably outdated by 1948. As a consequence of this sampling mistake, quota sampling fell out of favor with many social scientists.

For many social science research projects, adequate sampling frames can be constructed from existing listings (such as those mentioned earlier) that are available or can readily be created. Still, caution must be exercised when using such lists because they may inadvertently exclude some people. In fact, social science research is especially vulnerable to this when the populations we study are difficult to enumerate. For example, undocumented aliens are by definition not listed anywhere. We know they comprise a large segment of the population in such urban centers as Los Angeles, but a study of the poor in these areas that relied on a city directory would obviously miss large numbers of such persons. Early studies of gay men also fell prey to this problem (Bell & Weinberg, 1978; Hooker, 1957). In some of these studies, the sampling frame was homosexuals who were listed as patients by therapists who were participating in the research. The studies concluded that homosexuality was associated with personality disturbance. Yet, it does not take great insight to recognize that many gay men—those feeling no need to see a therapist—were not listed in the sampling frames, and the samples were thus strongly biased toward finding personality disorders among gays.

Sampling frames must therefore be assessed carefully to ensure that they include all elements of the population of interest. The remainder of Chapter 7 is a discussion of the different ways in which samples can be selected. First discussed is probability samples, for which we are most likely to have a sampling frame from which to draw the sample. Then nonprobability samples are discussed, which are often used when it is not possible to construct a sampling frame.

PROBABILITY SAMPLES

With luck, almost any sampling procedure could produce a representative sample. But that is little comfort to a researcher who wants to be as certain as possible that his or her sample is representative. Techniques that make use of probability theory can both greatly reduce the chances of getting a nonrepresentative sample and, what is more important, permit the researcher to estimate precisely the likelihood that a sample differs from the population by a given amount. In these samples, known as **probability samples**, each element in the population has some chance of being included in the sample, and the investigator can determine the chances or probability of each element's being included (Scheaffer, Mendenhall, & Ott, 1996). In their simpler versions, probability sampling techniques ensure that each element has an *equal* chance of being included. In more elaborate versions, the researcher takes advantage of knowledge about the population to select elements with differing probabilities. The key point is that, whether the probabilities are equal or different, each element's probability of being included in a probability sample is *nonzero* and *known*. Further, probability sampling enables us to calculate **sampling error,** which is an estimate of the extent to which the values of the sample differ from those of the population from which it was drawn. The major types of probability samples will be discussed, followed by an analysis of how to estimate the sample size needed to produce a given amount of sampling error.

Simple Random Sampling

The simplest technique for drawing probability samples is **simple random sampling** (SRS) in which each element in the population has an equal probability of being chosen for the sample. Simple random sampling treats the target population as a unitary whole. One begins with a sampling frame containing a list of the entire population or as complete a list as can be obtained. The elements in the sampling frame are then numbered sequentially, and elements are selected from the list using a procedure known to be random.

One method for drawing a random sample is to use a table of random numbers. Although today computers are normally used to select random samples, describing the use of a table of random numbers illustrates the random nature of the process. A table of random numbers consists of digits from zero to nine that are equally represented and have no pattern or

order to them. Appendix B contains such a table. To use the table for selecting a sample, first note the size of the population. This will determine how many of the random digits we need when selecting each element. For example, if the population had at least 1,000 elements but did not exceed 9,999 elements, we would use four columns of random digits for each of the selections from the sampling frame. If the population exceeded 9,999 elements, but not 99,999, we would take random digits five columns at a time.

The second step is to select a starting point in the table of random numbers. It is important that one not always start from the same place in the table, such as the upper-left corner. If that were done, every sample, assuming the same number of digits were used, would select the same elements from the population. That would, of course, violate the randomness we seek to achieve. This problem can be easily avoided by merely starting in the table at some point that is itself randomly determined (e.g., close your eyes and point).

Figure 7.2 implements this process by showing the selection of a sample of size 6 from a population of size 30. Each case in the population is assigned a number between 1 and 30. Then we proceed through the table from the randomly selected starting point, taking each set of two digits that is between 1 and 30 until we reach the desired sample size. We ignore any number between 31 and 99 and any number that is repeated. This list of random numbers is then used to identify the elements to be included in the sample. When using computers instead of tables to select random samples, each element in a sampling frame is assigned a number, and a computer program then generates a set of random numbers of the size of the sample needed. If there are 2,798 elements in the sampling frame and you need a sample size of 300, the computer will generate 300 random numbers between 1 and 2,798 (see Appendix B). Those 300 cases are your sample. If the whole sampling frame is in a

FIGURE 7.2 Using a Table of Random Numbers to Select a Sample

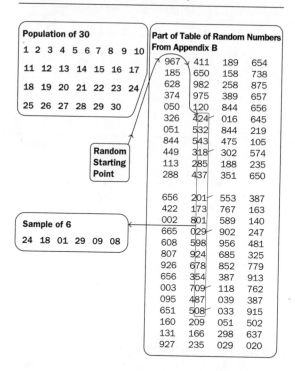

| Population of 30 |
| 1 2 3 4 5 6 7 8 9 10 |
| 11 12 13 14 15 16 17 |
| 18 19 20 21 22 23 24 |
| 25 26 27 28 29 30 |

Random Starting Point

| Part of Table of Random Numbers From Appendix B |
967	411	189	654
185	650	158	738
628	982	258	875
374	975	389	657
050	120	844	656
326	424	016	645
051	532	844	219
844	543	475	105
449	318	302	574
113	285	188	235
288	437	351	650
656	201	553	387
422	173	767	163
002	801	589	140
665	029	902	247
608	598	956	481
807	924	685	325
926	678	852	779
656	354	387	913
003	709	118	762
095	487	039	387
651	508	033	915
160	209	051	502
131	166	298	637
927	235	029	020

Sample of 6

24 18 01 29 09 08

computer file, the computer may be able to both assign the numbers to the sampling frame and select the 300 random cases.

Although simple random samples have the desirable feature of giving each element in the sampling frame an equal chance of appearing in the sample, they are often impractical. A major reason for this is the cost. Imagine doing a research project that calls for a national sample of 2,000 households. Even if one could obtain such a sample using SRS, which is unlikely, it would be prohibitively expensive to send interviewers all over the country to obtain the data. Further, alternatives to SRS may be more efficient in terms of providing a high degree of representativeness with a smaller sample. Simple random sampling is normally limited to fairly small-scale projects dealing with populations of modest size for which

adequate sampling frames can be obtained. The importance of simple random sampling lies not in its wide application but on the fact that it is the basic sampling procedure on which statistical theory is based, and it is the standard against which other sampling procedures are measured.

Systematic Sampling

A variation on simple random sampling is called **systematic sampling**, which involves taking every kth element listed in a sampling frame. Systematic sampling uses the table of random numbers to determine a random starting point in the sampling frame. From that random start, we select every kth element into the sample. The value of k is called the *sampling interval* and is determined by dividing the population size by the desired sample size. For example, if we wanted a sample of 100 from a population of 1,000, the sampling interval would be 10. From the random starting point, we would select every tenth element from the sampling frame for the sample. (If the starting point is in the middle of the list, we proceed to the end, jump to the beginning, and end up at the middle again.)

If dividing the population by the sample size does not produce a whole number, the decimal should be rounded up to the next largest whole number. This will provide a sampling interval that will take us completely through the sampling frame. If we rounded down, the sampling interval would be slightly too narrow, and we would reach the desired sample size before we had exhausted the sampling frame. This would mean that those elements farthest from the starting point would have no chance of being selected.

Systematic sampling is commonly used when samples are drawn by hand rather than via computer. The only advantage of systematic sampling over SRS is in clerical efficiency. In SRS, the random numbers will select elements that are scattered throughout the sampling frame. It is time consuming to search all over the sampling frame to identify the elements that correspond with the random numbers. In systematic sampling, we proceed in an orderly fashion through the sampling frame from the random starting point.

Unfortunately, systematic sampling can sometimes produce biased samples, although this is rare. The difficulty occurs when the sampling frame consists of a population list that has a cyclical or recurring pattern, called *periodicity*. If the sampling interval happens to be the same as that of the cycle in the list, it is possible to draw a seriously biased sample. For example, suppose we were sampling households in an apartment building. The apartments are listed in the sampling frame by floor and apartment number (1A, 1B, 1C, 1D, 1E, 1F, 2A, 2B, . . .). Further suppose that, on each floor, apartment F is a corner apartment with an extra bedroom and correspondingly higher rent than the other apartments on the floor. If we had a sampling interval of 3 and randomly chose to begin counting with apartment 2D, every F apartment would appear in the sample, which would mean that the sample is biased in favor of the larger and more expensive apartments and thus the more affluent residents of the apartment building or those with larger families (see Figure 7.3). So when using systematic sampling techniques, the sampling frame needs to be carefully assessed for any cyclical pattern that might confound the sample, and the list should be rearranged to eliminate the pattern. Alternatively, SRS could be used instead of systematic sampling.

Stratified Sampling

With simple random and systematic sampling methods, the target population is treated as a unitary whole when sampling from it. **Stratified sampling** changes this by dividing the population into smaller subgroups, called *strata*, prior to drawing the sample, and then separate random or systematic samples are drawn from each of the strata.

FIGURE 7.3 An Illustration of Periodicity in Using Systematic Sampling of Apartments in an Apartment Building

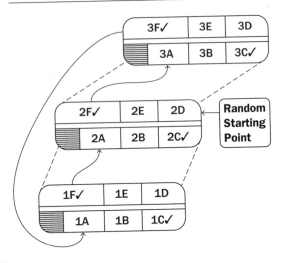

Reduction in Sampling Error

One of the major reasons for using a stratified sample is that stratifying has the effect of reducing sampling error for a given sample size to a level lower than that of an SRS of the same size. This is true because of a very simple principle: the more homogeneous a population on the variables being studied, the smaller the sample size needed to represent it accurately. Stratifying makes each stratum more homogeneous by eliminating the variation on the variable that is used for stratifying. Perhaps a gastronomic example will help illustrate this point. Imagine two large commercial-size cans of nuts, one labeled peanuts and the other labeled mixed nuts. Because the can of peanuts is highly homogeneous, only a small handful from it would give a fairly accurate indication of its contents. The can of mixed nuts, however, is quite heterogeneous, containing several kinds of nuts in different proportions. A small handful of nuts from the top of the can could not be relied on to represent the contents of the

TABLE 7.1   Hypothetical Proportionate Stratified Sample of University Students

Proportion in University	Percent	Stratified Sample of 200	Number
Seniors	20	Seniors	40
Juniors	20	Juniors	40
Sophomores	25	Sophomores	50
Freshmen	35	Freshmen	70
	100		200

entire can. If, however, the mixed nuts were stratified by type of nut into homogeneous piles, a few nuts from each pile could constitute a representative sample of the entire can.

Although stratifying does reduce sampling error, it is important to recognize that the effects are modest. One should expect approximately 10%–20% or less reduction in comparison to an SRS of equal size (Henry, 1990; Sudman, 1976). Essentially, the decision to stratify depends on two issues: the difficulty of stratifying, and the cost of each additional element in the sample. It can be difficult to stratify a sample on a particular variable if it is hard to access data on that variable. For example, it would be relatively easy to stratify a sample of university students according to class level because universities typically include class status as part of a database on all registered students. In contrast, it would be difficult to stratify the same sample on the basis of whether they had been victims of sexual abuse during childhood because these data are not readily available and obtaining it would require a major study in itself. Stratification therefore requires either that the sampling frame include information on the stratification variable or that the stratification variable be determined easily. The latter situation may be illustrated by telephone surveys and stratification by gender of respondent. Telephone interviewers can simply ask to speak to the man of the house to obtain the male stratum and request to speak to the woman of the house for the female stratum. If the desired gender is unavailable, the

household is dropped and another substituted. The process may require some extra phone calls, but the time and cost of doing this can pay for itself in the quality of the sample. As for the effect of cost issues on the decision of whether to stratify, if the cost of obtaining data on each case is high, as in an interview survey, stratifying to minimize sample size is probably warranted. If each case is inexpensive, however, stratifying to reduce cost may not be worth the effort unless it can be accomplished easily.

Proportionate Stratified Sampling

When stratification is used for reducing sampling error, *proportionate* stratified sampling is normally used, in which the size of the sample taken from each stratum is proportionate to the stratum's presence in the population. Consider a sample of the undergraduates at your college or university. Although students differ on many characteristics, an obvious one is their class standing in school. Any representative sample of the student body should reflect the relative proportions of the various classes as they exist in the college as a whole. If we drew an SRS, the sample size would have to be quite large in order for the sample to reflect accurately the distribution of class levels. Small samples would have a greater likelihood of being disproportionate. If we stratify on class level, however, the sample can easily be made to match the actual class distribution regardless of sample size. Table 7.1 contains the hypothetical class distribution of a university student body. If one wished a sample of 200 students with these proportions of students accurately represented, stratifying could accomplish it easily. One would begin by developing a sampling frame with the students grouped according to class level. Separate SRSs would then be drawn from each of the four class strata in numbers proportionate to their presence in the population: 70 freshmen, 50 sophomores, 40 juniors, and 40 seniors.

In actual practice, it is normal to stratify on more than one variable. In the case of a student population, one might wish to stratify on sex as well as class level. That would double the number of separate subsamples from four to eight: senior men, senior women, junior men, and so on. Even though stratifying on appropriate variables always improves a sample, it should be used judiciously. Stratifying on a few variables provides nearly as much benefit as stratifying on many. Because the number of subsamples increases geometrically as the number of stratified variables and their number of categories increases, attempting to stratify on too many variables can excessively complicate sampling without offering substantially increased benefits in terms of reduction in sampling error.

Disproportionate Stratified Sampling

In addition to reducing error, stratified samples are also used to enable one to make comparisons among various subgroups in the population when one or more of the subgroups are relatively uncommon. For example, suppose we were interested in comparing male firefighters to female firefighters. Nationally, only 3% of all firefighters are female (U.S. Bureau of the Census, 1998, p. 419). So if we selected a simple random sample of 500 from all firefighters nationally, we could end up with only around 15 women firefighters if the sample is representative. We might end up with even fewer. This number would be far too small to make meaningful statistical comparisons. Stratifying in this case would allow us to draw a larger sample of female firefighters to provide enough cases for reliable comparisons to be made. This is called *disproportionate* stratified sampling because the strata are not sampled proportionately to their presence in the population. This type of sample differs from most probability samples where representativeness is achieved by giving every element in the population an equal chance of appearing in the sample. With a disproportionate stratified sample, each element of a stratum has an equal chance of appearing in the sample of that stratum, but the elements in some strata have a better chance of appearing in the overall sample than do the elements of other strata.

Thus far I have said nothing about the selection of variables on which to stratify. This depends on the reason for stratifying. If stratifying is being done to ensure sufficient numbers of cases for analysis in all groups of interest, as in the example of female firefighters, then stratifying would be done on the variable that has a category with a small proportion of cases in it. This would often be an independent variable and would involve disproportionate stratified sampling. On the other hand, if the goal of stratifying is to reduce sampling error, as is the case in proportionate stratified sampling, then variables other than the independent variable may be used. Stratifying reduces sampling error only when the stratification variables are related to the dependent variables under study. Variables should therefore be selected that are known or suspected of having an impact on the dependent variables. For example, a study of the impact of religiosity on delinquency might stratify on socioeconomic status because this variable has been shown to be related to delinquency involvement. Stratifying on a frivolous variable, such as eye color, would probably gain little because it is unlikely to be related to delinquency involvement. It is worth noting, however, that stratifying never harms a sample. The worst that can happen is that the stratified sample will have about the same sampling error as an equivalently sized SRS, and our stratifying efforts will have gone for naught.

Area Sampling

Area sampling (also called **cluster** or **multistage sampling**) is a procedure in which the final units to be included in the sample are obtained by first sampling among larger units, called *clusters*, in which the smaller sampling units are contained. A series of sampling stages are involved, working down from larger clusters to smaller ones. Imagine, for example, that we wanted to conduct a needs assessment survey to determine the extent and distribution of preschool children with educational deficiencies in a large urban area. Simple random and systematic samples would be out of the question because there would likely be no sampling frame that would list all such children. We could turn to area sampling, which is a technique that enables us to draw a probability sample without having a complete list of all elements in the population. The ultimate unit of analysis in this needs assessment would be households because children live in households and we can create a sampling frame of households. We get there in the following way (see Figure 7.4). First we would take a simple random sample from among all census tracts in the urban area. The Census Bureau divides urban areas into a number of census tracts, which are areas of approximately 4,000 people. At the second stage, we would list all the city blocks in each census tract in our sample and then select a simple random sample from among those city blocks. In the final stage, we would list the households on each city block in our sample and select a simple random sample of households on that list. With this procedure, we have created what is called an *area probability sample* of households in that urban area. (Public opinion polling agencies, such as Roper, typically use area sampling or a variant of it.) Each household in the sample is interviewed regarding educational deficiencies among children in the household. If we were sampling an

FIGURE 7.4   Drawing an Area Probability Sample

Step 1:
Take a random sample of census tracts in an urban area (the shaded tracts are those sampled).

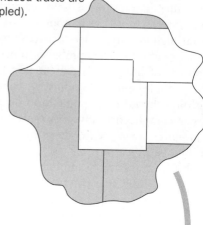

Step 2:
Identify city blocks in each census tract, and take a random sample from a list of those city blocks (the shaded blocks are those sampled).

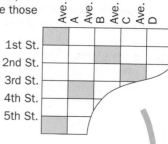

Step 3:
Using a table of random numbers, select a sample of five households from each city block sampled in each census tract sampled (addresses with an asterisk are those sampled).

1. 100 2nd St.*	10. 201 3rd St.
2. 110 2nd St.	11. 205 3rd St.
3. 120 2nd St.	12. 209 3rd St.
4. 130 2nd St.*	13. 213 3rd St.*
5. 140 2nd St.*	14. 217 3rd St.
6. 401 Ave. D	15. 400 Ave. C
7. 415 Ave. D	16. 410 Ave. C
8. 425 Ave. D	17. 420 Ave. C
9. 435 Ave. D*	18. 430 Ave. C

entire state or the entire country, there would be even more stages of sampling, starting with even larger areas, but eventually working down to the household or individual level, whichever is our unit of analysis.

A number of factors can complicate area sampling. For example, selected blocks often contain vastly different numbers of people—from high-density, inner-city areas to the low-density suburbs. The number of blocks and the number of households per block selected into the sample must be adjusted to take into account the differing population densities. Another complication involves the estimation of sampling error. With the simpler sampling techniques, there are fairly straightforward formulas for estimating sampling error. With area sampling, however, the many stages of sampling involved make error estimation more complex. (Procedures for doing so can be found in Kish [1965] or Scheaffer, Mendenhall, & Ott [1996].) However, error estimation is quite important for area samples because they are subject to greater error than other probability samples. The reason is that some error is introduced at each stage of sampling. The more stages involved, the more the sampling error accumulates. Other factors affecting sampling error are the size of the areas initially selected and their degree of homogeneity. The larger the initial areas and the greater their homogeneity, the greater the sampling error. This may seem odd because, with stratified sampling, greater homogeneity leads to less error. Remember, however, that with stratified sampling, we select a sample from *each stratum*. With area sampling, we draw samples only from a *few areas*. If the few areas in a sample are very homogeneous in comparison to the others, they will be unrepresentative. Small and more numerous heterogeneous clusters lead to more representative area samples. Despite the complexity, area sampling allows highly accurate probability samples to be drawn from populations that, because of their size or geographical spread, could not otherwise be sampled.

Estimating Sample Size

One of the advantages of probability samples is that, in part because of probability theory and the nature of sampling distributions, we can make some fairly precise estimates about how large a sample must be. People sometimes assume that a large sample is more representative than a small one, and thus one should go for the largest sample possible. Actually, deciding on an appropriate sample size is far more complicated than this. Five factors influence the sample size that a researcher will choose: the research hypotheses, the level of precision, the homogeneity of the population, the sampling fraction, and the sampling technique used.

Research Hypotheses

One concern in establishing desired sample size is that we have a sufficient number of cases to examine our research hypotheses. Consider a hypothetical study in which we have three variables containing three values each. For an adequate test of the hypotheses, we need a multivariate cross-tabulation of these three variables, and this would require a $3 \times 3 \times 3$ table or a table with 27 cells. If our sample was small, many cells would have few or no cases in them, and we could not test the hypotheses. Johann Galtung (1967, p. 60) suggests that there should be from 10 to 20 cases in each cell in order to provide an adequate test of hypotheses. Disproportionate stratified sampling could be used here to ensure an adequate number of cases in each cell. When that is not possible, Galtung suggests the following formula to determine sample size:

$$r^n \times 20 = \text{sample size}$$

where r refers to the number of values on each variable, and n refers to the number of variables. Thus, for our hypothetical study,

$$r^n \times 20 = 3^3 \times 20 = 540.$$

So we would need a sample of 540 in order to feel reasonably assured of having a sufficient number of cases in each cell. One weakness of this formula is that it is designed for situations where all the variables have the same number of values. If they don't, then a slight adjustment can be made by setting *r* at the largest number of values among the variables; however, this produces a sample that is probably larger than needed. Another weakness of this technique is that, while it provides a useful guide for deciding on a sample size, it does not guarantee an adequate number of cases in each cell. If some combination of variables is very rare in the population, then we may still find few cases in our sample.

Statistical procedures are often used in testing hypotheses, and most such procedures require some minimum number of cases in order to give accurate results. What is the smallest legitimate sample size? This depends, of course, on the number of variables and the values they can take, but generally 30 cases is considered the bare minimum, and some researchers conservatively set 100 as the smallest legitimate sample size (Bailey, 1987; Champion, 1981). Anything smaller begins to raise questions about whether statistical procedures can be applied properly.

Precision

Another factor influencing sample size is the level of precision, or the amount of sampling error, a researcher is willing to accept. Recall that sampling error refers to the difference between a sample value of some variable and the population value of the same variable. Suppose the average age of all teenagers in a city is 15.4. If we draw a sample of 200 teenagers and calculate an average age of 15.1, then our sample statistic is close to the population value, but there is an error of 0.3 years. Recall, however, that the ultimate reason for collecting data from samples is to draw conclusions regarding the population from which those samples were drawn. We have data from a sample, such as the average age of a group of teenagers, but *we do not have those same data for the population as a whole*. If we did, there

would be no need to study the sample because we would already know what we want to know about the population. If we do not know what the population value is, how can we assess the difference between our sample value and the population value? We do it in terms of the likelihood or probability that our sample value differs by a certain amount from the population value. This assessment is based on probability theory, which is discussed in more detail earlier in this chapter and in Chapter 14. One way this is done is by establishing a *confidence interval*, or a range in which we are fairly certain that the population value lies. I can illustrate this without showing the actual computations: If we draw a sample with a mean age of 15.1 years and establish a confidence interval of ± 1.2 years, we are fairly certain that the mean age in the population is between age 13.9 and 16.3. Probability theory also enables us to be precise about how certain we are. For example, we might be 95% certain, which is called the *confidence level*. (The way in which these confidence intervals and confidence levels are computed is discussed further in Chapter 14.) Technically, this means that if we draw a large number of random samples from our population and compute a mean age for each of those samples, 95% of those means would have confidence intervals that include the population mean and 5% would not. What is the actual population mean? We don't know because we have not collected data from the whole population. We have data from only one sample, but we can conclude that we are 95% sure that the population mean lies within the confidence interval of that sample.

Precision is directly related to sample size: Larger samples are more precise than smaller ones. I can demonstrate this with our earlier example in Figure 7.1 where 50 samples of size 50 were displayed. I instructed the computer to select 50 more random samples from the same population of 1,495 people, but these samples were set at size 10 rather than 50. The results are displayed in Figure 7.1B. You can

TABLE 7.2   Calculating Sample Size Based on Confidence Level, Sampling Error, Population Heterogeneity, and Population Size*

	Sample Size for the 95 percent confidence level					
	±3% Sampling Error		*±5% Sampling Error*		*±10% Sampling Error*	
Population Size	**50/50 Split**	**80/20 Split**	**50/50 Split**	**80/20 Split**	**50/50 Split**	**80/20 Split**
100	92	87	80	71	49	38
250	203	183	152	124	70	49
500	341	289	217	165	81	55
750	441	358	254	185	85	57
1,000	516	406	278	198	88	58
2,500	748	537	333	224	93	60
5,000	880	601	357	234	94	61
10,000	964	639	370	240	95	61
25,000	1,023	665	378	244	96	61
50,000	1,045	674	381	245	96	61
100,000	1,056	678	383	245	96	61
1,000,000	1,066	682	384	246	96	61
100,000,000	1,067	683	384	246	96	61

* How to read this table: For a population with 250 members whom we expect to be about evenly split on the characteristic in which we are interested, we need a sample of 152 to make estimates with a sampling error of no more than ± 5%, at the 95% confidence level. A "50/50 split" means the population is relatively varied. An "80/20 split" means it is less varied; most people have a certain characteristic, a few do not. Unless we know the split ahead of time, it is best to be conservative and use 50/50.

Numbers in the table refer to completed, usable questionnaires needed for various levels of sampling error. Starting sample size should allow for ineligibles and nonrespondents. Note that when the population is small, little is gained by sampling, especially if the need for precision is great.

Source   Adapted from Priscilla Salant and Don A. Dillman, *How to Conduct Your Own Survey*, p. 55. Copyright ©1994 by John Wiley & Sons. Reprinted by permission of John Wiley & Sons, Inc.

readily see that the smaller sample size produces sample statistics that spread much farther away from the population value—a spread of 22 years rather than 11 years with the larger samples. In other words, smaller samples produce results that, on average, diverge more from the population parameter. You can see in Figure 7.1 that the larger samples produce a sampling distribution with the samples bunched closer to the population parameter while the distribution of smaller samples is much more spread out. In other words, smaller samples diverge more from the true value of the population or have more sampling error.

Probability theory enables us to calculate the sample size that would be required to achieve a given level of precision. Table 7.2 does this for simple random samples taken from populations of varying sizes. As an example of how to read the table, with a population of 25,000 elements, you would need a sample size of 1,023 to obtain a sampling error of 3% or less, with 95% certainty and a relatively heterogeneous population (50/50 split identifies a heterogeneous population and an 80/20 split is a homogeneous population). Or, in other words, with that sample size, a 95% chance exists that the population value is within 3% of (above or below) the sample estimate. Again, to be technical, it means that if we draw many random samples of that size and determine a confidence interval of 3% for each, 95% of those confidence intervals will include the population value.

To summarize, the table shows that, other things being equal, sample size must increase when:

- you want to reduce sampling error (that is, increase precision),
- the population size is larger, or
- the population is more heterogeneous.

Although I have talked about what happens when we draw many samples, researchers actually draw only one sample; probability theory tells us the chance we run of that sample having a given level of error. There is a chance—5 times out of 100—that the sample will have an error level greater than 3%. In fact, there is a chance, albeit a very minuscule one, that the sample will have a very large error level. This is a part of the nature of sampling: Because we are selecting a segment of a population, there is always a chance that the sample will be highly unrepresentative of the population. The goal of good sampling techniques is to reduce the likelihood of that error. (Further, one goal of replication in science, as discussed in Chapter 2, is to protect against the possibility that the findings of a single study are based on a sample that unknowingly contains a large error.)

If the 95% confidence level is not satisfactory for your purposes, you can raise the odds to the 99% level by increasing the sample size (this is *not* shown in Table 7.2). In this case, only 1 out of 100 samples is likely to have an error level greater than 3%. However, a sample size large enough for this confidence level might be very expensive and time consuming from which to gather data. For example, at the 3% level of error, a sample from a very large population would have to increase from 1,067 to 1,843 in order to reach the 99% confidence level—potentially quite a bit more expensive (Backstrom & Hursh-Cesar, 1981). For this reason, professional pollsters are normally satisfied with a sample size that will enable them to achieve an error level in the 2%–4% range. Likewise, most social science researchers are forced to accept higher levels of error—often as much as 5%–6% with a 95% confidence level. At the other end of the spectrum, exploratory studies can provide useful data even though they incorporate considerably more imprecision and sampling error. The issue of sample size and error is therefore influenced in part by the goals of the research project.

Population Homogeneity

The third factor impacting on sample size is the variability of the population to be sampled. As noted, a larger sample is needed to achieve the same level of precision if a population is heterogeneous rather than homogeneous (see Table 7.2). Unfortunately, researchers may know little about the homogeneity of their target population. Accurate estimates of population variability can often be made only *after* the sample is drawn and data are collected and at least partially analyzed. On the surface, this would appear to preclude estimating sample size in advance. In fact, however, probability theory still allows sample size to be estimated by simply assuming maximum variability in the population. In Table 7.2, the assumption of "50/50 split" means that maximum variability is assumed. Such estimates are, of course, conservative. This means that the sample size estimates will be larger than needed for a given level of precision if the actual variability in the population is less than assumed.

Sampling Fraction

A fourth factor influencing sample size is the *sampling fraction*, or the number of elements in the sample relative to the number of elements in the population (or n/N, where n equals estimated sample size ignoring sampling fraction, and N equals population size). With large populations, the sampling fraction can be ignored because the sample will constitute only a tiny fraction of the population. In Table 7.2, for example, a population of 10,000 calls for a sample size of only 370 (5% sampling error and 50/50 split), which is less than 4% of the population. For such samples, the research hypotheses, sampling error, and population homogeneity would be sufficient to determine sample size. With smaller populations, however, a sample that meets these criteria may constitute a relatively large fraction of the whole population and in fact may be larger than necessary (Moser & Kalton, 1972). This is so because a sample that constitutes a large fraction of the population will contain less sampling error than if the sample were a small fraction. In such cases, the sample size can be adjusted by the following formula:

$$n' = \frac{n}{1 + \left(\dfrac{n}{N}\right)}$$

where

n' = adjusted sample size,

n = estimated sample size ignoring the sampling fraction, and

N = population size.

As a rule of thumb, this correction formula should be used if the sampling fraction is more than 5% percent. For example, suppose that a community action agency is conducting a needs assessment survey for an Indian tribal organization with 3,000 tribal members. On the basis of the research hypothesis, sampling error, and population variance on key variables, it is estimated that a sample of 600 is needed. The sampling fraction, then, is $n/N = 600/3,000 = 0.2$, or 20%. As this is well over 5%, we apply the correction:

$$n' = \frac{600}{1 + \dfrac{600}{3000}}$$

$$n' = 600/1.20$$

$$n' = 500$$

Thus, instead of a sample of 600, only 500 are needed to achieve the same level of precision. At costs that often exceed $50 per interview, the savings of this adjustment could be significant.

Sampling Technique

The final factor influencing sample size is the sampling technique employed. The estimates discussed thus far are for simple random samples. More complex sampling procedures change the estimates of sample size. Area sampling, for example, tends to increase sampling error in comparison to SRS. A rough estimate of sample sizes for area samples can be obtained by simply increasing the suggested sizes in Table 7.2 by one-half (Backstrom & Hursh-Cesar, 1981). That estimate will be crude and probably conservative, but it is simple to obtain. Stratified sampling, on the other hand, tends to reduce sampling error and decrease required sample size. Estimating sample sizes for stratified samples is complex, and procedures for doing so can be found in Kish (1965) or Scheaffer, Mendenhall, and Ott (1996).

This discussion of sample size, precision, and error points to some of the ways in which researchers exercise judgment regarding how scientifically sound a piece of research is and how much confidence we can have in its conclusions. As I have emphasized, single studies should be viewed with caution, irrespective of how low the sampling error is, because it is possible the conclusions are wrong and the sample drawn is one of the ones in error. As numerous studies begin to accumulate, however, and they all come to the same conclusion, then the likelihood that they are all in error declines. It is necessary to assess the samples in terms of how much error can be expected, given the sample size and the sampling technique. If the sampling errors appear to be quite low, then a few replications might confirm that the findings from these samples reflect the state of the actual population. With large sampling errors, however, the probability that the samples do not represent the population is increased. In such cases, confidence in the outcomes would be established only if a number of studies arrive at the same conclusions. More studies mean that more samples were drawn, which in turn reduces the likelihood that all the samples are in error.

NONPROBABILITY SAMPLES

Probability samples are not possible or even appropriate for all studies. Some research situations call for **nonprobability samples**—samples in which the investigator does not know the probability of each population element's being included in the sample. Nonprobability samples have some important uses. First, they are especially useful when the goal of research is to see whether there is a relationship between independent and dependent variables and there is no intent to generalize the results beyond the sample to a larger population. This is sometimes the case, for instance, in experimental research where generalizability would be established by future research in other settings (see Chapter 8). A second situation in which nonprobability samples are useful is in some qualitative research where the goal is to understand the social process and meaning structure of a particular setting or group (Maxwell, 1996). It is often the case in such qualitative research that the research goal is only to develop an understanding of one particular setting or group of people; issues of generalizing to other settings are either irrelevant or would be a matter for future research projects to establish. As we will see in Chapter 11 and Chapter 15, some qualitative researchers see probability samples as inappropriate for, or at best irrelevant to, conducting theoretically sound qualitative research.

A third situation in which nonprobability samples are useful is whenever it is impossible to develop a sampling frame of a population. Now that you understand what sampling frames and probability samples are, you can readily see that, with no complete list of all elements in a population, you cannot ensure that every element has a chance to appear in the sample. These populations are sometimes called "hidden populations" because at least some of their elements are hidden and difficult or impossible to locate. In fact, the members of hidden populations sometimes try to hide themselves from detection by researchers and others because they engage in illegal or stigmatized behavior, such as drug abuse or criminal activity. Rather than giving up on the study of such populations, researchers use nonprobability samples.

Although nonprobability samples can be very useful, they do have some serious limitations. First, without the use of probability in the selection of elements for the sample, no certain claim of representativeness can be made. There is often no way of knowing precisely which population, if any, a nonprobability sample represents. This question of representativeness greatly limits the ability to generalize findings beyond the level of the sample cases.

A second limitation is that the degree of sampling error remains unknown and unknowable. With no clear population being represented by the sample, there is nothing with which to compare it. The lack of probability in the selection of cases means that the techniques employed for estimating sampling error with probability samples are not appropriate. This means that the techniques for estimating sample size are also not applicable to nonprobability samples. Of the five criteria used in considering sample size among probability samples, the only one that comes into play for nonprobability samples is the first; namely, that sufficient cases be selected to allow the types of data analysis that are planned. Even population homogeneity and the sampling fraction don't come into play because you don't know exactly what the population is; this means you don't know either its size or its composition.

A final limitation of nonprobability samples involves statistical tests of significance. These commonly used statistics (discussed in Chapter 14) indicate to the researcher whether relationships found in sample data are sufficiently strong to be generalizable to the whole population. Some of these statistical tests, however, are based on various laws of probability and assume that a random process is utilized in selecting sample elements. Because nonprobability samples violate some basic assumptions of these tests, they should be used with caution on data derived from such samples. With an understanding of these advantages and disadvantages, we can now discuss the most widely used types of nonprobability samples.

Availability Sampling

Availability sampling (also called *convenience* or *accidental sampling*) involves placing into the sample whichever elements are readily available to the researcher. These samples are especially popular and appropriate for research in which it is very difficult or impossible to develop a complete sampling frame. Sometimes it is too costly to do so, whereas in other cases it is impossible to identify all the elements of a population. When Philippe Bourgois and his colleagues (1997) wanted to study homeless heroin addicts, they located some men who lived in a public park in San Francisco because these men were fairly readily available to them. It would obviously be impossible to develop a sampling frame of all homeless heroin addicts.

Using this availability sample made it possible to study the phenomenon, but at the same time the limitations on generalizability due to the sampling procedure reduces the utility of the findings. In an earlier study, Bourgois (1995) studied drug dealers by again using an availability sample: drug dealers known to him because they lived in the neighborhood he had moved to.

Availability samples are often used in experimental or quasi-experimental research. This is because it can be difficult to get a representative sample of people to participate in an experiment—especially one that is lengthy and time consuming. For example, Ronald Feldman and Timothy Caplinger (1977) were interested in factors that bring about behavioral changes in young boys who have exhibited highly visible antisocial behavior. Their research design was a field experiment calling for the youngsters participating in the study to meet periodically in groups over an eight-month period. Groups met an average of 22.2 times for two to three hours each. Most youngsters could be expected to refuse such a commitment of time and energy. Had the investigators attempted to draw a probability sample from the community, they probably would have had such a high refusal rate that the representativeness of their sample would have been questionable. They would have expended considerable resources and still had, in effect, a nonprobability sample. So they resorted to an availability sample. To locate boys who had exhibited antisocial behavior, they sought referrals from numerous sources: mental health centers, juvenile courts, and the like. In order to have a comparison group of boys who had not been identified as antisocial, they sought volunteers from a large community center association. Given the purpose of experimentation, representative samples are less important. Experiments serve to determine *if* cause-and-effect relationships can be found. The issue of how generalizable those relationships are becomes important only after the relationships have been established.

Snowball Sampling

When a snowball is rolled along in wet, sticky snow, it picks up more snow, becoming larger and larger. This is analogous to what happens with **snowball sampling**: We start with a few cases of the type we wish to study and have them lead us to more cases, who, in turn, are expected to lead us to still more cases, and so on. Like the rolling snowball, the snowball sample builds up as we continue adding cases. The researcher may begin with a convenience sample, and then ask each person in that sample for the name of another person who shares the same characteristic or activity. Douglas Heckathorn (1997) used a variant on this, which he called **respondent-driven sampling**, in a study of intravenous drug users and AIDS. He began with a small group of drug abusers called "seeds" who were known to the researchers and who were asked to contact three other drug abusers that he or she knew. The seeds were given a payment for being interviewed and an additional payment when each of the people they contacted came in for an interview. Each recruit who was interviewed then became a seed and was sent out to contact others. The procedure, especially the incentives, provided for successful recruiting of other drug abusers, and also protected their privacy because researchers didn't learn someone's name until they came in voluntarily to be interviewed.

Because snowball sampling depends on the sampled cases knowing other relevant cases, the technique is especially useful for sampling subcultures where the members routinely interact with one another. Snowball sampling can also be useful in the investigation of sensitive topics, such as child abuse or drug abuse, where the perpetrators or the victims might be hesitant to identify themselves if approached by a stranger, such as a researcher, but might be open to an approach by someone who they know shares their experience or deviant status (Gelles, 1978).

Snowball sampling allows the researcher to accomplish what Norman Denzin (1989b)

calls *interactive sampling;* that is, the sampling of persons who interact with one another. Probability samples are all noninteractive because knowing someone who has been selected for the sample does not change the probability of being selected. Interactive sampling is often theoretically relevant because many social science theories stress the impact of one's associates on one's behavior. In order to study these associational influences, researchers often combine snowball sampling with a probability sample. For example, Albert Reiss and Lewis Rhodes (1967), in a study of associational influences on delinquency, drew a probability sample of 378 boys between ages 12 and 16. They then had the members of this sample indicate their two best friends. By correlating various characteristics of the juveniles and their friends, the researchers were able to study how friendship patterns affect delinquency. In fact, the best snowball samples should begin with an initial group that is selected randomly. If that initial group is not random, then whatever biases it contains may get magnified as the sample snowballs through the initial individuals selecting others who are like themselves.

This interactive element, however, also points to one of the drawbacks of snowball sampling: Although it taps people who are involved in social networks, it misses people who may be isolated from such networks. Thus, a snowball sample of drug abusers would be limited to those who are a part of some social network, but would ignore those who abuse drugs in an individual and isolated fashion. It may well be that drug abusers involved in a social network differ from isolated users in significant ways. A second drawback of snowball samples is that they can produce "masking" where a respondent protects the privacy of others by *not* referring them to the researchers. Care must be taken in making generalizations from snowball samples, to ensure that we generalize only to those people who are like those in our sample. Actually, Heckathorn (1997) presents evidence to show that his respondent-driven sampling procedure substantially reduces the bias due to isolation, masking, and a nonrandom seed group.

With all the incentives available in Heckathorn's study, you can imagine that someone might try to take advantage of the situation. Given the overlapping networks among drug abusers, it is highly likely that one user might be contacted by more than one recruiter, creating the possibility that that one user could get interviewed twice by assuming a false identity at the second interview (and thus earn two rewards). Heckathorn reduced the likelihood of this occurring by recording identifying characteristics of each person interviewed: gender, age, height, ethnicity, scars, tattoos, or whatever characteristics would, in combination, identify a particular person. The concern here is less about money than about validity: If one person contributes multiple interviews, then the overall results become biased in the direction of that person.

Quota Sampling

Quota sampling involves dividing a population into various categories and setting quotas on the number of elements to be selected from each category. Once the quota is reached, no more elements from that category are put in the sample. Quota sampling is like stratified sampling in that both divide a population into categories, and then samples are taken from the categories; but quota sampling is a nonprobability technique, often depending on availability to determine precisely which elements will end up in the sample. As mentioned earlier in the chapter, quota sampling was at one time the method of choice among many professional pollsters. However, problems predicting the 1948 presidential election caused pollsters to turn away from quota sampling and embrace the newly developed probability sampling techniques. With its fall from grace among pollsters, quota sampling also declined in popularity among social science researchers.

Presently, use of quota sampling is best restricted to those situations in which its advantages clearly outweigh its considerable disadvantages. Its major positive attributes are that it is cheaper and faster than probability sampling. At times, these advantages can be sufficient to make quota sampling the logical choice. For example, if we wanted a rapid assessment of people's reactions to some event that had just occurred, quota sampling would probably be the best approach. In addition, quota sampling might be used to study crowds or other forms of behavior when the researcher cannot establish a sampling frame.

Normally, quotas are set with the idea of representativeness in mind. If the researcher knows the proportionate distribution of certain characteristics in the population, this information can be used to set the quotas for each category. Quotas are normally established for several variables. Typically included among these variables are common demographic characteristics such as age, sex, race, socioeconomic status, and education. In addition, it is common to include one or more quotas directly related to the research topic. For example, a study of political behavior would likely include a quota on political party affiliation to ensure that the sample mirrored the population on the central variable in the study.

In quota sampling, interviewers do the actual selection of respondents. Armed with the preestablished quotas, interviewers begin interviewing people until they have their quotas on each variable filled. The fact that quota sampling utilizes interviewers to do the actual selection of cases is one of its major shortcomings. Despite the quotas, much bias can enter quota sampling owing to interviewer behavior. Some people simply look more approachable than others, and interviewers may naturally gravitate toward the former. Interviewers are also not stupid. They realize that certain areas of major cities are less than safe places to go around asking questions of strangers. Protecting their personal safety by avoiding these areas can introduce obvious bias into the resulting sample.

Purposive Sampling

In some sampling procedures, the major concern is to select a sample that is representative of, and will enable generalizations to, a larger population. However, generalizability is only one goal, albeit an important one, of scientific research. In some studies, the issue of *control* may take on considerable importance and dictate a different sampling procedure. **Purposive** or **judgmental sampling** involves selecting elements for the sample that the researcher's judgment and prior knowledge suggests will best serve the purposes of the study and provide the best information. In some investigations, for example, this might take the form of choosing a sample that specifically *excludes* certain types of people because their presence might confound the research findings. For example, if one were conducting an exploratory study of a psychotherapeutic model of treatment for mental disorders, it might be desirable to choose people for the sample from among those who would be "ideal" candidates for psychotherapy. Because psychotherapy is based on talking about oneself and gaining insight into feelings, "ideal" candidates for psychotherapy are persons with good verbal skills and the ability to explore and express inner feelings. Because well-educated, middle-class people are more likely to have these characteristics, they might be chosen for the sample.

Although this may sound like "stacking the deck," it isn't—at least not for certain kinds of research questions. For example, suppose the basic question in the research is whether a particular type of psychotherapy can be effective on anybody. If we selected a random sample, we would have variation based on age, sex, education, socioeconomic status, and a host of other variables that are not of direct interest in this study but that might influence receptiveness to psychotherapy. Certainly, in a truly random sample, the

effects of this variation would be washed out. The sample, however, would have to be so large that it would not be feasible to engage that many people in psychotherapy. So, rather than use some other sampling technique, we choose a group that is homogeneous in terms of the factors that are likely to influence receptiveness to psychotherapy. This enables us to see whether psychotherapy works better than some other form of therapy. If it does not work with this "ideal" group, then we can probably forget the idea. If it does work, then we can generalize *only to this group*. Further research would be required among other groups to see how extensively the results could be generalized.

Targeted and Theoretical Sampling

Many variations on sampling have some similarities to both quota and purposive sampling. One variation is called **targeted sampling**, and involves strategies to ensure that persons or groups with specified characteristics have an enhanced chance of appearing in the sample. In other words, specified groups are targeted for special efforts to bring them into the sample. The targeted groups might be identified for theoretical reasons because they would be especially useful for gathering certain kinds of information; or they might be identified during the data-collection process, when it is found that some groups are not showing up in the sample. Targeted sampling is therefore interactive, with the results of data collection possibly changing the method of sampling. An example of this is a study of injecting drug users conducted by John Watters and Patrick Biernacki (1989). They began by constructing an "ethnographic map" of the city in which they were conducting the research. This told them where drug users gathered and drug activity tended to occur, in what amounts, and what were the users' characteristics in different locations. This information provided the basis for deciding where to start recruiting

APPLIED SCENARIO Sampling Homeless Populations

Over the past few decades, policymakers and social program managers have called for assistance from social science researchers in dealing with issues surrounding the homeless. In some cases, social science researchers have been asked to provide estimates of the numbers of homeless in particular locales for purposes of providing adequate services to them; in other cases, the researchers have focused on assessing the extent of severe mental illness among the homeless. In either case, the researchers confront a substantial challenge: coming up with a representative sample of homeless people. Exploring how some researchers met this challenge illustrates how applied researchers must often use a combination of the sampling strategies discussed in this chapter to come up with good samples.

For many reasons, homelessness is an extremely difficult phenomenon to study. In fact, difficulties in sampling the homeless probably account for the fact that studies have come to widely varying conclusions about the numbers of homeless and the degree and extent of mental illness among them (Burnam & Koegel, 1988). Clearly, conventional probability samples are out of the question because it would be impossible to develop a complete sampling frame of all homeless persons in a community. Most studies of the homeless have been limited to collecting data in one city at one type of location, such as at shelters for the homeless. Mental health researchers in Ohio developed a creative sampling procedure to try to overcome these problems (Roth, Bean, Lust, & Saveanu, 1985). First, they found that the concept of "homelessness" was too simple to reflect the complexities of life for people without permanent residences. They developed an operational definition that classified people into four types or levels of homelessness:

1. those with no shelter at all to spend the night in,
2. people living in public or private shelters or missions,
3. people living in cheap hotels or motels for 45 days or less, and
4. other unique situations, such as living with friends, in a tent city, or in jail.

Persons were classified into these categories based on where they had slept the night before the interview. (If the person became homeless on the day of the interview, then placement was based on where he or she intended to sleep during the coming night.) This illustrates once again the importance of conceptual development and the reconceptualization that often occurs in the process of operationalizing concepts (see Figure 5.1).

The goal of the sampling technique used in the Ohio study was to establish the highest probability that the sample was representative of the homeless in the entire state of Ohio, including both urban and nonurban areas. To achieve this, researchers used a combination of

people for participation in their study. This ethnographic map also told them that snowball sampling probably wouldn't work because the drug scene they found consisted of many nonoverlapping social networks. This meant that any network in which they were unable to find an initial informant would probably not show up in their sample.

Neighborhoods in the city were targeted for sampling based in part on their racial com-

stratified, purposive, and random sampling techniques. The state was stratified into five geographic regions, and four counties were selected from each region, one being the major urban center in the region. Additionally, two rural counties and one mixed type were selected from each region through a random process. Through this method, the researchers increased the likelihood that their sample would be representative of the homeless throughout the state.

To further ensure that sufficient data were collected from the different types of homeless people, the sample in each county was stratified according to the four types of homelessness mentioned above. Then equal numbers of each type of homeless person were interviewed in each county. Because of the small number of homeless in rural and mixed counties, all homeless persons identified in these counties were interviewed to ensure a sufficient number of cases for data analysis in each county. In urban counties, sampling frames were developed that listed sites at which homeless people of each type could be found, and sites were randomly selected for interviewing from these lists. At the sites themselves, the interviewers were trained to use random techniques, when possible, to select specific individuals to interview. For example, at shelters they interviewed the person at every tenth sleeping location. Had the interviewers selected whom they wished to interview at these sites, they might have been inclined to interview people who looked least dangerous, most approachable, or most articulate, and this would have biased the sample.

I described another study of the homeless in Chicago in Chapter 5. This research team used an additional sampling technique to reduce bias in the sampling of homeless people (Rossi et al., 1987). They began with a list of census blocks in the city and asked police and others in a position to know which blocks would have many homeless and which only a few. They then stratified blocks by the number of homeless, choosing a larger random sample among blocks believed to contain many homeless. Then the interviewers, accompanied by off-duty police officers, looked for homeless people to interview in all the nondwelling places on the sampled blocks to which they could gain access between midnight and 6 A.M., including hallways, roofs, abandoned buildings, and parked cars. This strategy is obviously far more likely to produce a sample that is representative of all homeless people than are the many studies that limit their sampling to people in shelters.

These studies, of course, are not free of sampling problems. For example, the use of a police escort in the Chicago study may have caused some homeless persons, understandably suspicious of someone seeking them out and possibly waking them up in the night, to flee, and this could have introduced a bias into the sample (Burnam & Koegel, 1988). Yet, the sampling procedures used in these needs assessments give us much more confidence that the results are accurate than does the more limited sampling typically done.

position. Watters and Biernacki wanted to ensure adequate numbers of African American and Latino drug abusers in their sample. To enhance Latino participation, for example, they sent two Latino males who were familiar with the drug scene into the community to inform people in the community of the research, to encourage their participation, and eventually to drive people to the center where data was to be collected. As data collection

EYE ON DIVERSITY   Sampling with Minority Populations

The key to selecting scientifically valid samples is to ensure their representativeness so that valid generalizations can be made. Accomplishing this can be an especially difficult challenge when research is conducted on racial or ethnic minorities. One problem is that some minorities can have "rare event" status; that is, they constitute a relatively small percentage of some populations. African Americans, for example, constitute approximately 12% of the U.S. population, Hispanics around 7% (slightly higher if those in the country illegally are counted), Native Americans perhaps 3% (U.S. Bureau of the Census, 1998). This means that a representative sample of 1,500 people nationwide would, if it included the proper proportions of minorities, contain 180 African Americans, 105 Hispanics, and 45 Native Americans. These numbers are too small for many data analysis purposes. The Native Americans, especially, are so few that any analysis that breaks down the sample into subgroups would result in meaninglessly small numbers in each subgroup. Further, these small numbers mean that the error rate will be much higher for the minorities than for nonminority groups because small samples are less reliable and contain more errors (A. W. Smith, 1987). These small sample sizes make it difficult to assess differences of opinion or behavior within a minority group, and thus it is easy to conclude that the group is more homogeneous than it actually is. As a consequence, we know little about gender, social class, regional, or religious differences among members of particular minorities. The outcome, according to one researcher, is "little more than a form of stereotyping, an *underestimation* of the variability of opinions among blacks. This leads to an *overestimation* of the contribution of race, per se, to black-white differences" in attitude and behavior (A. W. Smith, 1987, p. 445, italics in original). The Committee on the Status of Women in Sociology (1986) makes the same recommendations regarding gender: "Research should include sufficiently large subsamples of male and female subjects to allow meaningful analysis of subgroups."

Some minorities have "rare event" status in another way that can cause problems in sampling. Because of substantial residential segregation of minorities in the United States,

proceeded, Watters and Biernacki recognized that an insufficient number of female injecting drug users were showing up in the sample, so they revised the sampling techniques to enhance participation by women. One reason for the low participation by women, they learned, was that some of the female users were prostitutes, and going to the center to participate in the research meant lost time working the streets to earn money. To help alleviate this, they established a "ladies first" policy at the data-collection sites: If there was a wait, women were given precedence over men. This and other strategies they used were an effort to "target" women in order to include adequate numbers in the sample.

Douglas Heckathorn (1997), in his respondent-driven sampling, also used a form of targeted sampling to reduce bias problems that might arise because some groups of drug abusers are isolated from contact with other users. He used "steering incentives" in the form of bonus payments for contacting drug abusers with special characteristics. For example, female users were somewhat rare in his sample, so whoever contacted a female in-

minorities who live in largely Anglo areas are relatively small in number and can easily be missed by chance even in a well-chosen, representative sample. The result is a sample of minorities that is biased: It includes minorities living in largely minority communities but not minorities living elsewhere. Because minorities living in different communities probably vary in terms of characteristics, attitudes, and behavior, such a biased sample would give a deceptively homogeneous picture of the minority.

Efforts must therefore be made in sampling to ensure that those "rare events" have a chance to be selected for the sample. In some cases, this can be done with disproportionate stratified sampling, in which some individuals or households have a greater probability of appearing in the sample than do other individuals or households. Another way to avoid some of these problems is to use both probability and nonprobability sampling techniques when studying minorities. Some researchers suggest that purposive, dimensional, and snowball sampling can be combined effectively with some type of probability sample to ensure effective coverage of a minority population (Becerra & Zambrana, 1985). A dimensional sample of Hispanics, for example, might specify a series of dimensions that would all have to be covered by the sample. Thus, you might specify that certain age cohorts of Hispanic women (age 20–30, 31–40, and 41 +) would have to be included in the sample, or some minimum number of single-parent and two-parent Hispanic families would have to be included. This would ensure that there were sufficient people with certain characteristics in the sample for valid data analysis. A study of mental health among Asian immigrants used the snowball technique to ensure a complete sampling frame (Kuo & Tsai, 1986). Part of the sampling frame was developed by using local telephone directories and gathering names from ethnic and community organizations. However, given the dispersion of Asian Americans in the Seattle area where this study was done, researchers also used the snowball technique.

jector who then showed up for an interview received an extra $5. Barney Glaser and Anselm Strauss (1967) used a version of targeted sampling that they called **theoretical sampling** and which emerged from their grounded theory approach (to be elaborated on in Chapter 11 and Chapter 15). Glaser and Strauss argue that researchers should analyze their data and construct theories while they are still in the process of collecting the data. Then, as the theory is constructed, it helps the researcher decide from which units or individuals it would be most appropriate

to collect further data. This is similar to what Heckathorn and Watters and Biernacki did. Glaser and Strauss called it "theoretical" sampling because the emerging theory guides adjustments in the sampling procedures as data are collected.

So, the sampling procedure used in a particular study might involve a combination of a number of different sampling strategies. The Applied Scenario box describes how this has been done in applied research on homeless populations to come up with effective samples.

Dimensional Sampling

It is often expeditious, if not essential, that small samples be used. Small samples can be very useful, but considerable care must be exercised in drawing the sample. **Dimensional sampling** is a technique for selecting small samples in a way that enhances their representativeness (Arnold, 1970). There are basically two steps to dimensional sampling: (1) specify all the dimensions or variables that are important, and (2) choose a sample that includes at least one case representing each possible combination of dimensions. This can be illustrated with a study of the effectiveness of various institutional approaches to the control of juvenile delinquency (Street, Vinter, & Perrow, 1966). The population consisted of all institutions for delinquents. To draw a random sample of all those institutions, however, would have called for a sample size that would tax the resources of most investigators. As an alternative, the researchers used a dimensional sample. The first step was to spell out the conceptual dimensions that were important. In terms of juvenile institutions, this investigation considered three dimensions, each containing two values, as illustrated in Table 7.3: organizational goals (custodial or rehabilitative), organizational control (public or private), and organizational size (large or small). The second step was to select at least one case, or juvenile institution, to represent each of the eight possibilities that resulted.

Dimensional sampling has several advantages that can make it an attractive alternative in some situations. First, it is faster and less expensive than studying large samples. Second, it is valuable in exploratory studies where

TABLE 7.3 An illustration of Institutional Dimensions for a Dimensional Sample

	Custodial Goals		Rehabilitative Goals	
	Public	**Private**	**Public**	**Private**
Large Size				
Small Size				

there is little theoretical development to support a large-scale study. Third, dimensional sampling provides more detailed knowledge of each case than is likely to be gained from a large sample. With a large sample, data collection will necessarily be more cursory and focused (which is justified if previous research has narrowed the focus of which variables are important).

Despite their limitations, nonprobability samples can be valuable tools in the conduct of social science research. However, two points need to be reiterated. First, some research uses both probability and nonprobability samples in a single research project, and I have given some illustrations of this. The point is that the two types of samples should not be considered competitors for our attention. Second, findings based on nonprobability samples should be viewed as suggestive rather than conclusive, and opportunities to retest hypotheses using probability samples should be sought. The Eye on Diversity box suggests some ways in which nonprobability sampling strategies can be combined with probability sampling to develop effective samples.

REVIEW AND CRITICAL THINKING

In our everyday lives, we often draw conclusions from observations that have been made. People tend to assume that those observations provide an adequate basis from which to deduce something about how the world operates. Sometimes the observations do provide that solid basis, but often they do not. The observations made in scientific research differ from everyday observations because they are far more systematic in nature. An important dimension of making systematic observations has to do with sampling; that is, how you select which people or elements are to be observed. With what you have learned about sampling in this chapter, you have some tools with which to assess whether observations provide a sound basis for drawing conclusions. In assessing any conclusions that are based on observations, then, consider the following.

1. What is the whole group (or population) about which you want to draw conclusions?

2. What groups or people serve as the source of information? How were they selected? Are there people or groups who had no chance to show up in this sample? Could this lead to misunderstanding or inaccuracy?

3. Could the people observed be considered a probability sample? a nonprobability sample? What problems does this create as far as drawing conclusions to the whole group or population?

4. Are there other ways the sample could have been selected that could have produced a more accurate, or at least a different, conclusion or understanding?

Computers and the Internet

Computer software is now widely available to assist researchers with a number of tasks related to selecting a sample. Some software, for example, can help researchers determine an appropriate sample size by using information supplied by the researcher, such as anticipated data analysis procedures, time, money, projected response rate, and so on. Other programs lead the researcher through a structured decision-making system that reduces the possibility that some important considerations will be left out of the sampling decision.

In addition, computers are also used to actually select the sample. One sampling frame that has been rejuvenated thanks to computerization is the telephone directory. In addition to the limitations covered in the discussion on sampling frames, telephone directories are physically cumbersome to work with. Further, for anything other than a local survey, research would require amassing large numbers of directories, certainly an impractical approach to selecting a national sample. However, these problems are being overcome with the new high storage capacity available on personal computers, particularly with the development of CD-ROMs. CDs have the capacity to store vast amounts of data, such as encyclopedias, abstracts, guides to professional literature, as well as millions of names from telephone directories. The capacity to rapidly search the database and to apply various selection criteria make this kind of product a valuable sampling tool for many kinds of survey research. In addition to the U.S. database, similar products are available for Canada and European countries. Regular updates are available, in some cases as frequently as monthly, to assure that the sampling frame is current.

Because of the proliferation and frequent modification of computer products that are available for social research purposes, several professional publications, such as *Social Science Computer Review*, regularly report on new developments. Popular publications such as *Byte*, *PC Week*, *PC World*, and *PC Connection* often provide articles evaluating software, including survey, sampling, and statistical software products. You can also explore the Internet looking for Web sites related to the products discussed or other issues related to sampling. In a search engine, enter the name of a particular kind of sampling procedure or the word *sampling* combined with other related words (*statistical*, *survey*, etc.). The word *sampling* alone will produce far too many "hits."

Main Points

1. Social scientists study samples instead of whole populations because it is easier and less expensive to do so and because the information obtained from a sample is often better than information obtained from a population.

2. A sampling frame is a list of the population elements used to draw some types of probability samples. The representativeness of a sample is its most important characteristic, referring to the degree to which the sample reflects the population from which it was drawn.

3. Understanding sampling relies on having knowledge of probability theory and sampling distributions. When sampling is based on probability theory, then it is possible to

determine the likelihood that a sample is representative of a population.

4. Probability sampling techniques are the best for obtaining representative samples. The key characteristic of probability sampling is that every element in the population has a known chance of being selected into the sample. Probability sampling enables us to calculate sampling error, which is the difference between sample values and true population values.

5. Simple random, systematic, stratified, and area samples are all types of probability samples. Calculating the proper size for a probability sample is based on five factors: having sufficient data to test research hypotheses, level of precision required, degree of population homogeneity, the sampling fraction, and the type of sampling technique used.

6. Nonprobability samples do not assure each population element a chance of being selected into the sample and therefore lack the degree of representativeness of probability samples. Availability, snowball, quota, purposive, targeted, theoretical, and dimensional samples are all types of nonprobability samples.

Important Terms for Review

accidental sampling
area sampling
availability sampling
cluster sampling
convenience sampling
dimensional sampling
judgmental sampling
multistage sampling
nonprobability samples
population
probability samples
purposive sampling
quota sampling

representative sample
respondent-driven sampling
sample
sampling distribution
sampling error
sampling frame
simple random sampling
snowball sampling
stratified sampling
systematic sampling
targeted sampling
theoretical sampling

For Further Reading

Henry, Gary T. (1990). *Practical sampling*. Newbury Park, CA: Sage. This concise book is a handy guide to basic issues related to sampling. It also includes some interesting examples of research projects and the sampling procedures that were used in them.

Kish, Leslie. (1965). *Survey sampling*. New York: Wiley. Considered the mainstay regarding sampling issues in social research, this book assumes that the reader has an elementary understanding of statistics.

Maisel, Richard, & Persell, Caroline Hodges. (1996). *How sampling works*. Thousand Oaks, CA: Pine Forge Press. This book provides an excellent and detailed overview of scientific sampling as well as software that assists the student in learning through working problems.

Scheaffer, Richard L., Mendenhall, William, & Ott, Lyman. (1996). *Elementary survey sampling* (5th ed.). Belmont, CA: Wadsworth. As its name implies, this book is meant as an introductory text on the design and analysis of sample surveys. Limited to coverage of probability sampling techniques, it provides the information necessary to successfully complete a sample survey.

Stuart, Alan. (1987). *The ideas of sampling* (3rd ed.). New York: Oxford University Press. This book is another good review of sampling strategies that can be used in many social science research projects.

Sudman, Seymour. (1976). *Applied sampling*. New York: Academic Press. This book, along with the Kish work, will tell you almost all you need to know about sampling.

Wainer, Howard. (1986). *Drawing inferences from self-selected surveys*. New York: Springer-Verlag. This book focuses on the issue of self-selection into samples and the problems this can create in terms of making inferences from samples to populations.

Part III

COMMON RESEARCH DESIGNS

CHAPTER 8:
Experimental Research

CHAPTER 9:
Survey Research

CHAPTER 10:
Unobtrusive and Available Data Research

CHAPTER 11:
Field Research and Qualitative Methods

CHAPTER 12:
Evaluation Research

Part III describes the most commonly used research designs in the social sciences. The term "research design" refers to the specific ways in which scientists organize the collection of data. When possible, social scientists organize their observations into experiments (Chapter 8), which best enable us to conclude whether one factor was the cause of changes in another factor. Surveys (Chapter 9) involve gathering people's responses to questions and statements. In other cases, social scientists observe people's behavior directly or use documents or organizational records as a source of data (Chapter 10). The direct observation of behavior sometimes takes the form of field research—extended observations of people in their homes, places of business, or other parts of their natural environment—or of other kinds of qualitative research methods (Chapter 11). Finally, when doing applied research, social scientists bring one or more of these research designs to the task of solving some practical problem through the use of evaluation research (Chapter 12).

Many factors play a part in choosing a research design: the nature of the research problem itself, the resources available to the researcher, ethical considerations, the theoretical bent of the researcher (e.g., positivist versus interpretivist), and others. Over the course of these five chapters, how each research design is done and also what criteria are used to decide which design is most appropriate will be discussed.

Chapter 8

EXPERIMENTAL RESEARCH

When people think of "science" and "research," experiments are often the first things that come to mind. These terms can conjure up images of laboratories, white coats, and electronic instruments that are often associated with experiments. However, these images suggest a considerable misunderstanding of the nature of experimentation. In fact, we all engage in casual experimenting in the course of our everyday lives. For example, when a mechanical device malfunctions, we probe and test its various components in an effort to discover the elements responsible for the malfunction. If it is a lawn mower, we might put in a new spark plug and then see if it will work; if it doesn't, we replace another part and try the engine again. In essence, we are experimenting to find the component (or variable) that caused the device to malfunction. While these are illustrations of casual rather than systematic experiments, they do point to the essence of an **experiment**: It is a controlled method of observation in which the value of one or more independent variables is changed in order to assess its causal effect on one or more dependent variables.

The term "experimentation" as used in research, then, concerns a logic of analysis rather than a particular location, such as a laboratory, in which observations are made. In fact, experiments can be conducted in many settings. **Laboratory experiments** are conducted in artificial settings constructed in a way such that selected elements of the natural environment are simulated and features of the investigation controlled. **Field experiments**, on the other hand, are conducted in natural settings as people go about their everyday affairs. In this chapter, the underlying logic of experimentation and the major terms associated with experimental research are discussed first. This is followed by a presentation of the major factors that can lead to incorrect inferences in experiments and a discussion on how these problems can be avoided by properly designing the experiment. Finally, the problem of generaliz-ing experimental findings to other settings is analyzed and the advantages and disadvantages of experiments are reviewed.

THE LOGIC OF EXPERIMENTATION

Causation and Control

The strength of experiments as a research technique is that they are designed to enable us to make inferences about causality. The element that makes this possible is *control*: In experiments, the investigator has considerable control over determining who participates in a study, what happens to them, and under what conditions it happens. In order to appreciate the importance of this, let's look at some of the important terms in experimental research (Boruch, 1997; Kirk, 1982).

At the core of experimental research is the fact that the investigator exposes the persons in an experiment, commonly referred to as *experimental subjects*, to some condition or variable, called the *experimental stimulus*. The **experimental stimulus** or **experimental treatment** is an independent variable the value of which is manipulated in some fashion by the experimenter in order to assess its effect on a dependent variable. Recall from Chapter 2 that independent variables are those variables in a study that are hypothesized to produce change in another variable. The variable affected by the independent variable is the dependent variable—so called because its value is dependent upon the value of the independent variable. An **experimental group** is a group of subjects who are exposed to the experimental stimulus. It is also called the **experimental condition**.

I can illustrate the logic underlying experimentation by means of a series of symbols. The following symbols are commonly used in describing experimental designs:

O = an observation or measurement of the
dependent variable

X = exposure of people to the experimental stimulus or treatment (independent variable)

R = random assignment to conditions

In addition, the symbols constituting a particular experimental design are presented in time sequence, with those to the left occurring earlier in the sequence than those farther to the right. With this in mind, we can describe a very elementary experiment in the following way:

Pretest	Treatment	Posttest
O	X	O

In this experiment, the researcher measures the dependent variable at some starting point in time (the *pretest*), exposes the subjects to the experimental stimulus (the *treatment* or independent variable), and then remeasures the dependent variable at some point after the treatment (the *posttest*) to see if any change has occurred. One major yardstick for assessing whether the independent variable in an experiment has had an effect is a comparison of the pretest measures to those of the posttest. (Note that this is a different use of the term "pretest" from its use in Chapter 1 and some other chapters. In those contexts, the purpose of the pretest is to assess the adequacy of a data-collection instrument before actual data collection ensues. In this context, the pretest involves actual data collection, but before the introduction of the experimental stimulus.)

As an example of this, let's look at drug education programs in schools and see whether or not they have an effect on children's attitudes and behavior in relation to drugs. One such program that has been the focus of attention is Drug Abuse Resistance Education (D.A.R.E.) (Dukes, Stein, & Ullman, 1997; Wysong & Wright, 1995). Substantial resources have been expended on this program and many children have been exposed to it. It is sensible to ask: Does it work? These programs are based on the hypothesis that exposure to the program will increase students' negative attitudes toward drugs, increase their awareness of the risk of taking drugs, and reduce their use of drugs. One way to test this hypothesis, using the model just introduced, is to first measure the children's attitudes toward drugs so as to have a baseline against which to assess change. Then we would expose the children to the drug education program, the independent variable: For a specified period, children would be taught about drugs and their consequences. Finally, we would again measure their attitudes toward drugs—the posttest measure of the dependent variable—to see if it had changed since the first measurement.

This illustration shows one of the major ways that experiments offer researchers control over what occurs: The researcher manipulates the experimental stimulus. In this illustration, the researcher exposes a particular group of children to the drug education program. The experimenter specifies how and under what conditions the education program is delivered, including how long the program lasts. The researcher might use one short education program and then a longer program with another group to assess the impact of such variation. The key point here is that the "when" and "how much" of the experimental stimulus are *controlled* by the researcher.

The purpose of an experiment, again, is to determine what effects independent variables produce on dependent variables. Variation in the dependent variable produced by the independent variable is known as **experimental variability,** and is the focus of interest in experiments. However, variation in the dependent variable can occur for reasons other than the impact of the independent variable. For example, measurement error may affect the dependent variable (see Chapter 5). Or the individuals in the experiment may have peculiar characteristics that influence the dependent variable separately from any effect of the experimental stimulus. It is also possible that chance factors can affect the dependent variable. Variation in the dependent variable from any source other than the experimental stimulus is known as **extraneous variability** and makes inferences about change in the dependent variable a challenge. You can visualize the problem with the following formula:

total variation in the dependent variable
= experimental variation
+ extraneous variation

Every experiment contains some extraneous variability because there will always be things other than the experimental stimulus that influence the dependent variable. Researchers need some way of discovering how much variation is experimental and how much extraneous; experiments can be designed to provide this information through the use of *control variables* and *control groups*. **Control variables** are variables whose value is held constant in all conditions of the experiment. By not allowing these variables to change from one condition to another, any effects they may produce in the dependent variable should be eliminated. In the experiment on the impact of drug education programs mentioned above, for example, previous research might suggest that extraneous variation in the dependent variable can be caused by the gender of the students or by the time of day the education program is taught. We could control this variation by, say, giving the program only to boys and only at 10 A.M. in the morning. With such controls, we have greater confidence that changes in the dependent variable are caused by changes in the independent variable and not by the variables that are controlled.

A **control group** consists of research subjects who are provided the same experiences as those in the experimental condition with a single exception: The control group receives no exposure to the experimental stimulus. **Control condition** refers to the state of being in a group that receives no experimental stimuli. When we include the control group in the elementary experimental design just described, we end up with the following design:

	Pretest	Treatment	Posttest
Experimental Group	O	X	O
Control Group	O		O

The control group is very important in experiments because it provides the baseline from which the effects of the independent variable are measured. For example, in the study of the D.A.R.E. program, it is possible that the children's knowledge about and use of drugs would change between the pretest and posttest even if they hadn't participated in the education program, possibly because of something they saw on television. To test for this, it would be desirable to have a control group that is treated the same as the children in the experimental group but is not exposed to the drug education program. This is the way good evaluation experiments on the D.A.R.E. program have been designed. Because both experimental and control groups experience the same conditions, with the exception of the independent variable, we can more confidently conclude that any differences in the posttest value of the dependent variable between the experimental and control groups is due to the effect of the independent variable. So another yardstick for assessing whether an independent variable has an effect—in addition to the pretest-posttest comparison—is the posttest-posttest comparison between experimental and control groups (see Figure 8.1). By the way, the evaluations of D.A.R.E. have generally found it to be ineffectual, or at least less effectual than its proponents have claimed. Most studies show that D.A.R.E. participation does not seem to change the students' use of alcohol, cigarettes, marijuana, and other drugs, although a few comparisons suggest that the

FIGURE 8.1 Comparisons that are Made in Experimental Designs

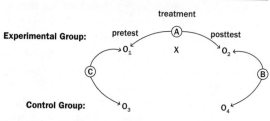

Comparison A, or $O_2 - O_1$:

This comparison will show change in the dependent variable. However, without a control group, it can't distinguish experimental from extraneous variability; change could be due to one of the validity threats, such as history or testing, rather than to the independent variable. If the comparison shows no change, then the treatment may not have an effect on the dependent variable.

Comparison B, or $O_4 - O_2$:

This comparison shows the difference between the experimental mental and control group. However, without a pretest, it can't show whether differences are due to the independent variable, to nonequivalent groups, or to one of the validity threats. If the comparison shows no change, then the treatment may not have an effect on the dependent variable.

Comparison C, or $O_3 - O_1$:

This comparison shows whether the random or other selection procedure actually produced equivalent groups on the dependent variable. If the comparison shows a difference, then there may be a selection bias producing nonequivalent groups.

program may lead to slightly lower levels of usage of some illegal drugs, such as cocaine and inhalants.

From the drug program example, you might get the impression that a control group is a group to which *nothing* is done (as in not being exposed to the drug education program). However, this is not always the case. For example, in a study of the impact on women of viewing media violence against women, the experimental treatment involved viewing a video clip that portrayed violence against a woman (Reid & Finchilescu, 1995). In this case, the control group, which the researchers referred to as the "alternate treatment," also watched a video clip of violence, but the violence was perpetrated against men. The key is that the control group needs to experience all the same things as the experimental group, which in

this case meant watching a video involving violence. What was new in the experimental condition—the independent variable—was that the violence was directed against women.

It is important that the ideas of a control variable and a control group be kept distinct because they serve quite different functions in research. The use of a *control variable* is an attempt to minimize the impact of a single known source of extraneous variability on the dependent variable. The use of a *control group*, on the other hand, is an effort to assess the impact of extraneous variability from any source—including variables that are not known to the researcher—on the dependent variable.

Equivalence of Experimental and Control Groups

When comparisons are made between the experimental and control groups to determine the effect of the independent variable, it is of crucial importance that the two groups be equivalent; that is, composed of persons who are as much alike as possible. If they are not, any comparison could be meaningless as far as the effects of the independent variable are concerned. For example, many of the assessments of the D.A.R.E. program have involved comparisons of students at a school with the D.A.R.E. program to students at a school that does not have the program. If D.A.R.E schools were to show a positive effect over the other schools, could we conclude that D.A.R.E. is effective? Not with complete certainty. The students in the two different schools may have differed from one another in some systematic ways, such as socioeconomic status, ethnicity, race, or motivational level. It may have been these differences between students, rather than exposure to D.A.R.E., that caused differences between the schools in the posttests. This possibility of nonequivalent groups is one of the criticisms that has been made of these evaluations of D.A.R.E.

Two methods are used to avoid the problem in these studies and increase our confidence that the experimental and control groups are equivalent: *matching* and *random assignment*. The first has a certain intuitive appeal, but on closer inspection, it proves to be the less desirable of the two.

Matching

As the name implies, **matching** involves matching individuals in the experimental group with similar subjects for a control group. People are matched on the basis of variables that we presume might have an effect on the dependent variable separate from the effect of the independent variable. By matching, we could produce an experimental group and a control group that are equivalent on the matching variables so that these variables could not account for any differences between the two groups on the dependent variable. For example, the assessment of the D.A.R.E. program might use socioeconomic status (SES) as a matching tool. A high SES student in one school would be matched with a high SES student in the other school. Any students that could not be matched with another student on SES would simply not be used in the data analysis. But other variables than SES could affect the outcome. Race, sex, study habits, grades, parents' involvement in the school— these and other factors could affect students' knowledge about and use of drugs. This illustrates one of the problems with matching: So many variables might be used in matching that it is usually impractical to consider more than a few at a time. Even though matching on SES would not be too great a problem, the difficulty of matching would increase geometrically with the addition of more variables. For example, the researchers might need to find a match for a female of middle-class background with good study habits but low grades and parents who are involved in the school. A second problem with matching is that the researcher needs to know in advance which factors might have an effect on the dependent variable so that these can be included in the

matching process, but often such information is not available.

On the positive side, matching is better than no attempt to control at all and may be the only type of control available. In our illustration, the schools may be prevented by school board policies from randomly assigning some students to participate in D.A.R.E. education while others are not allowed to participate. Another advantage of matching in this situation is that it does not require that all students in both schools participate in the study; the experimental and control groups could consist of only students for which appropriate matches can be made. These two groups can be compared on the dependent variable measures and the remaining students ignored.

Randomization

The second approach to assuring equivalent groups is *random assignment*. As its name implies, **random assignment** uses chance to reduce the variation between experimental and control groups. This can be done in a number of ways. For example, each person in the experiment can be assigned a random number, the numbers arranged in order, and every other person on the list placed in the experimental condition. Or, the names of all the subjects in the experiment can be listed alphabetically and assigned a number, and then a computer program could assign each number randomly to either the experimental or the control group. (See Chapter 7 and Appendix B on the use of random numbers tables.) The point is to be sure that each person has an equal chance of being placed into either the experimental group or the control group. In the long run, this technique offers the greatest probability that experimental and control groups will have no systematic differences between them. Chance rather than a priori knowledge of other variables is the foundation of random assignment.

The problem with relying on chance is that, even though in the long run and over many ap-

TABLE 8.1 Randomization and Blocking Illustration

Student	Sex	Height	True Test Score (Dependent Variable)
A	Male	Short	2
B	Male	Tall	4
C	Male	Tall	6
D	Male	Short	8
E	Female	Tall	12
F	Female	Short	14
G	Female	Short	16
H	Female	Tall	18
Total N = 8			Mean = 10

		Possible Groupings		
	Group 1	Group 1 mean	Group 2	Group 2 mean
Ideal groups	BCFG	10	ADEH	10
Worst random	ABCD	5	EFGH	15
Worst block on sex	ABEF	8	CDGH	12
Worst block on height	ABCD	5	EFGH	15

plications randomization generates equivalent groups, this approach can still occasionally yield nonequivalent groups. This is especially so when the study sample is very small. Table 8.1 illustrates the extreme between-groups difference that can occur by chance when using randomization. Let's suppose that the eight students in the table are the participants in the D.A.R.E. program and its evaluation. The dependent variable is their *knowledge of drug effects*. The students have been identified by sex and height and ranked in the table according to their score on a pretest of drug knowledge. The test has 20 questions and the student scores range from 2 correct to 18 correct. The mean score for the entire group is 10 correct. Thus, when divided into an experimental and a control group, the ideal situation (row labeled "Ideal groups") would be for each group to

have the same mean score of 10 correct, or at least very close to it, on the dependent variable. While probability theory tells us that, with random assignment, it is most likely that experimental and control group means will be close to the overall mean, it is possible to obtain groups with very different means because, by chance, any combination of subjects in the two groups is possible. Table 8.1 shows one possible arrangement of subjects into one group with a mean score of 5 correct and another with a mean score of 15 correct. It occurs when all males end up in one subgroup and all females cluster in the other subgroup. It is labeled "Worst random" because it is an arrangement that could result from random assignment and the two subgroups have mean scores that are as far as possible from the overall group mean.

Blocking

To reduce the likelihood of such occurrences, researchers often use a combination of matching and randomization known as *blocking*. In **blocking**, subjects are first matched on one or more key variables in order to form groupings, or blocks, where each group is homogeneous on the variables used to block. Members of each block are then assigned randomly to the experimental or control conditions. Blocking works by reducing the extreme range of groups that are possible. For blocking to be effective, it is necessary that the variable on which cases are blocked be associated with the dependent variable. Note in Table 8.1 that males have low scores and females have high scores; so the variable, *sex*, and the dependent variable are clearly associated. There is no such association between *height* and the dependent variable. When blocking is done on the basis of sex, two males are assigned randomly to the experimental group and the other two to the control group; females are likewise assigned randomly to each group. This ensures that the proportions of males and females in each subgroup are the same as their proportions in the overall group. The most extreme between-groups difference that is possible in our hypothetical illustration with *sex* as the blocking variable are group means of 8 and 12 (labeled "Worst block on sex"). This is a considerable improvement over the extremes of 5 and 15 that are possible with randomization alone. Note, however, that when the groups are blocked on the basis of *height* (labeled "Worst block on height"), there is no improvement; the worst possible group difference is still 5 and 15, as it was under randomization. This demonstrates that blocking on an appropriate variable can help reduce differences between experimental and control groups. Choosing a variable for blocking that is unrelated to the dependent variable won't make matters any worse than random assignment, but it also won't improve the chances of achieving equivalent groups. Whether or not it is worth it to use blocking basically comes down to the anticipated improvement in group equivalency versus the cost and complexity of carrying out the blocking procedure.

Internal Validity

The central issue in experimentation is its utility in enabling us to make statements about causal relationships between phenomena. This is the reason for such things as control variables, control groups, randomization, and matching. In this regard, what is of critical importance in experiments is their *internal validity* (Campbell & Stanley, 1963; Cook & Campbell, 1979; Kercher, 1992). Recall from Chapter 5 that the validity of a measure refers to whether it accurately measures what it is intended to measure. Likewise, **internal validity** in experiments refers to whether the independent variable actually does produce the effect it appears to have on the dependent variable; it is concerned with ruling out extraneous sources of variability to the point where we have confidence that changes in the dependent variable were caused by the independent variable. In the preceding discussion, I presented the basic logic involved in designing experiments that have internal validity. However, the problem is much more complex than I have so far stated because internal validity can be threatened in many ways. I will discuss the seven most serious threats to the internal validity of experiments.

History

The threat of history concerns events that occur during the course of an experiment, other than the experimental stimulus, that could affect the dependent variable. History is more of a problem for field experiments than for those conducted in the confines of a laboratory because field experiments typically last longer, allowing more time for events to occur that could affect the outcome. Again, using the example of drug education programs: In evaluating a drug education program, we might do a pretest and a posttest measure of levels of drug use and use the difference between the two as an assessment of the impact of the program. If the pretest and posttest measures were conducted a few months apart, a number of things, independent of the experiment itself, could happen between those two measures and affect the posttest measure. For instance, the price of drugs could increase and thus lead to lower usage because of financial considerations. Or some political, religious, or social service organization could initiate its own antidrug campaign in the media, and this could lead to lower usage. It should be evident that the effects of these historical events make it more difficult to ascertain what effect the drug education program itself has on drug usage rates.

Maturation

Maturation refers to changes occurring within experimental subjects that are due to the passage of time. Such things as growing older, hungrier, wiser, more experienced, or more fatigued are examples of maturation changes. If any of these changes are related to the dependent variable, their effects could confuse the effect of the independent variable. For example, a study of the impact of educational and social services on the development of language skills among very young children would have to take into account the fact that such skills typically show substantial improvements in young children because of their natural development. If this is not taken into account in the research design, then changes over time that are due to these natural developments could be confused with the effects of the educational and social services.

Testing

The threat of testing may occur any time subjects are exposed to a measuring device more than once. Because many experiments use paper-and-pencil measures and "before" and "after"

measurements, testing effects are often of concern. For example, individuals taking achievement or intelligence tests for a second time tend to score better than they did the first time. This improvement occurs even when alternative forms of the test are used. Similar changes occur on personality tests. It should be quite clear that these built-in shifts in paper-and-pencil measures could lead to changes in the dependent variable that are due to testing rather than to the impact of the independent variable.

Instrumentation

The threat of instrumentation refers to the fact that the way in which variables are measured may change in systematic ways during the course of an experiment, resulting in the measurement of observations being performed differently at the end from the way they were in the beginning. This has to do with the issue of reliability discussed in Chapter 5. To be useful in experiments, a measuring device must be reliable, or measure the same way at both pretest and posttest, if the thing being measured hasn't changed. In observing and recording verbal behavior, for example, observers may become more adept at recording and do so more quickly. This means that they could record *more* behaviors at the end of an experiment than at the beginning. If the observers do learn and become more skillful, then changes in the dependent variable may be due to instrumentation effects rather than to the impact of the independent variable. Or, suppose we were using teachers' judgments of students' aggressiveness as the dependent variable. If some teachers judged a given behavior as aggressive at the posttest but not at the pretest, then the change from pretest to posttest is due to changes on how behaviors are measured rather than to the independent variable.

Statistical Regression

The threat of statistical regression can arise any time subjects are placed in experimental or control groups on the basis of extremely high or low scores on a measure in comparison to the average score for the whole group. When remeasured, those extreme groups will, on the whole, tend to score less extremely. In other words, they will *regress toward the overall group average.* For example, suppose the bottom 10% of scorers on a standard achievement test are singled out to participate in a special remedial course. On completing the course, they are again measured with the achievement test and show improvement. Could you conclude that the remedial course was responsible for the higher scores? You would certainly be hesitant to make this inference because of the effects of testing and maturation already discussed. In addition, however, there could also be a regression effect. Some of those people scoring in the bottom 10% undoubtedly scored low in part for reasons other than their actual level of ability—because they didn't get enough sleep the night before the test, were ill the day of the test, or were upset over a quarrel. (How many times have you scored lower on an exam than you normally do because of factors such as these?) If these persons take the exam a second time, the conditions mentioned will have changed, and they will likely perform better—even without the special remedial course. Recognize that many, possibly most, people at the lowest 10% are performing at their normal level. However, in any such group based on a single test administration, a certain number of people will probably be scoring lower than their normal level. If the whole group repeats the exam—even without any intervening experimental manipulation—they can be expected to perform better and thus the average score of the group will improve.

Selection

Selection is a threat to internal validity when the kinds of individuals selected for one experimental condition differ from those who are selected for other conditions. The threat of selection derives from experimental and control groups being improperly constituted.

Recall the importance of random assignment or matching to equalize these groups. Suppose, for example, that an evaluation of a drug treatment program compared drug users in the program to drug users not in the program. Suppose, further, that the drug program was based on voluntary choice: People were accepted if they asked to join. In this case, any posttest differences between the two groups of drug users could be due to differences in SES or personality between drug users who decide to enter a program and those who do not.

Experimental Attrition

The attrition threat occurs when there is a differential dropout of subjects from the experimental and control groups. Especially in experiments that extend over long time periods, some people will fail to complete the experiment. People die, move away, become incapacitated, or simply quit. If subjects in the experimental group are asked to do something that takes time or effort, such as periodic interviews, and those in the control group are not asked to do this, then the attrition rate in the experimental group may well be higher. If there is a notable difference in attrition rates between the experimental and control groups, the groups may not be equivalent at the end of the experiment even though they were at the beginning.

With this lengthy list of threats to the internal validity of experiments, can one ever have confidence that changes in the dependent variable are due to the impact of the independent variable and not to some extraneous variability? In fact, such confidence can be established—and threats to internal validity controlled—through the use of a good experimental design, a topic that we will address shortly.

When to Choose an Experimental Design

Now that you have some idea of what the logic of experimentation is, you can appreciate that one significant indication that an experimental design would be a good choice for a research project is when the research problem focuses on demonstrating a causal link between an independent and dependent variable. This is one of the major strengths of experimental designs. This also means that experiments are appropriate to explanatory research, as discussed in Chapter 1, or evaluation research that has an explanatory component to it. Evaluation research in the form of program evaluations, for example, are often designed as field experiments.

In addition, to be appropriate for an experimental design, a research problem has to be such that the researcher has some control over the occurrence and measurement of the independent and dependent variables. These variables need to be things that occur in the present; we generally cannot manipulate the value of variables that occurred in the past. The variables also need to be precise, well developed, and quantitative. In other words, experiments can be conducted only in situations where the researchers have a fair amount of control: control to place people into experimental and control groups, control to vary the value of the independent variable (or have situations where it varies naturally), and control to measure the dependent variable both before and after treatment. There are many situations in which it is not possible or ethical or practical for the researcher to exercise this kind of control; in such situations, another research design must be chosen.

Researchers who adopt the positivist paradigm and search for nomothetic explanations often view the experiment as the preferred or ideal type of research design because it enables us to isolate the effects of one factor (the independent variable) and affords us the most confidence in making inferences about causality. For many positivists, then, an experiment is the research design of choice, unless factors dictate against it. Researchers who prefer this paradigm will try to develop research problems such that they can be studied with an experiment. For interpretivists and those seeking idiographic explanations, on the other hand, experiments do not have this special status and in fact may be seen as too artificial and limited to provide full explanations of phenomena.

EXPERIMENTAL DESIGNS

There are three categories of experimental designs. **Preexperimental designs** lack the random assignment to conditions and/or the control groups that are such a central part of good experimental designs. Although they are still sometimes useful, they illustrate some inherent weaknesses in terms of establishing internal validity. The better designs are called *true experimental designs* and *quasi-experimental designs*. True experimental designs are more complex and use randomization, control groups, and other techniques to control the threats to internal validity. **Quasi-experimental designs** are special designs that use procedures other than randomization to create experimental and control groups.

Preexperimental Designs

On the surface, Design P1 in Table 8.2 might appear to be an adequate design. The subjects are pretested, exposed to the experimental condition, and then posttested. It would seem that any differences between the pretest measures and posttest measures would be due to the

TABLE 8.2   Pre-Experimental Designs

P1 The One-Group Pretest-Posttest Design:

	pretest	treatment	posttest
Experimental Group:	O	X	O

P2 The Static Group Comparison:

	pretest	treatment	posttest
Experimental Group:		X	O
Control Group:			O

experimental stimulus. However, there are serious weaknesses in this design as it stands. Design P1 is subject to practically all the threats to internal validity. Without a control group, it is impossible to be sure if an observed change in a posttest is due to the independent variable (treatment) or to some historical event or maturation effect. If paper-and-pencil measures are used, a shift of scores from pretest to posttest could occur owing to testing effects. Regardless of the measuring process used, instrumentation changes could produce variation in the pretest and posttest scores. If the subjects were selected because they possessed some extreme characteristic, differences between pretest and posttest scores could be due to regression toward the mean. Even attrition could be a problem since a selective dropout from the experimental group between pretest and posttest could produce changes in the dependent variable that are not due to the treatment. In all these cases, variation on the dependent variable produced by one or more of the validity threats could easily be mistaken for variation due to the independent variable.

The other preexperimental design—the Static Group Comparison—involves comparing one group that experiences the experimental stimulus to another group that does not. In considering this design, it is important to recognize that the comparison group that appears to be a control group is not, in the true sense, a control group. The major validity threat to this design is selection. Note that no random assignment is indicated that would make the comparison groups comparable. In Design P2, the group compared to the experimental group is normally an intact group picked up only for the purpose of comparison. There is no assurance of comparability between it and the experimental group. For example, the evaluations of the D.A.R.E. program discussed earlier mostly used this sort of design. School districts implement drug education programs for a variety of educational and community reasons, and researchers are not normally free to randomly assign either schools or students to drug education programs and control groups. So, the experiments usually involve comparing schools that have implemented the program to schools that have not (the control group). All of these studies made efforts to compare the characteristics of the two groups (in terms of gender, race, social class, and so on) to see if there were any differences and didn't find them. This is done routinely when using this preexperimental design, but there is still the possibility that the two groups are not equivalent and that this is what accounts for any posttest differences found.

Despite their weaknesses, preexperimental designs are used when a lack of resources or other obstacles do not permit the development of true experimental designs. It should be evident, however, that conclusions based on such designs are to be regarded with the utmost caution and the results viewed as suggestive at best. These designs should be avoided if at all possible. If they are used, efforts should be made to assess the comparability of the two groups and to test the validity of those findings further by using one of the true experimental designs.

True Experimental Designs
The Classic Experimental Design

Diagrams of the true experimental research designs that are discussed are presented in Table 8.3. Probably the most common true experimental design is the Pretest-Posttest Control Group Design with random assignment, which is identical to the design on page 224 except for randomization. This design is used so often that it is frequently referred to by its popular name: the *classic experimental design*. This design utilizes a true control group, including random assignment to equalize the comparison groups, which eliminates all the threats to internal validity except certain patterns of attrition. Because of this, we can have considerable confidence that any differences between experimental and control groups on the dependent variable are due to the effect of the inde-

TABLE 8.3   True Experimental Designs

T1 The Pretest-Posttest Control Group Design with Randomization—the "classic" experimental design:

		pretest	treatment	posttest
Experimental Group:	R	O	X	O
Control Group:	R	O		O

T2 The Solomon Four-Group Design:

		pretest	treatment	posttest
Experimental Group 1:	R	O	X	O
Control Group 1:	R	O		O
Experimental Group 2:	R		X	O
Control Group 2:	R			O

T3 The Posttest-Only Control Group Design:

		pretest	treatment	posttest
Experimental Group:	R		X	O
Control Group:	R			O

T4 The Multiple Experimental Group with One Control Group Design:

		pretest	treatment	posttest
Experimental Group 1:	R	O	X_1	O
Experimental Group 2:	R	O	X_2	O
Experimental Group 3:	R	O	X_3	O
Control Group:	R	O		O

T5 The Factorial Design:

		pretest	treatment	posttest
Experimental Group 1:	R	O	$X_1 Y_1$	O
Experimental Group 2:	R	O	$X_1 Y_2$	O
Experimental Group 3:	R	O	$X_2 Y_1$	O
Control Group:	R	O	$X_2 Y_2$	O

pendent variable. In all true experimental designs, the proper test of hypotheses is the comparison of posttests between experimental and control groups (see Figure 8.1).

Let's take a closer look at how the classic design avoids the various threats. History is removed as a rival explanation of between-group differences on the posttest because both groups experience the same events except for the experimental stimulus. Because the same amount of time passes for both groups, maturation effects can be assumed to be equal and do not therefore account for posttest differences. Similarly, as both groups are pretested, any testing influences on the posttest should be the same for the two groups. Instrumentation effects are also readily controlled with this design because any unreliability in the measuring process that could cause a shift in scores from pretest to posttest should be the same for both comparison groups.

The classic experimental design deals with the regression threat through random assignment of subjects. This ensures that any extreme characteristics will be distributed randomly among experimental and control groups, and whatever regression does take place can be assumed to be the same for both groups. Regression toward the mean should not therefore account for any between-group differences on the posttest. Randomization also controls the validity threat of selection because randomization enables us to assume that the comparison groups are equivalent. Further, the pretest results can be used as a check on precisely how similar the two groups actually are. Because the two groups are very similar, attrition rates would be expected to be about the same for each group. In a lengthy experiment with a large sample, we would fully expect about the same number of subjects in each group to move away, die, or become incapacitated during the experiment. These can more or less be assumed to be random events. Attrition due to these reasons is therefore unlikely to create a validity problem.

Attrition due to voluntary quitting is another matter. It is quite possible that the experimental stimulus may affect the rate of attrition. That is, subjects might find something about the experimental condition either more or less likable than the subjects find the control condition, so the dropout rate could differ. If this occurs, it raises the possibility that the groups are no longer equivalent at the time of the posttest. It might be possible to put the people in the control group through something that is equally as difficult or unpleasant as the experimental group goes through but that is still different from the treatment. Recall the "alternate treatment" mentioned earlier where the control group watched a video clip showing violence against men while the experimental group saw one with violence against women. Efforts such as these try to reduce attrition by making the two groups as much alike as possible. If efforts such as these are not possible, there is no especially effective way of dealing with the attrition problem. About all one can do is watch for its occurrence and interpret the results cautiously if attrition bias appears to be a problem.

The Solomon Design

A second true experimental design—the Solomon Four-Group Design—is more sophisticated than Design T1 in that four different comparison groups are used. As should be evident by comparing Design T1 and Design T2 in Table 8.3, the first two groups of the Solomon design constitute Design T1, indicating that Design T2 is capable of controlling the same threats to internal validity as does Design T1. The major advantage of the Solomon design is that it can tell us whether changes in the dependent variable are due to some *interaction effect* between the pretest and the exposure to the experimental stimulus. Experimental Group 2 is exposed to the experimental stimulus but without being pretested. If the posttest of Experimental Group 1 differs from the posttest of Experimental Group 2, it may be due to an interaction effect of receiving both the pretest and the experimental stimulus—something we would not discover using the classic design.

One study that used a Solomon design was an investigation of whether women felt disempowered by watching media violence against women (Reid & Finchilescu, 1995). The concern was that the paper-and-pencil measure of *disempowerment* (the dependent variable), when administered in a pretest, might sensitize the subjects so that they reacted more strongly, or at least differently, to the treatment, which in this case was a film clip depicting violence against women. If we find higher levels of *disempowerment* in Experimental Group 1 than in Control Group 1, it might be due to the independent variable. But it could also be that filling out a pretest questionnaire on disempowerment sensitized people in the first group to these issues and they reacted *more strongly* to the experimental stimulus than they would have without such pretesting. If this is so, then Experimental Group 2 should show less change than Experimental Group 1. If the independent variable has an effect separate from its interaction with the pretest, then Experimental Group 2 should show more change than Control Group 1. If Control Group 1 and Experimental Group 2 show no change but Experimental Group 1 does show a change, then the change is produced only by the interaction of pretesting and treatment. (By the way, this study concluded that the pretest did not affect people's responses to the treatment and that exposure to media violence against women did lead women to experience heightened levels of *disempowerment*.)

It should be apparent that the Solomon design enables us to make a more complex assessment of the causes of changes in the dependent variable. In addition, the combined effects of maturation and history can be controlled (as with Design T1) as well as measured. By comparing the posttest of Control Group 2 to the pretests of Experimental Group 1 and Control Group 1, these effects can be assessed. However, our concern with history and maturation effects is usually only in *controlling* their effects, not *measuring* them.

Despite the superiority of the Solomon design, it is often bypassed for Design T1 because the Solomon design requires twice as many groups. This increases considerably the time and cost of conducting the experiment. Not surprisingly, many researchers decide that the advantages are not worth the added cost and complexity. If a researcher desires the strongest design, however, Design T2 is the one to choose.

The Posttest Only Control Group Design

There are times when pretesting is either impractical or unnecessary. For example, pretesting might sensitize subjects to the independent variable. In other cases, there may be nothing to measure with a pretest. For example, suppose we had a dependent variable that involved the number of months after release from prison until a first crime is committed. Treatment occurs at the point of release from prison, and there is nothing to measure with a pretest. In these cases, it is still possible to use a true experimental design, the Posttest Only Control Group Design, which consists of the two groups in the Solomon design that are not pretested. Despite the absence of pretests, Design T3 is an adequate true experimental design that controls validity threats as well as do the designs with pretests. It uses random assignment to conditions, which distinguishes it from the preexperimental Design P2. The only potential validity question raised in conjunction with this design is selection. The absence of pretests means that random assignment is the only assurance that the comparison groups are equivalent. Campbell and Stanley (1963), however, argue that pretests are not essential and that randomization reliably produces equivalent groups. They further argue that the lack of popularity of Design T3 stems more from the tradition of pretesting in experimentation than from any major contribution to validity produced by its use.

The Multiple Experimental Group Design

The preceding experimental designs are adequate for testing hypotheses when the independent variable is either present (experimental group) or absent (control group). Yet many hypotheses involve independent variables that vary in terms of *degree* or *amount* of something that is present. For example, *how long* does a drug education program have to last to change people's behavior? Or, *how severe* does a prison sentence have to be to reduce repeat offending? In cases such as this, Design T4, the Multiple Experimental Group with One Control Group Design is used. It is an extension of Design T1. The symbols X_1, X_2, and X_3 refer to different categories or amounts of a single independent variable or treatment. Comparing the posttests of the experimental groups enables us to determine the impact of the differing amounts of the independent variable.

The Factorial Design

The experimental designs considered thus far can assess the impact on the dependent variable of only a single independent variable at a time. We know, however, that many things can influence the dependent variables simultaneously. We also know that variables can *interact* with one another, and the combined effects of two variables may be quite different from the effects of each variable operating separately. For example, one experimental study investigated the impact of both social class and race on people's stereotypes of women (Landrine, 1985). People in the study were asked to describe society's stereotype of four different women: a middle-class African American, a middle-class Anglo, a lower-class African American, and a lower-class Anglo. Each of the four race-class combinations constituted a different experimental condition. They found that social class and race did affect stereotyping: Lower-class people received more negative stereotyping than did middle-class people, and African Americans were viewed less favorably than Anglos. They did not find an interaction effect, however. In other words, various class-race combinations did not produce more changes in stereotyping than the effects of class and race separately.

The experimental design used in this study is called a Factorial Design (see Table 8.3). Design T5 illustrates the simplest factorial design in which two independent variables (X and Y, or race and social class) each have only two values (X_1 or X_2 and Y_1 or Y_2). In factorial designs, each possible combination of the two independent variables constitutes one of the experimental conditions. Of course, one cannot be without some things, such as race or social class. So, technically, there may not be a control group, as in the example above, when there can be no "absence" of the factors involved. However, if variables involve the presence or absence of something, then a condition looking like a control group is found in the factorial design. In the Applied Scenario in Chapter 12, for example, I describe a field experiment with two variables: released inmates either receive financial assistance or they don't, and they either receive job placement services or they don't. If you look at Figure 12.2 on page 372 which outlines this experiment, you can see that one group receives neither financial assistance nor job placement services and looks like a normal control group in the other experimental designs.

Factorial designs can be expanded beyond the simple example used here by including more than two variables or by using variables that have more than two values. However, as the complexity of factorial designs increases, the number of groups required rapidly increases and can become expensive and time consuming. For example, with three independent variables, each with three values, we would need a $3 \times 3 \times 3$ factorial design, or 27 different groups. However, field experiments designed to assess the impact of social programs sometimes involve such complex designs.

The major types of true experimental designs have been covered in this section. However, circumstances often require researchers to use a variant of one of these designs. The Applied Scenario illustrates such a variation that, nonetheless, retains randomization.

Quasi-Experimental Designs

Sometimes it is impossible—for practical or other reasons—to meet the conditions necessary for the development of true experimental designs. The most common problems in this regard are an inability randomly to assign people to conditions and the difficulty of creating a true control group with which to compare the experimental groups. Although this can be a problem in experiments in any context, it is especially acute in field experiments. Rather than rule out experimentation in such settings, however, one may be able to use a *quasi-experimental design*, which are special designs that can control many threats to internal validity, with procedures other than random assignment to experimental and control groups. These designs allow the researcher to approach the level of control of true experimental designs in situations where the requirements of the latter cannot be met. Quasi-experimental designs afford considerable control but fall short of the true experimental designs and should only therefore be used when conditions do not allow the use of a true experimental design (Achen, 1986; Cook & Campbell, 1979).

The Time Series Design

One of the simplest and most useful of the quasi-experimental designs is the Time Series Design (see Table 8.4). This design involves a series of repeated measures, followed by the introduction of the experimental condition, followed by another series of measures. The number of pretest and posttest measurements can vary, but it would be unwise to use fewer than three of each. Time series designs can be analyzed by graphing the repeated measures and inspecting the pattern produced. We can conclude whether the independent variable produced an effect by observing when, over the whole series of observations, changes in the dependent variable occur. Figure 8.2 illustrates some possible outcomes of a time series experiment. Of major interest is what occurs between O_4 and O_5 because these are the mea-

TABLE 8.4 Quasi-Experimental Designs

Q1 The Time Series Design:

	pretests	treatment	posttests
Experimental Group:	$O_1\ O_2\ O_3\ O_4$	X	$O_5\ O_6\ O_7\ O_8$

Q2 The Different Group Time Series Designs:

	pretests	treatment	posttests
R	O_1		
R	O_2		
R	O_3		
R	O_4		
R		X	
R			O_5
R			O_6
R			O_7
R			O_8

Q3 The Multiple Time Series Design:

	pretests	treatment	posttests
Experimental Group:	$O_1\ O_2\ O_3\ O_4$	X	$O_5\ O_6\ O_7\ O_8$
Control Group:	$O_1\ O_2\ O_3\ O_4$		$O_5\ O_6\ O_7\ O_8$

sures that immediately precede and follow the experimental stimulus. The other measures are important, however, because they provide a basis for assessing the change that occurs between O_4 and O_5. As Figure 8.2 shows, some outcomes suggest that the stimulus has had an effect whereas others do not. In cases A, B, and C, stimulus X appears to produce an effect. In each case, the differences between measures O_4 and O_5 are greater than the differences between any other adjacent measures. Cases E and F illustrate outcomes where we can infer that changes result from something other than the stimulus because the differences between measures O_4 and O_5 do not substantially differ from those between some other adjacent measures. The results of a time series design are not always clear-cut, as illustrated

by case D. The sharp change between measures O_5 and O_6 could be a delayed effect of X, or it could be due to something else. More careful analysis would be required to come to a firm conclusion in this case.

In field studies involving a time series analysis, the experimental stimulus is often not something that the researcher manipulates; rather, it is something that occurs independently of the research. Thus any natural event presumed to cause changes in people's behavior could serve as an experimental stimulus. In a study of the mass media and violence, for example, the experimental stimulus was the heavyweight championship prizefights that occurred between 1973 and 1978 (Phillips, 1983). The dependent variable was the daily counts of all homicides in the United States

In the last 20 years, applied social scientists have conducted some ground-breaking field experiments to help criminal justice authorities find the most effective way of dealing with domestic violence. When police respond to a domestic violence call, they have three alternatives: mediate between the couple, separate the couple by sending the offender out of the home, or arrest the offender. In the past, police officers used their own experience and intuition in deciding which would be most effective. A series of creatively designed field experiments offers some scientific insight into which strategy works best.

The first in this series of studies was an investigation by criminologist Lawrence Sherman and sociologist Richard Berk (1984). These researchers randomly assigned 314 misdemeanor domestic assault cases in Minneapolis to one of the three types of police intervention (mediation, separation, and arrest), which was the independent variable. The dependent variable was whether or not the perpetrator was involved in another domestic violence incident in the six months following police intervention. For ethical and practical reasons, the study was restricted to misdemeanor assaults; that is, cases lacking severe injury or a life-threatening situation.

Random assignment to experimental conditions was achieved by requiring each police officer to use a pad consisting of color-coded forms corresponding to the three different treatments: mediation, separation, and arrest. The forms were randomly ordered in the pad, and the officers were instructed to select their intervention according to whatever color form came up when they intervened in a case that met the study guidelines. In this study, then, the officers doing the intervening were responsible for making sure that the randomization was done properly. In order for the study to work as planned, officers had to follow the instructions faithfully. In addition, the forms were sequentially numbered so that the researchers could monitor how well the officers followed the assigned interventions.

Sherman and Berk (1984) concluded that arrest was the most effective of the three options for dealing with a spouse abuser: Those arrested were significantly less likely to be involved in a repeat episode of domestic violence in the six months that followed. Encouraged by these findings, many communities developed more aggressive arrest policies toward domestic assault. However, researchers and policymakers cautioned against placing too much faith in the results of this one study because of some design flaws (Sherman, 1992). One criticism was that the study had a small sample size. In addition, a disproportionate number of cases had been submitted by a few officers. This may be because these officers patrolled a more violent section of town and, because of experience, were particularly effective at making arrest an effective deterrent. The other officers, who did not turn in many cases, may have been more effective at mediation or separation. If this was the case, then the study was really about variations in police officer skills rather than variations in the effectiveness of different treatments. This points to the importance of assessing internal validity: Is it variation in the independent variable that produces changes in the dependent variable, or is something else producing these changes?

Yet another criticism of Sherman and Berk's study is that there were inadequate controls over which treatments were actually delivered. There was also the possibility that testing effects from the multiple follow-up interviews distorted the impact of the treatment. Finally, if the sample was biased for any of the foregoing reasons, the results may not be generalizable to other settings.

A series of follow-up studies in various cities were designed differently from the Minneapolis project in several key respects (Dunford, Huizinga, & Elliott, 1989; Hirschel, Hutchison, & Dean, 1992). First, in some studies, in addition to evaluating the three treatments of mediation, separation, or arrest, the research included a unique facet. If the assailant was not present when officers arrived, but the case otherwise met study criteria, the suspect was assigned randomly to either receive or not receive an arrest warrant. Other projects addressed the criticism of the officers not properly delivering the treatments by having random assignment to treatment performed by a police dispatcher rather than the officers themselves. In this way, they assured that random assignment occurred.

For results to be valid, it was crucial that officers actually deliver the assigned treatment. This problem was controlled by three forms of monitoring. First, officers reported dispositions on a Domestic Violence Report form and forwarded it to the project. Second, victims were asked to describe the treatment that was delivered. Finally, official records of police, prosecuting attorneys, and courts were compared to the other forms of monitoring data. Analysis of these various case-monitoring systems indicated that 92% of the treatments were delivered as assigned.

The original Minneapolis study, based on 314 cases, concluded that arrest significantly reduced future assaults. Based on a total sample of thousands in six cities in the various replications, the general conclusion now is that the issue is far from clear-cut. Sherman (1992, p. 247) offers these conclusions based on these studies:

- arrest increases domestic violence among people who have nothing to lose, especially the unemployed;
- arrest deters domestic violence in cities with higher proportions of Anglo and Hispanic suspects;
- arrest deters domestic violence in the short run, but escalates violence later in cities with higher proportions of unemployed African American suspects; and
- offenders who flee before police arrive are substantially deterred by warrants for their arrest, at least in Omaha.

This research shows that field experiments can be designed to help us understand serious social problems and devise solutions to them. It also makes clear, however, that we need to rest policy decisions on repeated replications under varying conditions to ensure that we fully understand the complexity of things.

1. Look for other societal interventions that might be assessed with field experiments.
2. Describe what those field experiments might look like, addressing some of the issues that were dealt with in the spouse abuse experiments (how to randomize, and so on).

FIGURE 8.2   Possible Outcomes in a Time
Series Quasi-Experimental Design

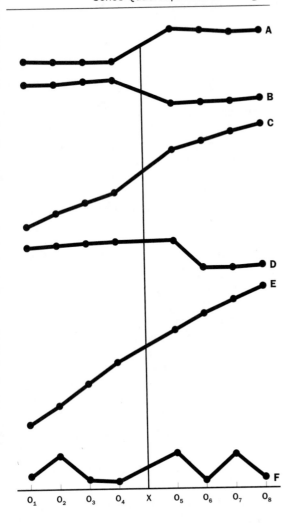

internal validity. With the exception of history, the other threats are controlled by the presence of the series of premeasures. Maturation, testing, instrumentation, regression, and attrition produce gradual changes that would be operating between all measures. As such, they could not account for any sharp change occurring between O_4 and O_5. Because this design does not use a control group, the issue of selection is not a factor. Thus history is the only potential threat to internal validity. It is always possible that some extraneous event could intervene between measures O_4 and O_5 and produce a change that could be confused with an effect of X. In many cases, there may not be an event that could plausibly produce the noted effect, and we could be quite sure it was due to X. Nevertheless, the inability of the time series design to control the threat of history is considered a weakness.

Different Groups Time Series Design

The time series design requires that we be able to measure the same group repeatedly over an extended period. Because this might require considerable cooperation from the subjects, there may be occasions when the time series design cannot be used. A design that avoids this problem and yet maintains the other characteristics of the time series is the Different Groups Time Series Design (see Table 8.4). Rather than making repeated measures on one group, this design substitutes several randomly selected groups, each of which is measured only once but at different times. Design Q2 produces the same type of data as the regular time series and can be analyzed using the same graphing procedure. In terms of internal validity, Design Q2 controls the same threats as the time series. Random sampling is relied on to equate the several comparison groups. Design Q2 also has the same weakness as the time series design: Historical events can intervene between O_4 and O_5, possibly confusing the results.

provided by the National Center for Health Statistics, so an assessment could be made of the daily homicide rates after championship prizefights. The study found a sharp increase in homicides after such fights, peaking on the third day after each fight. One possible explanation is that viewing heavyweight prizefights stimulates aggressive behavior in some Americans and turns fatal in a few cases.

The time series design fares quite well when evaluated on its ability to control threats to

Because of the many random samples required by this design, it is limited to situations where the cost of those samples is inexpensive. For example, if it is possible to draw the samples and collect data by telephone, this design is ideal. This is essentially the approach used by commercial pollsters when they repeatedly measure public opinion by conducting daily or weekly telephone surveys of different random samples. Although the pollsters do not conduct experiments, their repeated measures allow the assessment of the impact of events on public opinion. The event in question becomes the X in the design, and levels of opinion are compared before and after it occurred. For example, the 1991 Persian Gulf War against Iraq caused the popularity of President Bush to soar. His popularity later declined substantially, but because of the many before and after measurements, there is little doubt that it was the war that temporarily increased the president's popularity.

The Multiple Time Series Design

As noted, both of the preceding designs suffer from the validity threat of history. By the addition of a control group to the time series design, this last remaining threat is controlled. Design Q3 illustrates this and is known as the Multiple Time Series Design (see Table 8.4). Design Q3 controls all the threats to internal validity, including history. Events other than X should affect both groups equally, so history is not a rival explanation for between-group differences after the experimental group has been exposed to X. A design that uses a control group without random assignment might be thought suspect on the threat of selection. This is less of a problem than with preexperimental designs, however, because the series of before measures afford ample opportunity to see how similar the comparison groups are.

Data from Design Q3 are plotted and analyzed in the same way as with the two preceding designs. This time, however, there are two lines on the graph: one for the experimental group, one for the control group. The patterns

FIGURE 8.3   Example of Multiple Time Series Data: Percent of Alcohol-Related Accidents for "Ban" and "No Ban" Cities (California, 1981–1987)

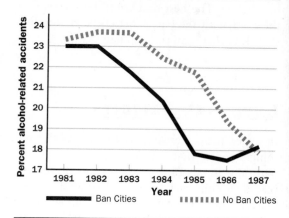

Source   Carol W. Kohfeld and Leslie A. Leip, "Bans on Concurrent Sale of Beer and Gas: A California Case Study." In *Sociological Practice Review,* Vol. 2 (April 1991), p. 109. Reprinted with permission of the American Sociological Association and the authors.

of the two groups are compared to see what effect X produced. As a demonstration of the utility of quasi-experimental designs, Carol Kohfeld and Leslie Leip (1991) looked at the impact of a policy in California that restricted the sale of beer and wine in stores that also sell gasoline. The idea was that people would be less likely to drink and drive if they could not purchase gasoline and alcohol at the same place. They measured the effectiveness of the policy in terms of the percentage of all automobile accidents that were alcohol related. They collected data on 77 cities from 1981 to 1987, 37 of which had instituted the ban on concurrent sales of alcohol and gasoline by 1986. As Figure 8.3 illustrates, the percentage of alcohol-related accidents did decline over the period, especially between 1983 and 1986, from a high of about 24% to a low of about 18%. This would suggest that the ban had an effect. But was the decline due to the ban?

Comparing cities with the ban to a control group of cities without such a ban over the same period, it appears that the ban had little effect since the "no ban" cities had a comparable decline. The best that could be said is that the cities with the ban experienced the decline slightly earlier than the "no ban" cities.

EXTERNAL VALIDITY

In experiments, researchers often make something happen that would not have occurred naturally. In laboratory experiments, for example, they construct a social setting that they believe simulates what occurs in the everyday world. In addition to being artificial, this setting is typically more simple than what occurs naturally because researchers try to limit the number of social or psychological forces operating in order to observe more clearly the influence of the independent variables on the dependent variable. Even many field experiments, which are considerably more natural than those in the laboratory, involve manipulation of an independent variable by the researcher—a change in the scene that would not have occurred without the researcher's intervention.

This "unnaturalness" raises a validity problem that differs from internal validity discussed earlier in the chapter. **External validity** concerns the extent to which causal inferences made in an experiment can be generalized to other times, settings, or groups of people (Cook & Campbell, 1979). The basic issue is whether an experiment is so simple, contrived, or in some other way different from the everyday world that what we learn from the experiment does not apply to the natural world. This is a part of the problem of generalizing findings from samples to populations discussed in Chapter 7. With experiments, however, some special problems arise that are not found with sampling in other types of research methods.

Indeed, resolving the problem of external validity can be more difficult and less straightforward than is the case with internal validity. Campbell and Stanley (1963, p. 17) state the problem succinctly:

> Whereas the problems of internal validity are solvable within the limits of the logic of probability statistics, the problems of external validity are not logically solvable in any neat, conclusive way. Generalization always turns out to involve extrapolation into a realm not represented in one's sample.

Four major threats to external validity and the ways in which these threats can be reduced are reviewed next.

Reactive Effects of Testing

In any design using pretesting, the possibility exists that experiencing the pretest can alter subjects' reactions to the independent variable. For example, items on a paper-and-pencil pretest measure might make subjects more or less responsive to the independent variable than they would have been without exposure to those items. The problem this raises for generalizability should not be difficult to understand. The populations to which we wish to generalize findings are composed of people who have not been pretested. Therefore, if the subjects are affected by the pretest, findings may not accurately generalize to the unpretested population.

Whether reactive effects of testing are a threat to generalizability depends on the variables involved in the experiment and the nature of the measuring process used. Paper-and-pencil measures may be quite reactive; unobtrusive observation is likely not to be reactive. When one is planning an experiment, it is therefore important to consider whether reactive effects of testing are likely to be a problem. If they are, then it would be desirable to choose a research design that does not call for pretesting, such as Design T3, or one that includes groups that are not pretested, such as the Solomon design. With the latter design, it is possible to measure the extent of any pretesting effects.

Unrepresentative Samples

As emphasized in Chapter 7, the representativeness of the people studied in any form of research is crucial to the issue of generalizability. Unfortunately, it is sometimes difficult to experiment on truly representative samples of any known population. For example, experimental subjects might be volunteers, such as college students, who are enticed in some way to participate in the research of an academic researcher. The implications of this for generalizing experimental findings are quite serious. For example, one review of a large number of studies concluded that persons who volunteer for experiments differ systematically from the general population in the following ways: The volunteers are better educated, come from a higher social class, have higher intelligence, have a greater need for social approval, and are more sociable (Rosenthal & Rosnow, 1975). Clearly these differences could be related to any number of the variables likely to be used in social science research. Generalizing findings from such volunteer subjects could be quite hazardous.

Although the use of coerced subjects may reduce the differences between experimental subjects and the general population from which they are drawn, it has problems of its own as far as generalization is concerned. Coerced subjects are likely to have little interest in the experiment and may resent the coercion used (Cox & Sipprelle, 1971). This effect has been found even where the nature of the coercion was quite mild, such as gaining extra credit in a college class for agreeing to participate in an experiment. It is reasonable to assume that these effects would be amplified where the level of coercion was greater, such as court-ordered participation in a treatment program under threat of incarceration. In fact, a study of a drug treatment program involving court-ordered clients found this anticipated pattern (Peyrot, 1985). The coerced clients were resentful, uncooperative, and unwilling to commit to the objectives of the program.

The threat to external validity created by unrepresentative samples is of great importance to applied research that focuses on evaluating the effectiveness of things like drug treatment programs, which often involve volunteers or coerced subjects. Volunteers may make programs look good because they are interested in the treatment and motivated to change in the direction promoted by the treatment. It should come as no surprise that treatment programs shown by research to work on volunteers may fail when applied to nonvolunteer groups. Alternatively, because of their lack of interest or resentment, coerced subjects may make programs appear ineffective when they might be effective on persons who are not coerced. The Eye on Diversity box explores issues of representativeness in regard to gender in experimental settings.

Reactive Settings

In addition to the reactivity produced by testing, the experimental setting itself may lead people to behave in ways that differ from their behavior in the everyday world. One reason for this is that experimental settings contain what researchers call **demand characteristics**: subtle, unprogrammed cues that communicate to subjects something about how they should behave. For example, people in experiments tend to be highly cooperative and responsive to the experimenter. In fact, the psychologist Martin Orne (1962) deliberately tried to create boring and repetitive tasks for subjects in experiments so that they would rebel and refuse to do them. One task was to perform a series of additions of random numbers. Each page required 200 additions, and each person was given 2,000 pages. After giving instructions, the experimenter told them to continue working until he returned; hours later the subjects were still working, and it was Orne who

EYE ON DIVERSITY Inclusion of Diverse Groups in Experimental Research

Feminist researchers have criticized conventional social science research for a lack of concern about gender and other forms of diversity in research. In fact, women and minorities are sometimes underrepresented in laboratory or field experiments. This is a problem that can affect external validity as well as internal validity. Researchers often choose as subjects for experiments people who are easily accessible to them. In basic research in psychology and sociology, for example, students in introductory psychology or sociology classes are often selected as research subjects. The sex ratio of college students is fairly even nationwide: 45% male, 55% female. However, African Americans constitute only about 10% of all college students while accounting for 13% of our total population (U.S. Bureau of the Census, 1998). This means that experiments using a representative sample of college students will tend to underrepresent African Americans, Hispanics, and some other minorities. And this underrepresentation is more severe at some colleges than at others: Some major research universities even today have only 1%–2% African American enrollment. So whenever experimental subjects are selected from a setting, care must be taken to ensure representation of various racial, ethnic, and gender groups in a sample if the generalizations made from the data will include these groups. Otherwise, mention should be made that no special procedures were used in the sampling to enhance minority involvement.

A related problem is what one sociologist calls *gender insensitivity*: "ignoring gender as an important social variable" (Eichler, 1988, p. 66). In experiments, this can occur when no mention is made of the sex ratio of the subjects in the experiment or data are not analyzed separately for each sex. Eichler reports one issue of a psychology journal in which the only article to mention the sex of the subjects was one that used rhesus monkeys; none of those using human subjects did so. This was true despite the fact that practically all those articles focused on variables (such as perception, verbal ability, and learning) on which people's gender might well have had an influence.

A final problem relating to minority participation in experiments is the failure to consider the gender or minority status of all participants in the experiment. In addition to the researcher and the subject, this might include interviewers, confederates of the experimenter, and any others who interact with the subject during data collection. We know, for example, that same-sex interaction differs from cross-sex interaction in many settings. Thus, a male interviewer will comport himself differently when asking questions of a female subject than he would when asking the same questions of a male subject. In the former case, he might be more friendly, attentive, or engaging without consciously realizing it, and this could influence the subject's response to questions. (These issues are discussed in more detail in the Eye on Diversity box in Chapter 9.)

These threats to internal and external validity can be reduced by reporting the sex or minority status of the persons involved in experiments and analyzing the data with sex or minority status as an experimental or control variable. There are, of course, reasons to have a homogeneous sample in an experiment: It can reduce extraneous variation when trying to establish a causal relationship between independent and dependent variables. However, if the results are to be generalized to all racial and ethnic groups and both sexes, then experiments need to be replicated using subjects from these groups.

gave up and ended the experiment! On the basis of this and other research, it has become evident that experimental settings exercise a powerful influence on subjects that can damage the generalizability of experimental findings. This has been labeled the problem of the *good subject*; that is, the subject in an experiment who will do whatever the investigator asks, even to the point of confirming the experimenter's hypotheses if they are communicated to the subject (Wuebben, Straits, & Schulman, 1974).

Subjects' reactions are not the only way in which experimental settings can be reactive. Experimenters themselves can introduce distortion into the results that reduce generalizability. Experimenters, of course, usually have expectations concerning the results of the experiment, wanting it to come out one way or another. These **experimenter's expectations** can be communicated to subjects in such a subtle fashion that neither experimenter nor subjects are aware the communication has taken place (Rosenthal, 1967). A classic illustration of how subtle this can be is recounted by Graham (1977) and involves a horse, not humans. The horse was called Clever Hans because of his seeming ability to solve fairly complex arithmetic problems. Hans and his trainer toured Europe in the early 1900s, amazing audiences and becoming quite famous. Hans would stand on stage pounding out the answers to problems with his hoofs. Amazingly, Hans was hardly ever wrong. Hans would perform his feats as well even when his trainer was not present. Could Hans really do arithmetic? After very careful observation, it was discovered that Hans was picking up the subtle cues given off by those who asked him questions. As Hans approached the correct number of hoofbeats, questioners would move or shift just enough to cue the horse that it was time to stop. People in experiments do the same thing. Subjects can be influenced by subtle cues, unconsciously given off by an experimenter, to behave in ways the researcher would like them to.

Reactive experimental settings can threaten external validity because changes in the dependent variable might be due to demand characteristics or experimenter expectancies rather than to the independent variable, and these factors don't exist in the real world to which we wish to generalize. One procedure for reducing the problem of reactivity in experimental settings is to conduct the experiment in a way such that subjects are *blind*—that is, unaware of the experimental hypotheses—so that this knowledge will not influence their behavior. In fact, subjects are commonly given a false rationale for what they are to do in order to reduce the reactions of subjects that might interfere with the generalizability of the findings. As an ethical matter, subjects should be informed of the experiment's true purpose during the postexperimental debriefing session.

The surest way of controlling reactivity due to the experimenter is to run what is called a *double-blind experiment*. In a **double-blind experiment**, neither the subjects nor the experimenter knows which people are in the experimental and which are in the control condition. This makes it impossible for the experimenter to communicate to subjects how they ought to behave because, for any given subject, the experimenter would not know which responses would confirm the experimental hypotheses. Although it is simple in theory, maintaining a double-blind procedure in practice can be difficult. Another layer of personnel must be added to assign subjects to conditions, issue sealed instructions to the experimenter, and keep track of the results. Further, all information relating to these activities must be kept from those actually running the experimental groups. It is easy for this structure to break down so that the double-blind feature is lost. Another way to control many forms of reactivity is to conduct a field experiment where those being observed are not aware they are part of an experiment. Many of the studies on domestic violence described in the Applied Scenario box did this by using police and court records as a source of data. The people who were the focus of the study were not aware of their involvement in the study, so their behavior could not have been a reaction to anything that was part of the study.

Multiple-Treatment Interference

In an experiment in which there is more than one independent variable, it may be the particular combination and ordering of experimental treatments that produce change in the dependent variable. If this same combination and this same ordering do not occur outside the experimental setting, the findings from the experiment cannot be generalized. Suppose, for example, that an experiment calls for subjects to experience four independent variables in succession. Also suppose that the last variable in the series appears to produce an interesting effect. Could the effect of that fourth variable be safely generalized on the basis of the experimental findings? The answer is no. The subjects in the experiment would have first experienced three other independent variables. Experiencing those other variables first might have affected the way they reacted to the fourth and last variable. If people outside the experimental setting do not experience all the variables in sequence, generalization con-cerning the fourth variable is risky. The problem of multiple-treatment interference is similar to reactive effects of testing in that the subjects in the experiment experience something that the people in the population at large do not.

The threat of multiple-treatment interference can be effectively eliminated through the use of complex designs in which the independent variables are experienced by the different experimental groups in every possible sequence. If a given variable produces a consistent effect regardless of the ordering of the variables, then multiple-treatment interference is not a threat to the generalizability of the findings. An alternative is to isolate the variable of interest in a multiple-treatment experiment and

conduct a follow-up experiment with that variable as the only treatment. If it produces an effect similar to that found when it was a part of a series of treatments, multiple-treatment interference can be ruled out.

Enhancing the External Validity of Experiments

Although both internal and external validity are important to experimental research, there is a tension between the two that produces a dilemma: increasing one type of validity often has the effect of reducing the other. So internal validity is enhanced through greater control. Consequently, the researcher seeking internal validity is attracted to the laboratory experiment and to precise testing. Yet we have just seen that the reactive effect of testing is a threat to external validity. As another way of seeking control, the researcher may use a homogeneous sample to avoid the confounding effects of other variables. In the effort to achieve internal validity, one might, for example, use a sample of 18-year-old Anglo males. Although the effects of age, race, and sex are now controlled, the external validity threat of unrepresentative samples is increased. I have suggested that one solution to this dilemma is the use of complex designs to reduce the threat to external validity of multiple-treatment interference. However, complex designs are more difficult to implement in a way that retains the integrity of the research design.

Another solution to the problem of external validity is, if possible, to conduct field experiments rather than laboratory experiments. While issues of external validity do arise in field experiments, field experiments on the whole have greater external validity than do laboratory experiments. The reason for this is that field experiments are conducted in the real world, which means that the results are more clearly generalizable to some settings in the real world—at least to the ones that are similar to the setting in which the field study was conducted.

However, the basic solution to this dilemma is not to seek a solution to both internal validity and external validity in a single study. As Thomas Cook and Donald Campbell (1979, p. 78) put it with regard to experiments: "In the last analysis, external validity . . . is a matter of replication." Confidence in generalizing from experimental findings increases as the same hypothesis is tested and supported in a variety of settings with a variety of designs. Often this takes the form of initial testing in the laboratory under ideal conditions to see if the hypothesis is at all supported. Then the same hypothesis can be tested under the less-than-ideal conditions of the field. Through replication, the dual objectives of internal and external validity can be achieved.

ASSESSMENT OF EXPERIMENTS

The Eye on Ethics box suggests some important ethical issues that need to be addressed in conducting experiments. I will close this chapter with a brief summary of the advantages and disadvantages of experiments.

Advantages
Inference of Causality
The major advantage of experimental research is that it places researchers in the most advantageous position from which to infer causal relationships between variables. Recall from Chapter 4 that causality can never be directly observed. Rather, we infer that one thing caused another by observing changes in the two things under the appropriate conditions. A well-designed and controlled experiment puts us in the strongest position to make that causal inference because it enables us to control the effect of other variables, which raises our confidence that it is the independent variable that is bringing about changes in the dependent variable. Experiments also permit us to establish the time sequence necessary for inferring

causality. We can measure the dependent variable to see if it has changed since the experimental manipulation. With other research methods, such as the survey, it is often not possible to directly observe whether the changes in one variable preceded or followed changes in another variable.

Control
Experimenters are not limited by the variables and events that happen to be naturally occurring in a particular situation. Rather, they can, to an extent, decide which variables will be studied, which values those variables will take, and which combination of variables will be included. In other words, they can create precisely what their hypotheses suggest are important. This is in sharp contrast to other research methods, such as the use of available data (see Chapter 10) or field techniques (see Chapter 11), in which hypotheses need to be reshaped to fit the existing data or observations.

The Study of Change
Many experimental designs are longitudinal, which means that they are conducted over a period of time, and measurements are taken at more than one time. This makes it possible to study changes over time. Many other research methods, such as surveys, tend to be cross-sectional—they are like "snapshots" taken at a given point in time. In cross-sectional studies, we can *ask* people how things have changed over time, but we cannot *directly observe* that change.

Costs
In some cases, experiments—especially those conducted in laboratory settings—can be considerably less expensive than other research methods. Because of the element of control, the sample size can be smaller and this saves money. The costly travel expenses and interviewer salaries necessitated by some surveys are eliminated. Some field experiments, however,

⚖️ **EYE ON ETHICS** **Deception and Informed Consent in Experiments**

Two of the most common ethical issues that arise in experimental research are deception and informed consent. Deception occurs in experiments when the subjects are not told about some key aspects of the experiment or are actually misled about them. Most commonly, subjects are not told about the true purpose of the research or they might actually be told something about it that is not true. In some cases, the subjects might be misled by being given some tasks or a questionnaire that are really unrelated to the experiment. This is more common in laboratory experiments when the researcher creates an artificial setting in which to test hypotheses, but it can also occur in field experiments.

Deception is engaged in to make it harder for the subjects to figure out the true purpose of the experiment. The reason this deception is necessary is that knowing the true purpose or the actual research hypotheses might lead the subjects to behave differently. For example, they might try to help the researcher by behaving in a way that helps confirm the hypotheses, or they might try to scuttle the experiment by behaving in a way to refute the hypotheses. Either way, the subjects' behavior would be a form of reactivity that throws into question the validity of the findings.

Researchers justify these deceptive practices in a number of ways. They argue that the research findings are important and that the research goals could not be achieved without the deception. They also argue that the deception in most cases is rather minor in nature. The subjects might be told, for example, that the study is about perception when it is actually about how friendships develop. Or the subjects might be asked to answer a series of mathematical questions when the researchers are really observing how the subjects interact with someone else in the room with them.

To alleviate some of the ethical concerns, researchers often have the subjects read and sign a consent form that informs them, among other things, that they may be deceived as to the true purpose of the experiment and may be asked to do some things to further that deception. Yet

can be expensive because they may require interviewing—possibly more than once—to assess changes in dependent variables.

Disadvantages
Inflexibility
Experimental designs generally require that the treatment, or independent variable, be well developed and that the same treatment be applied to all cases in the experimental group. If the nature of the independent variable changes as the experiment progresses, this extraneous variation makes it difficult to say what changed the dependent variable. Because of this inflexibility, experimental designs are

not suitable in the early stages of research when independent variables may not be well developed or precisely operationally defined. This requirement for consistently measuring the independent variable is especially true of large-scale longitudinal experiments, which are best suited for clearly designed treatments (Rossi, Freeman, & Lipsey, 1999).

Randomization Requirement
Some organizations or programs may be unwilling to accept random assignment to treatment and control conditions despite the value of this technique for controlling extraneous effects. This is especially true when the design calls for

another way to deal with the problem of deception is to conduct a **debriefing**, a session following the experiment when subjects are told the complete nature of the study, including any deceptions used, and subjects' reactions are assessed. Especially when deception is used, a debriefing is an ethical obligation of the researcher to ensure that people are in no way harmed by their participation. During the debriefing session, the researcher assesses whether the subject has had any kind of negative reaction to the deception or any other element of the experiment. It sometimes happens that things people read or do during an experiment or other people they come in contact with produce feelings of anger, depression, loneliness, or despair. In fact, almost any reaction is a possibility. The researcher's obligation is to assess these reactions and alleviate them. If necessary, a researcher might call in a counselor or other professional to assist in this. Such strong negative reactions are rare in most experiments, but milder reactions are not uncommon. The point is that a person should not walk away from an experiment feeling even a little angry or depressed because of the experience.

Ethical absolutists argue that no deception can be justified. They question whether subjects can truly give informed consent if they are misled about some aspects of the research. If those aspects were known to the subjects, might it lead some of them to withhold consent? That is, of course, impossible to say. It is certainly an appropriate ethical stance to inform the subjects that they might be deceived in unspecified ways, and it gives the subjects more information with which to make their decision about whether to participate. However, the subjects are still unable to decide for themselves whether the deception is "minor" because they don't know what the deception is until after the experiment has been completed.

Most social scientists agree that mild levels of deception, if done carefully, are necessary and ethically acceptable. However, the ethical absolutists, although a minority, are also a significant voice in the research community.

a control group to receive no treatment at all. Organizations may have ethical concerns about randomization, or may want to be sure that they "get something" for participating in a study and so are reluctant to serve as the control condition. In addition, with field experiments, there are times when it is impossible to find or create a control group that is truly comparable to the experimental groups and thus the true experimental designs are ruled out.

Artificiality

The laboratory setting in which some experiments are conducted is an artificial environment created by the investigator, which raises questions of external validity. We often do not know what the relationship is between this artificial setting and the real world in which people live out their daily lives. We cannot be sure that people would behave in the same fashion on the street or in the classroom as they do in the laboratory. To an extent, even field experiments are vulnerable to this criticism, especially when they create conditions that would never exist naturally. For example, the recipients of social services who are the subjects of research may be treated better by the research team than they would be in their everyday contacts with human service agencies.

Experimenter Effects

I just pointed out the artificiality of experimental settings in comparison to the real world. Yet, there is another side to this coin: The experimental setting itself is a real world—a social occasion in which social norms, roles, and values exist and shape people's behavior (Wuebben, Straits, & Schulman, 1974). Social processes arising within the experimental setting—and not part of the experimental variables—may shape the research outcome. Thus, changes in the dependent variable may be due to the demand characteristics of experiments, such as repeated measurement, or to the impact of experimenter expectations—factors that obviously do not influence people in the everyday world.

Generalizability

The logistics involved in managing an experiment often necessitate that experimenters use small samples. Further, the goal of holding measurement error to a minimum often compels researchers to make these samples as homogeneous as possible. Despite the advantages of random sampling, the experimenter must frequently settle for availability samples in order to obtain subjects who will participate in the study. The net result is that experiments are often conducted on small, homogeneous availability samples, and it may be difficult to know to whom such results can be generalized. It is often assumed that representative sampling is not as essential in experiments as in other types of research because of the randomization and control procedures used. Although these help, one must still exercise caution regarding the population to which generalizations are made.

Timeliness of Results

Although experiments may produce the strongest evidence of causation, they are often time consuming to conduct. The urgency of policy decisions may preclude using experimentation simply because it takes too long to acquire funding for the project, design the research, and carry it out. If results do not come in quickly enough, they may not be considered in the debate over a rapidly developing policy issue (Rossi, Freeman, & Lipsey, 1999).

REVIEW AND CRITICAL THINKING

Some scientists argue that experiments are the preferred or model research design because they are the strongest research design in terms of providing a solid logical foundation from which to infer causality. This also makes experimental designs a laudable model to serve as a guide for critical thinking in your everyday life. People constantly ask questions about causality: Why do people commit crimes? Why did I lose my job? Why did she treat me that way? When confronted with information, especially when it involves some causal judgment, ask yourself the following questions.

1. Can the observations or information available be organized into an experimental design? Can you identify independent and dependent variables?

2. Is there anything like random assignment or a control group involved? If not, how does this compromise your confidence in the conclusions that can be drawn?

3. Does the information look anything like a quasi-experiment, such as a time-series design?

4. Is there anything that threatens the internal validity or the external validity of the information available? What factors other than the independent variable might be producing changes in the dependent variable?

5. What about replication? Is there information available from other sources (other experiments?) that could increase your confidence in the causal inferences?

Computers and the Internet

If you've been taking advantage of Internet resources that apply to the previous chapters, then you've no doubt discovered the value of using the important terms in each chapter with various Internet search engines. For example, using such terms as "random assignment" or "randomized control group" with the AltaVista search engine will help you locate a number of Web sites that relate to experimental design in the social sciences. One site I located by doing this was The Hypertension Network (**http://www.bloodpressure.com**). The site is primarily devoted to health-related information and research on high blood pressure, so its relevance to experimental design is not immediately apparent. However, it includes a link (click on the "drug trials" button) to a page on clinical trials. Here you'll find a discussion of experimental design and how it is being applied to the study of hypertension.

The American Psychological Society maintains a Web site devoted to research experiments that are being conducted on the Internet titled "Psychological Research on the Net." Its URL is **http://psych.hanover.edu/APS/exponnet.html**. By connecting to this site, you will find links to experiments on the Internet that are related to psychological issues. They are organized by general topic area with the topic areas listed alphabetically. I recommend that

you explore some of these sites and participate in some of the projects that appeal to you. Keep in mind that not all the projects included there are true experimental designs complete with control groups, but many of the projects do employ features of experimental design. As an exercise, you should use the concepts discussed in this chapter to identify what type each experiment is and some of its features.

Main Points

1. Experiments are a controlled method of observation in which the value of one or more independent variables is changed in order to assess the causal effect on one or more dependent variables. Owing to the great control afforded by experiments, they are the surest method of discovering causal relationships among variables.

2. Experiments are designed so as to distinguish between experimental variability and extraneous variability. This can be done through the use of control variables or control groups. When comparisons are made between experimental groups and control groups, it is important that the two groups be equivalent. This can be accomplished through the mechanisms of matching, randomization, or blocking.

3. Internal validity refers to whether the independent variable does in fact produce the effect it appears to produce on the dependent variable. Numerous conditions may threaten the internal validity of experiments, including history, maturation, testing, instrumentation, statistical regression, selection, and experimental attrition.

4. Researchers are more likely to choose experimental designs when they want to be confident about inferring causality, when doing explanatory research, when independent and dependent variables are clear-cut and can be measured precisely, when they have sufficient control over the situation to

design an experiment, and when they are positivists seeking nomothetic explanations.

5. Preexperimental designs lack the random assignment to conditions or the control groups that are important to good experimental design. All of the threats to internal validity can be controlled by using one of the true experimental designs: classic design, Solomon four-group design, posttest only control group design, multiple experimental group with one control group design, and factorial design.

6. Quasi-experimental designs are useful for bringing much of the control of an experiment to nonexperimental situations. Three types of quasi-experimental designs are the time series design, the different group time series design, and the multiple time series design.

7. External validity concerns the degree to which experimental results can be generalized beyond the experimental setting. Threats to external validity include reactive effects of testing, unrepresentative samples, reactive settings, and multiple-treatment interference. Enhancing external validity often compromises internal validity, and vice versa. This problem can be alleviated through complex research designs, the use of field experiments, and replication.

8. Among the advantages of experimental designs are: ability to infer causality, control, the study of change, and lower costs. Disadvantages of experiments include: inflexibility, the requirement to randomize, artificiality, experimenter effects, difficulties in generalizing, and untimeliness of the results.

Important Terms for Review

blocking
control condition
control group
control variables

debriefing
demand characteristics
double-blind experiment
experiment
experimental condition
experimental group
experimental stimulus
experimental treatment
experimental variability
experimenter's expectations
external validity
extraneous variability
field experiments
internal validity
laboratory experiments
matching
preexperimental designs
quasi-experimental designs
random assignment
true experimental designs

For Further Reading

Adair, J. (1973). *The human subject: The social psychology of the psychological experiment.* Boston: Little, Brown. A good analysis of the many kinds of reactivity and experimenter expectancies that can be found in experimental settings.

Cook, T. D., & Campbell, D. T. (1979). *Quasi-experimentation: Design and analysis issues for field settings.* Chicago: Rand McNally. This is a classical work on quasi-experimental research designs. It is especially useful for designing experiments in field settings.

Fairweather, George W., & Davidson, William S. (1986). *An introduction to community experimentation: Theory, methods, and practice.* New York: McGraw-Hill. An interesting book ideally suited to those involved in community action programs. It outlines the reasons for experimentation and supplies the designs to evaluate the effectiveness of various intervention programs.

Gottman, John M. (1981). *Time-series analysis: A comprehensive introduction for social scientists.* New York: Cambridge University Press. *Comprehensive* is certainly the operative word in describing this book, as it details the mathematical analysis of time series designs. The author also argues persuasively against the common "eyeballing" approach to time-series analysis.

Kirk, R. E. (1982). *Experimental design: Procedures for the behavioral sciences* (2nd ed.). Belmont, CA: Brooks/Cole. This is a thorough overview of how to design both simple and complicated experiments. It goes well beyond the review in this chapter.

Pechman, Joseph, & Timpane, P. Michael. (Eds.). (1975). *Work incentives and income guarantees: The New Jersey income tax experiment.* Washington, DC: The Brookings Institution. An excellent discussion of methodological and political problems in conducting an applied experiment in a controversial field setting.

Ray, William, & Ravizza, Richard. (1993). *Methods toward a science of behavior and experience.* Belmont, CA: Wadsworth. This book covers many research topics, but also devotes considerable attention to experimentation, especially describing the logic behind and the value of experimental designs.

Chapter 9

SURVEY RESEARCH

A survey is a data-collection technique in which information is gathered from individuals (*respondents*) by having them respond to questions or statements. Surveys have been conducted for eons, although those done before the twentieth century would be considered primitive by today's standards (Rossi, Wright, & Anderson, 1983). The ancient civilizations of Rome and Athens conducted surveys in the form of population censuses to provide information for their rulers. The Constitution of the United States mandated a population survey (census) over 200 years ago. The earliest surveys of modern times were also often population surveys: a block-by-block household survey of the poor in London in the late 1800s, for example, or a complete household survey of the African American community in Philadelphia in the same era. A needs assessment survey was conducted in Pittsburgh in 1907 that focused on assessing the extent of problems accompanying industrialization, such as poor living conditions and industrial accidents. Surveys are still probably the most widely used research method in the social sciences. In the 1995 volume of the *Journal of Health and Social Behavior* (to pull just one journal off my shelf), 84% of the articles were based, at least in part, on survey data. This illustrates a major attraction of surveys—their flexibility. Surveys can be used for all types of studies: exploratory, descriptive, explanatory, and evaluative. They can even play a part, as we will see, in research designs, such as experiments, evaluation research, and field research, all of which are discussed in other chapters.

The many techniques available for conducting a survey make it a versatile tool. However, all surveys share certain characteristics. First, surveys typically involve collecting data from large samples of people; they are therefore ideal for obtaining data representative of populations too large to be dealt with by other methods. Indeed, the generalizability of survey findings is another major attraction of the method. Second, all surveys involve presenting respondents with a series of questions to be answered. These questions may tap matters of fact, attitudes and opinions, future expectations, or virtually any other kind of information that can be elicited through people's responses to questions or statements. The questions may be simple, single-item measures or complex, multiple-item indexes and scales. In whatever form, however, survey data are basically what people say to the investigator in response to a question.

An important point to emphasize about surveys is that they only measure what people *say* about their thoughts, feelings, and behaviors. Surveys do not measure those thoughts, feelings, and behaviors directly. For example, if people tell us in a survey that they do not take drugs, we have not measured actual drug-taking behavior but only people's reports about that behavior. This is very important in terms of the conclusions that can be drawn: We can conclude that people report not taking drugs, but we cannot conclude that people do not take drugs. This latter would be an inference we draw from what people say. So, surveys always involve data on what people say about what they do, not what they actually do.

Data are collected in survey research in two basic ways: with *questionnaires* or with *interviews*. A **questionnaire** contains written questions that people respond to directly on the questionnaire form itself, without the aid of an interviewer. A questionnaire can be handed directly to a respondent, or it can be sent (via mail or Internet) to the members of a sample or population who then fill it out on their own and return it to the researcher. An **interview** involves an interviewer reading questions to respondents and recording their answers. Interviews can be conducted either in person or over the telephone. Some survey research uses both questionnaire and interview techniques, with respondents filling in some answers themselves and being asked other questions by interviewers. Because questionnaires and interviews involve asking people to respond to questions, a problem central to both is the type of question to be asked. Chapter 6 discusses this topic: how to design valid questions, indexes, and scales. This chapter focuses on how the questions and statements described in Chapter 6 are organized into questionnaires and interviews and how these surveys are conducted.

DESIGNING QUESTIONNAIRES

Questionnaires are designed so that they can be answered without assistance. Of course, if a researcher hands a questionnaire to the respondent, as is sometimes done, the respondent then has the opportunity to ask the researcher to clarify anything that is ambiguous. A good questionnaire, however, should not rely on such assistance. In fact, questionnaires are often mailed or sent over the Internet to respondents, who thereby have no opportunity to ask questions. In other cases, questionnaires are administered to many people simultaneously in a classroom, auditorium, or organizational setting. Such modes of administration make questionnaires quicker and less expensive than most interviews. But this also places a burden on researchers to design questionnaires that can be properly completed without assistance (Dillman, 1991).

Structure and Design
Directions
One of the simplest, but also most important, tasks of questionnaire construction is the inclusion of precise directions or instructions for respondents. Good directions can go a long way toward improving the quality of data generated by questionnaires. Some of these directions have to do with how to answer questions (discussed in Chapter 6). Other directions have to do with how the respondent should move through the questionnaire and which questions should be answered or skipped. Some of these directions are described below, but it is always

desirable to pretest the questionnaire on groups with similar characteristics to those who will be the ultimate respondents. This makes it possible to detect any confusion in the directions.

Order of Questions
An element of questionnaire construction that takes careful consideration is the proper ordering of questions. Careless ordering can lead to undesirable consequences, such as a reduced response rate or biased responses to questions. Generally, questions asked early in the questionnaire should not bias answers to questions that come later. For example, if we asked several factual questions regarding poverty and the conditions of the poor and later asked a question concerning which social problems people consider to be serious, more respondents will likely include poverty in their answer than would otherwise have done so. These potentially biasing effects can sometimes be avoided by placing opinion questions first when a questionnaire contains both factual and opinion questions and when the factual questions might bias the opinion questions.

Ordering of questions can also increase a respondent's interest in answering a questionnaire—this is especially helpful for boosting response rates on mailed questionnaires. Questions dealing with particularly intriguing issues should be asked first. The idea is to interest the recipients enough to get them to start answering. Once they have started, they are more likely to complete the entire questionnaire. If the questionnaire does not deal with any topics that are obviously more interesting than others, then opinion questions should be placed first. People like to express their opinions and, for the reasons mentioned earlier, opinion questions should often be first anyway. A pitfall generally to be avoided is beginning a questionnaire with the standard demographic questions of age, sex, income, and the like. People are so accustomed to those questions that they may be disinclined to answer them again and may promptly file the questionnaire in the nearest wastebasket.

Filter and Contingency Questions

Some questions on a questionnaire may apply to some respondents and not to others. These questions are normally handled by what are called *filter questions* and *contingency questions*: A **filter question** is a question whose answer determines which question the respondent goes to next. In Table 6.1 on p. 151, questions 2 and 3 are both filter questions. In question 2, the part of the question asking about "how many items they have taken" is called a **contingency question** because whether a person answers it depends on (is contingent upon) their answer to the filter question. Note the two ways the filter question is designed. With question 2, the person answering "Yes" is directed to their next question by the arrow and the question is clearly set off by a box; also in the box, the phrase "If Yes" is included to be sure the person realizes that this question is only for those who answered "Yes" to the previous question. With question 3, the answer "No" is followed by a statement telling the person which question he or she should answer next. Either format is acceptable for filter questions, but the point is to provide clear directions for the respondent. By sectioning the questionnaire on the basis of filter and contingency questions, the respondent can be guided through even the most complex questionnaire. The resulting path that an actual respondent follows through the questionnaire is referred to as the *skip pattern*. As is true of many aspects of questionnaire design, it is important to evaluate the skip pattern by pretesting the questionnaire to be sure that all appropriate sections are completed and that minimal respondent frustration occurs.

Matrix Questions

In some cases, a number of questions or statements may have identical response alternatives. A very efficient way of organizing such questions is in the form of a **matrix question**, where the response alternatives are listed only once and each question or statement is followed only by a box to check or a number or letter to circle. Table 6.3 on page 155 and Table 6.6 on page 161 are examples of matrix questions. They are a very compact way of presenting a number of items and are often used with multiple-item indexes and scales.

Matrix questions should, however, be used cautiously because they contain a number of weaknesses. One weakness is that, with a long list of items in a matrix question, as in Table 6.3, it is easy for the respondent to lose track of which line is the response for which statement and indicate their answer on the line above or below it. This can be alleviated by following every third or fourth item with a blank line so it is easier to keep track of the proper line on which to mark an answer. A second weakness of matrix questions, especially those that are poorly designed, is that they may produce response set. The problem of response set and techniques for alleviating it are discussed at length in Chapter 6. A third weakness of matrix questions is that the researcher may be tempted, in order to gain the efficiencies of the format, to force the response alternatives of some questions into that format when some other format would have been valid. The response format of any question or statement should be determined by theoretical and conceptual considerations of what is the most valid way to measure a variable.

Response Rate

A major problem in many research endeavors is to gain people's cooperation so that they will provide whatever data are needed. In surveys, this cooperation is measured by the **response rate**; that is, the proportion of a sample or population who complete and return a questionnaire or agree to be interviewed (Bridge, 1974; Groves & Couper, 1998). With interviews, response rates are often very high—in the area of 90%—largely because people are

TABLE 9.1 Items to be Included in the Cover Letter of a Questionnaire or the Introduction to an Interview

Item	Cover Letter	Interview Introduction
1. Sponsor of the research	yes	yes
2. Address/phone number of the researcher	yes	if required
3. How the respondent was selected	yes	yes
4. Who else was selected	yes	yes
5. The purpose of the research	yes	yes
6. Who will utilize or benefit from the research	yes	yes
7. An appeal for the person's cooperation	yes	yes
8. How long it will take the respondent to complete the survey	yes	yes
9. Payment	if given	if given
10. Anonymity/confidentiality	if given	if given
11. Deadline for return	yes	not applicable

reluctant to refuse a face-to-face request for cooperation. (In fact, with interviews, the largest nonresponse factor is often the inability of the interviewers to locate respondents.) With mailed questionnaires, however, this personal pressure is absent, and people feel freer to refuse. This can result in many *nonreturns*—people who refuse to complete and return a questionnaire. Response rates with questionnaires, especially mailed ones, vary considerably, from an unacceptable low of 20% to levels that rival those of interviews.

Why is a low response rate of such concern? This has to do with the issue of the representativeness of a sample (discussed in Chapter 7). If we selected a representative sample and obtained a perfect 100% response, we would have confidence in the representativeness of the sample data. However, as the response rate drops below 100%, the sample may become less and less representative. Those who refuse to cooperate may differ from those who return the questionnaire in some systematic ways that affect the results of the research. In other words, any response rate less than 100% may result in a biased sample. Of course, a perfect response rate is rarely achieved, but the closer the response rate is to that level, the more confidence we have that the data are representative. A number of things can be done to improve response rates. Most of these apply only to questionnaires, but a few can also be used to increase response rates in interviews.

The Cover letter

A properly constructed *cover letter* can help increase the response rate. A **cover letter** accompanies a questionnaire and serves to introduce and explain it to the recipient. Because with mailed questionnaires the cover letter is the researcher's only medium of communicating with the recipient, it must be carefully drafted to include information recipients will want and to encourage them to complete the questionnaire (see Table 9.1).

Prominent in the cover letter should be information about the sponsor of the research project. Recipients are understandably interested in who is seeking the information they are asked to provide, and research clearly indicates that the sponsoring organization influences the response rate (Goyder, 1985; Rea & Parker, 1992). The highest response rates are for questionnaires sponsored by governmental agencies. University-sponsored research generates somewhat lower response rates, and commercially sponsored research produces the lowest rate of all. Apparently, if the research is at all associated with a governmental agency, stressing that in the cover letter may have a beneficial effect on the response rate. The response rates of particular groups can be increased if the research is sponsored or endorsed by an organization that people in the group believe has legitimacy. For example, response rates of professionals can be increased if the research is linked to relevant professional organizations. In a study of social workers, it would be helpful to be able to state in the cover letter that the National Association of Social Workers supported or sponsored the study, whichever is actually the case (Sudman, 1985).

Attractive and informative letterhead stationery should be used for the cover letter. It should include the address and telephone number of the researcher as well as an e-mail address. In some cases, it might be useful to include a Web site, if the organization sponsoring or conducting the research maintains one and if the information there would be helpful to potential respondents in making their decision about whether or not to participate. Especially if the sponsor of the research is not well known, some recipients may desire further information before making their decision. Although relatively few respondents will ask for more, including additional information gives the cover letter a completely open and aboveboard appearance that may further the general cooperation of recipients.

The cover letter should also inform the respondent about how people were selected to receive the questionnaire. It is not necessary to go into great detail on this matter, but those who receive an unanticipated questionnaire are quite naturally curious about how and why they were chosen to be part of a study. A brief statement that they were randomly selected or selected by computer (if this is the case) should suffice.

Recipients will also want to know the purpose of the research. Again, without going into great detail, the cover letter should explain such things as why the research is being conducted, why and by whom it is considered important, and the potential benefits anticipated from the study. Investigations have clearly shown that the response rate can be significantly increased if the importance of the research, as perceived by the respondent, is emphasized. This part of the cover letter must be worded very carefully, however, so that it does not sensitize respondents such in a way that their answers to questions are affected. Sensitizing effects can be minimized by keeping the description of the purpose very general—certainly, do not suggest what the research hypotheses might be. Regarding the importance of the data and anticipated benefits, the researcher should resist the temptation of hyperbole and instead make honest, straightforward statements. Exaggerated claims about "solving a significant social problem" or "alleviating the problems of the poor" are likely to be seen as precisely what they are.

The preceding information provides a foundation for the single most important component of the cover letter: a direct appeal for the recipient's cooperation. General statements about the importance of the research are no substitute for a more personal appeal to the recipient as to why he or she should take time to complete the questionnaire. Respondents must believe that their responses are very important to the outcome (as, in reality, they are). A statement to the effect that "your views are important to us" is a good approach. It emphasizes both the importance of each individual respondent and that the questionnaire will allow the expression of opinions, which people like.

The cover letter should indicate that the respondent will remain anonymous or that the data will be treated as confidential, whichever is the case (see Chapter 3). "Anonymous" means that *no one*, including the researcher, can link a particular respondent's name to his or her questionnaire. "Confidential" means that, even though the researcher can match respondents to their questionnaires, the information will be treated collectively, and no individuals will be publicly linked to their responses.

With mailed questionnaires, two techniques assure anonymity (Sudman, 1985). The best technique is to keep the questionnaire itself completely anonymous, with no identifying numbers or symbols, and provide the respondent with a postcard, including the respondent's name, to be mailed at the same time as, but separately from, the completed questionnaire itself. This way, the researcher knows who has responded and need not send reminders to respond; and no one can link a respondent's name with a particular questionnaire. A second way to ensure anonymity is to attach to the questionnaire a cover sheet with an identifying number and assure the respondents that the cover sheet will be removed and destroyed once the receipt of the questionnaire has been recorded. This second procedure provides less assurance to the respondent because an unethical researcher might retain the link between questionnaires and their identification numbers. The first procedure, however, is more expensive because of the additional postcard mailing, so the second procedure may be preferred in questionnaires that do not deal with highly sensitive issues that would make respondents very concerned about anonymity. If the material is not very sensitive, assurances of confidentiality can be adequate to ensure a good return rate. No evidence indicates that assuring anonymity rather than confidentiality will increase the response rate in nonsensitive surveys (Moser & Kalton, 1972).

Finally, the cover letter should include a deadline for returning the questionnaire, calculated to take into account mailing time and a few days to complete the questionnaire. The rationale for a fairly tight deadline is that it will encourage the recipient to complete the questionnaire soon after it is received and not set it aside where it can be forgotten or misplaced.

Payment

Research consistently shows that response rates can be increased by offering a payment as a part of the appeal for cooperation, and the payment need not be large to have a positive effect. Studies find that, depending on the respondents, an incentive of $2 or less can add almost 10% to the response rate, and an additional $2 can add another 5%–10% (James & Bolstein, 1990; Warriner, Goyder, Gjertsen, Hohner, & McSpurren, 1996). For the payments to have the greatest effect, they should be included with the initial mailing and not promised on return of the questionnaire. One study found that including the payment with the questionnaire boosted the return rate by 12% over promising payment on return of the questionnaire (Berry & Kanouse, 1987). Researchers have used other types of incentives, such as entering each respondent in a lottery or donating to charity for each questionnaire returned, but these have shown mixed results as far as increasing response rate.

Follow-ups

The most important procedural matter affecting response rates is the use of follow-up letters or phone calls. A substantial percentage of nonrespondents to the initial mailing will respond to follow-up letters. With two follow-ups, 15%–20% increases over the initial return can be achieved (James & Bolstein, 1990). Such follow-ups are clearly essential and can be done by telephone if budget permits and speed is important. With aggressive follow-ups, the difference in response rates between mailed questionnaires and interviews declines substantially (Goyder, 1985).

In general, two follow-ups should be used. When response to the initial mailing drops off, follow-up letters should be sent to the nonrespondents, encouraging them to return the questionnaire. The follow-up letter should include a restatement of the points in the original cover letter with an additional appeal for their cooperation. When response to the first follow-up declines, a second follow-up is then sent to the remaining nonrespondents, including another copy of the questionnaire because people may have misplaced the original. After two follow-ups, the remaining nonrespondents should be considered a pretty intransigent lot, and additional follow-ups will generate relatively few responses.

Length and Appearance

The rate of response to a mailed questionnaire is also affected by the length of the questionnaire and its appearance. The problem is that as length increases, response rate declines. Most researchers realize that there are no hard-and-fast rules about the desirable length for a questionnaire or interview. I know one researcher who won't design a questionnaire any longer than three pages (he sometimes fits more questions in by reducing the font size, which strikes me as a form of cheating). What is most important, of course, is not the physical length or even the number of questions but how long it takes a respondent to complete it. The danger is that a long questionnaire either will not be returned or the respondent will answer quickly without giving sufficient thought and thus throw into question the validity of the responses. Much depends, of course, on the intelligence and literacy of the respondents, the degree of interest in the questionnaire topic, and other such matters.

One guideline is that a questionnaire should not exceed five pages in length or take more than 30 minutes to complete unless there are compelling reasons why additional questions need to be asked (Fowler, 1988). Certainly, questionnaires should be kept as short as possible while still collecting all the data necessary to answer the research questions at hand. In addition, any questionnaire approaching or exceeding 30 minutes to complete should be scrutinized very closely to determine if any questions can be deleted. Although keeping the questionnaire short is important, this should not be achieved by cramming so much material onto each page that the respondent has difficulty using the instrument—the appearance of the questionnaire is also important to generate a high response rate. As discussed earlier, the use of boxed response choices and smooth transitions through filter and contingency questions make completing the questionnaire easier and more enjoyable for the respondent, which in turn increases the probability that it will be returned.

Other Influences on Response Rate

Other factors can work to increase response rates. In telephone surveys, for example, the voice and manner of the interviewer can have an important effect (Oksenberg, Coleman, & Cannell, 1986). Interviewers with higher-pitched, louder voices and clear and distinct pronunciation have lower refusal rates. The same is true for interviewers who sound more competent and upbeat toward the respondent. But reminders of confidentiality can adversely affect the response rate (Frey, 1986). People who are reminded of the confidentiality of the information partway through the interview are more likely to refuse to respond to some of the remaining questions than are people who do not receive such a reminder. It may be that the reminder works to undo whatever rapport the interviewer has already built up with the respondent.

Response rates are also affected by mailing procedures. It almost goes without saying that a stamped, self-addressed envelope should be supplied for returning the questionnaire, making things as convenient as possible for the respondent. The type of postage utilized also affects the response rate, with stamps bringing about 4% higher return than bulk-printed postage (Scott, 1961). Presumably, the stamp makes the questionnaire appear more personal and less like unimportant junk mail. A regular stamped envelope also substantially increases the response rate in comparison to a business reply envelope (Armstrong & Luck, 1987).

A survey following all the suggested procedures should yield an acceptably high response rate. Specialized populations may, of course, produce either higher or lower rates. Because so many variables are involved, only rough guidelines can be offered for evaluating response rates with mailed questionnaires. The desired response rate is, of course, 100%. Anything less than 50% is highly suspect as far as its representativeness is concerned. Unless some evidence of the representativeness can be documented, one should generalize from such a sample with great caution. In fact, it might be best to treat the resulting sample as a nonprobability sample from which one cannot make very confident generalizations. In terms of what a researcher can expect, response rates in the 60% range are considered good, and anything more than 70% is very good. However, even with these response rates, the researcher must be cautious about generalizing and make attempts to check for bias due to nonresponse. Whether the response rate is high or low, the bottom line is to report it honestly so that those reading the research can judge for themselves about its generalizability.

Checking for Bias Due to Nonresponse

Even if a relatively high rate of response is obtained, possible bias due to nonresponse should be investigated by determining the extent to which respondents differ from nonrespondents (Groves, 1989; Miller, 1991; Rea & Parker, 1992). One common way of doing this is to compare the characteristics of the respondents to the characteristics of the population from which they were selected. If a database on the population exists, this task can be simplified. For example, if you are studying a representative sample of welfare recipients in a county, the Department of Social Services is likely to have data on age, sex, marital status, level of education, and other characteristics for all the welfare recipients in the community. You can compare your respondents to this database on the characteristics for which data have already been collected. A second approach to assessing bias resulting from nonresponse is to locate a subsample of nonrespondents and interview them. In this way, the responses to the questionnaire by a representative sample of nonrespondents can be compared to those of the respondents. This is the preferred method because the direction and the extent of bias due to nonresponse can be directly measured. It is, however, the most costly and time-consuming approach.

Any check for bias due to nonresponse informs us only about those characteristics on which we make comparisons. It does not prove that the respondents are representative of the whole sample on any other variables—including those that might be of considerable importance to the study. In short, even though we can gather some information regarding such bias, in most cases it is not possible to *prove* that bias due to nonresponse does not exist.

An Assessment of Questionnaires
Advantages

As a technique of survey research, questionnaires have a number of desirable features. First, they can be used to gather data far more inexpensively and quickly than interviews. Only a month to six weeks is needed for mailed questionnaires, whereas obtaining the same data by personal interviews would likely take several months—at a minimum. Mailed and online questionnaires also save the expense of interviewers, interviewer travel, and other such costs.

Second, questionnaires distributed via mail or Internet enable the researcher to collect data from a sample that is dispersed geographically. It costs no more to mail or e-mail a questionnaire across the country than across a city. Costs of interviewer travel rise enormously as distance increases, making interviewing over wide geographic areas expensive.

Third, with questions of a personal or sensitive nature, mailed questionnaires may provide more accurate answers than interviews. Individuals may be more likely to respond honestly to such questions when they are not face to face with a person they perceive as possibly making judgments about them. Some evidence indicates, for example, that questions about premarital sex are answered more honestly on a questionnaire than during an interview. However, this does not appear to be true for all topics, so the choice of questionnaire versus interview on these grounds would depend on the subject of the research (Knudsen, Pope, & Irish, 1967; Moser & Kalton, 1972).

Finally, mailed and online questionnaires eliminate the problem of interviewer bias. Interviewer bias occurs when an interviewer influences a person's response to a question through what the interviewer says, his or her tone of voice, or demeanor. Because no interviewer is present when the person fills out a questionnaire, it is not possible for an interviewer to bias the respondent's answers in a particular direction.

Disadvantages

Despite their many advantages, questionnaires have important limitations that may make them less desirable for some research efforts (Moser & Kalton, 1972).

First, questionnaires require a minimal degree of literacy and facility in English that some respondents do not possess. Substantial nonresponse is, of course, likely with such people. With most general population surveys, the nonresponse due to illiteracy will not seriously bias the results. Self-administered questionnaires are more successful among persons who are better educated, motivated to respond, and involved in issues and organizations. However, some groups of interest to social scientists often do not possess these characteristics. If the survey is aimed at a special population in

which less than average literacy is suspected, personal interviews are the better choice.

Second, the questions must all be sufficiently simple to be comprehended on the basis of printed instructions. Third, there is no opportunity to probe for more information or evaluate the nonverbal behavior of the respondents. The answers they mark on the questionnaire form are final. Fourth, there is no assurance that the person who should answer the questionnaire is the one who actually does so. Fifth, responses cannot be considered independent, for the respondent can read through the entire questionnaire before completing it. Finally, mailed and online questionnaires, more so than other types of surveys, raise the problem of nonresponse bias.

INTERVIEWS

In an interview, the researcher or an assistant reads the questions directly to the respondents and records their answers. The comments made in the previous section regarding good directions for questionnaires also apply to interviews. Even though interviewers will be familiar with the content of the interview, they still need good directions and devices such as contingency questions to ensure that they collect all information and do so quickly. Interviews offer the researcher a degree of flexibility that is not available with questionnaires. One area of increased flexibility relates to the degree of structure built into an interview.

The Structure of Interviews

The element of structure in interviews refers to the degree of freedom the interviewer has in conducting the interview and respondents have in answering questions. Interviews are usually classified in three levels of structure: (1) the *unstandardized*, (2) the *nonschedule-standardized*, and (3) the *schedule-standardized*.

The *unstandardized interview* has the least structure. All the interviewer typically will have for guidance is a general topic area, as

FIGURE 9.1   Examples of Various Interview Structures

The Unstandardized Interview

Instructions to the interviewer: Discover the kinds of conflicts that the child has had with the parents. Conflicts should include disagreements, tensions due to past, present, or potential disagreements, outright arguments and physical conflicts. Be alert for as many categories and examples of conflicts and tensions as possible.

The Nonschedule-Standardized Interview

Instructions to the interviewer: Your task is to discover as many specific kinds of conflicts and tensions between child and parent as possible. The more *concrete* and detailed the account of each type of conflict the better. Although there are 12 areas of possible conflict which we want to explore (listed in question 3 below), you should not mention any area until after you have asked the first two questions in the order indicated. The first question takes an indirect approach, giving you time to build up rapport with the respondent and to demonstrate a nonjudgmental attitude toward teenagers who have conflicts with their parents.

1. What sorts of problems do teenagers you know have in getting along with their parents?
   (Possible probes: Do they always agree with their parents? Do any of your friends have "problem parents"? What other kinds of disagreements do they have?)
2. What sorts of disagreements do you have with your parents?
   (Possible probes: Do they cause you any problems? In what ways do they try to restrict you? Do you always agree with them on everything? Do they like the same things you do? Do they try to get you to do some things you don't like? Do they ever bore you? Make you mad? Do they understand you? etc.)
3. Have you ever had any disagreements with either of your parents over:
   a. Using the family car
   b. Friends of the same sex
   c. Dating
   d. School (homework, grades, activities)
   e. Religion (church, beliefs, etc.)
   f. Political views
   g. Working for pay outside the home
   h. Allowances
   i. Smoking
   j. Drinking
   k. Eating habits
   l. Household chores

(continued on next page)

---

illustrated in Figure 9.1. By developing his or her own questions and probes as the interview progresses, the interviewer explores the topic with the respondent. This approach is called "unstandardized" because different questions will be asked by each interviewer and different information obtained from each respondent. There is a heavy reliance on the skills of the interviewer to ask good questions and to keep the interview going, and this can only be done if experienced interviewers are available. This unstructured approach makes unstandardized interviewing especially appropriate for some exploratory research. As discussed in Chapter 11, the unstandardized interview is also often used in qualitative research, where it is called an *in-depth interview*. It permits the respondents to create the structure and meaning of their responses rather than having them imposed by the researcher by closed-ended or highly structured, open-ended questions and response options. In Figure 9.1, for example, the interviewer is guided only by the general topic of parent-child conflicts. The example also illustrates the suitability of this style of interviewing for exploratory research because the interviewer is directed to search for as many areas of conflict as can be found.

FIGURE 9.1 *(continued)*

The Schedule-Standardized Interview

Interviewer's explanation to the teenage respondent: We are interested in the kinds of problems teenagers have with their parents. We need to know how many teenagers have which kinds of conflicts with their parents and whether they are just mild disagreements or serious fights. We have a checklist here of some of the kinds of things that happen. Would you think about your own situation and put a check to show which conflicts you, personally, have had and about how often they have happened. Be sure to put a check in every row. If you have never had such a conflict then put the check in the first column where it says "never."

(Hand him the first card dealing with conflicts over the use of the automobile, saying, "If you don't understand any of those things listed or have some other things you would like to mention about how you disagree with your parents over the automobile let me know and we'll talk about it.") (When the respondent finishes checking all rows, hand him card number 2, saying, "Here is a list of types of conflicts teenagers have with their parents over their friends of the same sex. Do the same with this as you did with the last list.")

Automobile	Never	Only Once	More Than Once	Many Times
1. Wanting to learn to drive				
2. Getting a driver's license				
3. Wanting to use the family car				
4. What you use the car for				
5. The way you drive it				
6. Using it too much				
7. Keeping the car clean				
8. Putting gas or oil in the car				
9. Repairing the car				
10. Driving someone else's car				
11. Wanting to own a car				
12. The way you drive your own car				
13. What you use your car for				
14. Other				

Source From Raymond L. Gorden, *Interviewing Strategy, Techniques, and Tactics*, 4th ed. Copyright © 1987 by the Dorsey Press. Reprinted by permission of the author

Nonschedule-standardized interviews add more structure, with the topic narrower and specific questions asked of all respondents. However, the interview remains fairly conversational, and the interviewer is free to probe, rephrase questions, or use the questions in whatever order best fits that particular interview. Note in Figure 9.1 that specific questions are of the open-ended type, allowing the respondent full freedom of expression. As in the case of the unstandardized form, success with this type of interview requires an experienced interviewer.

The *schedule-standardized interview* is the most structured type. An **interview schedule** is used that contains specific instructions for the interviewer, specific questions in a fixed order, and transitional phrases for the interviewer to use. Sometimes the schedule also contains acceptable rephrasings for questions and a selection of stock probes. Schedule-standardized interviews are fairly rigid, with neither interviewer nor respondent allowed to depart from the structure of the schedule. Although some questions may be open-ended, most will likely be closed-ended. In fact, some schedule-standardized interviews are very similar to a questionnaire except that the interviewer asks the questions rather than having the respondent read them. Note in Figure 9.1 the use of cards with response alternatives that are handed to the respondent. This is a popular way of supplying respondents with a complex set of closed-ended alternatives. Note also the precise directions for the interviewer as well as verbatim phrases to be read to the respondent. Schedule-standardized interviews can be conducted by part-time interviewers with relatively little training or experience because nearly everything they need to say is contained in the schedule. This makes schedule-standardized interviews the preferred choice for studies with large sample sizes requiring many interviewers. The structure of these interviews also ensures that all respondents are presented with the same questions in the same order. This heightens reliability and makes schedule-standardized interviews popular for rigorous hypothesis testing.

Contacting Respondents

As with mailed questionnaires, interviewers face the problem of contacting respondents and eliciting their cooperation. Many interviews are conducted in respondents' homes or other locations, and locating and traveling to those places is sometimes the more troublesome and costly aspect of interviewing. It has been estimated that as much as 40% of a typical interviewer's time is spent traveling (Sudman, 1965). Because so much time and cost

are involved and because high response rates are desirable, substantial efforts are directed at minimizing the rate of refusal. The way prospective respondents are first contacted has substantial impact on the refusal rate.

Two approaches to contacting respondents that might appear logical to the neophyte researcher have, in fact, an effect opposite of that desired. It might seem that *telephoning* to set up an appointment for the interview would be a good idea. In reality, it greatly increases the rate of refusal. In one experiment, for example, those who were telephoned had nearly *triple* the rate of refusal of those contacted in person (Brunner & Carroll, 1967). Apparently, it is much easier to refuse over the relatively impersonal medium of the telephone than in a face-to-face encounter with an interviewer. *Sending a letter* asking individuals to participate in an interview has much the same effect (Cartwright & Tucker, 1967). The letter seems to give people sufficient time before the interviewer arrives to develop reasons they do not want to cooperate. Those first contacted in person, on the other hand, have only the excuses they can muster on the spur of the moment. Clearly, then, the lowest refusal rates are obtained by contacting interviewees in person.

Additional factors can affect the refusal rate (Gorden, 1987). For example, information regarding the research project should blanket the total survey population through the news media. The purpose of this is to demonstrate general community acceptance of the project. The information provided should be essentially the same as that of a cover letter for a mailed questionnaire, with a few differences (see Table 9.1). Pictures of the interviewers and mention of any props they will be carrying such as clipboards or zipper cases should be included. This information assists people in identifying interviewers and reduces possible confusion with salespeople or bill collectors. In fact, in some cases, it may be a good idea to equip the interviewers with photo identification badges clipped to their clothing in a visible location so they are easily recognizable and not mistaken for others who go door to door. When the interviewers go into the field, they should take along copies of the news coverage. If they encounter an interviewee who has not seen the media coverage, the clippings can be shown during the initial contact.

The timing of the initial contact also affects the refusal rate. It is preferable to contact interviewees at a time convenient for them to complete the interview without the need for a second call. Depending on the nature of the sample, predicting availability may be fairly easy or virtually impossible. For example, if the information required can be obtained from any household member, almost any reasonable time of day will do. On the other hand, if specific individuals must be contacted, timing becomes more critical. If the breadwinner in a household must be interviewed, for example, then contacts should probably be made during evening hours or on weekends, unless knowledge of the person's occupation or schedule suggests a different time of greater availability. Whatever time the interviewer makes the initial contact, it still may not be convenient for the respondent, especially if the interview is lengthy. If the respondent is pressed for time, it is better to use the initial contact to establish rapport and set another time for the interview, even though callbacks are costly. This is certainly preferable to the rushed interview that collects inferior data.

When the interviewer and potential respondent first meet, the interviewer should include certain points of information in the introduction. One suggestion is the following (Smith, 1981):

> Good day. I am from the Public Opinion Survey Unit of the University of Missouri [shows official identification]. We are doing a survey at this time on how people feel about police-community relationships. This study is being done throughout the state, and the results will be used by local and state governments. The addresses at which we interview are chosen entirely by chance, and the interview only takes 45 minutes. All information is entirely confidential, of course.

Potential respondents in interviews will be looking for much the same basic information as is found in the cover letter of a questionnaire. As the example illustrates, respondents should also be informed of the approximate length of the interview. After giving the introduction, the interviewer should be prepared to elaborate on any points the interviewee questions. Care must be exercised, however, when discussing the purpose of the survey, to avoid biasing responses.

Conducting an Interview

A large-scale survey with an adequate budget often turns to private research agencies to train interviewers and conduct interviews. A smaller research project, however, may not be able to afford this and will have to train and coordinate its own team of interviewers, possibly with the researchers themselves doing some of the interviewing. It is important, therefore, to know how to conduct an interview properly.

The Interview As a Social Relationship

The interview is a social relationship designed to exchange information between respondent and interviewer. The quantity and quality of information exchanged depends on how astute and creative the interviewer is at understanding and managing that relationship (Bradburn & Sudman, 1979; Fowler & Mangione, 1990). A few elements of the research interview are worth emphasizing because they have direct implications for conducting the interview.

The research interview is a secondary relationship in which the interviewer has a practical, utilitarian goal. It is easy, especially for an inexperienced interviewer, to be drawn into a more casual or personal interchange with the respondent. For example, if the respondent is friendly and outgoing, the conversation could drift off to sports, politics, or children. That, however, is not the purpose of the interview. The goal is not to make friends or give the respondent a sympathetic ear but rather to collect complete and unbiased data following the interview schedule.

We all recognize the powerful impact that first impressions can have on our perceptions of other people. This is especially true in interview situations, where the interviewer and respondent are likely to be total strangers. The first things that affect a respondent are the physical and social characteristics of the interviewer. Therefore, considerable care needs to be taken to ensure that the first contact enhances the likelihood of cooperation by the respondent (Warwick & Lininger, 1975). Most research suggests that interviewers are more successful if they have social characteristics similar to those of the respondents. Thus, characteristics such as socioeconomic status, age, gender, race, and ethnicity might influence the success of the interview—especially if the subject matter of the interview relates to one of these characteristics. The Eye on Diversity box explores at greater length this issue in regard to race, ethnicity, and gender. In addition, the personal demeanor of the interviewer plays an important role; interviewers should be neat, clean, and businesslike, but friendly. The concern is with **interviewer bias**, that the characteristics of the interviewer may influence how respondents answer questions. If this occurs, it is a form of reactivity and reduces the validity of the data collected.

After initial pleasantries have been exchanged, the interviewer should begin the interview. The interviewee may be a bit apprehensive during the initial stages of an interview. In recognition of this, the interview should begin with fairly simple, nonthreatening questions. If a schedule is used, it should be designed to begin with these kinds of questions. The demographic questions that are reserved until the later stages of a mailed questionnaire should begin an interview. Respondents' familiarity with these questions makes them nonthreatening and a good means of reducing tension in the respondents.

Another issue in interviews is whether third parties, such as a spouse or a friend, should be present during the interview. The problem, again, is that the interviewee's responses might be shaped in part by the presence of other persons and thus not be entirely valid. In some cases, ethical or other considerations might dictate that another person be present; for example, when interviewing a child, a prison inmate, or a psychologically incompetent person, a third party might be appropriate to protect the interests and well-being of either the interviewer or the interviewee. In other cases, the general wisdom is to interview the respondent alone in order to avoid any potential effect. Yet, the problem is complex and the research available on this topic is not entirely consistent (Aquilino, 1993; Pollner & Adams, 1997; T. W. Smith, 1997). Some research suggests that, with most interviews, the presence of a third party, including a spouse, has little if any effect on data gathered in an interview. Other research does suggest that having a spouse present makes a difference: When spouses are present, for example, respondents give more positive assessments of their marriages, report higher levels of marital conflict, and give higher estimates of their spouse's contribution to the housework. However, the complication is to determine which portrayal is more accurate—the one with the spouse present or the one with spouse absent. Each interview situation requires careful consideration as to whether the presence of a particular person might have a negative impact.

Probes

If an interview schedule is used, the interview will progress in accordance with the schedule. As needed, the interviewer will use **probes**, or follow-up questions, that are intended to elicit clearer and more complete responses. In some cases, suggestions for probes will be contained in the interview schedule. In less structured interviews, however, interviewers must be prepared to develop and use their own probes. Probes can take the form of a pause in conversation that encourages the respondent to elaborate. Or a probe could be an explicit request to clarify or elaborate something. A major concern with any probe is that it not bias the respondent's answer by suggesting how he or she should answer (Fowler & Mangione, 1990).

Recording Responses

A central task of interviewers is, of course, to record the responses of respondents. The four most common ways are: classifying responses into predetermined categories, summarizing the "high points" of what is said, taking verbatim notes, or making an audio or video recording of the interview.

Recording responses is generally easiest when an interview schedule is used. Because closed-ended questions are typical of such schedules, responses can simply be classified into the predetermined alternatives. This simplicity of recording is another factor making schedule-standardized interviews suitable for use with relatively untrained interviewers since no special recording skills are required.

With nonschedule interviewing, the questions are likely to be open-ended and the responses longer. Often, all that needs to be recorded are the key points made by the respondent. The interviewer condenses and summarizes what the respondent says. This requires an experienced interviewer, familiar with the research questions, who can accurately identify what should be recorded and do so without injecting his or her own interpretation, which would bias the summary.

Sometimes it may be desirable that everything the respondent says be recorded verbatim in order to avoid the possible biasing effect of summarizing responses. If the anticipated responses are reasonably short, competent interviewers can take verbatim notes. Special skills such as shorthand, however, may be necessary, or one could use laptop or notebook computers for recording. If the responses are lengthy, verbatim note-taking can cause difficulties such as leading the interviewer to fail to monitor the respondent or to be unprepared to probe when necessary. It can also damage rapport by making it appear that the interviewer is ignoring the respondent. Problems such as these can be eliminated by recording the interviews, but this increases costs substantially. Although individual cassettes and recorders are not very expensive, the number needed for a large-scale survey would certainly drive up costs considerably. The really big cost comes, however, when transcribing the tapes. Vast amounts of secretarial time are required for this (Gorden, 1987).

The fear of some researchers that tape recorders will increase the refusal rate appears unwarranted (Gorden, 1987). If use of the recorder is explained as a routine procedure that aids in recording complete and accurate responses, few respondents object. You should avoid asking the respondents if they mind if you record the interview. The question itself implies legitimate reasons for objecting and almost invites the respondent to object. The interviewer should assume that the tape recorder will be accepted unless the respondent raises the issue.

Controlling Interviewers

Once interviewers go into the field, the quality of the resulting data is heavily dependent upon them. It is a naive researcher indeed who assumes that, without supervision, they will all do their job properly, especially when part-time interviewers who have little commitment to the research project are used. Proper supervision begins during interviewer training. The importance of contacting the right respondents and meticulously following established procedures should be stressed. Interviewers should be informed that their work will be checked carefully and failure to follow procedures will not be tolerated.

One particularly serious problem with hiring people to conduct interviews is a practice

EYE ON DIVERSITY The Effect of Diversity on the Interview Relationship

In a diverse world, it may often be the case that an interviewer may differ from those being interviewed in terms of race, ethnicity, gender, or in some other way. Does it make a difference? It appears to, at least in some cases. In survey research, three elements interact to affect the quality of the data collected: the characteristics of the interviewer, the characteristics of the respondent, and the content of the survey instrument. The interrelationships among these elements need to be considered carefully to ensure the least reactivity and bias in the data-collection process.

Substantial research documents a tendency for persons to choose more desirable or socially acceptable answers to questions in surveys (DeMaio, 1984). This stems in part from the desire to appear sensible, reasonable, and pleasant to the interviewer. In all of our contacts with individuals, including an interview relationship, people typically prefer to please someone rather than to offend or alienate them. In cases in which interviewer and respondent are from different racial, ethnic, or gender groups, respondents tend to give answers they perceive to be more desirable, or at least less offensive, to the interviewer; this is especially true when the content of the questions is related to racial, ethnic, or gender issues. In addition to social desirability, social distance between interviewer and respondent can also affect responses in interviews. Social distance refers to how much people differ from one another on important social dimensions such as age, race, or ethnicity. Generally, the less social distance between people, the more freely, openly, and honestly they will talk. Racial, sexual, and ethnic differences often indicate a degree of social distance.

The impact of cross-race interviewing has been studied extensively with African American and Anglo respondents (Anderson, Silver, & Abramson, 1988; Bachman & O'Malley, 1984; Bradburn & Sudman, 1979; Cotter, Cohen, & Coulter, 1982; Schaeffer, 1980). African American respondents, for example, express more warmth and closeness for Anglos when interviewed by an Anglo person and are less likely to express dissatisfaction or resentment over discrimination or inequities against African Americans. Anglo respondents, in turn, tend to express more positive attitudes toward African Americans when interviewed by an African American than by an Anglo. This race-of-interviewer effect can be quite large and occurs fairly consistently. Some research concludes that it plays a part mostly when questions involve race or other sensitive topics, but recent research suggests its effect is more pervasive, affecting

variously known as "curbing," "curbstoning," or "shade-treeing." These terms refer to interviewers who fill in the responses themselves without contacting respondents (Frey, 1989). This produces false data and threatens the validity of the research. Although not widespread, it can be a problem if adequate supervision is not provided. To prevent this, completed interviews should be scrutinized for any evidence of falsification. Spot checks can be made to see if interviewers are where they are supposed to be at any given time, and respondents can be telephoned to see if they have in fact been interviewed.

people's responses to many questions on a survey and not just the racial or sensitive questions (Davis, 1997).

Gender also has an effect on interviews. Women are much more likely to report honestly about things like rape, battering, sexual behavior, and male-female relationships in general when they are interviewed by a woman rather than a man (Eichler, 1988; Reinharz, 1992). Both men's and women's responses are affected by the gender of the interviewer when they are being asked about issues of gender inequality (Kane & Macauley, 1993). Recent research suggests that, on topics regarding sexual behavior in general, matching the interviewer's gender to that of the respondent provides the most complete and valid responses (Catania, et al., 1996).

Some researchers recommend routinely matching interviewer and respondent for race, ethnicity, and/or gender in interviews on racial or sensitive topics, and this is generally sound advice. Sometimes, however, a little more thought is called for. The problem is that we are not always sure in which direction bias might occur. If Anglo respondents give different answers to Anglo as opposed to African American interviewers, which of their answers most accurately reflect their attitudes? For the most part, we aren't sure. It is generally assumed that the same-race interviewer will gather more accurate data (Fowler & Mangione, 1990). A more conservative assumption is that the truth falls somewhere between the data that the two interviewers collect. This issue goes back to topics of measurement and measurement error first discussed in Chapter 5. When individuals' responses are affected at least in part by the characteristics of the interviewer, then we are not getting a true measure of their attitudes. The problem lies in measuring the extent and the direction of the impact of that error on our measurement.

When minorities speak a language different from that of the dominant group, the quality of data collected can be affected if the interview is conducted in the dominant group's language (Marin & Marin, 1991). For example, a study of Native American children in Canada found that these children expressed a strong Anglo bias in racial preferences when the study was conducted in English; the bias declined significantly when the children's native Ojibwa language was used (Annis & Corenblum, 1986). So, when interviewing groups where a language other than English is widely used, it would be appropriate to consider conducting interviews in that other language.

An Assessment of Interviews
Advantages

First, interviewing is a more *flexible* form of data collection than questionnaires. This flexibility makes interviewing suitable for a far broader range of research situations. It can be used in highly quantitative research projects that are favored by positivists who might use highly structured interviews (the focus of Chapter 5 and Chapter 6). However, interviews can also be adapted to the research problems commonly approached by the interpretivist, feminist, and critical paradigms, who seek research methods that allow respondents to create their own meaning and structure. The style of interviewing can be tailored to these needs by using more unstructured, in-depth interviews, and a free conversational style with much probing. This approach is explored in more detail in Chapter 11 and Chapter 15.

Second, interviews can help *motivate* respondents to give more accurate and complete information. There is little motivation for respondents to be especially accurate or complete when responding to a mailed questionnaire. They can hurry through it if they want to. The control afforded by an interviewer encourages better responses. This becomes especially important as the information sought becomes more complex.

Third, interviewing affords an opportunity to *explain* questions that respondents may not otherwise understand. Again, if the information sought is complex, this can be of great importance, and the literacy problem that was a limitation of mailed questionnaires is virtually eliminated. Even lack of facility in English can be handled by using multilingual interviewers. (When I assisted with a needs assessment survey in some rural parts of Upper Michigan some years ago, we employed one interviewer who was fluent in Finnish because a number of people in the area spoke Finnish but little or no English.)

Fourth, the presence of an interviewer allows *control* over factors not controlled with mailed questionnaires. For example, the interviewer can ensure that the proper person responds to the questions and that questions are responded to in sequence. Further, the interviewer can arrange for the interview to be conducted such that the respondent does not consult with, or is not influenced by, other people before responding.

Finally, the interviewer can add *observational information* to the responses of the respondent. What was the respondent's attitude toward the interview? Cooperative? Indifferent? Hostile? Did the respondent appear to be fabricating answers? Did he or she react emotionally to some questions? This additional information helps to better evaluate the responses given, especially when the subject matter is highly personal or controversial (Gorden, 1987).

Disadvantages

The first disadvantage is *cost*. Interviewers must be hired, trained, and equipped; and their travel must be paid for. Altogether, these can be very expensive. As one example, the General Social Survey involves nationwide interviews of about 1,600 people conducted just about every year since 1972. It is probably the most important survey conducted in the United States for research purposes, but it currently costs nearly $1 million annually to conduct this survey.

The second disadvantage is *time*. Traveling to respondents' homes requires much time and limits each interviewer to only a few interviews each day. If particular individuals must be contacted, several time-consuming callbacks may be needed to complete many of the interviews. Considerable time is also required for start-up operations, such as developing questions, designing schedules, and training interviewers.

A third disadvantage of interviews is the problem of *interviewer bias*. Especially in unstructured interviews, interviewers may misinterpret or misrecord something because of their own personal feelings about the topic. Further, just as the respondent is affected by the interviewer's characteristics, so the interviewer is similarly affected by the characteristics of the respondent. Sex, age, race, social class, and a host of other factors may subtly shape the way in which the interviewer asks questions and interprets respondents' words.

A fourth disadvantage of interviews, especially with less structured interviews, is the fact that the *meanings of words and phrases can vary* from person to person and setting to setting. A part of this problem is the possibility of significant but unnoticed variation in wording from one interview to the next or among interviewers. We know that variations in wording can produce variations in response, and the more freedom interviewers have in this regard, the more of a problem it becomes. Wording variation can affect both reliability and validity (see Chapter 5). A deeper issue, however, is that meaning arises not from words and phrases but from a whole context and from the process of interaction between people. In other words, each interview is a unique social situation, and the possibility exists that the meaning constructed in one interview situation—because it is a unique situation—can be quite different from the meaning constructed in another interview. Non-positivists especially have been critical of interviews on these grounds, arguing that interview data is not nearly as objective as positivists claim and is certainly much more complicated.

SPECIALIZED SURVEY FORMATS

Telephone Surveys

Face-to-face interviews tend to be a considerably more expensive means of gathering data in comparison to mailed questionnaires or telephone surveys (Rea & Parker, 1992). As Table 9.2 shows, face-to-face interviews can cost more than twice as much as a survey conducted via phone or mail. The table shows that with face-to-face interviews, there are substantially higher costs for such things as locating and contacting respondents, conducting interviews, traveling, and training interviewers. With mail or telephone surveys, there is no travel time and fewer interviewers are needed; in addition, fewer supervisory personnel are needed. Although such costs as telephone charges are higher in telephone surveys, they are far outweighed by the other savings. The cost advantage of the less expensive types of surveys makes feasible research that otherwise would be prohibitively expensive.

The speed with which a telephone survey can be completed also makes it preferable at times. If we wanted people's reactions to some event, for example, or repeated measures of public opinion, which can change rapidly, the speed of telephone surveys makes them preferable in these circumstances.

Response bias can also be reduced by using the more anonymous technique of a telephone survey rather than an interview. For example, a study of tobacco, alcohol, marijuana, and cocaine use showed that people were more willing to admit using these substances and admit to higher levels of use, when interviewed over the phone rather than in person (Aquilino & LoSciuto, 1990).

Certain areas of the country and many major cities contain substantial numbers of non-English-speaking people. These persons are difficult to accommodate with mailed questionnaires and personal interviews unless we know ahead of time what language a respondent speaks. Non-English-speaking people can be handled fairly easily with telephone surveys, however. All that is needed are a few multilingual interviewers. (Spanish-speakers account for the vast majority of non-English-speaking persons in the United States.) If an interviewer contacts a non-English-speaking respondent, that respondent can be conveniently transferred to an interviewer conversant in the respondent's language. Even though multilingual interviewers can be and are used in personal interviews, this is far less efficient, probably involving at least one callback so that an interviewer with the appropriate language facility can be sent out. A final advantage of telephone interviews is that supervision of the interviewers is far easier. The problem of curbing is eliminated because supervisors can monitor the interviews any time they wish. This makes it easy to ensure that specified procedures are followed and that any problems that might arise are quickly discovered and corrected.

Despite these considerable advantages, telephone surveys have several limitations that make the method unsuitable for some research purposes. First, they must be quite short in duration. Normally, the maximum length is about 20 minutes, with most being even shorter. This is in sharp contrast to personal interviews, which can extend to an hour or more. The time

TABLE 9.2   Cost Comparison of Telephone, Mail, and Face-to-Face Surveys, With a Sample Size of 520

A. Mail Survey	Total Cost (dollars)		Total Cost (dollars)
Prepare for survey		Postage for return envelopes, 960 @ $.52 each	500
Purchase sample list in machine-readable form	375	Sign letters, stamp envelopes	100
Load database of names and addresses	17	Prepare mail-out packets	118
Graphic design for questionnaire cover (hire out)	100	**Third mail-out (960)**	
Print questionnaires: 4 sheets, legal-size, folded, 1,350 @ $.15 each (includes paper) (hire out)	203	Prestamped postcards, 4 bunches of 250 @ $.19 each	190
Telephone	100	Address postcards	25
Supplies		Print message and sign postcards	50
Mail-out envelopes, 2,310 @ $.05 each, pre-addressed but no return address	116	Process, precode, edit 390 returned questionnaires, 10 min each	546
Return envelopes, 1,350 @ $.05 each, with return address	68	**Fourth mail-out (475)**	
Letterhead for cover letters, 2,310 @ $.05 each	116	Print cover letter	25
Miscellaneous	200	Address envelopes	25
First mail-out (960)		Sign letters, stamp envelopes	25
Print advance-notice letter	25	Prepare mail-out packets	168
Address envelopes	25	Postage for mail-out, 475 @ $.52 each	247
Sign letters, stamp envelopes	50	Postage for return envelopes, 475 @ $.52 each	247
Postage for mail-out, 960 @ $.29 each	278	Process, precode, edit 185 returned questionnaires, 10 min each	260
Prepare mail-out packets	134		
Second mail-out (960)		**Total,** excluding professional time	4,883
Print cover letter	25	Professional time (120 hrs @ $35,000 annual salary plus 20% fringe benefits)	2,423
Address envelopes	25		
Postage for mail out, 960 @ $.52 each	500	**Total,** including professional time	7,306

limitation obviously restricts the volume of information that can be obtained and the depth to which issues can be explored. Telephone surveys work best when the information desired is fairly simple and the questions are uncomplicated.

A second limitation stems from telephone communication's being only voice-to-voice. Lack of visual contact eliminates several desirable features characteristic of personal interviews. The interviewer is unable to supplement

TABLE 9.2   *(continued)*

B. Telephone Survey	Total Cost (dollars)	C. Face-to-Face Survey	Total Cost (dollars)
Prepare for survey		**Prepare for survey**	
Use add-a-digit calling based on systematic, random sampling from directory	84	Purchase map for area frame	200
Print interviewer manuals	37	Print interviewer manuals	29
Print questionnaires (940)	84	Print questionnaires (690)	379
Train interviewers (12-hour training)	700	Train interviewers (20-hour training session)	1,134
Miscellaneous supplies	25	Miscellaneous supplies	25
Conduct the survey		**Conduct the survey**	
Contact and interview respondents; edit questionnaires; 50 minutes per completed questionnaire	2,786	Locate residences; contact respondents; conduct interviews; field edit questionnaires: 3.5 completed interviews per 8-hour day	9,655
Telephone charges	3,203	Travel costs ($8.50 per completed interview; interviewers use own cars)	4,420
		Office edit and general clerical	
		(6 completed questionnaires per hour)	728
Total, excluding professional time	6,919	**Total,** excluding professional time	16,570
Professional time (120 hours @ $35,000 annual salary plus 20% fringe benefits)	2,423	Professional time (160 hrs @ $35,000 annual salary plus 20% fringe benefits)	3,231
Total, including professional time	9,342	**Total,** including professional time	19,801

Source   Adapted from Priscilla Salant and Don A. Dillman, *How to Conduct Your Own Survey* (New York: John Wiley and Sons, 1994), pp.46–49. Used with permission.

responses with observational information, and it is harder to probe effectively without seeing the respondent. Further, the use of cards with response alternatives or other visual stimuli is precluded. The inability to present complex sets of response alternatives in this format can make it difficult to ask some questions that are important.

Finally, as noted in Chapter 7, surveys based on samples drawn from telephone directories may have considerable noncoverage because some people lack telephones and others have unlisted numbers. Although modern telephone sampling techniques, such as random digit dialing, eliminate the problem of unlisted numbers, the approximately 6% of households without

telephones remain unreachable, and they are concentrated among the poor and transient segments of the population. So some sampling bias remains even if random digit dialing is used.

Answering machines have come to play a significant part in telephone surveys (Piazza, 1993; Xu, Bates, & Schweitzer, 1993). Approximately one-quarter to one-third of households have an answering machine, although the amount of time they are used varies across households. Answering machines affect response rates: Households with answering machines are more likely to complete an interview and less likely to refuse participation than households where there is no answer at an initial call for an interview. So, reaching an answering machine is better news for the telephone interviewer than getting no answer because the former is more likely, ultimately, to complete the interview. However, it also takes more time and resources to complete an interview when the initial contact was a message on an answering machine: It takes twice as many calls to complete an interview with answering-machine households than with non-answering-machine households. In addition, households with answering machines tend to be better educated and have higher incomes than households without them, so there is a potential for a biased sample if the answering-machine households have higher response rates. There are two things about the impact of answering machines that telephone surveyors scrutinize closely: costs (do you spend more to reach an answering-machine household) and bias (is the response rate skewed because of the answering machines).

Computer-Assisted Interviewing

Computer-assisted interviewing (CAI) refers to using computer technology to assist in the completion of questionnaires and interviews. This originally took the form of computer-assisted telephone interviewing (CATI) where the interview is conducted over the telephone as the interviewer reads questions from a computer monitor instead of a clipboard, and records responses directly into the database via the computer's keyboard, instead of using a paper form (Buetow, Douglas, Harris, & McCulloch, 1996; Groves, Blemer, Lyberg, Massey, Nicholls, & Waksberg, 1988). Superficially, CATI replaces the paper-and-pencil format of interviewing with a monitor-and-keyboard arrangement, but the differences are far more significant. Some of the special techniques that are possible with CATI include personalizing the wording of questions based on answers to previous questions, and automatic branching for contingency questions. For example, in a study on sibling relations, if a respondent reports that she has a son named James who is

14 and a daughter named Virginia who is 11, the computer can automatically insert the name "Virginia" into questions to refer to the younger child. If one subset of questions concerns same-sex siblings, and a different set addresses boy-girl relations, the computer will automatically select the latter subset and skip the former. These features speed up the interview and improve accuracy because the interviewer can concentrate fully on the questions at hand instead of searching through pages of items that do not apply. The CATI program may also include enforced probing when respondents give incomplete answers, editing of responses, and automatic call scheduling.

In computer terminology, such programs are "interactive"; that is, the interview schedule changes and presents customized instructions for the interviewer, depending on the responses that are recorded. Although these CATI features are expensive and thus have traditionally been available only to large-scale survey organizations, advances in computer software and hardware now make the technology feasible for smaller organizations that need to do periodic surveys.

Computer-assisted interviewing involves two steps. The first is designing the interview schedule, which can be done using specialized software that can produce either hard copy or online versions of the interview. The second step is conducting the interviews. The online survey forms are interactive, and the software enters the data from respondents directly into a data file for analysis. If the printed version is used, the filled-out survey forms can be scanned into the data file.

Besides customizing questions and utilizing branching routines, the program helps prevent errors from entering the data during the collection phase. For example, with a question that requires numeric data such as "How old are you?" the program can require that only numeric characters be entered. If the interviewer accidentally enters a letter, the program responds with an error message and requires the interviewer to reenter the response. Range checks can also be used to catch errors. Assuming one is interviewing adults, the age range might be set to 18–99. Any response outside that range would result in an error message or a request to recheck the entry.

Data entry can be simplified by prerecording responses to which most people will give the same answer. For example, when asking about health status, very few respondents may have had heart attacks. A "No" response can be prerecorded for this item so that data need be changed only for those few individuals who respond "Yes."

The interview schedule can be further refined to warn of logical inconsistencies. If a respondent lists three family members as being employed during the year, an error message would result if the interviewer attempts to enter zero annual income for one of those persons at a later point in the survey. The program will also catch errors such as a February 29 birthday in a non-leap year.

In some situations, the order in which questions are asked can affect the responses in a survey. Because of this, it is common to administer scale items in a random order in a questionnaire. Some software now includes a special feature to take the principle of randomized question ordering a step farther. In administering a scale, it will randomly order the items separately for each respondent to eliminate possible bias due to question order.

Although telephone interviewing was the first application of computerized interviewing, it has now been extended to what is called **computer-assisted personal interviewing** (CAPI) where face-to-face interviewers use computers in much the same way that the telephone interviewers did. Laptop and notebook computers have proved especially versatile in this regard, as online, interactive questionnaires can be taken into the field. CAPI has all the advantages of CATI, and is also more versatile since it can be used in many field settings. An extension of this is called **computer-assisted self-interviewing** (CASI) where the respondents read questions on the computer screen and enter their responses at the keyboard themselves. CASI was used for clinical tasks by psychologists and social workers before it was taken up by researchers for use in large-scale surveys. In some cases, computers may be left in the homes of respondents so that they can periodically respond to survey questions on a prearranged schedule. In other cases, CASI is used in combination with CAPI, with sensitive questions being answered through CASI, which gives people a greater sense that their privacy is being protected. One disadvantage of CASI is its effect on response rates: Older and less educated respondents, as well as those with less computer experience, are less likely to give answers to CASI portions of interviews (Couper & Rowe, 1996; Tourangeau & Smith, 1996).

Two evaluations suggest that CATI clearly has some advantages over traditional methods but is not without problems (Catlin & Ingram, 1988; Weeks, 1988). The costs of the two approaches proved to be about equal, but CATI had a lower overall response rate because CATI interviews took longer, especially when conducted by interviewers who were unfamiliar with the procedure. However, as the interviewers gained practice, interview time was reduced dramatically. Another problem was computer downtime. If the computer system is inoperable for a period of time, which happens, then interviewing is delayed and some interviews may be lost entirely. CATI had its greatest benefit in reducing the error rate. It was 50% less overall with CATI, and 60% less in the portion of the questionnaire that required complex branching. The researchers concluded that studies that require complex branching would benefit the most from CATI.

Online Surveys

The growth of the Internet in the 1990s has led to the possibility of conducting surveys using that medium rather than in person or through the mail or telephone (Nesbary, 2000). Online surveys are conducted in two ways. One involves sending the survey to respondents via e-mail or as an attachment to e-mail. E-mail surveys are fairly straightforward, requiring only that one have access to the e-mail addresses of all the respondents.

In the second type of online survey, the survey is made available at a Web site. These surveys are designed using Web design software and uploaded to an Internet server, such as a university or a private Internet Service Provider. Respondents are directed in some fashion to go to the Web site where the survey is displayed. If necessary, access to the survey can be controlled by requiring a password to log onto the Web site where it is located. Web design software, such as Microsoft Front Page and Adobe PageMill, now enables one to design surveys for use online. They are designed such that the respondents simply click the options they wish to choose for closed-ended questions or keyboard in their answers to open-ended questions in a text box provided. When done, the respondent clicks a "submit" button and the survey is returned to the researcher, often directly to a data file that is ready for analysis.

Online surveys have many advantages. They are quick, inexpensive, and can reach audiences around the world. One study found, for example, that a survey sent to 100 people cost $72 to distribute and collect via e-mail versus $503 via postal mail—quite a difference (Mavis & Brocato, 1998). The same survey took an average of one day to return via e-mail versus 13 days via postal mail. While research on using online surveys is sparse at this point, earlier research has shown that administering questionnaires by computer rather than by mail or in person generates responses that are more extreme and less affected by the social desirability effect discussed in Chapter 6 (Kiesler & Sproull, 1986). The anonymity and impersonal nature of interacting with a machine rather than a human interviewer may reduce respondents' concerns about how their responses appear to others. It also makes it a good way to contact groups that are difficult to access in other ways, possibly because they are involved in deviant interests or activities that they wish to hide from others (Wysocki, 1999).

There are, of course, disadvantages to online surveys. One major concern is the issue of sampling and representativeness (Kaye and Johnson, 1999). The population of people who use the Internet tends to be skewed toward the affluent, well educated, young, and male. So, unless you have a clearly defined population, all of whose members have access to and actually use the Internet, questions of the representativeness of online respondents are difficult to resolve. Even with a clearly defined population, this could be a problem. For example, an online survey of the faculty members at a university would probably involve a population where all members have access to the Internet. However, it may be the younger faculty or those from particular disciplines who are most likely to respond. So the issues of response rate and representativeness must be scrutinized as carefully as they are with other types of surveys. Data available at this point suggest that postal surveys achieve a considerably higher response rate than do e-mail surveys (Mavis & Brocato, 1998). However, for needs assessment surveys and some kinds of qualitative research where probability samples are not critical, online surveys can be very useful.

Another difficulty with online surveys is formatting: Different computer systems sometimes change the formatting of the survey in unpredictable ways. A survey that looked clear on the designer's computer screen may be partially unintelligible when e-mailed to a respondent's computer. In addition, the design features available in some Web page design software may not be supported by all Internet servers. We have seen earlier in this chapter that the formatting and design of surveys are both very important in terms of achieving high response rates and gathering complete and valid responses. If respondents with various computers receive differently formatted surveys, this may influence their willingness to participate, or their responses, and introduce error into measurement. This is a serious concern, although new approaches and improvements in software will reduce and possibly eliminate this problem in the near future.

Technology like the telephone and the Internet have made it easier than ever to conduct surveys. However, as the Eye on Ethics box argues, this ease of administering surveys has raised ethical issues regarding whether too many surveys are being conducted.

Focus Groups

Research situations arise in which the standardization found in most surveys and interviews is inappropriate and researchers need more flexibility in how they elicit responses to questions. One area where this is likely to be true is exploratory research. Here, research questions cannot be formulated into precise hypotheses, and our knowledge of some phenomenon is too sketchy to allow precise measurement of variables. Another research situation where less structure and more flexibility is called for is qualitative research that explores very personal and subjective experiences that are unlikely to be tapped adequately by asking the same structured questions of everyone.

In such research situations, a flexible strategy for gathering data is the **focus group,** or **group depth interview** (Krueger, 1994; Morgan, 1994). It is similar to an unstandardized interview but, as the name implies, it is an interview with a whole group of people at the same time. Originally, focus groups were intended as an early step in the research process, to help researchers develop hypotheses and questionnaire items, and they are still sometimes used this way. Survey researchers, for example, sometimes use focus groups as a tool, for developing questionnaires and interview schedules. However, focus groups are now also used in applied research as a strategy for collecting data, especially when doing qualitative research to tap people's subjective experiences. Today, focus groups are widely used in applied

research, marketing research, and political campaigns. One example of this is a study of the barriers women confront in obtaining medical care to detect and treat cervical cancer, a potentially fatal ailment that can be readily detected and treated if women obtain Pap smears on a regular basis and return for follow-up care when necessary. The researchers decided that a focus group "would allow free expression of thoughts and feelings about cancer and related issues" and would be the most effective mechanism to probe women's motivations for not seeking appropriate medical care (Dignan, Michielutte, Sharp, Bahnson, Young, & Beal, 1990, p. 370).

A focus group is led by one or more moderators whose job is to direct the discussion by following an outline of the main topics of inquiry. The group itself contains anywhere from 5 to 15 members, who are selected on the basis of their usefulness in providing the data called for in the research. The women for the study on cervical cancer, for example, were chosen, among other reasons, because they had had some previous experience with cancer. Focus group membership is not normally based on probability samples, which Chapter 7 points out as the most likely to be representative samples. This can therefore throw the generalizability of focus group results into question. However, in exploratory and qualitative research, such generalizability is not as critically important as it is in other research. In addition, most focus group research enhances its representativeness and generalizability by collecting data from more than one focus group. The cervical cancer study involved four separate focus groups of 10 to 12 women, and some research projects use 20 + focus groups.

The moderator in a focus group encourages discussion by asking questions, probing areas that are not clear or sufficiently discussed, and pursuing new lines of inquiry that emerge from the discussion and seem relevant. In addition, the moderator uses a knowledge of group dynamics to elicit data that might not have been obtained in an in-depth interview. For

EYE ON ETHICS   When Is It Wrong to Conduct a Survey?

It sometimes seems that almost everyone wants to survey practically everyone else about almost everything. In fact, surveys can be so easy to administer that they are subject to abuse by overuse. One of the consequences of this is that people can react against this constant intrusion into their lives by phone or mail by refusing to respond to surveys. I know that I have done this—especially when a telephone survey hits me just as I am preparing dinner! The danger is that this saturation by survey will make it more difficult for survey researchers to obtain acceptably high response rates.

It is notoriously difficult and controversial to decide which surveys should be done and which not, but there are some criteria that can be applied. A good survey is one that has been prepared with care and rigor, using the kind of measurement, sampling, and other criteria discussed in this book. I know from my own experience that some people who have done surveys, especially those without any special technical training in the field, tend to quickly write up the questions that will constitute the survey, with, at best, a vague use of face validity to decide whether the questions are any good. There is no pretesting or more sophisticated assessments of validity and reliability, or any concern with such matters as the sequencing of questions. If these poorly designed surveys are not valid measures—which they likely are not—then it is ethically questionable to distribute them. The conclusions from such surveys are of little use and may actually be deceptive or misleading. In addition, they contribute to the overall saturation of survey respondents, whose disinclination to respond seems to increase with each additional survey.

Another criterion that can be used to assess whether a survey is appropriate is its purpose, which should be solely that of gathering respondents' opinions on some topic. The survey

example, natural leaders in the group will be encouraged by the moderator to help others speak out and express their views. Group members often respond more readily to other group members than to the moderator. People in a focus group will make side comments to one another—obviously not possible in a one-person interview—and the moderator notes these comments and possibly encourages group members to elaborate on them. In fact, in a well-run focus group, the members may interact among themselves as much as with the group moderator. In a standard interview, the stimulus for people's responses is the interviewer's questions; by contrast, focus group interviews provide a second stimulus for peoples' responses—the group experience itself.

The typical focus group session begins by exploring more general topics and then is gradually directed by the moderator toward more specific issues (Krueger, 1994). For example, in the focus group study of cervical cancer, the moderators began with questions about general life concerns and the perceived value of health and ended with specific questions about cancer, cancer screening, and Pap smears. The general questions provided a foundation and a context without which the women might not have been as willing or as able to come up with useful answers to the more specific questions. Group moderators take great care in using questions to transition from the general to the specific. Moderators also try to ensure that all members of the group have an opportunity to participate. For

should not include an effort to change the respondent's opinion, persuade them on some issue, or sell something to them. In fact, some professional survey researchers have coined the term "sugging" to refer to "selling under the guise" of research. This occurs when a telephone survey is ostensibly to gather someone's opinion but then changes into an effort to sell a product. The survey nature of the sales effort is presumably used as a way to gain cooperation and legitimacy and get a foot in the door in the hopes of increasing the chance of a sale. This is not much of a concern in academic or basic research where there is normally no commercial product to sell. However, it could become a problem in areas like marketing research where companies do have products to sell. There is nothing unethical about doing surveys in marketing research, as long as the purpose of the survey itself is to gather opinion about a product or service. But there is an ethical line to be crossed if the opinion-gathering context transforms into a sales pitch. In fact, the federal Telemarketing and Consumer Fraud and Abuse Prevention Act makes selling under the guise of research, or "sugging," illegal by requiring telemarketers to promptly disclose their name and that the purpose of the call is sales related—including the nature and price of the product the caller is attempting to sell. These issues are discussed on the Web page of the Council for Marketing and Opinion Research (**www.cmor.org**).

Professional organizations, such as the American Association for Public Opinion Research and the Council for Marketing and Opinion Research, strive to maintain both the quality and the ethical standards of those conducting surveys. Survey researchers address other ethical issues, of course, such as privacy and informed consent. However, the ease of conducting surveys has led to their abuse.

example, naturally quiet people may have to be encouraged to join in, whereas outgoing people may have to be politely restrained. In short, focus group moderators must possess a knowledge of group dynamics as well as skills in understanding and working with people. During a focus group, the moderator or his or her assistant records the data in field notes or on a video or audio recording for later analysis. What results is mostly qualitative data, the analysis of which is discussed in Chapter 15.

Focus groups have major advantages over more structured, single-person interviews: The former are more flexible, cost less, and can provide quick results. In addition, focus groups have the advantage of using the interaction between people to encourage people to participate. The fact that the individuals being studied participate so heavily in the production of data in focus groups has made focus groups a preferred research method for some feminist and other nonpositivist researchers (Wilkinson, 1998). Focus groups can be organized such that the relationship between the moderator and the participants is fairly equal, and the participants share in both the production of the data and the analysis of its meaning. Unfortunately, focus groups also have disadvantages: The results are less generalizable to a larger population, and the data are more difficult and subjective to analyze. Focus groups are also less likely than interviews to produce quantitative data; in fact, focus group data may more closely resemble the field notes produced in field research discussed in Chapter 11.

WHEN TO CHOOSE A SURVEY

From what I have described in this chapter, it should be clear that surveys are an appropriate research technique when variables can be accurately measured by people's responses to questions. This is the core of survey research. If the variables of interest refer to individuals' feelings, values, expectations, or some other internal state, then surveys would be very appropriate. Asking people questions about how they feel or what they think or believe is one of the key ways we have of getting at these internal or subjective phenomena. It is also reasonable to ask respondents about their behavior in the past or at other times, especially when the researcher has no way of directly observing those behaviors. A survey, for example, can be a quick and inexpensive way of determining whether a person has ever been married, whether he or she graduated from high school, or whether that person has been convicted of a crime in the past year. Other ways of gathering this data would be far more costly. However, as noted, care must be taken to ensure that people's answers are accurate. It is also reasonable to ask people survey questions about their recollection of the behavior of others or about events and circumstances, but we cannot assume that their recollections are accurate.

Also, surveys are chosen when the research calls for collecting data from large numbers of people or those widely dispersed geographically. The survey format is sufficiently flexible in that it can accommodate quantitative (questionnaires and some interviews) as well as qualitative (some in-depth interviews and focus groups) forms of data collection. Surveys are also efficient, in that they can, in a short period of time, collect data on many variables, including multiple indicators of some phenomena.

Surveys can be used in research that pursues any of the four goals of research discussed in Chapter 1 (description, explanation, prediction, evaluation). As for the unit of analysis, it is most often used when the individual is the focus of attention, but surveys are very flexible. We can use them to collect data about groups or organizations, for example, although the questions about these units of analysis would normally have to be asked of some individual. (See the discussion in Chapter 4 on the difference between the unit *about which* data is collected and the unit *from which* the data is obtained).

REVIEW AND CRITICAL THINKING

The research techniques discussed in this chapter involve observations of what individuals say about their own thoughts, feelings, or behaviors; or what they say about others. This kind of research technique has many advantages, as we have seen, but it also has many drawbacks, which leads social scientists to be very careful when using this research strategy. You also need to be cautious in your everyday life when confronted with information or conclusions based on similar data. Consider the following.

1. Are the topic and the conclusions best addressed by what persons *say* about their thoughts or behavior (a survey) or by direct observation? What do the conclusions focus on: what people have said or their

actual behavior? Is it legitimate to conclude something about people's behavior from what they have said?

2. What questions were asked and how were they asked? Do they contain any of the flaws discussed in this chapter that could produce bias or misinterpretation?

3. What about reactivity? Could the manner in which the information was gathered have influenced what people said?

4. What about sampling? Could the manner in which the information was gathered, such as by telephone or online, have influenced on whom the observations were made?

Computers and the Internet

Some of the major uses of computers in survey research, such as CATI, CASI, and CAPI, are discussed in Chapter 9. A rich variety of resources also awaits anyone who seeks information about survey research on the Internet. Most major survey research centers maintain Web sites, some of which are extremely useful. At many sites, you can find a basic overview of survey research, a discussion of ethics in surveys, and information on how to plan a survey. Some of these sites contain mini courses on all aspects of conducting surveys. Some sites even provide the opportunity of examining questions that are used in actual surveys. It is also possible to download and read entire questionnaires and survey instruments. By reviewing these surveys, you can explore how the instrument is structured, the ordering of questions, skip patterns, and other features that enhance the quality of the research instrument. Not only will you become more familiar with major survey projects around the world, you can also learn a great deal about how good survey questions are designed.

Many survey Web sites have a search field that enables you to locate questions on almost any topic imaginable. Try one of the search fields by entering a topic of interest to you, such as domestic violence, health care, or poverty. Determine if the questions that result are open ended or closed ended. Examine how the questions are worded and consider how you might improve them or adapt them to a project of interest to you. Depending on the source, you may be able to use or adapt questions from these surveys for use in a survey instrument that you are designing.

Four of the best sites that I have found are listed below.

- The Odum Institute for Research in Social Science (IRSS):**http://www.irss. unc.edu**
- The General Social Survey: **http://www. icpsr.umich.edu/gss/**
- The Centre for Applied Social Surveys in England: **http://www.scpr.ac.uk/cass/**
- The Survey Research Center at Princeton University: **http://www.princeton.edu/~abelson/index.html** (This site has some of the best links to organizations involved in survey research throughout the world that I have seen.)

Use the search engine on your Web browser to look for other survey research centers like these.

You can also locate commercial survey research organizations on the Internet and learn quite a bit about survey research and conducting online surveys from them. Two that I am familiar with are eLISTEN (**http://www.surveyssay.com**) and The Survey System (http://www.surveysystem.com). Search the Web for other such commercial firms.

Main Points

1. Surveys are of two general types: questionnaires, which are completed directly by respondents; and interviews, in which the questions are read and the responses recorded by an interviewer. A key point

about surveys is that they measure what people *say* about their behavior; they don't measure behavior directly.

2. Clear directions must be provided on questionnaires, to indicate what respondents are to do and to guide them through the questionnaire. Questions should be ordered so that early questions maximize the response rate but do not affect the responses to later questions. Contingency questions and filter questions are often used to guide people through questionnaires.

3. Obtaining a high response rate is very important for representativeness in survey research. Central to efforts to maximize the response rate with the mailed questionnaire are the cover letter, the use of payments, follow-up letters and telephone calls, and the length and appearance of the questionnaire. A variety of procedures exist for checking whether nonresponse produces a biased sample.

4. Interviews are classified by their degree of structure as unstandardized, nonschedule-standardized, or schedule-standardized. Contacting respondents to elicit their participation in interviews is important, and some procedures are more effective at this than others.

5. An interview is a social relationship and a secondary relationship. The interviewer's goal in the relationship is to collect valid data. Probes are used to elicit clearer and more complete responses during interviews. Interviewers need to be trained well and then supervised to ensure that they collect data properly and completely.

6. Telephone surveys are a suitable alternative in many cases and offer significant time and cost benefits compared to interviews or mailed questionnaires.

7. Computer-assisted interviewing has also become a useful way of conducting telephone and personal interviews. Computer assistance can make interviewing more efficient and accurate by enabling responses to be input into the computer directly, as well as checking for errors or inconsistencies in responses as the interview is being conducted.

8. Surveys are now also conducted online, either via e-mail or on a Web site. While these surveys are useful, they can raise concerns about the possibility of bias due to nonresponse.

9. Focus groups are often used in situations where the standardization of question format found in most questionnaires and interviews is inappropriate. Focus groups can be used to provide data to develop hypotheses or when the group dynamics of the focus group will help produce data that would not be discovered with a standard questionnaire or interview format.

Important Terms for Review

computer-assisted interviewing
computer-assisted personal interviewing
computer-assisted self interviewing
computer-assisted telephone interviewing
contingency question
cover letter
filter question
focus group
group depth interview
interview
interview schedule
interviewer bias
matrix question
probes
questionnaire
response rate
survey

For Further Reading

Dillman, Don A. (2000). *Mail and Internet surveys: The tailored design method* (2nd ed.). New

York: John Wiley & Sons. This is an excellent introduction to survey research in general, and it also provides the most up-to-date overview of how to conduct surveys via mail and Internet.

Gorden, Raymond. (1992). *Basic interviewing skills.* Itasca, IL: Peacock. A very useful "how-to" book on interviewing, it covers everything from how to develop questions, to how to motivate good responses, to how to evaluate respondents' nonverbal behavior.

Rosenberg, M. (1968). *The logic of survey analysis.* New York: Basic Books. A presentation of the logic of surveys, understanding relationships between variables, and analyzing the findings of surveys that requires very little statistical background to comprehend.

Salant, Priscilla, & Dillman, Don. (1994). *Conducting surveys: A step-by-step guide to getting the information you need.* New York: John Wiley & Sons. As the title states, this is a very useful guide to all the steps in conducting sound survey research.

Schuman, Howard, & Presser, Stanley. (1996). *Questions and answers in attitude surveys: Experiments on question form, wording, and content.* Thousand Oaks, CA: Sage. This is a comprehensive handbook on the rules, problems, and pitfalls of designing questions for surveys. It goes far beyond what this chapter is able to cover on this important topic.

Schwarz, N., & Sudman, S. (Eds.). (1996). *Answering questions: Methodology for determining cognitive and communicative processes in survey research.* San Francisco: Jossey-Bass. This collection discusses the methods involved in observation and/or taping and the subsequent coding and analysis of interviews; explores how to obtain, code, and analyze verbal protocols from respondents; and examines other survey techniques, including sorting tasks and response latency measures.

Stewart, David W., & Shamdasni, Prem N. (1990). *Focus groups: Theory and practice.* Newbury Park, CA: Sage. This book presents a thorough discussion of the role that focus groups can play in survey as well as other types of research. It presents detailed coverage of the design and conduct of focus groups as well as techniques for analyzing the data generated from them.

Weisberg, Herbert F., Krosnick, Jon A., & Bowen, Bruce D. (1996). *An introduction to survey research, polling, and data analysis.* Thousand Oaks, CA: Sage. A comprehensive guide to conducting surveys and large-scale public opinion polls, this also provides the reader with a cautious approach toward the interpretation of survey results.

Weiss, Robert S. (1993). *Learning from strangers: The art and method of qualitative interview studies.* New York: Free Press. This is the definitive work on qualitative research interviewing. Weiss provides examples and running commentary on how social interaction during interviews either inhibits or promotes trust and alliance. Used as a reference, handbook, or text, this book is appropriate for novices and professionals.

Chapter 10

UNOBTRUSIVE AND AVAILABLE DATA RESEARCH

Consider the following research studies done by various social scientists:

1. A study of the lyrics in modern rap music, showing that it is a cultural form with messages of resistance and empowerment for minorities in the United States and serves as subcultural social criticism expressing distrust and anger at what is perceived as a racist and discriminatory dominant Anglo culture (Martinez, 1997).

2. A study of prime-time television programs that explores what messages about alcohol use are presented in this medium, especially as it affects adolescents (Mathios, Avery, Bisogni, & Shanahan, 1998).

3. A study of magazine advertisements to determine whether the behavior displayed in the ads differs for men and women and to assess whether stereotypical portrayals of men and women have changed over the past two decades (Kang, 1997).

4. A study of how emotions are expressed and interpreted in different settings by reviewing audiotapes of emergency 911 calls (Whalen & Zimmerman, 1998).

All four of these studies involve a research technique that is the focus of this chapter. In all cases, the raw data—the music lyrics, the television programs, the advertisements, and the 911 audiotapes—were produced in such a way that the people involved were not aware that their behavior or products would be the focus of research attention. In experiments, surveys, and much field research (chapters 8, 9, and 11), the people being observed are usually aware that they are the focus of research attention and may even know exactly what data is being gathered about them. For some research questions, however, there is a danger that this awareness might change people's behavior in ways detrimental to the research question. I discuss this in Chapter 4 as the issue of *reactiv-*

ity. Most of the research in this chapter would be classified as **unobtrusive** or **nonreactive research**: Those under study are not aware that they are being studied, and the investigator does not change their behavior by his or her presence (Sechrest & Belew, 1983; Webb, Campbell, Schwartz, Sechrest, & Grove, 1981). Most of the research covered in this chapter is also called **available data research**: the researcher uses data that was collected by someone other than the investigator himself or herself for purposes that differ from the investigator's but that are nonetheless available to be analyzed.

This chapter begins with an exploration of some different sources of data used in unobtrusive research. Then the conduct of available data research is described, covering both the analysis of statistical data and of content. Finally, comparative and historical research is discussed as a combination of both unobtrusive and available data research. The chapter closes with an assessment of these various types of research.

SOURCES OF UNOBTRUSIVE DATA

Unobtrusive data can be found in naturalistic or contrived settings and can involve both quantitative or qualitative observations. A vast array of this type of data is available for scientific analysis.

Statistical Data

One form of unobtrusive data is *statistical data,* or quantified observations of some element of human behavior. Large amounts of statistical data are collected by the various branches of government, for example, and provide quantified information about crime, health, birthrates, death rates, and the like. In fact, virtually all large organizations collect statistical data that have to do with the operation of the organization and its membership and clients or consumers. Another form of available statistical data involves surveys that are conducted by someone and then made available to other people to be analyzed. Specific examples of this type of data will be discussed in the next section.

Hidden and Disguised Observation

In some research projects, it is possible to make *hidden observations*; that is, to observe behavior from a vantage point that is obscured from the view of those under observation. This might be done by observing people through a one-way mirror or by videotaping them with a hidden camera. For example, studies of aggressive behavior among children have utilized such hidden observations by videotaping the children at play in schoolyards with the camera in a hidden location (Pepler & Craig, 1995). With some types of behavior, it is possible to observe people in a natural setting, but without participating and without revealing that one is a researcher making observations. Any setting in which one can be present and not participate without calling attention to oneself is a potential scene for such *disguised observation*. One setting that might lend itself to such disguised observation is a public establishment like a bar or tavern. A researcher could pretend to be simply another patron in the bar and use that position to make observations. This would still be considered unobtrusive observation, even though the other patrons can see the researcher, because they are unaware of the researcher's identity as researcher or the fact that research observations are being made. In such a setting, the researcher's impact on the behavior of the clientele would probably be minimal. Hidden and disguised observations of this sort are a type of field research and are discussed at more length in Chapter 11.

Physical Traces

Another type of unobtrusive data, **physical traces**, are the physical objects or evidence that result from people's activities and that can be used as data to test hypotheses. There are two types of physical traces. *Erosion measures* in-

volve the degree to which some materials are worn, eroded, or used up. For example, nurses making home visits commonly observe the level of medicine in medicine bottles as an unobtrusive indicator of whether the patient is taking the prescribed amount. *Accretion measures* are those involving materials that are deposited or accumulated because of human activity. For example, the archaeologist William Rathje directed a project devoted to the study of modern household refuse as an accretion measure (Rathje & Murphy, 1992). Sometimes called "garbology," this research was designed to learn more about contemporary civilization by studying the things we throw out as garbage. Such a study can serve a number of purposes. First, it can serve as a check on what a respondent tells an interviewer. If a person claims to consume a few cans of beer a day, inspection of that person's refuse can validate or refute that claim. Second, it can measure people's responses to various social and economic changes. As food prices rise, one can measure the changing patterns of food consumption by people in varying social classes.

Another example of accretion analysis involves graffiti. Dismissed by some lay people as either meaningless or destructive, social scientists recognize that graffiti can have considerable cultural significance. Ferrell (1995), after a four-year study of "hip hop" graffiti writing, suggests that it represents an important form of youthful communication and expression. Graffiti writing, known as "tagging," is, for many youths, a form of protest against many of the unpleasant things they face in their urban lives. The most frequent targets are symbols of authority, such as public buildings or large businesses. The "tagging" crews are even looked upon as surrogate families who significantly supplement the often fragmented and dysfunctional urban family.

Documents

Documents are another form of available data and refer, in their broadest sense, to any form of communication that is nonquantitative in format, such as books, music, magazines, letters, diaries, and other communication media such as radio, television, movies, and plays. Documents can also make up a part of the records maintained by institutions, such as military records, police records, court records, or social agency records. Some such records, of course, contain statistical data as well as information in nonstatistical form.

AVAILABLE STATISTICAL DATA

Sources of Statistical Data

Who among us hasn't complained about the mountains of paperwork associated with organizational and bureaucratic life in modern societies. Yet the facts and figures generated by these organizations can be a rich source of data for social science research. However, great care must be taken in doing available data research because those who collect the organization's data are not researchers. Rather, they collect the data for the purposes of the organization, and those purposes may not produce data that are useful to researchers. These people may also not be aware of the considerations that are necessary to produce valid data for research.

Statistical data are collected for many reasons, and much of these data are available to social science researchers. First, some statistical data are collected as part of research projects. Many research organizations, such as the Institute for Social Research at the University of Michigan or the Institute for Research on Poverty at the University of Wisconsin, collect large amounts of highly useful data, as do the national public opinion polling organizations, such as the Gallup Organization and the Roper Center. These organizations make some of this data available to be reanalyzed by others with different research questions in mind. In this way, data collected for one project may be reanalyzed by a number of different people in the

years that follow. Table 10.1 lists just a few of these datasets and describes the kinds of data they contain. This reanalysis of data collected for some other research project is called **secondary data analysis**. In fact, some research organizations, such as universities or governmental agencies, have organized a national system of **data archives**, which are essentially libraries that lend or sell datasets much as libraries or bookstores lend or sell books. Among the better known are the Inter-university Consortium for Political and Social Research (ICPSR) at the University of Michigan, the Roper Center at the University of Connecticut, the National Institute of Child Health and Human Development, and the National Institute of Justice. Also, some private organizations have created data archives in specialized areas. The Sociometrics Corporation, for example, offers numerous datasets from studies focusing on adolescent pregnancy issues. Many individual researchers also make datasets available to those with legitimate secondary uses of the data (Young, Savola, & Phelps, 1991). Some underwriters of research, such as the National Science Foundation, now require, as a stipulation for receiving research funds, that the data eventually be delivered to a public data archive. Available statistical data from these sources can now sometimes be downloaded from the Internet (Clark & Maynard, 1998).

Data collected for research purposes and made available for secondary analysis tend to be of fairly high quality because they are collected by professional researchers. In fact, when these data are originally collected, the research would often be classified as survey research, and the issues discussed in Chapter 6 and Chapter 9 would have been relevant in their design; however, they are treated as available data here because the data were not collected by the researchers using them as secondary data and are thus subject to many of the problems discussed in this chapter.

A second source of available statistical data is the federal, state, and local social service

TABLE 10.1   Illustrations of Available Statistical Data Sets

Name	Location	Types of Data Available
National Survey of Families and Households	Center for Demography and Ecology, University of Wisconsin-Madison. **http://ssc.wisc.edu/ nsfh/avail.htm**	Based on a national probability sample of 13,000 households, this data set contains data on parents and their children, relationships, income, education, health, marriage and divorce, child rearing, religion, and a host of other variables.
KIDS COUNT	The Annie E. Casey Foundation, Baltimore, MD **http://www. aecf.org/kc1998/toc.html**	This data set focuses on variables relating to the status of children and youth in society: infant mortality, infant and child health, teen pregnancy rates, suicide, school drop-outs, poverty, and so on.
General Social Surveys	National Opinion Research Center, University of Chicago **http:// www.icpsr.umich.edu/gss99/**	This is a survey of a national cross-section of adults that has been conducted almost every year since 1972. It contains data on marriage, divorce, crime, education, and questions on a host of attitudes from abortion to homosexuality.
American National Election Studies (ANES)	Inter-university Consortium for Political and Social Research, University of Michigan **http:// www/icpsr.umich.edu**	Based on a national multi-stage area probability sample, this data set focuses on variables relating to voting, gender, race, education, political affiliation and attitudes, political efficacy, and other factors that might affect voting patterns.
Monitoring the Future: A Continuing Study of American Youth	Inter-university Consortium for Political and Social Research, Survey Research Center, University of Michigan **http://www/ icpsr. umich.edu**	Surveys of 8th and 10th grade students that focus on values, behaviors, lifestyle orientations, and drug use.
Current Population Survey (CPS) and Public Use Microdata Samples (PUMS)	U.S. Census Bureau **http://www.cenus.gov http://www.icpsr.umich.edu/**	The CPS is a national sample of households with data on labor force participation, employment status, income, education, and other variables. The PUMS are a variety of data sets based on a sampling of households that were included in the decennial census.

agencies that collect data for either administrative purposes or for purposes of client service. Community mental health centers, Head Start programs, departments of social services, and health and educational institutions are repositories of vast amounts of available data. Figure 10.1 illustrates some of the data regarding clients that are routinely collected by social

FIGURE 10.1 An Example of a Client Information Form Used in a Human Service Agency

1. Primary Recipient Name										Date Received
2. Case Number			3. Referral Date		**MANAGEMENT INFORMATION REPORT** Department of Social Services CHILDRENS PROTECTIVE SERVICES					Reason Issued
4. County | District | Unit | Worker			5. Referral Number							Transaction Number
6. Action	7. Referral Source		8. No. Prefer Ref.	9. Hours to First Contact	10. Date Invest. Comp.		11. Living Arr. – Invest			12. ADC Status
ADDRESS OF PRIMARY RECIPIENT	13. In Care Of				Comments:					18. Document Number
	14. Number and Street									
	15. City			16. State	17. Zip Code					

19. Recipient Update Action	20. NAME (Last, First, Middle Initial)	21. PLANS CLIENT IDENTIFIER	22. BIRTH DATE	23. Sex	24. Race	25. Role	26. Abuse	27. Neglect	28. Living Arrangement at Closing

CLOSING DATA (Also Complete Item 28)

29. Close Date	30. Closing Code	If Item 30 – 6, Complete Item 31	31. Court Involvement/ Disposition	If Item 31 – 1, Complete Item 32	32. Non-Court Disposition	33. Greatest No. Days In Temp. Foster Care	34. Reason in Foster Care Over 21 Days	35. Reason Case Open Over 6 Months
Signature of Worker							Date	

CENTRAL OFFICE

6. ACTION
1 • Investigation
2 • Close
3 • Invest/Close
4 • Change/Update

7. REFERRAL SOURCE
Mandated
01 • Pvt. Physician
02 • Hosp./Clinic Phys.
03 • Coroner/Med. Examiner
04 • Dentist
05 • Audiologist
06 • Nurse (Not school)
11 • School Nurse
12 • Teacher
13 • School Administrator
14 • School Counselor
21 • Law Enforcement
31 • Child Care Provider
41 • Hosp./Clinic Soc. Wkr.
42 • DSS Facil. Soc. Wkr.
43 • DMH Facil. Soc. Wkr.
44 • Other Pub. Soc. Wkr.
45 • Pvt. Agy. Soc. Wkr.
46 • Court Soc. Wkr.
47 • Other Soc. Wkr.

Non-Mandated
51 • Hosp./Clinic Personnel
52 • DSS Facil. Personnel
53 • DMH Facil. Personnel
54 • Other Pub. Soc. Agy. Pers.
55 • Pvt. Soc. Agy. Pers.
56 • Court Personnel
57 • Other School Pers.
61 • Victim
62 • Relative
63 • Sibling
64 • Parent/Sub in Home
65 • Parent/Sub out of Home
66 • Anonymous
67 • Friend/Neighbor
68 • Other

11. LIVING ARR. AT INVESTIGATION
01 • Own Home
02 • Foster, Shelter, Group Home
03 • Shelter Facility
04 • DSS Facility
05 • DMH Facility
06 • Court Facility
07 • Pvt. Facility
08 • Resid. Ed. Facility
09 • Other Child Care Inst.
10 • Other Out of Home
11 • Mult. Plcmts.

12. ADC STATUS
1 • ADC
2 • GA
3 • ADC & GA
4 • None

19. RECIPIENT UPDATE
1 • Update Existing Recipient
2 • Add New Recipient
3 • Delete Recipient

23. SEX
M • Male
F • Female

24. RACE
Migrant
A • Caucasian
B • Negro
C • Indian
D • Other
E • Unknown
F • Sp. Surname

Non-Migrant
1 • Caucasian
2 • Negro
3 • Indian
4 • Other
5 • Unknown
6 • Sp. Surname

25. ROLE
11 • Victim
21 • Perp-Parent in Home
22 • Perp-Sibling
23 • Perp-Other Relat.
24 • Perp-Other Household Memb.
25 • Perp-Day Care
26 • Perp-Fost. Parent
27 • Perp Inst. Staff
28 • Perp-Parent Out of Home
29 • Perp-Other
31 • Uninvolv. – Parent
32 • Uninvolv. – Sibling
33 • Uninvolv. – Other

26. ABUSE (Victim only)
1 • Physical Injury
2 • Congen. Drug Addict
3 • Rape
4 • Incest
5 • Molestation
6 • Exploitation
7 • Unnatural Acts

27. NEGLECT (Victim Only)
1 • Physical
2 • Social
3 • Abandonment
4 • Inapp. Use of Funds
5 • Unlicensed Home/ Improp. Guardianship

28. LIVING ARR. AT CLOSE (Victim and uninvolved siblings only)
1 • In Home
2 • Out of Home

30. CLOSING CODE
1 • Unable to Locate
2 • No Evidence
3 • Not Sufficient
4 • Other Agency Contact
5 • Invest. Only
6 • Services Provided

31. COURT INVOLVEMENT/DISPOSITION
1 • No Court
2 • Accept PS Recom – Adjudication and Disposition
3 • Accept PS Recom – Adjudication Only
4 • Not Accept PS Recommendation
5 • Court – Other

32. NON-COURT DISPOSITION
1 • Satisfact. – No Referral
2 • Satisfact. – Refer to DSS
3 • Satisfact. – Refer Comm. Agency
4 • Unsatisf. Family Resp. – Court Not Feasible
5 • Unsatisf. – Comm. Agency Not Avail. PS Not App.
6 • Unsatisfact. – Refer to DSS
7 • Other

34. REASON FOSTER CARE OVER 21 DAYS
1 • Pending Court
2 • Other

35. REASON OPEN OVER 6 MO.
1 • Await Court Action
2 • Protective Payment
3 • Court Ward in Home
4 • Still Danger
5 • New Danger
6 • Other

service agencies. This form is completed by a county social service agency for each suspected case of child abuse or neglect that is referred to it. The upper portion of the form provides space for various specific bits of information on cases to be recorded; the lower portion provides a coding scheme to ease the entry of information. Some of these data are in quantitative form (such as the age of the client, the number of prior referrals, and the number of days the client is in foster care), and these variables are amenable to statistical manipulation. Agencies collect such data for reasons that may have nothing to do with their eventual research uses. Sometimes agencies are required by law to collect such information, which is then stored in databanks and may be available to researchers. I used the data on these forms, combined with data from other organizational records, in a study to assess the impact of a school social work program on the referral of cases of child abuse and neglect. Unfortunately, data from social service agencies are often not cataloged and indexed efficiently and may therefore be difficult for investigators to locate and use. Such data have, nonetheless, grown voluminously during the past few decades. This has occurred because social service agencies have increasingly been required to engage in needs assessment, planning, accountability, and evaluation. In addition, the strides made in computer technology, along with the widespread use of management information systems in social agencies, have all contributed to the growth and increased availability of data produced by social agencies.

A third source of available statistical data is the organizations and governmental agencies that collect data as a public service or to serve as a basis for social policy decisions. Such agencies as the Federal Bureau of Investigation (FBI) and the National Center for Health Statistics (NCHS), for example, collect vast amounts of data, as do state and local governments. The U.S. Census Bureau also collects enormous amounts of data to be used for establishing and changing the boundaries of political districts and allocating government funds that are based on population size. Often data from this source can be used in conjunction with other types of data—from questionnaires, for example, or from agency records—to test hypotheses. One study of social workers' accuracy hypothesized that caseworker errors might be influenced by the political and economic characteristics of the local community (Piliavin, Masters, & Corbett, 1977). Data were based on interviews, questionnaires, and other sources. In addition, the political and economic characteristics of each community were measured with data from the Bureau of the Census regarding the percentage of poor people in the county, how the county voted in the most recent gubernatorial election, and the percentage of impoverished families that were headed by women. Thus data from multiple sources can be combined to measure variables and test hypotheses in a single study.

A fourth source of available statistical data is the private organizations and corporations that collect data for purposes of running their organizations or providing services to their members. Professional associations, such as the American Medical Association, the American Bar Association, and the National Association of Social Workers, produce statistics relating to their membership and issues of concern to their members. A research question that relates to such organizations or their members might make use of such data. Data from these private organizations is generally distributed less widely than governmental statistics and for that reason may be more difficult to locate.

Issues in Using Statistical Data

Most available statistical data were not collected for research purposes—or at least not for the specific research questions for which you now intend to use them. They were collected to meet the needs of whatever agency, organization, or researcher originally collected them, and you are limited by the form in which the data were collected. This leads to some special issues that call for attention.

Missing Data

For a variety of reasons, a dataset may not include complete data for every person studied or may fail to collect data from the entire population or sample of interest. For example, an individual may refuse to answer certain questions, which results in a gap in his or her dataset. Or data may not be collected in a particular neighborhood because it is considered too dangerous for interviewers to enter. Such gaps in a data set are referred to as **missing data** and occur to some degree in practically all studies. For example, one of the most widely used sources of available data about crime is the FBI's annual publication *Uniform Crime Reports (UCR)*. In order for an event to be recorded as a crime in the *UCR*, it must come to the attention of police officials. In some cases, police chance across criminal activity taking place so it gets officially recorded. Most often, however, police must rely on citizens who are victimized by crime to report it to them, and herein lies one of the problems with *UCR* data. A considerable percentage of even serious crimes are never reported to the police. One estimate made by Callie Rennison (1999) of the Bureau of Justice Statistics suggests that only around 45% of all violent crimes and only one-third of other crimes are reported to the police and make their way into the *UCR* data. Clearly this reporting gap causes the *UCR* data to significantly understate the true impact of criminal activity in the United States. These estimates of unreported crime are made by comparing *UCR* results to the National Crime Victimization Survey (NCVS), which is conducted every year on a representative sample of over 100,000 citizens, inquiring if they have been victims of crimes. The disparity between the NCVS data and the *UCR* data provides an indicator of the amount of crime that goes unreported in the *UCR* and thus gives a clearer picture of actual criminal activity.

The problem with using available data files is that you have no control over this failure to collect a complete set of data. Missing data result in incomplete coverage, which, if extensive, can throw into question the representativeness of the data. Further, because statistical procedures are based on the assumption of complete data, missing data can result in misleading statistical conclusions. As we dig back into data from the past, it is not uncommon to discover that data for whole periods of time are missing. This can occur for many reasons, such as data destroyed by fires, lost data, changes in policy, and the like. Finding data that cover only a portion of one's target population is also quite common. For example, the researchers in a study of marriage rates among older Americans in the United States were forced to do without data from three states— Arizona, New Mexico, and Oklahoma— because those states did not maintain central marriage files (Treas & VanHilst, 1976). Given the high concentration of retirees in Arizona, however, it is possible that the exclusion of data from that state might have affected their results. Another example of this is with the FBI's *UCR* crime data. Over the years that the *UCR* has been published, the percentage of police agencies that report their crime statistics to the FBI has steadily increased (such reporting is voluntary for police jurisdictions). This of course means that any comparison over time has a built-in problem. Even if crime has not increased, it may appear to have increased over time due to the higher rate of reporting in more recent years. When working with data that suffer from noncoverage, one should assess the implications that noncoverage has for the results of the research.

Inductive versus Deductive Analysis

In research investigations in which data are collected firsthand, a positivist deductive approach is often used. That is, hypotheses are deduced from a theory, variables in the hypotheses are operationally defined, and data are collected

FIGURE 10.2   The Measurement Process with Available Statistical Data

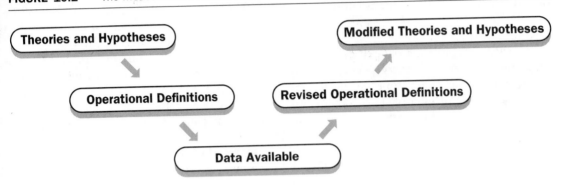

based on these operational definitions. In short, research moves from the abstract to the concrete: The kind of data collected is determined by the theory and hypotheses being tested (see Chapter 2). In the analysis of available data, however, such a deductive approach is often impossible because the data necessary to measure the variables derived from a theory may not have been included when the original data were collected. If this difficulty is encountered, a compromise is often made. As shown in Chapter 5, the measurement process often calls for modifying nominal and operational definitions (see Figure 5.1). With available data research, not only operational definitions but, in some cases, hypotheses and theories may have to be revised so they can be tested with the data available (see Figure 10.2).

In other words, we have modified the research to fit the data. When this occurs, research takes on a somewhat inductive character. In inductive research, we move from the concrete to the abstract: Starting with the data collected, we develop hypotheses and theories to explain what we find in the data. The situation we have described is actually somewhere in limbo between induction and deduction. We have begun with theories and hypotheses, but we have also been forced to let the available data influence how we test the hypotheses.

This is a frequent problem in the analysis of available statistical data and has led some to argue that available data are most appropriately used in purely inductive research that has an exploratory purpose (Hoshino & Lynch, 1981). While this may be true in some cases, especially when the data available does diverge substantially from the theories and hypotheses being tested, other researchers argue that available data can be used in many areas of deductive research, provided that care is exercised in how extensively hypotheses and operational definitions are changed to fit the data. When extensive changes are made, the major danger is one that has been discussed in relation to other data collection techniques: validity.

Validity and Reliability

Validity refers to whether a measurement instrument actually measures what it is intended to measure (see Chapter 5). Many data in existing statistical files can be considered valid indicators of certain things that they directly describe. The age, sex, and racial profiles of clients, along with the amounts of different kinds of services provided, would be examples of these. However, validity problems frequently arise in three areas.

First, many elements of the operations of organizations or agencies—such as achievement of goals, success of programs, or satisfaction of clients—may not be measured directly by any data normally collected by the organizations. To study these, one would have to search for indirect measures among the organization's data that might enable one to infer such things as goal achievement, success, or satisfaction. These situations create the same kinds of validity problems that confront researchers using other methods except that, when available data are used, validity problems cannot be considered and resolved *before* the data are collected. Because the data have already been collected, the problems are entrenched.

A second area in which validity problems frequently arise is data analysis that becomes inductive, as I have just discussed; that is, when operational definitions are changed so that variables can be measured with the data available. The more the definitions are changed, the more the validity of the measures is called into question. The operational definitions may be changed so drastically that they no longer measure the theoretical concepts they were first intended to measure.

Finally, validity questions arise when procedures used in gathering data by an agency or organization have changed over the years. Changing modes of collecting data, such as dropping some questions or developing new definitions of something, are quite common in any agency or organization. These procedural changes can, however, affect the comparability of data collected at various times and the validity of using the same operational definitions. An example of this can be found in the FBI's *Uniform Crime Reports* discussed earlier in the context of missing data. In 1979, the FBI added an additional crime, arson, to the six crimes that had made up the FBI's "crime index," or index of more serious crimes. Obviously, comparisons of the crime index before 1979 with the crime index after 1979 must be done with care so that the result is not misleading.

There are a number of ways to tackle the problem of validity in available data research.

One is to search for tests of validity and reliability that have already been conducted. Especially when you are doing a secondary analysis of data collected by other researchers or research organizations, some of the procedures for assessing validity and reliability discussed in Chapter 5 may well have been carried out. This is less likely to be the case if you are using data collected by a nonresearch organization or agency. A second attack on the validity problem would be to conduct your own tests of validity and reliability. Even though the measuring devices have already been used and the data collected, you can still (a) assess face or content validity or (b) use the measuring devices on some other population to evaluate criterion validity, construct validity, or one or more of the tests of reliability. A third attack on the validity problem would be to test hypotheses by using more than one measuring device, if the available data provides you with more than one way to measure variables. If measuring the variables in different ways produces the same results, it gives you more confidence in the validity of the results. A final attack on the validity problem is replication: If possible, test the hypotheses on more than one set of available data. If all tests produce the same results, then it increases confidence in the validity of the findings.

Unit of Analysis

In some cases, available data may be collected from a unit of analysis that differs from the unit of analysis posited in the research hypotheses. The most common occurrence along these lines is to have hypotheses referring to the individual as the unit of analysis while the data available is at the group or organizational level. The Census Bureau, for example, often collects data on the characteristics of census tracts, cities, or other aggregations of individuals. Such data would not be the most appropriate when hypotheses have to do with the characteristics of individuals. As with other measuring problems stemming from available data, you cannot normally change the unit of analysis once the data have been collected. One alternative is to change the hypotheses so that they are stated at the level available in the data. Another alternative is to be cautious about drawing conclusions: Relationships tested with the inappropriate unit of analysis provide some evidence for relationships between variables but do not provide the strongest inference about the relationship. Recall the ecological and reductionist fallacies discussed in Chapter 4: It is possible to find relationships at one level of analysis and not at another, even when you appear to be focusing on the same variables at the two levels of analysis. The fact that you are measuring at different levels of analysis means that you are measuring different variables.

The amount of available statistical data from various sources is likely to grow astronomically in the future. In addition, the increasing availability and sophistication of computer technology are likely to make these data more readily available for research. Such data represent vast, but currently highly underutilized, sources for research. When contemplating a research project, you would be wise to consider whether there are some available data that might serve as legitimate sources of information.

CONTENT ANALYSIS

While available statistical data is common, available data often exist in a more qualitative form. Whenever activity is recorded in some document or format—whether a book, diary, case record, film, tape recording, or any other medium—it is amenable to scientific analysis. Although the data discussed in the previous section were already quantified when made available for research, the documents now being considered have basically *qualitative* data that have to be analyzed in its qualitative form or that must be quantified for analysis. As mentioned in the beginning of the chapter, the term *document* is used very broadly here to refer to any recorded form of communication that is nonquantitative in nature, whether it be a book, letter, audio or video recording, or some other format. This chapter focuses on some of the procedures used in quantifying the content of documents; the qualitative analysis of the data in documents is discussed in Chapter 15.

Quantifying the content of documents involves the issue of measurement, a topic discussed at length in Chapter 5 and Chapter 6. By quantifying the content of documents, we are trying to extract some precise meaning from the documents. **Content analysis** refers to a method of transforming the symbolic content of a document or medium, such as words or other images, from a qualitative, unsystematic form into a quantitative, systematic form (Krippendorff, 1980; Roberts, 1997; Weber, 1990). Content analysis is a form of coding. **Coding** refers to categorizing behaviors or elements into a limited number of categories. I discuss coding as a way of transforming data in a number of chapters. Coding is done in some surveys (Chapter 9), in some structured field observations (Chapter 11), in quantitative data analysis (Chapter 13), and in qualitative data analysis (Chapter 15). In content analysis, coding is performed on documents that are produced for purposes other than research and then made available for research purposes.

Coding Schemes

A major step in content analysis is to locate or develop a coding scheme that can be used in analyzing the documents at hand. Coding schemes in content analysis, like those in coding observed behavior, can be quite variable, and their exact form depends on the documents being studied and the hypotheses being tested in the research project.

Existing Coding Schemes

In some cases, it is possible to find coding schemes that have been previously developed and used by others. For example, a general coding scheme has been developed for use in categorizing themes in fiction and drama. The basic themes are: (1) love, (2) morality, (3) idealism, (4) power, (5) outcast, (6) career, and (7) no agreement (McGranahan & Wayne, 1948). There are also categories for classifying endings, temporal settings, spatial settings, and patterns of love in the love theme. As another example, a content analysis study of fiction in women's magazines categorized stories in terms of the characteristics and roles of the main character in each story (Peirce, 1997). So the goal of the main character was categorized into one of six categories: romantic love, career, personal fulfillment, community betterment, material accumulation, or family happiness. Existing coding schemes such as these could be used to code the content of books, short stories, plays, movies, and even music videos—if the coding scheme is a valid measure of the researcher's variables.

Coding schemes have also been developed to analyze the content of conversations, such as the dialogue between a therapist and a client during a therapeutic interview. In a study of the impact of a clinician's responses on whether clients continued with treatment, a coding system was developed that used three categories.

1. Substantive congruent responses: clinician responses that referred to the client's immediately preceding response or contained some elements of that response.

2. Nonsubstantive congruent responses: clinician responses, such as "I see" or "Yes," which indicate that the clinician is aware of and paying attention to the client's verbalizations.

3. Incongruent responses: clinician responses that appear unrelated to what the client has said.

Such a coding scheme could be used by other researchers who are interested in similar aspects of any interview relationship or in elements of the process of social interaction in general.

In conducting content analysis, then, it is beneficial to search for an existing coding scheme that might be applicable to the research problem at hand. Not only does the use of an existing coding scheme result in considerable savings for the researcher in terms of time, energy, and money, it also serves to make the research project comparable with other studies that have used the same coding system. For example, a study of gender stereotypes in popular culture wanted to explore changes in tendencies to engage in such stereotyping over a number of decades (Brabant & Mooney, 1997). The aspect of popular culture selected for scrutiny was newspaper comic strips. The study used the same coding scheme as the earlier studies to ensure comparability. For each comic strip in the Sunday newspaper, the researchers coded whether or not each character appeared inside or outside the home, what major activities each character engaged in (home or child care, work, leisure, and so on), and whether each character was shown reading. The study concluded that the trends were mixed: There has been movement away from stereotyped imagery by some measures but movement toward greater stereotyping by other measures. These researchers might have been inclined to develop a new, possibly more contemporary, coding scheme, but this would have made it more difficult to draw conclusions about one of their major concerns: changes over time.

Coding schemes can be found in the many journals that report social science research or in books, such as Holsti's (1969) or Krippendorff's (1980), that are devoted to the study of content analysis.

Characteristics of Coding Schemes

If a suitable coding scheme cannot be located, then one must be developed. Since a coding scheme is a measure, or operational definition, of the content of a document, many of the principles relating to measurement presented in Chapter 5 and Chapter 6 would also apply to developing coding schemes. In particular, like categories in any measuring process, those used in content analysis should be exhaustive and mutually exclusive. Categories are *exhaustive* when a category is available for every relevant element in the documents. If there are only a few possibilities and these can be clearly defined, an exhaustive set of categories will not be difficult to develop. If what we are trying to measure is rather open-ended with many possibilities, however, developing an exhaustive set of categories may well be difficult. For example, presidential speeches have been analyzed in terms of the values expressed. Obviously, there are so many values that an exhaustive list is unlikely to be developed. One such list contained 14 value categories, which undoubtedly cover the most important or commonly mentioned values (Prothro, 1956). It is doubtful whether even that many categories are really exhaustive, however. If it is impossible to be exhaustive, then the most common or most important categories are used, and a residual category ("others") is made available for those items that do not fit any of the other categories.

Coding categories should also be *mutually exclusive*, which means that each coded item can fall into one—and only one—category. This requirement forces one to have precise definitions for each category so that there is no ambiguity concerning which items it includes and which it does not. Failure to meet this requirement can threaten the validity of the measuring process because items will be placed every which way by coders confused with overlapping categories. Lack of mutual exclusiveness will likely show up in low levels of reliability as the coders disagree on the placement of items into categories.

In addition to these two characteristics, coding categories also need to be *theoretically relevant*. This means that they should derive from some theoretical analysis of the phenomenon being studied.

Units of Analysis

With the categories established, the next research decision in content analysis concerns precisely which aspects of the documents will be recorded. Generally, there are four units of analysis in document coding: a word, a theme, a major character, or a sentence or paragraph. (The units of analysis commonly found in research other than content analysis are discussed in Chapter 4.) An often convenient unit of analysis is a *single word*. Coding the presence of certain words in documents can be accomplished easily and with a high degree of reliability. If a single word qualifies as a valid indicator of what you wish to measure, it is a good choice for the unit of analysis. For example, as a part of their study of colonial families, Herman Lantz and his colleagues (1968) counted the frequency with which the word *power* was associated with men or women in colonial magazines. In this context, the single word was used as a measure of the perceived distribution of power between men and women during colonial times. When using single words as the unit of analysis, it is often helpful to make use of a *context unit*, which is the context in which a single word is found. The words surrounding the word used as the unit of analysis modify it and further explain its meaning. This contextual information is taken into account when coding the unit of analysis. For example, in the study just mentioned, the investigators needed to know whether the word *power* referred to men or to women. The context surrounding the word supplied this crucial information. The amount of context needed to explain the use of a given word adequately, of course, varies.

The *theme* as a unit of analysis refers to the major subject matter of a document or part of a document. An entire document can be characterized as having a primary theme. Novels, for example, can be described as mysteries, science fiction, historical, and so on. In a study of the images of women during World War II and their treatment by governmental propaganda agencies, one researcher analyzed themes relating to the role of women in the workforce found in fiction pieces in two magazines that were popular during that war, *Saturday Evening Post* and *True Story* (Honey, 1984).

Themes can, however, be difficult to delineate. The overall theme may or may not be clear, or there may be multiple themes. Coder reliability is likely to be lower than when easily identifiable words are the unit of analysis.

A third unit of analysis in documents is the *main character*. The use of this unit of analysis is, of course, limited to documents that contain a cast of characters, such as plays, novels, movies, or television programs. Studies of books, magazine stories, and television programs, for example, have often used the main character as the unit of analysis in studies of sexism. The Eye on Diversity box describes some of this research and suggests how beneficial nonreactive available data is in the study of sensitive topics such as racism and sexism.

The fourth unit of analysis is a *sentence* or *paragraph*. The study described earlier regarding congruency in clinician-client interaction used a similar unit of analysis: Each clinician response, preceded and followed by a client comment, was considered the unit to be coded as congruent or incongruent. A paragraph or even a single sentence, however, often contains more than one idea, and this may make these larger units more difficult to classify while maintaining mutually exclusive categories and intercoder reliability. Indeed, reliability may well be lower than with the word or main character units. Yet the larger units are often more theoretically relevant in the social sciences. In clinician-client interaction, for example, it would make little conceptual sense to characterize an interchange as congruent on the basis of one word. Meaning in social interaction normally arises from a whole block of words or sentences. So, as I emphasized when discussing units of analysis in Chapter 4, the primary consideration in selecting a unit of analysis is theoretical: which unit seems to be preferable given theoretical and conceptual considerations.

Manifest versus Latent Coding

In content analysis, a distinction is made between *manifest* and *latent* coding. The term

manifest coding refers to coding the more objective or surface content of a document or medium (Holsti, 1969). One example of manifest coding would be to note each time a particular word appears in a document. Another example of manifest coding is found in a study of the portrayal of African Americans in children's picturebooks. This study observed whether African Americans, Anglos, or people of other racial groups appeared in the pictures in a book (Pescosolido, Grauerholz, & Milkie, 1997). It may in some cases be difficult to judge the race of a particular character—certainly more difficult than deciding whether a particular word appears in a document. However, this is still manifest coding because there is a fairly direct link between the image in the document and the coding category. **Latent coding,** on the other hand, inherently involves some inference in which the coder has to decide whether the representation in the document is an instance of some broader or more abstract phenomenon. The representations in the document are assessed for what they say about some more abstract or implicit level of meaning. The study just mentioned also used latent coding by judging the nature of the African American representations in the picture books and of the African American–Anglo interactions portrayed. So the coders made judgments about how *central* the African American characters were in the pictures and how *intimate* or *egalitarian* the African American–Anglo interactions were. "Centrality," "intimacy," and "egalitarianism" are qualities that a coder must infer on the basis of a judgment or assessment of the materials.

Another example where both manifest and latent coding were used was a study of how alcohol use is portrayed in prime-time television programs (Mathios, Avery, Bisogni, & Shanahan, 1998). The manifest coding involved recording incidents where alcohol was a part of such shows and noting the type of alcohol used and the gender of the characters involved.

EYE ON DIVERSITY   **Studying Sexism Using Available Data**

When studying emotionally charged topics such as racism or sexism, reactivity can be a very serious problem because people may be inclined to disguise their true feelings or motives if these would meet with disapproval. With research methods such as surveys, for example, it is always possible for people to lie or at least to distort their true positions on issues. One of the major benefits of available data is that they are less reactive when used to study behaviors such as racism and sexism.

One way to study racism and sexism is to look for stereotypes in the portrayal of people in various cultural productions, such as books, magazines, or movies. Decades ago, the portrayals of women and men in children's books, televisions programs, and movies were highly stereotyped, and things have not changed as much as you might think. Recent studies of children's picture books, for example, find that more women are portrayed than in the past but still fewer men (only one-third of the illustrations are of women). In addition, almost no women are portrayed as working outside the home, women are still shown in fewer occupations than men, and women are portrayed as less brave and adventurous and more helpless (Peterson & Lach, 1990; Purcell & Stewart, 1990; Williams, Vernon, Williams, & Malecha, 1987). Another study of newspaper photographs found that men appear in these photos far more frequently than women and that men are typically pictured in professional roles while women are in domestic roles (Luebke, 1989). Recent studies of prime-time television documented that only one out of three roles in prime time is written for women and that women are more likely to be portrayed as preoccupied with romance, dating, and personal appearance rather than with work or education (Gunther, 1993; D. Smith, 1997). Even college textbooks are not immune to these influences. Studies of the pictorial content of texts for college-level psychology and sociology courses found that women are shown less often than men and are portrayed more passively and negatively than men (Ferree & Hall, 1990; Peterson & Kroner, 1992). For example, the psychology texts portray women as the victims of mental disorders and the clients in therapy while men are pictured as the therapists. These are examples of how cultural products, like books or newspapers, can serve as available data to detect gender stereotypes that might not have been detected by more reactive research methods.

All of this can be fairly easily observed, with little need to make judgments or inferences. The latent coding involved judging something about the personality characteristics of those portrayed consuming alcohol. For each such character, for example, the coders had to rate them on a four-point scale from "Very powerful" to "Very powerless" and from "Very admirable" to "Very despicable." To do this latent coding, the observers were required to make some fairly sophisticated and subjective judgments about personality from the behaviors and overall portrayal of the characters on the television shows. As these two examples illustrate, manifest coding is generally more reliable than latent coding, but latent coding is often a more valid way to get at some fairly complex and theoretically important social processes and characteristics.

Validity

In content analysis, validity refers to whether the categories we develop and the aspects of

Some differences in the portrayal of men and women in cultural products can be very subtle. In a creative series of studies involving an intriguing application of available data, Dane Archer and colleagues (1983) measured the amount of space in photographs devoted to people's faces. They divided the distance from the top of the head to the bottom of the chin by the distance from the top of the head to the lowest point of the body visible in the depiction. The result is a proportion with an upper limit of 1.00 (face fills entire picture) that indicates the degree to which the face is prominent in the picture. The first application of this measure was on 1,750 photographs drawn from *Time*, *Newsweek*, and *Ms.* magazines, and from the *San Francisco Chronicle* and the *Santa Cruz Sentinel*. Before even measuring the photos, the researchers discovered something interesting: More than 60% of the photos depicted men! When they applied their facial prominence indicator, they found another difference between the sexes: The average "face-ism" value was .67 for men and .45 for women, indicating that the pictures of males tended to feature the face more prominently, whereas pictures of women featured more of the body; and more recent studies have confirmed this (Nigro, Hill, Gelbein, & Clark, 1988). Archer and colleagues also found that this pattern held across pictures in publications from 11 other countries. Applying the indicator to paintings going back as far as 600 years showed that the pattern they found had a long history and the difference seems to be increasing. Just how deeply ingrained this tendency is was illustrated in another part of the study, in which people were instructed to draw a picture of a man or a woman. Consistently, and regardless of their sex, people drew men with faces more prominent than for women.

Cultural practices such as face-ism are important to study because they may represent mechanisms for perpetuating stereotypes about gender. Archer and his colleagues (1983) present evidence showing that those depicted with more prominent faces are rated as more intelligent, more ambitious, and as having a better physical appearance. In other words, the faces of men are more prominently displayed in the pictures we are surrounded with every day, and a more prominent facial display leads people unwittingly to the perception of the person as more intelligent and ambitious. So, face-ism may contribute to unconscious stereotyping of women as less intelligent and capable and could contribute to discrimination against women.

the content coded are meaningful indicators of what we intend to measure. Developing coding schemes that are valid indicators can be a challenging task. Anne Fortune (1979), for example, reported on an effort to study communication patterns between social workers and their clients in which recordings of interviews were content analyzed. The goal of the research was to discover which interview techniques used by social workers would bring about cognitive and affective change in clients. Fortune was concerned particularly with how techniques varied when clients were adults rather than children. First, a typology of communication techniques was developed. As there are many ways to execute a particular verbal technique in a therapeutic setting, it was impossible to base the content analysis on a single word or phrase. Instead, coders were provided with a description of each technique and examples of the verbal forms that each technique might take. For example, the communication technique of "exploration" was described as "communication intended to elicit information, including questions and restatements or 'echoes' of client's communications." Examples of this technique provided to the coders were such phrases as "What class was that?" and "You said your son misbehaved . . ." The communication technique of "direction" was exemplified by "I think the first step would be to talk this over with your daughter" (Fortune, 1979, p. 391). By the provision of specific examples of how to code comments, it was hoped that the resulting coding would have greater validity than if coders were left to their own devices.

Depending on the nature of the research and the documents to be analyzed, any of the methods of assessing validity discussed in Chapter 5 may be applicable. The logical approaches of face validity, content validity, and jury opinion are the most generally used assessments of validity in content analysis, although criterion validity is sometimes used. For example, in a content analysis of suicide notes, an attempt was made to identify aspects of the content of real suicide notes that would differentiate them from simulated ones written by people who had not attempted suicide (Ogilvie, Stone, & Shneidman, 1966). The researchers compared half of the genuine notes with half of the simulated ones and found that genuine notes made more references to concrete things and greater use of the word *love*. The simulated notes, on the other hand, contained more references to the thought processes that went into the decision to kill oneself. Using these observations to predict which of the remaining notes were genuine, the researchers were able to do so at a 94% level of accuracy. The ability to predict at such a level suggests that the content aspects of the notes they discovered were valid indicators of genuine suicide notes.

Researchers must be able to argue convincingly that indicators are valid if scientific outcomes are to be accepted. Although most content analysis does not go beyond face or content validity, it is important to remember that these are the weakest demonstrations

of validity. Whenever possible, more rigorous tests should be attempted.

Reliability

Reliability refers to the ability of a measure to yield consistent results each time it is used. In content analysis, reliability relates to the ability of coders to consistently apply the coding scheme that has been developed. The question is: Can several people code the documents according to the coding scheme and obtain consistent results? Reliability in content analysis depends on many factors, including the skill of the coders, the nature of the categories, the rules guiding the use of the categories, and the degree of clarity or ambiguity in the documents (Holsti, 1969; Scott, 1990; Weber, 1990). The clarity of the documents in any given study is largely fixed, so our discussion of control over reliability is limited to the coders and the categories utilized.

First, reliability can be enhanced by a thorough training of the coders and by practice in applying the coding scheme. Coders who in their coding continually deviate from others during this practice period should not be relied on as coders. One must be cautious, however, about eliminating coders without assessing whether their deviation indicates poor performance or an ambiguous coding scheme.

Second, the nature of the categories to be applied to the document is also important for reliability. The simpler and more objective the categories—manifest coding—the higher the reliability will be. Vaguely defined categories or those requiring substantial interpretation—latent coding—will decrease reliability because they create greater opportunity for disagreement among the coders. However, latent coding categories should not be avoided just because of this; if theoretical considerations suggest that latent coding is the most appropriate measurement of variables, then it should be used and special effort devoted to training the coders.

Measuring reliability in content analysis is much the same as measuring reliability when coding observed behavior (see Chapter 11). One simple way to do this is to calculate the percent of judgments on which coders agree out of the total number of judgments that they must make:

$$\text{percent of agreement} = \frac{2 \times \text{no. of agreements}}{\text{total no. of observation recorded by both observers}}$$

However, this measure of intercoder reliability does not take into account the extent to which intercoder agreement can occur by chance, and there are other ways of measuring intercoder reliability. Whichever particular procedure is used, accepted levels of reliability are 75% or better agreement between coders. A content analysis of children's television advertisements, for example, achieved an overall intercoder reliability of 94%, with some coding categories achieving 99% (Furnham, Abramsky, & Gunter, 1997). Well-trained coders using well-constructed coding schemes should achieve better than 85% agreement.

A certain tension exists between validity and reliability in content analysis. The simplest coding schemes—such as those employing word frequency counts—produce the highest reliability because they are very easy to apply in a consistent manner. As noted, however, word frequency counts may not validly measure what we want to measure. Sacrificing some degree of reliability for the sake of validity may be necessary. Researchers often end up performing a balancing act between the dual requirements of validity and reliability.

Level of Measurement

Like other methods of collecting data, content analysis involves a decision about the level of measurement to be used (see Chapter 5). The level of measurement achieved depends on the variable being measured and on the process used to measure it. The variable being measured puts an upper limit on the level of measurement that can be reached. For example, if we rate main characters in a book according to marital status, the highest level of measurement would be nominal because marital status is a nominal variable. No amount of measurement finesse can change this once the key variables have been selected. However, researchers have some control over the level of measurement, depending on the process used to measure the variables. The important factor is how the unit of analysis is quantified in the coding process. Coding systems in content analysis generally fall into one of four categories: (1) presence or absence of an element, (2) frequency of occurrence of an element, (3) amount of space devoted to an element, and (4) intensity of expression.

The simplest rating system is to indicate merely *the presence or absence of an element* in a document. For example, as mentioned above, we could determine the presence or absence of people of different races in a pictorial representation. This simple system would yield nominal data and convey a minimum of information about content. Several important questions would remain unanswered. We would not know whether the portrayals were positive or negative in nature. Also unknown would be how much of the representation was devoted to particular racial characters. Consideration of these factors might make the study more informative.

Frequency counts are very common methods of rating. We simply count how often an element appears. For example, Jack Levin and James Spates (1970) sought to compare the dominant values expressed in middle-class publications to those in the so-called underground press. The investigators rated the frequency with which statements of various values appeared in the two types of publications. On the basis of the frequency counts, the study concluded that the underground press emphasized values related to self-expression, whereas the middle-class publications stressed values related to various types of personal achievement. Frequency counts reveal more information about the document than does the simple present or absent approach. They also open the way for more sophisticated statistical analysis because frequency counts can, with the appropriate theoretical concepts, achieve an interval or ratio level of measurement.

Coding systems based on *amount of space devoted to an element* have proved useful for analyzing the mass media. In newspapers or magazines, the normal approach is to measure column inches. The equivalent for films or television would be time. For example, we could measure the amount of space devoted to a topic, such as family violence or the problems of lesbian mothers, over the years in order to assess the impact of popular trends or political events on public concern about various issues. The major attraction of space/time measures is their ease of use. The amount of space devoted to a given topic can be measured far more rapidly than word frequency counts or even whether key words are present or absent. As each document takes only a short time to analyze, space/time measures allow a larger sample size, possibly leading to greater representativeness. Space/time measures can also yield interval or ratio level measurement with theoretical concepts that are amenable to such a level of measurement.

Unfortunately, space/time measures are still somewhat crude. Other than the volume of space devoted to the particular topic in question, they reveal nothing further about the content. For example, time measures applied to the network news could tell us which types of news are allotted the most coverage, but nothing about more subtle issues such as whether the news coverage was biased. As Ole Holsti (1969, p. 121) notes, "A one-to-one relationship between the amount of space devoted to a subject and the manner in which it is treated cannot be assumed."

The most complex rating systems involve a *measure of intensity*—the forcefulness of expression in the documents. However, developing intensity measures is difficult. Intensity of expression can be quite subtle and dependent upon many aspects of word usage. This makes it extremely difficult to clearly specify the conditions for coding content elements. The coders will have to make many judgments before deciding how to categorize the content, and this leads to disagreements and low reliability.

Developing intensity measures that will produce reliable results is quite similar to the construction of measurement scales discussed in Chapter 6. We could, for example, have coders rate documents along a scale (as in semantic differential or Likert scales). Newspaper editorials dealing with public assistance programs might be rated as (1) Very unfavorable, (2) Unfavorable, (3) Neutral, (4) Favorable, or (5) Very favorable. The study mentioned earlier on portrayals of alcohol use on television used an intensity measure: Coders had to rate characters on a four-point scale from "Very powerful" to "Very powerless" and from "Very admirable" to "Very despicable" (Mathios, Avery, Bisogni, & Shanahan, 1998). The options for intensity scales are nearly endless given the vast variety of intensity questions that can arise concerning the contents of documents. Despite their complexity, intensity measures are often the most revealing concerning a document's contents.

Sampling

In document analysis, the number of documents is often too vast for all to be analyzed; it is then necessary to take a sample from a group of documents. For example, we might wish to study the extent to which concern for child abuse has changed among social scientists between 1950 and 2000. We could do this by studying the extent of coverage of the topic in social science journals during those years. Considering the number of journals and thousands of pages involved, sampling would clearly be necessary to make such a project feasible. As with other types of sampling, a critical issue is representativeness. In order to generalize the findings of our document analysis, the sampling procedure must be likely to yield a representative sample, which is often difficult to achieve with documents. One problem is that the elements of the population of documents may not be equal. In studying child abuse, for example, some journals would be more likely to publish articles on that topic than would others. Such journals as *Social Problems* or the *Journal of Interpersonal Violence*, for example, would be more likely to receive and publish an article on child abuse. A more general journal, like the *American Sociological Review* would include a much smaller proportion of articles dealing with child abuse. Likewise, a journal specializing in a different area, such as the *Journal of Gerontology*, would probably publish few if any articles on child abuse. A random sample of all social science journals, then, might include only a few of those most likely to publish articles on this topic. Therefore, the sample would be dominated by journals having a small likelihood of publishing family violence articles, making it more difficult to measure changes in focus on family violence. This problem could be solved by choosing our sample from among those journals specializing in areas related to family violence. Or we could stratify the journals by type and then select a stratified random sample, taking disproportionally more journals that are likely to publish articles on family violence (see Chapter 7).

A second issue in sampling documents is the difficulty in defining the population of documents. For example, for a study of the role of experts in the social construction of the crime problem, Michael Welch, Melissa Fenwick, and Meredith Roberts (1997) used quotes by crime experts as their unit of analysis. These quotes could be found in many different places but are most prominently featured in newspaper articles. The researchers decided to limit their sample to four newspapers (the *New York Times*, the *Washington Post*, the *Los Angeles Times*, and the *Chicago Tribune*) on the grounds that these newspapers had large circulations, offered national news coverage, and provided a geographic spread across the country. These charac-

teristics would provide the best data to test their hypotheses. Within these newspapers, they sampled only articles identified in the index as "feature" articles because these are the ones most likely to include actual quotations from experts.

Assuming that the population to be sampled can be defined adequately, the normal procedures of sampling discussed in Chapter 7 can be applied to documents. Of those, simple random sampling is probably most common and most generally applicable (Scott, 1990). As we have seen, however, stratified sampling may be required to avoid bias when sampling from a population of unequal elements. Systematic sampling should be approached cautiously as it is easy to be trapped by periodicity in documents. *Periodicity* is the problem in which elements with certain characteristics occur at patterned intervals throughout the sampling frame. For example, the size and content of newspapers vary substantially with the days of the week. If the sampling interval were seven or a multiple of seven, a systematic sample would include only papers published on the same day of the week and could be very biased depending on the aspects of content under study.

Sampling documents often involves multistage sampling. As in all multistage sampling, we start with large units and work down through a series of sampling stages to smaller and smaller units. Levin and Spates's sample of the underground press (1970) illustrates multistage sampling. The first stage was selecting a sample of publications from the membership of the Underground Press Syndicate. This was done by choosing the top five in terms of circulation. The second stage involved randomly selecting a single issue of each publication for every other month from September 1967 through August 1968. This produced six sample issues of each of the five publications for a total of 30 issues. The final stage was to select every other nonfiction article appearing in the sample issues for analysis. The resulting sample contained 316 articles to represent the underground press.

HISTORICAL-COMPARATIVE RESEARCH

Comparative and historical research are two distinct but closely linked types of social science research. **Comparative research** involves the comparison of two or more events, settings, societies, or cultures in order to determine the similarities and differences among them (Bollen, Entwisle, & Alderson, 1993). The fundamental impetus behind comparative research is the recognition that oftentimes we can learn something only by seeing things compared to other things. If we restrict our focus, for example, to one culture, then it would be impossible to assess how cultures are similar to or different from one another. Only by comparing cultures can we be in a position to assess cultural variability. Comparative research is cross-sectional in nature in that it focuses on one point in time.

Historical research is inherently longitudinal; it involves an examination of the records and other evidence that has survived from the past (Alford, 1998; Gottschalk, 1969). The researcher uses this evidence to reconstruct the past and infer some abstract knowledge from it. Historical research involves the imaginative interpretation of the historical record based in part on the historian's or researcher's sophisticated knowledge of history. It focuses on one or more events, societies, or cultures over a period of time as events and processes unfold. Historical research always involves an assessment of the past, although it might be the recent past, but it can also include in its analysis an assessment of and comparison to the present. Historical research can be very broad or narrow in focus. *Macrolevel historical research* focuses on broad issues that have to do with whole social institutions or with society as a whole. For example, Barrington Moore (1966) conducted a historical analysis whose focus was "the transformation from agrarian societies . . . to modern industrial ones" (p. xi) in Europe, Asia, and the United States over the span of centuries. In addition to being historical, this research was also comparative in focus since Moore was looking at how the transformation occurred in a variety of different societies. *Microlevel historical research* focuses more narrowly on a single event or possibly even the life of a particular individual. For instance, Arthur Thurner (1984) conducted a historical analysis of a labor strike that occurred in the Copper Range of Michigan's Upper Peninsula in 1913 and 1914. While Thurner had to discuss societal conditions in general in order to place the strike and subsequent events into a context, the primary focus of the research was the events surrounding the strike itself.

Historical and comparative research are often combined when the focus is both on multiple societies or cultures and on an assessment of past occurrences, and I will discuss them together as historical-comparative research. Historical-comparative research is discussed in this chapter because, as will be seen, it relies heavily on the analysis of available data, often in the form of documents. In addition, historical-comparative research typically has an important qualitative dimension to it, although it also often involves some quantitative data analysis.

Some of the classic works in the social sciences are historical-comparative in nature. Karl Marx, for example, studied social stratification, social inequality, and political and economic change in societies throughout history; Max Weber studied the development of rationality, bureaucracy, and religion in Asian, African, and Western societies as well as in ancient and modern times; Emile Durkheim compared rates of suicide in a number of European societies and forms of religion in societies from the ancient to the modern. So, historical-comparative research, while not nearly as common today as other research methods such as surveys, has a rich and substantial foundation in the social sciences.

Goals of Historical-Comparative Research

Historical and comparative research can pursue any of the goals of research discussed in Chapter 1 and take the approach of any of the paradigms discussed in Chapter 2 (Skocpol, 1984). For many such researchers, the ultimate goal is explanation: the development and testing of theories and hypotheses. Even with historical research, the goal is not just to know what happened in history but rather to develop an abstract knowledge about how human societies operate and evolve. Some historical approaches involve the application of a *single theoretical model to a number of historical cases*. For example, Kai Erikson (1966) used a Durkheimian and functionalist model of how communities regulate deviant behavior and applied it to the actions of people in the Puritan community of Massachussets Bay in the 1600s. Erikson's goal was not so much to learn about the Puritans of the 1600s as it was to use the experience of those Puritans as evidence to test the general model of deviant behavior. In fact, this is one criterion used by some to distinguish sociology from history: History's primary interest is in the Puritans (or any other historical peoples) in their own right, whereas sociology's primary interest is in developing some broader theoretical knowledge by using the historical data as evidence.

A different and somewhat less ambitious approach to historical analysis is to *search for causal regularities in history*. The differences and similarities among a variety of historical cases can be explored to see if any commonalities can be found. This is more akin to a search for empirical generalizations among a limited number of variables (as discussed in Chapter 2) than to the testing of an elaborate theory. So, if a similar historical outcome is found in a number of societies, we might be able to conclude that all those societies share something else in common that can serve as an explanation of those common outcomes. For example, Barrington Moore (1966) took this approach in his study of the transition of agrarian monarchies into either dictatorships or democracies. He compared the experiences of many different societies in terms of such things as the role of the landed upper classes, the strength of the peasantry, and forces toward commercialization. The basic logic of the approach was to search for those factors shared in common by all the societies that also shared a common outcome. To simplify the logic: If all the societies that transitioned to democracy shared a particular social, political, or economic structure and none of the societies that became dictatorships shared that same structure, then it is a plausible assumption that that structure played a causal role in producing the outcome. Or at least, a historical analysis such as this would enable Moore to formulate a hypothesis that could then be tested with yet further historical analysis (Smith, 1984).

These first two approaches are generally consistent with the positivist approach discussed in Chapter 2. The first one, the application of the general model, is more deductive in nature and the second is more inductive. Some historical-comparative researchers adopt an approach that is more consistent with the interpretive framework: *the development of interpretations of historical events and processes* that have meaning in terms of the people involved and that have significant meaning to the people who are using the history to understand their own lives. This interpretive approach downplays the importance or even the reality of generalizing from broad models or finding deterministic, causal patterns that are applicable to diverse historical settings. Instead, there is a sense that each historical setting involves a set of unique meanings and that we must understand how these meanings are perceived and understood by the people involved. In fact, interpretive historical sociologists often use a comparative approach to show what is common to a number of cases but also what is distinct. Anthropologist Clifford Geertz (1971), for example, used this approach in his study of the role of religion in societal modernization. He selected two countries for study, Morocco and

Indonesia, because they shared a religion in common, Islam, but also were vastly different in terms of cultural heritage, one Western and the other Eastern. The idea is that understanding arises by grasping the commonalities of the two societies as well as their uniqueness.

From this discussion of the goals of historical-comparative research, you may have concluded that historical-comparative research sometimes seeks nomothetic explanations, at other times idiographic explanations, and often a combination of the two. So, the Erikson and Moore studies just described are more nomothetic in nature because they attempt to isolate some specific elements and look at their causal impact. Geertz's approach, on the other hand, seems more akin to the idiographic approach in that he attempts a detailed study of a wide range of dimensions of the societies investigated. Actually, historical-comparative research often involves elements of both nomothetic and idiographic explanations.

Sources of Data

Historical-comparative data tends to be, by its very nature, available data; that is, data that was accumulated or collected for reasons other than the research purpose. Because of this, some of the research techniques already reviewed in this chapter are often used in historical-comparative research, including secondary analysis, data archives, content analysis, and coding schemes.

One key source of data in historical-comparative research are the writings or other accumulated materials of people about historical events. These materials are usually divided into primary and secondary sources. *Primary sources* are the records of people who were eyewitnesses to historical or societal events: their letters, diaries, books, autobiographies, or even handwritten notes. In the modern age, such primary sources also include audio and video recordings as well as e-mail messages. Basically, primary sources consist of any kind of record left by someone who was an eyewitness to

events. *Secondary sources* are the records of people who were close to an event, either temporally or geographically, and who, for that or some other reason, are in an especially good position to have valid knowledge of the event.

Another source of historical-comparative data are the official records of government, business, or other organizations of a particular time period. These provide the researcher with information on exactly what happened in politics or commerce during a particular time period. These official records can sometimes provide highly quantitative data. For example, the records might show the percentages of people voting for various politicians, changes in the gross domestic product, or shifts in the import and export of various products.

An additional important source of data are the various cultural products, such as newspapers, books, novels, poems, and the like. Newspapers often serve as a semi-official record of historical events. Depending upon the research topic, researchers may use such nationally recognized newspapers as the *New York Times*, the *Wall Street Journal*, and the *Washington Post* as key sources of information about events. A historical-comparative study that focuses on a small community might use a local newspaper in a similar fashion.

Finally, a source of data that is sometimes used is oral histories—verbal or written accounts from someone who experienced an event or an era but that are gathered long after the event or era. While these are subject to problems of memory and recall, oral histories are nonetheless considered important because they provide eyewitness assessments.

The range of materials that are relevant to a particular historical research might be very extensive—so extensive that the researcher could not possibly review them all and must consider issues of sampling. In some cases, time sampling, discussed earlier in this chapter, might be a relevant approach. In other cases, where it is not possible to establish a sampling frame of all the materials relevant to the study, a purposive sampling procedure could be used. In that case, the researcher would use his or her judgment in deciding which of the materials would provide the most useful and valid data. The Applied Scenario box describes an area of applied research and social policy development where historical-comparative research has been especially useful.

Validity and Reliability

Historical-comparative research confronts many of the problems that other qualitative, available data research confronts, but there are also some special considerations that come into play. One of these considerations is the recognition that all historical documents and records are prepared from a particular perspective or point of view that may bias its content. Even firsthand accounts of primary sources may contain such a bias. The people who write the diaries, autobiographies, personal notes, or e-mails that constitute the primary sources may desire to present themselves in a positive light. Even though they may not explicitly falsify, they may leave out of the account incidents or events that might suggest a different interpretation. So, one challenge for the historical researcher is to assess how and why a particular record was produced and determine the perspective of the person who produced it.

Even the quantitative data found in the historical record are subject to concerns about validity and reliability since the data may not be the best measures of the phenomenon under study. As discussed earlier in this chapter, one always has to be careful that the measures used from available data are valid measures of the phenomenon of interest.

The data used as evidence in historical-comparative research is sometimes much less clearly subject to validity checks than are other sources of data. Conclusions, for example, might be based on the researcher's imaginative interpretation of multiple sources of data. A conclusion might be based on an interpretation of material from diaries, autobiographies, fiction, newspa-

pers articles, and data about commerce all at the same time. It might also be based in part on the researcher's general knowledge of a historical epoch achieved through a wide reading of many sources. In fact, just as other researchers conduct a literature review in refining a research problem, historical researchers often conduct broad reviews of written histories of the period under consideration in order to develop a familiarity with social, economic, and cultural trends of the era. This understanding provides the historical researcher with a framework from which to assess evidence and events. Of course, the researcher recognizes that these histories also are produced from a particular perspective that may color their presentation or interpretation of events.

In this regard, historical-comparative research has some parallels with field research, which is explored in depth in Chapter 11. One parallel is the highly subjective nature of some historical-comparative analysis. Another parallel is contextualism: Both field research and historical-comparative research attempt to describe a complex context in which many forces are operating simultaneously. Unlike some survey or experimental research that attempts to isolate one or a few variables to study their effects, most field and historical-comparative research assume that the whole context must be comprehended, with all its complexity and interlinkages. In other words, historical-comparative research often focuses more on idiographic explanations, and less on nomothetic ones, than do surveys and experiments.

Social scientists have developed procedures that can improve the validity of historical-comparative research, as they have with other types of qualitative research. One such procedure is to be as thorough as possible in describing and interpreting events and situations. Things that may seem unimportant to you at one time may be recognized as important by others who later review your data. Obviously, one could never attend to all the information uncovered in the historical record,

APPLIED SCENARIO   What Causes Race Riots to Occur?

An unwanted, yet enduring, feature of the cultural landscape for the United States during the twentieth century was race riots. Sometimes the riots were sporadic, as in Detroit in 1943, Miami in 1980, and Los Angeles in 1992; at other times, riots came in waves as city after city exploded into violence in quick succession in the 1960s. Social science research has focused on trying to ferret out the causes of these riots, and historical-comparative research has proved useful in this endeavor.

Riots are a very difficult phenomenon to study firsthand, for at least two reasons. One reason is that, since we cannot predict when and where a riot will occur, it is difficult for social scientists to be on hand to make observations when they do occur. Actually, there have been a few cases where sociologists happened to be in the right place at the right time, but this is not common. A second reason why it is difficult to study riots firsthand is that riots are notoriously dangerous places to be! Most social scientists would be disinclined to place themselves in such danger for the sake of data collection.

Beyond these issues, however, firsthand data collection at the scene of a riot is not necessarily the best way to learn about some aspects of riots. In fact, one of the major concerns about riots that has held the attention of applied researchers is the underlying societal and community structures that cause riots to occur. Data that can be collected at the scene of the riot does not always measure variables appropriate to this structural level of analysis. Researchers have instead turned to historical-comparative research methods to address these issues. One advantage of historical-comparative methods in studying riots is that they make possible a larger sample size. Since riots are (fortunately!) not common occurrences, a sufficiently large sample of riots can be obtained only if we look at riots that have occurred over some historical period, say over the past 40 years. Historical methods enable us to include in our sample events that occurred long in the past.

Many studies of riots compare cities that experience riots to those that do not. So, the researcher must find some way to measure whether a riot has occurred in particular cities and in some cases the number or severity of riots. One common way of measuring this is to use the *New York Times* as essentially a public record of important events. Coders read through this newspaper for reports of disturbances that might be race riots. Then, for each such disturbance found, certain criteria are applied to see whether it passes the test of being a race riot. One study, for example, required for inclusion as a race riot that a disturbance involve grievances against racial discrimination or racial injustice, that at least 30 people be involved, and that the event involve violence that lasted for several hours (Olzak & Shanahan, 1996). Some studies

but observers try to err on the side of being too complete rather than too skimpy. A second procedure that enhances validity is for researchers to carefully assess their own desires, values, and expectations to see if these might bias their observations or their interpretations of the record. If you expect to find something, you will tend to find or interpret evidence in support of it; so look especially rigorously for the opposite of what you expect and be critical if what you expect to find actually seems to be happening. A third check on validity is to have other researchers review the same historical or comparative evidence to see if they come to

also measure the severity of the riot by using such criteria as the number of deaths and injuries, the amount of property damage, or the kinds of violence that occur (for example, looting, arson, and so on) (Lieberson & Silverman, 1965; Spilerman, 1976). You can see how a newspaper such as the *New York Times* provides a permanent historical record of race riots that have occurred. It is not perfect, of course; it may not report some riots that occur and may misreport details about numbers of participants or the source of the grievances in other riots. Yet, given the comprehensive reporting style of that newspaper, it probably provides a more complete record than most other sources. In some studies, this newspaper source is supplemented by other compilations of riots, such as the Congressional Quarterly's *Civil Disorder Chronology* and the *Report of the National Advisory Commission on Civil Disorders* (Myers, 1997).

The independent variables in these studies are the structural conditions in the riot communities that presumably cause riots to happen. These independent variables often include such phenomena as social disorganization in the community (measured with data on in-migration and out-migration); levels of deprivation (data on income, occupation, and education); or degree of racial competition for jobs. Data to measure these variables is collected annually by various agencies of the government—such as the Department of Education, the Census Bureau, and the Department of Labor—and is available going back many years.

Historical-comparative research of this sort has advanced our understanding of the underlying causes of race riots in the United States and produced policy recommendations for improving the situation. For example, many of these studies conclude that race riots are not merely a function of poverty or racial discrimination, although these factors do play a part. Rather, riots occur when there is increasing competition between different racial groups for housing and jobs. To prevent such riots in the future, these findings suggest, policies should be implemented that increase the availability of both housing and jobs, especially in situations where population is growing and competition is increasing.

With this understanding of the role of historical-comparative research, you should do the following.

1. Look for other examples of historical-comparative research that have policy implications and identify what those implications are.

2. Go to the sources of data described above (such as the *New York Times* or the Census Bureau) and identify other possible research topics with a historical-comparative approach that could be addressed using data from those sources.

the same conclusions. Fourth, validity can be assessed by comparing whether the results of historical-comparative research are consistent with the results of other research. If they are at wide variance with other research, consider the possibility that there are validity problems in the interpretation of the data.

Many of the procedures for analyzing qualitative data and assessing its validity and reliability discussed in Chapter 15 are relevant in historical-comparative research. The Eye on Ethics box analyzes some of the ethical issues that arise in the research methods discussed in this chapter.

EYE ON ETHICS Privacy, Confidentiality, and Informed Consent

One of the advantages of nonreactive, available data, and historical-comparative research is that the researcher confronts fewer ethical dilemmas than when gathering data directly from people who are still alive. However, ethical issues can arise, depending on how the data are collected. In hidden and disguised observation, for example, even in public places, the researcher needs to consider whether the people might prefer that their behavior not be recorded or that it not be used for research purposes. In this type of research, it is not generally possible to obtain informed consent because that would compromise the "hidden" or "disguised" nature of the observations. We saw this as an issue in Chapter 3 in Laud Humphreys' (1970) research on men having sex in public bathrooms. One of the criticisms of Humphreys' research was that he did not obtain informed consent and had violated the men's privacy because the men were clearly trying to hide their behavior from the prying eyes of others. Even the observations of people's behavior in bars, described in this chapter, might confront this issue since some individuals would prefer that their antics in bars not be recorded for discussion or use in other settings.

This is admittedly a very judgmental consideration since, as Humphreys argued in his own defense, the behaviors are on public display. Yet, recall from Chapter 3 the weight given to people's right to privacy and self-determination. Critics of Humphreys argue that his justification might be technically or logically correct but not morally correct. The morally correct option, they argue, would be to give the people the benefit of the doubt in a situation where they never have the opportunity to make the choice themselves. So, researchers conducting this kind of research need to weigh carefully the rights of privacy and self-determination against the benefits of the research and any possible harm that could come to the subjects of the research. If it is decided that obtaining informed consent is the ethical route to go, then some research method other than disguised or hidden observation will probably have to be used since in most settings it would be impossible to both obtain consent and retain the "hidden" or "disguised" nature of the observations.

Available data can also raise privacy and informed-consent issues since you must determine whether a particular document or record is in the public domain and can be used without getting permission. Often this is not the case. Researchers who gather data from school records, hospital records, corporate databanks, or any other organizational source must determine whose permission needs to be obtained. In some cases, it might only be that of the organization (in the case of many corporations), whereas in other cases permission might have to be gotten from both the organization and the individuals whose information is kept by the organization (in the case of hospitals, schools, and some social service agencies). In the case of schools, for example, informed consent may have to be obtained from both the school administration and the parents of the students. When using this kind of available data, careful investigation must be done about whose consent must be gathered. This is partly a legal issue, of course, since you may violate the law by using data to which you don't have legal access. However, it is also an ethical issue in terms of ensuring that people's rights to privacy and self-determination are not violated. In some situations, a researcher may have legal access to data without requesting permission but still be ethically bound to obtain informed consent.

ASSESSMENT OF UNOBTRUSIVE AND AVAILABLE DATA RESEARCH

Advantages
Lower Costs

Document analysis can be one of the least costly forms of research, and using statistical available data can help reduce the costs of a research project. Even historical-comparative analysis, which relies on existing documents, can be relatively inexpensive. The major expense of data gathering is borne by the producers of the documents and statistics rather than by the researcher. The sheer volume of data collected by the U.S. Census Bureau, the National Association of Social Workers, or the many local and state government and agencies is so massive that even the best-endowed research projects could not possibly duplicate their data collection efforts. Document analysis can become expensive if the documents of interest are widely scattered and difficult to obtain or if very large samples are employed. Also, the more complex the coding process, the more expensive the study will become. Overall, however, available data offer an opportunity to conduct valuable research at reasonable cost.

Nonreactivity

Available data are nonreactive. Unlike the case of surveys or experiments, in which the participants are usually aware that they are being studied, producers of documents do not normally anticipate a researcher coming along at some later date to analyze those documents. The contents of the documents are therefore unaffected by the researcher's activities. Of course, those producing the documents may well be reacting to something that might result in biased documents. For example, police officers might be more likely to record an instance of physical or verbal assault in an arrest report when the suspect is non-Anglo than when the suspect is Anglo, thus generating data that make non-Anglos appear more abusive and introducing bias into arrest records. In addition, those who compile documents may be reacting to how people other than researchers, such as a supervisor or politician, may respond to their document. Likewise, the preparation of a document might be shaped by the preparer's hope for a "place in history." Despite all this, the data are still considered nonreactive because it is not the researcher who is introducing the bias.

Inaccessible Subjects

Properly cared for, documents can survive far longer than the people who produce them. Document analysis allows us to study the ways of society long ago and the behavior of people who are long dead. Several of the examples cited in this chapter illustrate the use of document analysis for this purpose.

Longitudinal Analysis

Many statistical data and documents are collected routinely over a period of years—even centuries. This contrasts sharply with the typical "one-shot," cross-sectional survey data. With such longitudinal data, trend analysis—looking for changing patterns over time—can be accomplished. For example, a longitudinal analysis of available data was used in an evaluation of the progress of African Americans in the United States since the 1950s by Reynolds Farley (1984). Using census data from the 1950s, 1960s, and 1970s, Farley compared the position of African Americans relative to Anglos over three decades. He found mixed results. The census data revealed that African Americans as a whole had made substantial gains in the quality of employment, family earnings, and educational attainment. On indicators of residential integration, unemployment rate, and school integration, however, African Americans' position relative to Anglos had not improved during 30 years. Obviously, this type of longitudinal analysis of past trends can be done only through the analysis of available data. Available statistical data often lend themselves to the kind of time series analysis described in Chapter 8. In fact, the illustration of the multiple time series analysis by Kohfeld and Leip (1991) in that chapter utilized existing data on traffic accidents in California that were alcohol-related.

Sample Size

Many types of documents are abundant. As mentioned when discussing sampling, a researcher is likely to confront far more documents than can be analyzed, rather than too few. This means that large samples can be employed to increase one's confidence in the results. The low cost associated with document analysis also contributes to the researcher's ability to use substantial sample sizes without encountering prohibitive costs.

Disadvantages
Variable Quality

Because documents are produced for purposes other than research, their quality for research purposes is quite variable. Unless the limitations of the documents are known, as with crime statistics, researchers may have little idea of the conditions under which the data were collected or the amount of attention to quality that went into them. Researchers should, of course, investigate the issue of quality when possible so that any deficiencies in the data are discovered. When this cannot be done, we simply have to draw conclusions cautiously.

Incompleteness

Documents, especially those of a historical nature, are frequently incomplete. Gaps of weeks, months, or even years are not uncommon. Data may also be missing from available statistical data. The effect of these gaps on a study is often impossible to know except that confidence in the findings is reduced. Incompleteness is unfortunately a common characteristic plaguing available data, and researchers have to work around it if they can.

Lack of Comparability Over Time

Even though documents are commonly used in longitudinal analysis, care must be taken in doing so. Things can change over time that create statistical artifacts in the data and render comparisons useless. I illustrated this with crime statistics, where changes in how the data are collected by the FBI can make it misleading to compare crime statistics over long periods of time. Researchers need to be aware of similar kinds of changes in other data.

Bias

Because documents are produced for purposes other than research, there is no assurance that they are objective. Data from private sources, for example, may be intentionally slanted to present a particular viewpoint. A researcher blindly accepting these data could be walking into a trap. Nonstatistical documents may suffer from biased presentation as well. A good example is the corporate annual reports to stockholders, which are filled with glowing praise for management, impressive color photographs, and bright prospects for the future, with anything negative presented in the most favorable light or camouflaged in legalese.

Sampling Bias

Bias may creep into otherwise objective data during the sampling process. As I noted, sampling documents is often difficult owing to unequal population elements and hard-to-define populations. I also commented that document sampling is frequently complex, requiring several stages. If errors or bad decisions are made in the sampling process, it is quite possible to end up with a highly biased sample that will produce misleading results.

When to Choose Unobtrusive or Available Data Research

As this chapter should make clear, a number of key criteria would lead a researcher to choose unobtrusive or available data research. One criterion is the need for nonreactivity. If the research topic is on some sensitive issue where other data collection techniques are likely to produce biased results, then some form of unobtrusive observation or available data would be a good choice.

A second criterion that comes into play, especially in terms of choosing available data, is whether variables can be measured with the data found in some document or existing data set. This is not always the case, or the available data may be of poor quality or incomplete. Often, however, data on a topic have been collected by someone else, such as the U.S. Census Bureau, or the human behavior you are interested in naturally leaves a record in the form of documents. So, movies, music, newspapers, and the many other cultural products are the natural outgrowth of human social activity and can often be used in measuring variables.

A third criterion is whether the research question calls for data about the past or for longitudinal data. While other research methods can produce longitudinal data, available data naturally lends itself to this because the sources of the data—the documents—usually persist over time once they have been created. In addition, since documents from the past persist into the present, this provides us with an eye on events in the past that surveys, experiments, and field research cannot offer.

REVIEW AND CRITICAL THINKING

While this chapter introduces some new research methods, some of the issues discussed, such as measurement and reactivity, are familiar because they have been discussed in other chapters. This illustrates the variety of forms that these issues can take and their centrality to the conduct of research. It also shows how important these issues are to evaluating information that you may come upon in your everyday life. When evaluating information and conclusions, consider the following.

1. Are the conclusions based on information that is nonreactive? In other words, might the people observed behave differently because the observer is present?

2. If the information is based on available data (e.g., records or documents), what kind of measuring problems does this create? Is the data limited or distorted in some fashion? Is the sampling limited?

3. If the information is qualitative in nature, is it transformed into a quantitative form? If so, how does this affect or change the meaning of the information?

Computers and the Internet

A number of issues related to doing research with available data can be explored on the Internet. To begin, explore the site of the

Inter-university Consortium of Political and Social Research (ICPSR) at their home page: **http://www.icpsr.umich.edu/.** If you click on "Topical Archives," you will find a link to the National Archive of Criminal Justice Data (NACJD). The NACJD currently holds over 500 data collections relating to criminal justice. This Web site provides browsing and downloading access to most of the data and documentation.

The site also provides a link to the National Archive of Computerized Data on Aging (NACDA), which is funded by the National Institute on Aging. NACDA's mission is to advance research on aging by helping researchers to profit from the underexploited potential of a broad range of data sets. NACDA acquires and preserves data relevant to gerontological research, processes them as needed to promote effective research use, and disseminates them to researchers. By preserving and making available the largest library of electronic data on aging in the United States, NACDA offers opportunities for secondary analysis on major issues of scientific and policy relevance.

To learn more about the ICPSR, I suggest that you select "Contents" for an overview and then try "FAQ" (Frequently Asked Questions). Next, select the "Archive" page and use the search field to locate studies that relate to a topic of interest to you. For example, you might enter "domestic violence" or "poverty"; the archive search function returns a descriptive listing of data sets that include a focus on that concept.

Julian Faraway of the University of Michigan maintains an extensive list of data sources at **http://www.stat.lsa.umich.edu/~faraway/ data.html.** Exploring just a few of the many sites listed here will give you a sense of the wide array of data sites. Some of the better sites are the Roper Center, the Gallup Organization, the U.S. Census Bureau, and Statistics Canada.

A recently established Web site is very useful for those interested in issues related to content analysis research. Called "Content Analysis Resources: For Qualitative Analyses of Texts, Transcripts, and Images," its Internet address is **http://www.gsu.edu/~www.com/ content.html.** It includes buttons that can lead you to summaries of recent publications on content analysis, analyses of software used for content analysis, and links to related Internet sites. You might explore this Web site by identifying materials found on this site that have been discussed in this chapter.

Main Points

1. Unobtrusive research and available data research are forms of research that, in one way or another, avoid the problem of reactivity. Sources of unobtrusive data include statistical data collected by others, hidden and disguised observation, physical traces, and documents, which include any form of communication.

2. Problems confronted in analyzing available statistical data are missing data, switching from an inductive to a deductive approach, validity and reliability, and units of analysis.

3. Content analysis quantifies and organizes the qualitative and unsystematic information contained in documents; it is essentially a form of measurement making the issues of validity and reliability paramount. The data in documents are analyzed by using coding schemes, either existing coding schemes developed by others or a newly developed coding scheme.

4. In content analysis, the researcher must identify the unit of analysis, decide on whether manifest or latent coding is most valid, assess the validity and reliability of the coding, and assess the level of measurement. Content analysis will normally include some form of sampling procedure, which must be performed carefully for the sake of representativeness.

5. Historical and comparative research often use some form of available data; the former analyzes the past, whereas the latter compares different situations, societies, or cultures in order to make generalizations about human social behavior.

6. Historical-comparative research can focus on any of the goals of research. For data, it uses the writings and other accumulated materials of people; the official records of governments, businesses, or other organizations; cultural products, such as newspapers, books, movies, and the like; and oral histories.

7. Available data analysis offers the advantages of low cost, nonreactivity, ability to study otherwise inaccessible subjects, easy longitudinal analysis, and often large samples. Problems in using available data include the variable quality of the data, incomplete data, changes in data over time, possible bias in data, and possible sampling bias.

Important Terms for Review

available data research
coding
comparative research
content analysis
data archives
historical research
latent coding
manifest coding
missing data
nonreactive research
physical traces
secondary data analysis
unobtrusive research

For Further Reading

Davies, James A., & Smith, Tom W. (1994). *The NORC general social survey: A users guide.* Newbury Park, CA: Sage. This book describes a body of available statistical data based on surveys conducted since the 1970s. It includes data on a wide range of topics of interest to people in the social sciences.

Jacob, H. (1984). *Using published data: Errors and remedies.* Beverly Hills, CA: Sage. An excellent guide to the problems and pitfalls of using existing data sets. The book focuses primarily on issues of validity and reliability.

Price, J. (1986). *Handbook of organizational measurement.* Marshfield, MA: Pitman. This volume describes a diversity of ways of measuring organizational variables, many of which are based on available data.

Rutman, L. (Ed.). (1984). *Evaluation research methods* (2nd ed.). Beverly Hills, CA: Sage. Although devoted to evaluation in general, it also contains an excellent discussion of how to make the information systems of social service agencies and other organizations of maximum utility for research purposes.

Stewart, David W., & Kamins, Michael A. (1993). *Secondary research: Information sources and methods.* Thousand Oaks, CA: Sage. This useful guide to the many sources of data that are available for secondary analysis also reviews some of the methodological issues that are important to consider when doing such data analysis.

Webb, J., Campbell, D., Schwartz, R., Sechrest, L., & Grove, J. (1981). *Nonreactive measures in social research* (2nd ed.). Chicago: Rand McNally. A discussion of the advantages and problems in the use of nonreactive measures, with two chapters devoted to archival records.

Weber, R. P. (1990). *Basic content analysis* (2nd ed.). Thousand Oaks, CA: Sage. This is a basic introduction to the procedures used in conducting content analysis research.

Chapter 11

FIELD RESEARCH AND QUALITATIVE METHODS

ASSESSMENT OF FIELD RESEARCH AND QUALITATIVE METHODS
Advantages
Disadvantages

REVIEW AND CRITICAL THINKING
Computers and the Internet
Main Points
Important Terms for Review
For Further Reading

The research methods introduced in this chapter, in many ways, bring social scientists into the most intimate, and some would say most exciting, contact with the people being observed. As you will see, field research and qualitative methods involve social scientists becoming a more intimate part of the everyday lives and routines of the people being studied than do most other research methods. Social scientists doing field research, for example, have joined white supremacist groups, run with joyful and sometimes violent crowds of sports fans, and observed children in the classroom and in the schoolyard. Many field researchers over the years have hung around with drug users as well as drug dealers and joined in the lives of various religious cults. Field researchers have even been witness to a wide range of human sexual activities. However, none of this observation was done for the excitement or the titillation; it was done because it provided a unique perspective and insight into some domain of human behavior—insight that could not be obtained by other research methods.

As discussed in Chapter 1, qualitative research basically involves research in which the data comes in the form of words, pictures, narratives, and descriptions rather than in numerical form. Field research, which is a type of qualitative research, involves observations made of people in their natural settings as they go about their everyday lives. Surveys involve people's reports to the researcher about what they said, did, or felt. With surveys and available data, the researcher does not directly observe what will be the focus of the research. With field techniques, on the other hand, the researcher actually sees or hears the behavior that is the data for the research. Field research is most closely associated, in many people's minds, with the work of anthropologists who live among indigenous peoples for extended periods and write ethnographic reports that summarizes the group's way of life. Yet, other social scientists also engage in field research.

This chapter begins with a brief overview of some of the special characteristics of qualitative research methods. Next, different examples and types of field research as well as some other types of qualitative research methods are reviewed. Then I explore how field research is conducted and, finally, an assessment of field research and qualitative methods is made.

CHARACTERISTICS OF QUALITATIVE METHODS

We begin with a quick overview of some of the special characteristics that set qualitative research methods apart from quantitative methods.

Contextual Approach

In Chapter 2 and Chapter 4, I discuss in detail the distinction between qualitative and quantitative research and explore the positivist and interpretive paradigms in the social sciences. Although qualitative research methods might at times be conducted by people who take a positivist approach and might sometimes involve collecting some quantitative data, qualitative research has been closely associated over the years with the interpretive paradigm. Proponents of this approach argue that qualitative and contextual approaches offer access to a very valuable type of data: a deeper and richer understanding of people's lives and behavior, including some knowledge of their subjective experiences (Benton, 1977; Guba & Lincoln, 1994; Gubrium & Holstein, 1997). By observing people in their natural settings, researchers have access to a view of group life that cannot be gained in an experiment or questionnaire. It is a unique view because it is seen from the perspective of the group—a perspective that is especially valuable to anyone who works with groups that are stigmatized or commonly misunderstood by both researchers and laypeople. These qualitative approaches stress the idea that knowledge best emerges when researchers understand the full context in which people behave.

Grounded Theory Methodology

Many qualitative methods are also very closely "grounded" in the data in that they let meaning, concepts, and theories emerge from the raw data rather than being imposed by the researcher. In many social science research methodologies, the role of theories parallels what is discussed in Chapter 2 as deductive reasoning and the positivist model. Theories are abstract explanations containing a variety of concepts and propositions. From these theories, the researcher derives testable hypotheses and operational definitions of concepts. Then observations are made to determine whether the hypotheses are true. This deductive approach, however, is not always appropriate or useful. One reason that it may not be useful is that some research projects are exploratory in nature, meaning that there is little existing theory, concepts, or propositions from which to shape hypotheses and develop operational definitions. Another reason such a deductive approach may not be useful, at least for some research projects, is that it involves the scientists, through the theory and measurement, imposing structure, categorization, and meaning onto reality rather than letting the structure and meaning emerge from reality.

Proponents of more inductive approaches to theory development argue that deducing hypotheses from existing theories can sometimes be limiting, especially in the early stages of theory development when the theory may not include some relevant variables (Strauss & Corbin, 1994). If the variables are not in the theory, they cannot be part of hypotheses and thus may be ignored. In other words, strict adherence to deductive hypothesis construction might blind researchers to some key phenomena. One of the more widely used approaches to these issues is called *grounded theory*. **Grounded theory** is a research methodology for developing theory by letting the theory emerge from the data, or be "grounded" in the

data. With this method, there is a continual interplay between data collection, data analysis, and theory development. In the positivist model, these three elements are sequenced: first data collection, followed by data analysis, followed in turn by theory development. Using a grounded theory method, these three elements can occur simultaneously as one goes constantly back and forth among them. So theory development occurs in the midst of data collection rather than following it.

In the absence of theory, we begin by making observations; those who use grounded theory often do qualitative research by making direct observations or conducting interviews in field settings. Without the restrictions of a preexisting theory, they describe what happens, try to identify relevant variables, and search for explanations of what they observe. Beginning with these concrete observations, they develop more abstract concepts, propositions, and theoretical explanations that would be plausible given those observations. Inductive research of this sort can serve as a foundation for building a theory, and the theory that emerges can later serve as a guide for additional research, possibly even as a source of testable hypotheses through deductive reasoning.

Proponents of qualitative methods argue that concepts and theories produced by such grounded approaches provide a more valid representation of some phenomena because they emerge so directly from the phenomena being studied. In fact, in positivist science, one can engage in theory development without engaging in any data collection or data analysis. This would be impossible in grounded theory, since the theory emerges from the data or observations. However, theory produced in this manner could also be subjected to further verification by deducing and testing hypotheses. In fact, one common misconception of grounded theory is that it is entirely inductive in nature. To the contrary, grounded theory does at times use existing theory to understand and explain data, and it does include procedures for verifying theories. Thus some of the considerations of more positivist approaches are relevant to grounded theory: Grounded theory uses evidence to verify theories, it rigorously follows precise procedures, it makes efforts at replication, and one of its goals is to generalize about social processes across a variety of social settings.

Grounded theory will be discussed in more detail in Chapter 15 in the context of analyzing qualitative data. However, I wanted to review these ideas about the contextual and grounded nature of qualitative research so that you can recognize the commonalities among the different types of qualitative methods that we will explore.

FIELD RESEARCH

Field research is one of the more common types of qualitative methods used by social scientists. I begin with some examples of good field research. This will help you to understand about how and why field research is done and some of the issues involved in doing it. These studies will also be used to illustrate points later in the chapter.

Examples of Field Research
Mental Patients in the Community

Anthropologist Sue Estroff studied the daily lives and problems of mental patients released into the community. She joined for two years in the lives of a group of deinstitutionalized mental patients, experiencing the drudgery and degradation of their daily routine (Estroff, 1981). She worked at low-paying jobs (such as slipping rings onto drapery rods) that were the lot of these expatients. She took the powerful antipsychotic medications that were routinely administered to them and that had distinctive side effects such as hand tremors and jiggling legs. And she experienced the extreme depression and despair that result when one suddenly stops taking these potent drugs. From Estroff's position as participant in the subculture of these mental patients, she could observe the con games that characterized the relationships between the patients and mental health professionals.

Teenage Drug Dealers

Sociologist Terry Williams (1989) did field research among teenage cocaine dealers in New York City. He spent two hours a day, three days a week, hanging out with these teens in the bars, discos, restaurants, parties, crack houses, and street corners where they spent their days. He got to know them, their friends and acquaintances, and their activities intimately. Through this exposure, he was able to see why they couldn't get jobs—good entry-level jobs were disappearing; and why they dealt cocaine—the opportunity was there and they could make a good deal of money. Williams also learned that these kids were shrewd businesspeople who worked very hard, and most of all sought the same kind of respect and reward that motivates people pursuing more conventional careers. By using such field methods, Williams was able to observe processes and grasp deep and intimate levels of meaning that other methods, such as surveys or available statistical data, would not have been able to uncover.

Poverty and Marginalization in the Inner-City

In the mid-1980s, anthropologist Philippe Bourgois (1995), recently married and looking for an affordable apartment, moved into "El Barrio," the name of an East Harlem neighborhood in New York City. Although fieldwork was not his intent in moving there, he eventually used his residence in El Barrio as a springboard for doing research on the underground economy, social marginalization, and how the impoverished families in that community managed to survive. He lived there for three and one-half years and got to know many of the community members. He visited their homes often and attended parties and celebrations with them. Bourgois was able to observe how people managed to get by when they could obtain no job at all or only a poorly paying one. Most of El Barrio's residents were law-abiding; however, a small but publicly visible portion of them were involved in the drug trade. Bourgois became friends with two dozen street dealers and their families. He spent many nights on the streets and in the crack houses both with dealers and with addicts. He also was able to observe how the residents who had nothing to do with drugs coped with the vibrant and sometimes violent drug world that pervaded their community.

Learning Gender Differences

Sociologist Barrie Thorne (1993) was interested in the sociology of childhood, and especially how childhood differed for boys and girls. In part because she had been asked to assess gender equity in the early grades of school, she was able to observe elementary school students in a number of schools for many months. Thorne observed students in the classroom as well as at lunch and in the schoolyard at play. In fact, she was free to make observations almost wherever the schoolchildren went. Sociologist Karin Martin (1998) also did field research on gender in schools, but with a somewhat more narrow focus: the role that schools play in teaching children about gender differences. She and her assistants made observations in five preschool settings three times a week for eight months; they sat in classrooms and made observations without materially affecting the interactions occurring in the classroom. Some of these observations were very unstructured, while at other points they structured their observations to ensure that certain kinds of children or their behaviors were noted. Through these observations, they were able to assess how the schools taught different things to boys and girls about how to use their bodies in interacting with others. They could

see how clothing and other types of adornment of boys and girls were treated differently by the teachers and other children. They could also observe how gender influenced the formal and informal behaviors displayed as well as how boys' and girls' yelling received different responses from educational authorities. By directly observing these things rather than asking students and teachers about them in a survey, Martin was able to learn things that would probably not be revealed in a survey.

Poverty, Violence, and Death in a Brazilian Shantytown

Nancy Scheper-Hughes (1992) was a young women when she first went to the shantytown in Brazil in the 1960s as a Peace Corps volunteer. The routine and daily death and despair that she found among the impoverished residents with whom she worked left an indelible mark on her. Her concern about the plight of the children and their mothers remained with her after she left Brazil and pursued her education to become an anthropologist. As an anthropologist, she returned to the same shantytown on four different field trips beginning 15 years later. She and her family spent a total of 14 months living among the residents of the shantytown. While there as a Peace Corps volunteer, she had undertaken the role of community activist, once being arrested because her activities were defined as "subversive" by the authorities. When she returned as a field researcher, she struggled to balance her professional role as a scientific observer with her personal inclination to again be the activist who helps people improve their lives. Through her experience living in the community and interviewing the women who live there, she was able to develop an understanding of how they cope with the despair, violence, and death that are part of their daily existence.

With these examples fresh in mind, we can begin discussing the types of field research that can be conducted. A number of different approaches to conducting field research exist.

While there is much overlap among them, each introduces some special techniques of data collection.

Participant Observation and Ethnography

One technique for doing field research is **participant observation**, where the researcher observes people in their natural environment (the "field"), and the researcher is a part of, and participates in, the activities of the people, group, or situation that is being studied (Burgess, 1984; Lofland & Lofland, 1995). The researcher has some degree of firsthand involvement in the lives of the people being studied. A related form of field research is called *ethnographic research* (Vidich & Lyman, 1994). The Greek word *ethnos* refers to a people, a race, or a cultural group; the suffix -graphy means "knowing something" or "a knowledge of something." So, **ethnography**, in its broadest sense, refers to the social science description and knowledge of a people and the cultural bases of their lives. Ethnography is often associated with anthropology, especially early anthropological work, which focused on producing detailed descriptions of a whole way of life and its associated meaning system as seen from the point of view of the people being described. However, other social scientists conduct ethnographies, and participant observation research can come close to being ethnographic in nature.

If there is a distinction between participant observation and ethnography, it is that the latter is a broader and more complete description of a culture and its meaning system; it also does not necessarily have to involve participation in the group, although it usually does. In fact, a volume could be written on different thoughts about and variations on ethnography as well as the theoretical bases that underlie them. However, the distinction between them is more a matter of degree than kind, and many of the issues and techniques involved in field research are relevant for both participant

observation and ethnography. So I will use the two terms interchangeably rather than distinguish between them.

The Interpretive Paradigm and Empathic Understanding

A brief review of some of the points made in Chapter 2 regarding the interpretive paradigm and qualitative research will be helpful in appreciating the benefits of field research and participant observation. Recall that positivism argues that the world exists independently of people's perceptions of it and that scientists can use objective techniques to discover what exists in the world (Durkheim, 1938; Halfpenny, 1982). Astronomers, for example, use telescopes to discover stars and galaxies, which exist regardless of whether we are aware of them. So, too, human beings can be studied in terms of behaviors that can be observed and recorded using some kind of objective techniques. Recording people's gender, age, height, weight, or socioeconomic position in a survey are legitimate and objective measuring techniques—the equivalent of the physicist measuring the temperature, volume, or velocity of some liquid or solid. For the positivist, quantifying these measurements—assessing the average age of a group or looking at the percentage of a group that is male—is merely a precise way of describing and summarizing an objective reality. Such measurement provides a solid and objective foundation for understanding human social behavior. Limiting study to observable behaviors and using objective techniques, positivists argue, is most likely to produce systematic and repeatable research results that are open to refutation by other scientists.

The interpretive approach, on the other hand, argues that these "objective" measures miss a very important part of the human experience: the subjective and very personal meanings that people attach to themselves, what they do, and the world around them (Glaser &

Strauss, 1967; Wilson, 1970). Max Weber (1925/1957), an early proponent of this view, argued that we need to look not only at what people do but also at what they think and feel about what is happening to them. This "meaning" or "feeling" or "interpretive" dimension cannot be adequately captured through objective, quantitative measuring techniques. Researchers need to gain what Weber called **verstehen,** or a subjective understanding. They need to view and experience the situation from the perspective of the people themselves. To use a colloquialism, the researchers need "to walk a mile in the shoes" of the people being studied. They need to talk to the people at length and to immerse themselves in people's lives so they can experience the highs and lows, the joys and sorrows, the triumphs and the tragedies as seen from the perspective of the people being studied. Qualitative research methods are the best method for gaining this *verstehen*; for interpretivists, quantitative research, by its very nature, misses this very important dimension of social reality. Participant-observers gain *verstehen* by immersing themselves in the lives and daily experiences of the people they study. By experiencing the same culture, the same values, the same hopes and fears—as anthropologist Estroff did among mental patients in the community, for example, or as sociologist Williams did when hanging out with teenage cocaine dealers—researchers are in a better position to take on the points of view of these people.

It is important to recognize that, in participant observation, knowledge is gained from two distinct, although linked, kinds of observations (see Figure 11.1). One kind of observation is based on *participation*, where the researcher learns about the social world of those being observed by personally experiencing that world. This is the *verstehen* method proposed by Weber and provides understanding through empathic experience. This empathic experience is an important source of knowledge that offers an intersubjective understanding of other people's lives. The second kind of observation

is produced by *observation*, noting and recording how others behave and what occurs at a social setting. This observation provides understanding through a deep appreciation of the full context within which people live their lives. It enables social scientists to make rich descriptions of everyday social life.

Despite its focus on subjective experiences, however, field research is still empirical in the sense that it is grounded in observation, and those who use this method are also concerned about issues of reliability and validity. Field researchers consider their research no less systematic or scientific than the more positivistic research techniques.

Researcher Roles and Intervention

In many types of research, the relationship between the researcher and those participating in the research is fairly clear-cut. In surveys, for example, participants know who the researchers are and that they as respondents are providing data to the researchers. In field research, the researcher-participant relationship can become more problematic in that it can take a number of different forms. These various forms revolve around two key issues: Will the observer's identity be known to those being observed? and, Will the observer intervene in and possibly change the setting that is being observed?

In many cases, the researcher takes what is called the *participant-as-observer role* where the researcher's status as observer is revealed to those who are being studied (see Figure 11.2). In this role, the observer enters a group and participates in their routines but is known to be doing so for research purposes. The participant-as-observer spends a considerable amount of time observing in the group being studied. In the examples mentioned earlier, Bourgois, Thorne, Martin, and Scheper-Hughes made their identities as researchers known to those being observed.

However, the participant-as-observer role needs to be further clarified. Will the researcher emphasize the *participant* dimension

FIGURE 11.1   Sources of Data in Participant Oberservation and Ethnographic Research

Participation

researcher learns about the social world of those being observed by actually personally experiencing that world

Observation

noting and recording how others behave and what occurs at a social setting

Researcher

Interviewing

informal and unstructured interviews which can explore a wide range of topics and may last for a long time and look more like a rambling conversation

Researcher

(take an active role in the group, as any other group member would) or the *observer* dimension (remain detached, passive, and observe what others do)? Let's look at each side of this issue. Those who emphasize the importance of the participation of the observer argue that the investigator plays two roles: scientist and group member. In order to comprehend fully the activities of the group and the dynamics of the situation, the researcher must become fully involved in the group. Otherwise, group members may not confide in the researcher. In addition, the researcher who does not participate will be hampered in achieving *verstehen*; that is, an empathic understanding of the deep meanings and experiences important to the

group. In order to become fully involved, the researcher must act like any other group member—and this means intervening in those situations in which other group members might do so. The situation of Scheper-Hughes in this regard was complicated because she had participated and intervened as a social activist in her first prefield research visit to the Brazilian shantytown. When she returned as a field observer years later, she at first remained somewhat detached lest she might lose objectivity. However, the residents put considerable pressure on her to resume those roles and she succumbed to the pressure. She argues that her active involvement actually benefited her field research: as the residents "pulled me toward

FIGURE 11.2   Researcher Roles in Field Research

Researcher Roles   **Examples of Field Research**

Participant as Observer   Participant Observation

Researcher's identity as
researcher is known to
those being observed

Complete Observer   Hidden Observation

Researcher does not
participate in group and
identity is not known to
those being observed

Complete Participant   **Participant Observation**
Disguised Observation

Researcher joins group
but keeps identity as
researcher hidden

the 'public' world of Bom Jesus [the shanty-town], . . . the more my understandings of the community were enriched and my theoretical horizons . . . expanded" (1992, p. 18).

On the other side of the issue, those who emphasize observation over participation argue that the more fully one becomes a group member, the less objective one becomes. The real danger is that researchers might "go native": become so immersed in the group that they completely take on the perspective of the group and can no longer view the situation from a less interested perspective (Shupe & Bromley, 1980). This is always a danger in

field research. In fact, some researchers caution against the use of field research for this very reason. Most field researchers attempt to strike a balance between total immersion and loss of objectivity on the one hand, and total separation with its consequent loss of information on the other.

However, beyond the problem of objectivity, critics of the participation approach point to another problem it may create: It may represent an unnatural intrusion into the natural setting. If the researcher-participant has some skills, talents, or whatever that are not possessed by other members of the group, then the group may develop in ways that would not occur naturally. This is especially a problem if the group is one that the researcher would not normally belong to. Scheper-Hughes, for example, undoubtedly brought some intellectual, organizational, and other resources to the Brazilian town and may have helped to produce outcomes that would not have occurred without the intervention of an outsider. Although this was probably not a problem in the Scheper-Hughes study, it could become a problem if it interfered with the observer's ability to understand the natural development of the group.

So, any field setting needs to be assessed carefully to determine whether participation would tend to be unacceptably unnatural and interfere with research goals. We will see in Chapter 17 that some approaches to social science research would not view such researcher intervention as that much of a problem. Proponents of participatory or empowerment research argue that researchers should work together with the subjects of research to try to improve social conditions. (Scheper-Hughes would agree with this as she viewed helping the residents of the shantytown as part of her obligation.) In such an effort, any special or unique skills that the researcher brings to the endeavor would be seen as an advantage.

Unobtrusive Observation

Some field research questions can best be addressed if the investigator refrains from participation in the group being investigated. This is the case when the intrusive impact of an outsider might change the behavior of group members in ways detrimental to the research question. In such cases, the relationship the investigator has with those being observed is that of a nonparticipant, or what is called the *complete observer role*: the observer has no direct contact with, or no substantial influence on, those being observed (see Figure 11.2). One way of doing such nonparticipant observation is to use an observational technique called **un-**

obtrusive or **nonreactive observation:** Those under study are not aware that they are being studied, and the investigator does not change their behavior by his or her presence (Sechrest & Belew, 1983; Webb, Campbell, Schwartz, Sechrest, & Grove, 1981). This is discussed briefly as one form of unobtrusive research in Chapter 10. Unobtrusive observation can involve both quantitative or qualitative observations. However, the focus in this chapter is on the use of unobtrusive observation in qualitative field research, and it can take two forms: hidden observation or disguised observation.

Hidden Observation

In some field research, it is possible to observe behavior from a vantage point that is obscured from the view of those under observation. This might be done by observing people via a one-way mirror or by videotaping them with a hidden camera. For example, studies of aggressive behavior among children have utilized observations in field settings by videotaping children at play in schoolyards with the camera in a hidden location (Pepler & Craig, 1995). The camera can be set up in a classroom window, out of sight of the children because modern cameras with zoom lenses make possible clear observation of children's behavior from long distances. As long as the children are not aware of the cameras and of the fact that they are being observed, the research is truly nonreactive in nature. This naturalistic, hidden observation is important in the study of aggression among children because children behave more naturally under these conditions; in an interview or if children know they are being observed, it would probably be impossible to make the same honest, natural observations that are possible when children are just being themselves among their peers. In fact, one group of researchers of such behavior noted that "the use of knives on our school playground tapes was so covert that it often took several passes through the audiovisual tapes to discern their presence" (Pepler & Craig, 1995,

p. 550). Because the children are clearly attempting to hide the use of knives from observation by others, they would be unlikely to display such aggressive behavior to researchers in surveys or other more obtrusive research formats. Naturalistic and hidden observations are key to making such discoveries.

A major problem with hidden observations is ensuring that the observations are in fact hidden and truly unobtrusive. The unobtrusive nature of the study of children's aggression just mentioned was compromised in two ways. First, the school principal and the teachers and other adults supervising the children at play were aware that observations were being made. All people have a tendency to react differently when they are being observed—they put their best foot, or at least a different foot, forward, or behave in a fashion that will be acceptable to the observer. The adults supervising these children may well have done the same thing—watching the children more closely or reacting more quickly to behaviors that hint of aggression. If these adults behaved differently with the children than they would have were the children not being observed, this would compromise the unobtrusive nature of the observations. A second compromise occurred because the researchers also wanted an audiotape of the children's conversations in order to fully understand the nature of their aggressive behavior. So the children had a wireless microphone clipped to their clothing while they played in the schoolyard, thus making them aware that they were being observed. The researchers decided that both of these compromises were essential and that they had only minor effects on the validity of the observations made. The adults, for example, did not know exactly which children on the playground were being observed at a given time, and the children, after a short period of being aware of the microphone, seemed to ignore it and play naturally. Other research supports this idea that people often tend to forget they are being observed and to behave normally, especially if the

observation occurs over a long period. This research on aggression among children does illustrate, however, that what appears at first to be hidden observation is sometimes not completely unobtrusive.

Disguised Observation

Another way to conduct field observations is through *disguised observation*, where researchers observe people in natural settings but without revealing that they are researchers making observations. This is sometimes called the *complete participant role* (see Figure 11.2). The observer enters a group under the guise of being just another member and essentially plays that role while conducting the study.

One setting, for example, that might lend itself to such disguised observation is a public establishment like a bar or tavern. Exactly this kind of disguised observation was done by a group of investigators interested in alcohol-related aggressive behavior (Graham, LaRocque, Yetman, Ross, & Guistra, 1980). They decided that information could best be gathered through an unobtrusive field observation of the behavior of people as they consumed alcohol in several bars in Vancouver, British Columbia. Teams of observers spent from 40 to 56 hours per week making observations in drinking establishments. Each team consisted of a male-female pair who would enter an establishment, locate a table with a good view of the saloon, and order drinks. They made every effort not to influence the people in any way. We would consider such research unobtrusive, rather than participant, because the investigators made no effort to have contact with, or influence on, the behavior of the patrons. At times, this was impossible—a patron would wander over to their table and engage them in conversation. In those cases, they quickly terminated the observations and left the bar. In some other ways, one might question the unobtrusive character of their observation: The female observer was often the only Anglo female in the bar (the others being Native Americans), the observers were sometimes a small island of sobriety in a sea of drunkenness, and there were often social class differences between the dress and behavior of the observers and the observed. Nonetheless, their impact on the behavior of the clientele was probably minimal, so we can consider this as disguised, nonparticipant observation.

The complete participant role has proved valuable in studying groups that otherwise might be closed to research if the observer's true identity were known. Drug abusers (Becker, 1953), people engaging in sex in public restrooms (Humphreys, 1970), drug dealers (Adler, 1985), and members of the Satanic Church (Moody, 1976) are among such groups that have been studied by using the complete participant role. The complete participant role has also proved valuable in research on sensitive topics, such as racism and discrimination. For example, discrimination in the rental and sale of dwelling units has been studied by sending Anglo, African American, and Hispanic buyers to the same housing unit or realtor, without the renter or seller aware that they are a part of an observational study (Yinger, 1995). They found that African American and Hispanic buyers are charged more for the rental and purchase of homes, are more likely to be told an advertised housing unit is no longer available, and are more likely to be shown units in predominantly minority neighborhoods. Only field research using the complete participant role is likely to uncover these kinds of social phenomena.

Informed Consent

Unobtrusive observation raises a critical ethical issue in the researcher-participant relationship: To what extent should the people studied be informed of the investigator's research purposes? Unobtrusive observation, by its very nature, means that participants are neither told

of the research purposes, nor given the opportunity to make informed consent. At the same time, this research is often conducted in public settings where, one could argue, the researcher has the right, along with anyone else, to observe people's behavior. This can be an especially troubling problem in field research because in some cases fully informing people would undermine the researcher's ability to gather accurate data. For example, a study of staff treatment of patients in a mental hospital was conducted by having researchers admitted to the hospital as patients without informing the staff of their research purposes (Rosenhan, 1973). Undoubtedly, hospital staff would have behaved quite differently had they known they were under surveillance. Therefore, some researchers take the position that concealment is sometimes necessary in order to conduct scientific work, and that researchers must judge whether the scientific gain justifies the deception of human beings and any potential injury—social or psychological—they might suffer. Others, however, hold adamantly to the position that any research on human beings must include "informed consent": The people involved should be fully informed concerning the purposes of the research, any possible dangers or consequences, and the credentials of the researcher. Anything less, they argue, would be unethical and immoral because it tricks people into cooperation and may lead to undesirable consequences of which they are not aware. This complex ethical dilemma is dealt with at greater length in Chapter 3.

OTHER QUALITATIVE METHODS

In addition to field research, a number of other important qualitative research methods are used by social scientists. These methods are sometimes used on their own and at other times as a part of a field research study.

In-depth Interviewing

Chapter 9 discusses the use of interviews in survey research, but those interviews are generally more structured and focused than the *in-depth interviewing* often used in qualitative research. **In-depth interviews**, sometimes called **ethnographic interviews**, are informal and unstructured interviews that can explore a wide range of topics and may last for a long time, even days or weeks. Often, they are more like a rambling conversation, with the interviewer being relatively nondirective while the person being interviewed is fairly unconstrained in what he or she talks about (Fontana & Frey, 1994; Patton, 1990). Actually, there is some overlap between participant observation and in-depth interviewing since some data are often gathered in participant observation through informal interviews with people in the field. This interview data is a third source of data collected in field research, in addition to gathering data through participation and observation (see Figure 11.1). Yet, the two methods are distinct since true participant observation research also gathers data through participation and through observation. In addition, interviews can be a mechanism for collecting qualitative data apart from field research. For example, Sean Gilmore and Alicia Crissman (1997) studied the link between gender and violence among teens who play video games. Their main data-gathering tools were unstructured interviews that enabled the researchers to explore issues of gender identity in great breadth and depth. This topic had been the focus of little previous research, so there was no existing knowledge base from which to derive meaningful concepts or develop hypotheses. Instead, the researchers wanted the concepts and hypotheses to emerge from the freewheeling and friendly conversations that were their unstructured interviews. These interviews were supplemented with some direct observation of the teens' conversations and behavior as they played the video games. As another example of in-depth interviews, Betty Dobratz and Stephanie Shanks-Meile (1997) conducted a study of the white separatist movement in the United States by conducting in-depth interviews with leaders and members of the movement. However, they also supplemented the interview data with some field observations, such as by attending white separatist rallies, dinners, congresses, and cross lightings. While their data were primarily from the interviews, the field observations provided additional important information and insights that the interviews could not provide.

The researcher approaches an in-depth interview with some general topics of interest and asks questions that probe into those areas. However, the whole process is much more interactive and collaborative and much less directive than in survey interviews (Holstein & Gubrium, 1995). It is more like a conversation. The researcher will talk, rather than just ask questions, as a way of keeping the conversation going. The researcher may talk about herself and her life and even respond to questions asked by the respondent—something not normally done in a survey interview. The researcher will also permit the interview to take some unexpected directions if these appear rewarding in terms of the research question or essential to maintain the rapport and interest of the respondent. In fact, the relationship between interviewer and respondent is likely to become one more of equals conversing than of the "expert" gathering data from the "subject." Yet there is a structure of sorts to the interview that is imposed by the general topics the researcher needs covered. This is especially apparent when more than one person is interviewed because the researcher may have to be sure that the same general topics are covered with each person interviewed.

The goal of the in-depth interview is similar to participant observation: to explore how the world appears to the respondent, without imposing inappropriate structure on the views they express by using preestablished categories or overly restrictive direction on what is explored and how it is expressed. The respondent's perspective should unfold and be framed in ways that make sense to him or her rather than being limited and constrained by the researcher's preconceived category systems or structures of meaning.

Case Studies, Life Histories, and Narratives

Another qualitative research technique is variously called *narrative inquiry, life histories*, or *case studies* (Clandinin & Connelly, 2000; Marshall & Rossman, 1995). While there are some differences among these approaches, they all involve a detailed descriptive account of part or all of a particular individual's life or, with some case studies, of an organization or an event. The goal is to gain understanding through the depth and richness of detail achieved with this method in comparison to other methods. The description in case studies and life histories is a detailed and what is sometimes called "thick" description: A complete and literal accounting of the person or setting under study. While some quantitative data might be presented as a part of this, the emphasis is on telling a story in prose or narrative. Case studies and life histories can be based on direct observation, interviews, document analysis, organizational records, or some combination of these. Basically, any data that would contribute to a description of the case under study could be used.

Case studies and life histories share with participant observation a desire to understand how the social world looks from the perspective of the person being studied. Anthropologist Bourgois, for example, used a life history approach in his participant observation research in El Barrio: he tape-recorded long conversations with his informants in order to see how their current circumstances and behavior were a part of the flow of their lives. One of the strengths of case studies and life histories is that they permit the people being studied to play a large part in framing and providing meaning for their lives rather than having meaning and interpretation imposed by the observers. However, case studies take a much longer time perspective since they typically explore a good part of, if not the whole of, a person's life. Case studies and life histories have proven useful methodologies in feminist research and with interpretivist and critical approaches (Lawless, 1991).

The primary goal of most case studies and life histories is an idiographic explanation that focuses on an in-depth understanding of a particular case. Such an understanding might enhance one's comprehension of other cases and situations, but the primary focus is description, not generalization. Their advantage is the richly detailed descriptions they provide of people's lives, experiences, and circumstances. In addition, the ability of these methods to allow individuals to speak in their own voices make them valuable sources of data. However, being based on one person's life, case studies and life histories are criticized on the grounds that their results are not generalizable beyond that one case. The data from such studies also contain a considerable element of subjectivity because of its dependence on the accounts of the individual being studied. It can also suffer from the normal errors of people's memories as well as from selective recall on the part of the individual. In fact, the data produced in narratives is sometimes the result of a collaboration between researcher and subject where the final story is one that they each find acceptable. To overcome some of these problems, those producing case histories sometimes check for errors or misinterpretations with others knowledgeable of people or events in the case history. Of course, errors in memory or selective recall may themselves produce valuable data in that we may learn as much about people from how they remember or reconstruct their past as from their actual past. In fact, proponents of case studies and life histories argue that a person's own "story" is important to understand, irrespective of what the objective facts may be.

Focus Groups

I discuss focus groups as a form of survey research in Chapter 9 because they involve interviewing people, but they also deserve mention here as a qualitative research method. Focus groups are also called *group depth interviews* because they are like in-depth interviews, each with a number of people at the same time. The advantage of focus groups is that they are very flexible forms of data collection, and leave the participants free to frame their answers and construct meaning as they wish. While some quantitative data is collected as a part of focus groups, people are free to talk as much as they wish, and their complete responses serve as data for the research.

WHEN TO CHOOSE FIELD RESEARCH

If the research question calls for exploring the contextual or the subjective, then qualitative research methods may well be the most appropriate research methodology. If a person's research, for example, focuses on the interpretive paradigm, then qualitative methods would be a way to get at that interpretive and subjective dimension of reality. For this reason, feminist researchers have found qualitative research methods to be very useful in studying social phenomena. In fact, I have laid out some of the linkages between feminist approaches and qualitative research methods in Table 11.1 (DeVault, 1996; Wilkinson, 1998; Wolf, 1996). However, even some positivist researchers may recognize that a particular social phenomenon is far too complex to be captured through existing quantitative measurements.

TABLE 11.1   Linkages Between Feminist Approaches and Qualitative Research Methods

Feminist Approaches	Qualitative Research Methods
▪ stress that people are inextricably linked to their social context	▪ avoid data collection techniques that isolate people from their social context
▪ are sensitive to issues of power and control	▪ place researcher and research subject into a more egalitarian relationship
▪ are concerned about the imposition of meaning by powerful people	▪ leave research subjects open to create their own meaning
▪ posit that social reality is a product of social interaction and negotiation	▪ consider the researcher and research subject to be co-creators of meaning in the research data; research results are a product of interaction between researcher and research subject
▪ stress the importance of connectedness between people rather than separation	▪ provide opportunities for social connections and empathy to develop between researcher and research subject

Or the researcher may be searching for an idiographic explanation, which means looking for all the factors that affect a situation or outcome rather than trying to isolate only a few (see Chapter 2). So, any time the complete and natural context is important to understanding people's attitudes or behaviors, qualitative methods may be the most appropriate.

Qualitative research may also be a good choice when studying behavior that people are likely to wish to hide from public view or that people are sensitive about. This is true of many forms of deviant and criminal behavior, and research projects on these topics have often turned to field research methods. In addition, there are many attitudes and behaviors whose expression is shaped by social desirability: People say things or behave in ways that are socially acceptable. For example, questions about such topics as racism, sexual behavior, or alcohol consumption may produce socially desirable responses. These topics may be candidates for field research.

Qualitative methods are also often selected when the research question is exploratory in nature and theoretical development does not enable researchers to spell out relevant concepts or develop precise hypotheses. Qualitative methods permit researchers to view human behavior as it occurs in the natural environment, without the restrictions of preconceived notions or explanations. Through observation, the researcher can begin to formulate concepts, variables, and hypotheses that seem relevant to the topic and grounded in the actual behavior of people.

It is worth repeating that, whereas qualitative research has a strong link with interpretive approaches, this does not mean that quantitative research cannot be done with some of the qualitative methods. Many field researchers, for example, focus on both qualitative and quantitative data in order to get a more complete picture than either approach would provide alone.

STEPS IN FIELD RESEARCH

I next describe in some detail the steps taken in conducting field research. Other qualitative methods would also include some of these steps. The exact steps taken in conducting field research will vary depending on whether it is participant observation, unobtrusive observation, or some other variant. I describe the steps taken in participant observation, recognizing that some parts of what follows may be unnecessary in other types of field research.

Problem Formulation

The first step in field research is to engage in problem formulation, including a literature review and conceptual development. From this assessment, the researcher establishes the specific goals of the research and decides whether field research is the most appropriate research strategy. The elements that influence this decision have already been discussed at length in the preceding section. A part of this problem formulation includes learning as much as one can about the people, groups, and settings that will be the focus of observation. This could include reviewing previous research on the groups as well as any historical and literary materials that might shed light on the group and its culture. The more you can learn before going into the field, the more effective you will be at making observations in the field.

Selecting a Field Setting

The second step is to decide which specific group to study. Whether you are studying deinstitutionalized mental patients, cocaine dealers, or gangs of motorcycle riders, you have to decide exactly which group you will join and observe. One way to decide this is by finding a group that is accessible to you. Sometimes a field setting presents itself as a by-product of some research or other activities. Thorne's (1993) participant observation in elementary schools arose because she had been asked to assess issues of gender equity in schools. Her field observations were designed to achieve that goal, but she also obtained permission to make additional field observations to be used for her own research purposes. Clifford Stott (Stott & Reicher, 1998) was interested in how violence escalates among fans of British soccer teams, so this focus determined the necessity of making field observations of fans before, during, and after soccer matches. Being a soccer fan himself, Stott joined other fans of the team he supported at international soccer matches, and these matches and fans became the focus of his participant observation. As we have seen, Scheper-Hughes selected a particular shantytown in Brazil as her field setting because she had lived there before becoming an anthropologist and had continuing ties with some people who lived there.

Entering the Field

The third step is very challenging: gaining entry into the group to be studied. In the complete participant role, this step may be less of a problem because the people do not know they are being studied. However, you must be sufficiently like those to be studied to gain access. In the other participant roles, however, where they know that you are an outsider and a researcher, you must find some way to convince the people to agree to accept your involvement as a researcher. Several methods increase the likelihood that people will cooperate (Johnson, 1975; Jorgensen, 1989). One way to gain their cooperation is to win the support of those with more status or influence in the group and use your relationship with them to achieve access to others. It would be best, for example, to approach the directors of a mental health center and enlist their aid before contacting caseworkers and ward staff.

Another way to increase cooperation is to present your reasons for conducting the research to people in a way that seems plausible and makes sense in their frame of reference. Esoteric or abstract scientific goals are unlikely to be very appealing to drug dealers, deinstitutionalized mental patients, or single parents struggling on welfare. You should emphasize that your major concern is understanding their thoughts and behaviors as legitimate, acceptable, and appropriate. Nothing will close doors faster than the hint that you intend to evaluate the group.

Cooperation is also enhanced if you have some means of legitimizing yourself as a researcher. This might be done through an affiliation with a university that supports the study or an agency that has an interest in the research. In a study of a sexually transmitted disease clinic, for example, Joseph Sheley (1976)

gained entrance because he was involved with a larger community study of sexually transmitted disease (STD). Such legitimation can backfire, of course, if the group you hope to study is suspicious of, or hostile toward, the organization with which you are affiliated. Sheley, for example, although allowed into the clinic, found himself having a degree of "outsider" status because of his association with the community STD study.

Finally, it may be necessary to use informants to gain entry into some groups. An *informant* is an insider who can introduce you to others in the group, ease your acceptance into the group, and help you interpret how the group views the world. Especially with informal subcultures, the informant technique can be a valuable approach.

Sometimes it takes a bit of creativity, or perhaps even courage, to gain entry into some field settings. Sociologist Ruth Horowitz (1987) was interested in studying Latino gangs in a particular city. Being a Jewish woman, she did not have easy entry into these groups. So she sat on a bench in a park frequented by gang members until they approached her. Eventually, one of the leaders of the gang asked who she was—an obvious outsider on their turf. She told him that she wanted to write a book about Latino youth and was able to convince him to introduce her to other members of the gang.

In some cases, the researcher may know members of the group being observed, and these personal contacts can ease entry into the group. For example, in a study of violence among British soccer fans, the anthropologist doing the study selected a particular group of fans to join and observe because he had grown up and gone to school with some of them (Armstrong, 1993). These personal contacts made it easier to gain their cooperation.

Attitude of the Researcher

In positivist research, the researcher's attitude is presumed to be one of objectivity and detachment. In field research, however, researcher-subject relationships are likely to be more extensive, complex, and personal. As Bourgois (1995) put it, "[I]n order to collect 'accurate data,' ethnographers violate the cannons of positivist research; we become intimately involved with the people we study" (p. 13). Another anthropologist put it this way: "The ethnographer must be intellectually poised between familiarity and strangeness, while socially, he or she is poised between 'stranger' and 'friend'" (Powdermaker, 1966, p. 20). So the researcher's role may be that of friend or stranger or somewhere in between. This complexity and ambiguity means that each field observer needs to address the issue of what his or her attitude should be toward those being observed.

In most field settings, the researcher's attitude should probably be one of *openness* to a wide range of types of behavior and *respect* for the dignity of the research subjects. As the examples at the beginning of this chapter show, field researchers sometimes find themselves observing behavior that they find morally offensive or politically unpalatable. However, the researcher's goal is not to judge but to observe, record, and learn. Moral or political reactions can interfere with these goals and may threaten the rapport necessary to achieving good field research. In addition, ethical principles (see Chapter 3) dictate that the people who let us into their lives should not be harmed by that, and taking moral or political offense can be a form of attack. Researchers who cannot achieve this attitude of openness and respect toward particular groups might be best served by not conducting field research on those groups.

The researcher's attitude also needs to be a *reciprocal* one, recognizing that the subjects may make as significant a contribution to the production and interpretation of knowledge as does the researcher. A dangerous attitude is one that Thorne calls "studying down," where the researcher assumes that it is the researchers who produce or "discover" knowledge while the subjects are less informed and less able to

contribute to the production of knowledge (1993, p. 12). Thorne ran across this attitude on research on children where many researchers seemed to dismiss the possibility that the children being observed could teach the researchers something and that the children could help produce knowledge and understanding. Thorne approached her field observations "with an assumption that kids are competent social actors who take an active role in shaping their daily experiences. I wanted to sustain an attitude of respectful discovery, to uncover and document kids' points of view and meanings" (1993, p. 12).

Finally, the researcher's attitude needs to be *balanced* in terms of identifying both the positive and negative aspects of the settings or cultures being observed. Bourgois, in his field research on inner-city street culture, points out that

> [T]he methodological logistics of participant-observation requires researchers to be physically present and personally involved. This encourages them to overlook negative dynamics because they need to be empathetically engaged with the people they study and must also have their permission to live with them (1995, p. 14).

This can produce an unwitting self-censorship where the researcher fails to note or report on some of the negative aspects of the behavior or lifestyle of the subjects. Bourgois found aspects of the street culture he studied to be violent, dangerous, and abhorrent. Yet a balanced attitude requires that the researcher's overall picture is neither unrealistically negative nor unrealistically positive.

Developing Rapport in the Field

A critical step in field research is to develop rapport and trust with the people being studied so that they will serve as useful and accurate sources of information. This can be time consuming, trying, and traumatic. In the initial stages of the research, people may be distant if not outright distrustful. You are likely to commit errors and social gaffes that offend the people you have joined. More than one participant observation effort has had to be curtailed because the investigator inadvertently alienated the people being studied.

Many elements are involved in developing rapport or trust among one's informants (Johnson, 1975; Jorgensen, 1989). Rapport can emerge if the informants and group members view the investigator as a basically nice person who will do them no harm. It matters little if the informants know of or agree with the research goals—only that they develop a positive attitude toward the investigator. Trust and rapport can also emerge if the investigator shows through behavior that he or she agrees with, or at least sympathizes with, the perspective of the people being studied. If you join your field contacts in some of their routine activities, such as drinking beer or playing cards, they are likely to view you as one who accepts them and can be trusted. Here, of course, the researcher must balance the need for acceptance against personal and professional standards of behavior. Another way to enhance trust and rapport is by reducing the social distance between the researcher and those being observed. In field studies of schoolchildren, this can often be achieved by avoiding positions of authority or avoiding intervening with the children in ways that other adults, teachers, or school officials might do. For example, field researchers observing children often avoid intervening in fights or other behavioral problems unless some serious accident or injury might occur (Fine, 1987; Mandell, 1988; Thorne, 1993). Finally, rapport can be enhanced if the relationship between investigator and group members is reciprocal; that is, both the observer and the group members have something the other needs and/or wants. You might, for example, gain scientific data from your informants while they in turn hope to gain some publicity and attendant public concern from the publication of your results.

The Eye on Ethics box explores some of the ethical complexities of the relationships with people that can occur in field research.

Becoming "Invisible"

In field research, being "invisible" means that the observer is perceived by those present to be a natural part of the setting, not an outsider or in any way unusual. Even in the participant-as-observer role when the people know you are a researcher, you can become invisible if the others begin to see you as "just one of them." Partly, this has to do with rapport: As rapport develops, the observer comes to be seen as a natural part of the setting by the people being observed. Becoming invisible is also in part a matter of time: The longer a person is in the scene, the less he or she is noticed as unusual or as an outsider. Initially, the presence of the observer may change behavior, but this effect often dissipates as time passes. Observers also become invisible when they join in the routine activities in the setting, whether that be working side by side at some task, drinking beer or smoking marijuana with the others at the scene, or joining in some illegal behavior. By doing these things, the observer comes to be viewed as a routine part of the setting. Another way to become invisible is to develop friendships with those being observed. When this happens, the friendship role comes to be seen as more important in the eyes of those being observed than does the researcher role. As the friendship role becomes more salient, it calls for people to be more open and honest and less guarded in their interactions with the observer.

Observing and Recording in the Field

The center of attention in field research is, of course, the observing and recording of what occurs. However, this cannot truly begin until you have accomplished the other steps just described. While field researchers will keep track of what they observe while entering the field and gaining rapport, the best observations depend on those things having been achieved. For example, those being observed may hide many important feelings and activities from

EYE ON ETHICS — The Moral Complexity of Working With Deviant Subjects in Field Research

Field researchers sometimes find themselves observing people who are doing something illegal, disreputable, or stigmatizing. As examples, field research has been conducted on drug dealers, drug abusers, striptease dancers, hustlers, shoplifters, pickpockets, and police officers who harass or brutalize civilians. On occasion, the subjects of the observation have asked or expected the researcher to participate in or in some fashion assist them in their illegal or deviant behavior, at least to the point of not interfering or not bringing any harm or punishment to them. What ethical obligations does a researcher have in such situations?

One solution to such ethical dilemmas is, of course, not to initiate field research among groups for whom such issues are likely to arise, and certainly some researchers have chosen this route. However, this is an extraordinarily limiting strategy in that it places many forms of social behavior beyond the reach of field research. In addition, one does not always know ahead of time when such ethical dilemmas will be an issue. In choosing to study police detectives, for example, one might not predict that those detectives would harass people they had arrested. Should one avoid all field research on police? Again, that would remove extraordinary amounts of social life from the view of social science research.

Another solution to the dilemma rests on the purported value of the research and weighs that against the harm that might be caused by the deviant behavior of those being observed. By refusing to participate or by interfering, the researcher would likely threaten the rapport with the observed to the point where the research project becomes untenable. The observed may no longer cooperate or may become so restrained and inauthentic in their behaviors that few valid observations can be made. Or, even worse, they might reject the researcher outright. In either event, the argument goes, the researcher is justified in ignoring, or even in some cases supporting, the deviant behavior in order to protect the integrity of the research project. This assumes, of course, that the benefits of the research are considerable and the harm of the deviant behavior is not too substantial. Judgments about this must be considered carefully, possibly including consultations with others who have no stake in the research project. And there are certainly lines that a researcher should never cross. No one would support, for example, the ignoring of a police officer's murder of a suspect in order to maintain rapport for the research project.

Carl Klockars (1979) suggests a third solution to the dilemma, which is based on the idea that the people being observed have, by virtue of permitting the researcher into their lives and space, placed a confidence in the researcher:

the researcher until after rapport has emerged. What observations to record and how to record them are discussed later in this chapter.

Going Native

Because of the deep immersion in the group that can occur in field research, along with the strong relationships that can develop, one danger that can arise is what some ethnographers call "going native": becoming so involved in and identified with the group or individuals being observed that the researcher takes on their perspective and finds it difficult to see the group from more objective, or at least different, perspectives. Any semblance of a scientific

This confidence and the access which is granted based upon it is a complex set of understandings, some explicit and some tacit, some formal, some taken for granted. In any given field research relationship it is very difficult to describe the details of such understandings as roles evolve and the rules and responsibilities proceeding from them evolve with them. However, in all field research with deviant subjects that confidence begins with a promise by the researcher not to "blow the whistle" on his subjects . . . (Klockars, 1979, p. 275).

Klockars argues that field researchers develop complex relationships with those they observe, and those relationships involve a plethora of moral obligations. The fieldworker is placed into a particular role by the agreements—either implicitly or explicitly—that are forged between researcher and subject. Police officers, for example, may cast the observer into the role of "fellow undercover cop," and fully expect the researcher to live up to that role, just as another officer would. A fellow cop would not blow the whistle on a cop who harasses or brutalizes suspects, and the police expect the fieldworker to live up to that role. In fact, Klockars argues, the police have let the fieldworker into their lives precisely because of the moral assumption that the researcher will live up to those role obligations. Even though nothing was explicitly said about it, such a belief is at the foundation of the relationship.

So, what does a researcher do if the police harass or brutalize someone in his or her presence? For Klockars (1979), the resolution of the ethical dilemma is clear:

It is the immediate, morally unquestionable, and compelling good end of keeping one's promise to one's subjects. In particular, it is the keeping of that minimal promise which every fieldworker makes explicit or implies to deviant subjects in the process of gaining first-hand access to their deviance (pp. 275–276).

The researcher ignores and goes along. A civilian who is trying to be a "decent human being" might not ignore the deviance of the subjects. But Klockars's answer to that is: "Decent human beings do not; morally competent fieldworkers do" (Klockars, 1979, p. 276). Not everyone would agree with Klockars's conclusion, and certainly police could cross a line that could not be ignored by a field researcher. However, whether one agrees with Klockars or not, his analysis does suggest the complexity of the relationships forged in the field.

perspective or judgment can be lost. For example, one British social scientist who was a strong soccer fan has admitted to strong feelings of partisanship toward a particular team and its fans, who were also the subjects of his field observation (Hughson, 1998). On at least one occasion this went so far that he joined, although reluctantly, in physical confrontations when "his" fans came into physical conflict with fans supporting opposing teams. So going native is something to be concerned with, but it is a delicate balance. After all, going native can assist in developing the empathic understanding that can be one of the goals of field research. As stated above, the field researcher must be somewhere between a stranger and a friend, and it requires judgment to detect the point beyond which friendliness has more negative than positive consequences for the research.

Exiting the Field

In most field research, the period of observations in the field ends because sufficient data has been collected (or the researcher's grant money runs out). When this time comes, the researcher must leave the field in a way that brings no negative consequences to the people who have been observed. Since field research involves the researcher living in some degree of intimacy with those being studied, it is not uncommon for some level of personal relationship to develop. Acquaintances, friendships, and perhaps even more intense relationships can emerge. The researcher needs to sever these relationships so that people don't experience significant social or emotional loss. This may mean, throughout the period of being in the field, that the researcher remains somewhat socially or emotionally distant so that complicating personal relationships don't develop to the point where difficulty exiting the field occurs. This is one way to avoid expectations of excessive intimacy or permanency in a relationship. Of course, there is a balance here because some level of friendly involvement may be essential to developing rapport. However, from the very beginning, the researcher has an eye on the exit in terms of developing relationships with respondents.

RECORDING OBSERVATIONS IN FIELD RESEARCH

While field research is most closely associated with qualitative research, observations made in the field could involve either qualitative or quantitative data or both. Accordingly, the manner in which observations are recorded depends on whether observations are primarily quantitative or qualitative in nature. Qualitative observation typically makes use of less-structured *field notes*, whereas quantitative observation calls for more structured recording of data using *coding sheets*.

Field Notes

Field notes contain the detailed record of the observations made in the field and constitute the raw data in qualitative field research. Although the precise nature of field notes can vary greatly from study to study, field notes should generally include six elements (Bernard, 1994; Lofland & Lofland, 1995):

1. A *running description* makes up the bulk of the field notes and is pretty much self-explanatory. It is simply the record of the day's observations. The primary concern of the running description is to record accurately the concrete events that are observed. The observer should avoid *analyzing* persons or events while in the field because of time constraints and because it would interfere with observation of the ongoing scene. Instead, one should concentrate on faithfully recording what occurs.

2. Field notes also include *accounts of previous episodes that were forgotten or went unnoticed* but were remembered while the investigator was still in the field. When the field notes from any observation session are being prepared, it is likely that certain events may be forgotten and left out. During subsequent observations, events may transpire that bring the forgotten episodes back to mind. These events should be recorded when remembered, with the proper notation concerning when they originally occurred.

3. *Analytical ideas and inferences* refer to spur-of-the-moment ideas concerning such things as data analysis, important variables, speculation regarding causal sequences, and the like. "Flashes of insight" regarding any aspect of the study should be recorded when they occur. Reviewing these ideas after the completion of observations can be of great benefit to

the final data analysis and writing of the report. Although most data analysis is reserved until after the observation period is over, you would not want to forget whatever analytical ideas occurred to you while in the field.

4. *Personal impressions and feelings* should be noted because the possibility for bias to color one's observations is always present. Recording personal impressions and feelings helps to minimize bias by giving a sense of the perspective from which the observer is viewing various persons, places, or events. Does the observer just plain dislike a certain individual in the setting who is being observed? If so, the observer should honestly record such a feeling when it first occurs. This can prove beneficial when reviewing accounts relating to this person to see if one's personal feelings may have influenced the description.

5. *Methodological notes* refer to any ideas that relate to techniques for conducting field research in this setting. Any difficulties you have in collecting data, any biases that might be introduced by the data-collection techniques, or any changes in how you make and record observations should be noted. The purpose of this is to better prepare you to assess validity (to be discussed below) and to provide insight for future researchers who might be making observations in similar settings.

6. *Notes for further information* are those notes observers write to and for themselves: plans for future observations, specific things or persons to look for, and the like. It is risky to rely on memory for anything important relating to the study.

How to Record

Field notes can actually take a number of different forms (Bernard, 1994; Lofland & Lofland, 1995). For example, parts of field notes might consist of brief jottings where researchers make note of something that is happening or something that occurs to them. Many field researchers carry a note pad at all times, even when not in the field, to note things as they occur or as they think of them.

Other parts of field notes may contain a more detailed and complete written recording of what is happening in some setting. In some cases, instead of writing field notes, researchers use laptop or notebook computers or audiotapes to make their recordings. In other cases, conversations or interviews conducted in the field might be tape-recorded or videotaped for later analysis. Which of these formats to use depends on what will most accurately preserve the record without interfering with rapport.

Recording field notes can be particularly problematic for participant-observers whose status as observers is disguised. Because such observers must constantly guard against having their true identity revealed, note-taking must be accomplished surreptitiously. In some settings, a bit of ingenuity on the part of the researchers can handle the problem quite nicely. For example, in her study of people's behavior in bars, Sheri Cavan (1966) solved the note-taking problem by making frequent trips to the rest room and recording her observations there. Given the well-recognized effect of alcoholic beverages on the human body, her trips probably raised little suspicion among the other bar patrons. The study mentioned earlier, conducted by researchers who were admitted as patients to mental hospitals, found another wrinkle on note-taking in participant observation: They could take notes openly because the staff defined note-taking as part of the meaningless activities that people who were "crazy" engaged in (Rosenhan, 1973).

In some participant observation settings, no amount of innovation will allow the researcher to record observations on the scene. In these situations, there is no alternative but to wait and record observations after leaving the observational setting. Relying on memory in this fashion is less than desirable because memory is quite fallible. The observer should record observations as soon as possible to minimize the likelihood of forgetting important episodes.

What to Record

For someone who has never conducted field research, collecting data through field notes can be a particularly frustrating and confusing affair. What to watch for? What to include in the field notes? These can be very difficult questions even for veteran observers. In addition, because field research may be exploratory, the observer is often only partially aware of what might be relevant. It is possible, nevertheless, to organize one's thoughts around some general categories of things to be observed and recorded (Bogdan & Biklen, 1992; Lofland & Lofland, 1995; Runcie, 1980).

1. *The setting.* Field notes should contain some description of the general physical and social setting being observed. Is it a bar, restaurant, or ward of a mental institution? Are there any physical objects or barriers that might play a role in the social interaction in the setting? Bogdan and Taylor (1975) suggest beginning each day's field notes with a drawing of the physical layout being observed. Such things as time of day, weather, or the presence of others who are not the focus of your observations might be useful information in some field research. In short, the field notes should serve to remind you—when you review them weeks, months, or years later—of the characteristics of the setting in which you observed behavior.

2. *The people.* Field notes should include physical and social descriptions of the main characters who are the focus of your observations. How many people? How are they dressed? What are their ages, genders, and socioeconomic characteristics (as well as you can observe from physical appearance)? Again, field notes should tell you, for each separate day of observation, who was present, who entered and exited the setting during observation, and how the cast changed from one day to the next.

3. *Individual actions and activities.* The central observations in most studies are the behaviors of the people in the settings. How do people relate to one another? Who talks to whom, and in what fashion? What sequences of behavior occur? In addition, you may want to record the duration and the frequency of interaction among people. Are there repetitive cycles of behavior that occur? Is there a particular sequencing of behavior?

4. *Group behavior and relationships.* In some cases, the behavior of groups may be an important bit of information. How long does a group of people remain on the scene? How does one group relate to another? It might, for example, be useful to know what cliques have formed in a setting. What is recorded also has to do with describing the social structure of the setting: the statuses and roles that various people occupy and the relationships between those people.

5. *Meanings and perspectives.* Field researchers are very sensitive to the subjective *meanings* that people give to themselves and their behavior (which is one of the reasons for doing qualitative field research). So field notes should contain observations about these meanings and what words or behaviors are evidence of those meanings. *Perspectives* refers to general ways of thinking that people exhibit, and evidence of this should appear in the field notes.

It should be clear that an enormous amount of information might be gathered in any setting. In addition, the more exploratory the research, the more information must be recorded because it can be difficult to be sure what is relevant. With experience in the field and the development of narrowly focused hypotheses, it is often possible to reduce the amount of information collected.

Coding Sheets

When it is possible to do quantitative observation in field research, this often involves a process of coding, or categorizing behaviors into a limited number of preordained categories. The process of developing coding schemes was outlined in Chapter 10 when I discussed content analysis. The process would be the same in field research. One should specify as clearly as possible the behaviors to be observed or counted during data collection. When this can be done, the use of coding sheets is desirable. A *coding sheet* is a form designed to facilitate categorizing and counting behaviors. For example, a typical coding sheet lists various behaviors with blanks following in which the behaviors could be checked as they occur. If duration of a behavior is also important, additional blanks would be provided to record the timing of the behaviors.

Table 11.2 presents a coding scheme for coding different elements of face-to-face social interaction that was used in a study of residents of homes for the aged. You can see that the coding categories describe various strategies that people use when interacting with one another. The coding scheme includes the symbols that will be recorded on the coding sheet to indicate that a given strategy was used and a definition of each strategy to assist the observers in deciding whether a particular strategy had occurred. In this study, observations were made at group meetings where residents of the home interacted with social workers. At each meeting, two observers sat adjacent to the group where they could observe social interaction without interfering with the group. (Does this qualify as unobtrusive observation?) The actual coding sheet used in this research is presented in Figure 11.3. Behaviors were recorded on the coding sheet at one-minute intervals. In this fashion, the researchers had a running account of the verbal and nonverbal interaction between the group members and the social workers. The researchers could then analyze the relationship between interactional strategies used by social workers and the behavior of the home residents.

The coding scheme and coding sheet for a particular research project are likely to be somewhat unique and highly specific, reflecting the special concerns of that project. However, Table 11.2 and Figure 11.3 give you an idea of what they look like. The development of an efficacious coding scheme requires considerable care. The coding categories should derive from the theories and hypotheses being tested in the research. If it is possible to specify concepts and hypotheses very precisely—in other words, if the research is clearly hypothesis testing rather than exploratory—then it may be possible to develop a precise coding scheme for data collection. In their disguised observation study of alcohol-related aggression, Kathryn Graham and colleagues (1980) were able to do this. Their basic hypothesis was that aggressive behavior among men when they drink is due to situational factors as much as psychological predispositions toward violence. They hypothesized that *aversive stimuli in bars* (their independent variable) serve as cues that allow or encourage *aggressive behavior* (the dependent variable). The dependent variable was coded by using a dichotomous coding scheme: nonphysical aggression (swearing or other forms of abusive language) and physical aggression. Within the physical category, behavior was coded as physical threats or challenges to fight but no actual contact, aggressive but noninjurious physical contact (for example, grabbing and pushing), and actual physical violence (punching and kicking).

The independent variable—*situational factors eliciting aggression*—was, needless to say, more complex to code. First, the researchers spent some weeks in the field observing and developing precise definitions and coding schemes for the situational variables. In other words, they used qualitative observation as a means of developing more quantitative observations. A few of the coding categories they developed follow:

time of day: 1 = 9 A.M. to noon

2 = noon to 3 P.M.

3 = 3 P.M. to 6 P.M.

. . .

noise level: 1 = very quiet

2 = medium quiet

3 = medium loud

4 = loud

TABLE 11.2 A Coding Scheme for Recording Data on Strategies Used in Social Interactions

Behavior	Recording Symbols *	Definition
		Social Group Worker Behaviors
Questions	G, I	Verbal behavior that demands or suggests a response from one or more group members, indicated by words that suggest a question (i.e., why, how) or a direct request or demand for a response
Statements	G, I	Verbal behavior that gives information and does not call for a response from group members and is not a direct consequence of a previous behavior of individual or group residents. Includes reading to residents
Positive Comments	G, I	Verbal behavior that followed the behavior of one or more group members and relates to this behavior to encourage similar responses. Suggests recognition, approval, or praise.
Negative Comments	G, I	Verbal behavior that followed the behavior of one or more group members and relates to this behavior to discourage similar responses. Suggests disapproval or displeasure.
Listening	✓	Silence on the part of the worker either while a group member verbalizes or while waiting for resident response in the absence of other worker behavior.
Demonstration/Participation	✓	Demonstrating equipment or activity or participating in activity.
Attending to External Events	✓	Watching, listening to, or talking to a stimulus outside the activity.
		Group Member Behaviors
Appropriate Verbal Behavior	O	Verbal behavior related to current group task (subject under discussion, activity, or relating to activity stimuli, e.g., phonograph recording)
Verbal Behavior Related to Environment	O	Verbal behavior related to another person present at activity or related to the room or other aspects of the environment, but not related to the current group task.
Inappropriate Verbal Behavior	∅	Verbal behavior that does not relate to group task or other residents or staff present or the environment. Verbalizations not audible to the entire group or observers.
Appropriate Attention	Δ	Visual or apparent listening attention, indicated by head orientation or other observable response, that is directed toward social worker, a resident who is making or has just made an appropriate verbal response, or activity stimulus.
Appropriate Activity	☐	Manipulating equipment related to activity as worker has demonstrated or similar appropriate use, helping another resident to do so, nodding or head-shaking appropriately, raising hand for recognition.
Inappropriate Activity	/	Repetitive actions, aggressive actions, manipulating materials not related to group task or activity stimuli, leaving activity, sleeping, talking to oneself, any attention direct away from worker, group task, activity stimuli, or a resident making a verbal response.

* Questions, statements, and comments were judged as to whether directed to an individual (I) or to the group (G).

Source Adapted from N. Linsk, M. W. Howe, and E. M. Pinkston, "Behavioral Group Work in a Home for the Aged." Copyright © 1975, National Association of Social Workers, Inc. Reprinted with permission from *Social Work*, Vol. 20, No. 6 (November 1975), p. 456, Fig. 1; p. 457, Fig. 2.

FIGURE 11.3 Sample Observation Form (Used to Record Data During 3-Minute Observation Periods at Group Meetings)

Group Activity Study—Observation Form—Observer I

Date_____ Page_____
Observer_____

Leader Behavior:

	1		2		3		4		5	
Questions	G									
Statement			G		G		G			
Comment +			I							
Comment −										
Listening					✓					
Demo/Part										
Att/Ext										

Residents Present:

1. _____   6. _____
2. _____   7. _____
3. _____   8. _____
4. _____   9. _____
5. _____   10. _____

Resident Behavior:

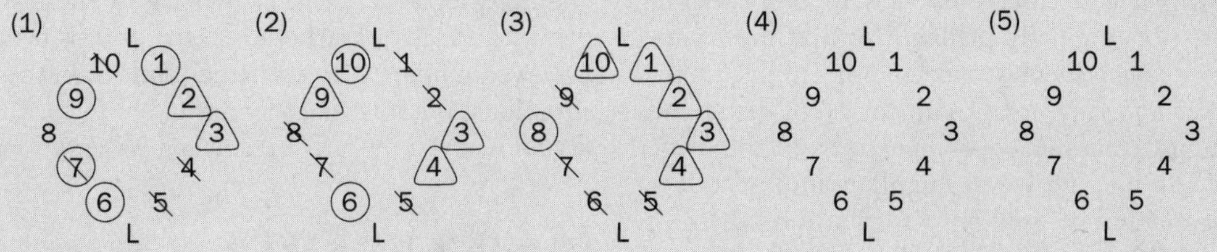

Source   Adapted from N. Linsk, M. W. Howe, and E. M. Pinkston, "Behavioral Group Work in a Home for the Aged." Copyright © 1975, National Association of Social Workers, Inc. Reprinted with permission from *Social Work,* Vol. 20, No. 6 (November 1975), Fig. 1, p. 456, Fig. 2, p. 457.

sexual contact:

1 = none, very casual

2 = discreet necking

3 = heavy necking, touching

4 = flagrant fondling

friendliness to strangers:

1 = open, lots of conversation with strangers

2 = closed, people talk only to members of their own group

Each two-person observation team would spend from two to two and one-half hours in an establishment. Most recording of observations was done after leaving the establishment so that note-taking would not attract attention.

This unobtrusive observation illustrates the manner in which a precise, quantifiable coding scheme can be used in field observation if the hypotheses to be tested are developed sufficiently. The investigators also found, however, that coding, although the major form of data collection, was not sufficient by itself. It was also necessary to collect qualitative data because, while they were in the field, more variables of importance began to emerge. They found it useful to record descriptive accounts of aggressive incidents in order to ensure a complete record. This provides an illustration of a grounded theory approach where data collection changes while still in process because evaluation suggests that some new theoretical concepts not originally planned for are important.

Coding schemes can become highly complex, involving many categories of behavior, timing of the behaviors, measures of intensity, and the like. However, highly complex coding schemes can be used only in situations where it is possible to record accurately all that is necessary. In Graham's (1980) study, this would have been considerably more difficult because the group (the clientele of a bar) was large and shifting in composition, much was going on, and actual recording had to be reserved until after the observation was concluded. This research setting necessitated a less complex coding scheme. Further, an investigation using complex coding schemes that require intense concentration on the part of the observer often require a number of observers, each of whom records for a short period and then is relieved by another observer. This reduces error due to observer fatigue or fluctuations in concentration. In some investigations, group behavior is recorded on videotape in order to reduce error and to allow researchers to view the group as often as needed to code behavior properly. In developing a coding scheme, in short, one must be sure that the scheme does not become more complex than is usable, given the constraints and resources at hand.

Coding schemes are not appropriate forms of recording observations in all field research. In some cases, as with exploratory research, hypotheses cannot be developed with sufficient precision to make possible the operationalization of concepts through coding schemes because the research is intended to explore and discover rather than explain and predict. In other cases, the interpretivist and critical approaches suggest that the nature of some phenomena does not lend itself to quantification; in fact, they argue, quantification can lead to distortion and misunderstanding of what is going on. In these cases, hypotheses involve variables and relationships that require considerable interpretive effort on the part of the observers in the field and cannot be easily condensed to a few coding categories. In still other cases, the complexity and lack of structure in the groups being observed render coding schemes useless. In these situations, the investigators are likely to turn to field notes as a means of recording observations.

ISSUES IN FIELD RESEARCH

Sampling

In most field research, it is difficult, if not impossible, to use probability samples because no adequate sampling frame can be established. In addition, the research questions sometimes addressed by field research do not call for probability samples. So nonprobability samples are widely used in field research. Especially common are snowball sampling, targeted sampling, and theoretical sampling. In fact, in Chapter 7, I described a number of field studies that used one of these sampling strategies or some combination of them. These samples make possible sampling procedures in situations where sampling frames do not exist, and also encourage the researcher—especially with targeted and theoretical sampling—to avoid samples that are biased because some group is inadvertently missed by a particular sampling procedure.

Sampling in field research must often address another challenge: the fact that observations are made throughout the length of the study, which may involve a considerable period of time. Participant-observers record things as they happen, and more quantitative observers mark coding schemes for as long as an interchange or social setting persists. However, in some of these research projects continuous data collection is costly and unnecessary. In addition, as discussed with other forms of sampling in Chapter 7, it is often unnecessary to collect data from *all* elements of a population. In studies of child development, for example, it may not be necessary to record all that occurs during an hour, a day, or a week. Instead, we can gather valid data through **time sampling**; that is, make observations only during certain selected time periods (Irwin & Bushnell, 1980). For example, if one was doing an observational study of adolescents in a group home, the prime hours for observation would be those when the residents were most likely to be at the home, such as weekdays from 3:00 P.M. until lights out at 11:00 P.M. (40 hours per week) and Saturdays and Sundays from 7:00 A.M. until 11:00 P.M. (32 hours per week). The researcher could construct a sampling frame comprised of a weekly list of these 72 one-hour time segments and then select a random sample of these elements. To be sure that both weekdays and weekends are in-

cluded, one might specify eight hours during the week and four hours from the weekend. The resulting sample of 12 one-hour time segments would provide sufficient time coverage for observation while reducing biases that might occur if, say, all observations were made on weekdays after 8:00 P.M. when many youths would be tired.

When conducting time sampling, there are several guidelines to keep in mind. The length of each time-sampling interval and the distance between intervals depend on the nature of the behaviors being observed: They should occur with sufficient frequency that they are likely to be seen during the sampled time periods. Very infrequent behaviors might call for continuous observations. The more frequently a behavior occurs, the smaller the number of intervals that will need to be sampled. Further, the time interval should be long enough for the behavior to occur and for the observer to make whatever recordings are called for.

Many of the considerations in time sampling are like the issues in sampling subjects or respondents discussed in Chapter 7. If the primary concern of the researcher is to be assured of observing *some* occurrences of the behavior under study, it is advisable to use the equivalent of a purposive sample. For example, in studying domestic violence, a researcher would want to observe instances of family quarreling. Conducting observations at mealtime would be one way to increase the probability of witnessing the desired events. On the other hand, if the researcher is concerned with accurately estimating the frequency of occurrence of a particular event or with studying the pattern of responses over a time period, it would be necessary to use the equivalent of a probability sample. For example, in observing nursing home residents for frequency of contacts with nonresidents, the week could be divided into hourly segments and a random selection of hours could then be used as the basis for doing the observations.

Particularly when the observation process is highly complex and difficult to sustain for long periods of time, some form of time sampling can be of considerable help in improving the quality of data collected.

Validity and Reliability

Field research techniques, like other forms of data collection, need to be assessed in terms of how valid and reliable they are (Kirk & Miller, 1986). Observation rests on human sense organs and human perceptions—both of which are notoriously fallible. This is a particularly difficult and insidious problem because we are so often totally unaware of the ways in which our senses and our perspective can lead us to misperceive situations. Especially with field methods, people are inclined to resolutely say, "I was there. I saw it. I comprehend what was going on." Yet, as any trial lawyer will readily attest, eyewitnesses are often very unreliable spectators to events, and considerable experimental evidence indicates that firsthand accounts of events are often partially inaccurate (Buckhout, 1974). Given these problems, we need to consider the validity and reliability of observations carefully.

Little question exists that observational techniques have greater face validity as measures of some behaviors and events than do many techniques that rely on secondhand accounts. Surveys or questionnaires depend on someone else's perception and recollection, which can be shaped and clouded by many factors beyond the control of the researcher. Observational techniques, on the other hand, provide firsthand accounts of occurrences under conditions that are partially controlled by the investigator. Misperception may, of course, still occur, but the researcher is in a position to recognize its impact and possibly control its magnitude. For these reasons, field observations are considered to possess greater face validity than many other data-collection techniques.

As shown in Chapter 5, it is sometimes possible to measure the validity of an instrument through such means as correlating the results of the instrument with the results achieved by some other instrument known to be a valid measure of the variable. Often, such direct measures of validity are not possible in qualitative field research; nevertheless, there are certain procedures that can be employed in field research in order to enhance the validity of the observations.

1. *Be as thorough as possible in describing and interpreting situations.* This increases the likelihood that you will make important observations and produce a valid assessment of a situation. Observations that may seem unimportant while in the field may be recognized as important by yourself or others who later review the data. Observations not recorded, of course, are lost forever as data. Obviously observers cannot record everything that happens in a situation, but it is preferable to err on the side of being too complete than too skimpy.

2. *Carefully assess your own desires, values, and expectations to see if these might bias your observations.* People's perceptions are drastically shaped by their expectations or lack of them. If we expect something to occur, we are far more likely to observe it—whether or not it actually occurs. If we expect welfare recipients to be lazy, then we will be acutely aware of all those behaviors among welfare recipients that might be interpreted as laziness. Thus validity of observations will be reduced to the extent that our expectations—whether or not recognized—mold our perceptions. Observers should assiduously look for the *opposite* of what they expect to happen, and should be careful and critical if what they expect to find seems to be found. On the other side of the coin, a lack of expectations may lead us to miss something of importance in a setting. The Eye on Diversity box suggests some ways in which observers' values and perspectives can influence observations and conclusions drawn in field research.

3. *Have other observers visit the same group or setting to see if they arrive at the same conclusions.* If they do, this provides validation that the conclusions are a response to

the actual setting rather than to the expectations or biases of the observers—especially if the other observers had different expectations or biases.

4. *Compare the conclusions reached through field observations with conclusions reached by other research methodologies,* whether it be observational research in other settings or surveys, available data, or experimental research (Weinstein, 1982). This is a variation on criterion validity discussed in Chapter 5. If the various methodologies yield the same conclusions, then we have greater confidence that the field observations have validity. Field research, the conclusions of which are at wide variance with the results of other research, must be reviewed very carefully, especially if none of these other checks on validity are available.

5. *Consider how the condition of the observer might influence observations and conclusions.* Hunger, fatigue, stress, and/or personal problems can lead to very distorted perceptions and interpretations. Likewise, physical characteristics, such as the lighting in an establishment, may lead to invalid observations. (This is another good reason for keeping complete field notes—field conditions affecting validity can be assessed at a later point.) If a number of these conditions exist, it may be judicious to terminate observation and resume when conditions are more favorable.

6. *Look for behavior that is illegal, stigmatizing, potentially embarrassing, or risks punishment.* If people engage in these kinds of behaviors, especially when they know they are being observed, then they are probably acting naturally and not putting on a performance for the benefit of the observer. In a study of police officers in a juvenile division, for example, Irving Piliavin and Scott Briar (1964) made this argument:

> While these data do not lend themselves to quantitative assessments of reliability and validity, the candor shown by the officers in their interviews with

the investigators and their use of officially frowned upon practices while under observation provide some assurance that their behavior accurately reflects the typical operations and attitudes of the law-enforcement personnel studied (p. 207).

Likewise, in his study of an STD clinic, Sheley (1976) argued that the validity of his data was quite strong because "staff members dropped their professional masks and displayed quite unprofessional behavior and ideas in the company of the researcher" (p. 116). Under such conditions, you can assume that people are reacting to environmental stimuli that normally guide their behavior rather than shaping a performance for the benefit of the investigator. If the people being observed do not have *anonymity,* their behavior may not be a true reflection of how they behave normally. Especially when controversial, sensitive, or potentially embarrassing issues are investigated, validity will decline substantially if anonymity has not been ensured. For this reason, hidden or disguised observation and observation in which the researcher takes the complete participant role are more valid than other types of observation.

7. *If possible, make a video or audio recording of the scene.* While such recordings have their weaknesses as records of what occurred, they do provide another way for observers to check and validate their observations and conclusions. These recordings can also be reviewed by others, which offers yet a further check on possible bias or misinterpretation.

As many of these guidelines as possible should be followed in designing field research; the more that are incorporated, the more confidence we have in the validity of the results. Although many of the conditions influencing validity may be beyond the control of the investigator, it is important to assess their impact on the research honestly so that accurate appraisal of the results can be made.

EYE ON DIVERSITY | **Gender in Field Observations: The "Big Man Bias"**

In most cultures, gender is an important determinant of people's behavior, and this often means that men and women have distinctively different ways of interacting with people. In general, men tend to be louder, more aggressive, and more domineering in their interactional styles. This often produces a problem in field research that one anthropologist labeled the "Big Man bias": the tendency for the more powerful, more assertive, and louder individuals (who are also more likely to be male) to gain the attention of ethnographers and to have their behaviors and lives recorded in ethnographies (Ortner, 1984). As I have mentioned repeatedly in this chapter, a field researcher cannot observe and record everything that occurs in the field. Selectivity is inevitable. However, selectivity can also lead to bias when some important things are ignored. So, a field researcher overwhelmed with stimuli will have a natural inclination to find ways to choose. In the remote village of some indigenous tribe, the ethnographer may choose to pay more attention to the village leaders, who are also likely to be elder males. This means that the perspectives of younger tribal members and of females will be less represented in the ethnographic understanding of the village. To the extent that age and gender produce significant variations in perspectives in this village, the ethnographer's choices can produce a biased outcome.

Historians have for some time noted the danger that history can become more an account of the lives of societal elites than of nonelites. It is the elites, after all, who leave more of the historical record that historians use as data in writing histories: the elites write books and memoirs, pass legislation, make speeches, and in other ways come to the attention of historians. To the extent that these elites are also mostly male, then histories written primarily from these sources will reflect the activities, interests, concerns, and perspectives of male elites.

Sociologist Barrie Thorne (1993) detected a Big Man bias in field studies of preschool and elementary school children:

As for reliability, in the case of an individual researcher who is studying a single group or setting through participant observation, there is no practical way to assess it (Kirk & Miller, 1986). When observations are more structured, as when using a coding scheme, reliability can be readily assessed through tests of *intercoder reliability*; that is, the ability of observers to code behaviors consistently into the same categories of the coding scheme. Two or more observers code the same behavior, and the resulting codes are correlated to determine the degree of agreement between them. For example, in their study of barroom aggression,

Graham and her colleagues (1980) did this by correlating the coding results from the two observers who visited each bar. They achieved reliabilities ranging from $r = .57$ to $r = .99$. Many experts suggest that structured observation should achieve an intercoder reliability of $r = .75$ or better (Bailey, 1987).

Reactivity

A major concern in any research is **reactivity**; that is, the degree to which the presence of the researcher influences what is being observed (Webb, Campbell, Schwartz, Sechrest, & Grove, 1981). To take an extreme example,

Large, bonded groups of boys who are physically assertive, engage in "tough talk," and actively devalue girls anchor descriptions of "the boys' world" and themes of masculinity. Other kinds of boys may be mentioned, but not as the core of the gender story. (p. 98)

In other words, it was not just a bias in favor of males, but of a limited range of males. The voices of females and the less visible or dominant males are silenced or marginalized. Visibility could come from being a part of a large group of boys who were more likely to be noticed by an observer. Or visibility could come from the loud and aggressive behavior of the boys. Or visibility might come from popularity or athletic prowess. Thorne also argues that ethnographers have a bias toward noticing things that are consistent with ethnographers' stereotypes or hypotheses. Like everyone else, ethnographers carry with them cultural assumptions and stereotypes about masculinity (that males are tough, competitive, and tend to form dominance hierarchies). Boys who conform to this stereotype in the classroom or schoolyard get noticed by the ethnographers, whereas quiet boys who hang out with only a few other friends, and possibly even some girls, are less visible.

Thorne summarizes the problem by stating that "socially constructed contours of visibility skew ethnographic reports" (1993, p. 97). In her own and other field research, Thorne found a much richer and more complex social field among children's play groups than the gender stereotypes would suggest, both for boys and for girls. She found boys who were quiet, shy, and cooperative and girls who were aggressive and hostile. Further, given the diversity actually found, it is not clear whether the stereotype even portrays the experiences of the majority of boys and girls. When the ethnographer begins with the stereotype that the play worlds of boys and girls are different, then the ethnographer tends to notice the differences in the field. It takes a special effort to notice instances that do not conform to the stereotype.

suppose a researcher enters a group for the purpose of studying it through participant observation. Also suppose that the researcher takes a very active role in the group's proceedings by talking a great deal, offering suggestions, and the like. It should be clear that an observer behaving in this fashion will exert considerable influence on what occurs in the group, making the observer's presence highly reactive. This affects the validity of the observations because you do not know whether you have measured the group's *natural* activities or their *reactions* to the observer. One is never sure if events very different from those observed may have taken

place had the observation been conducted in a less reactive manner. Reactivity also relates to the generalizability of findings. If the observer's presence is reactive, it is difficult to generalize findings to similar groups that have not had an observer in attendance.

It is generally agreed that participant observation generates the best results when reactivity is kept to a minimum. This is a major argument in favor of the complete participant role, where the observer's true status is concealed (Johnson & Bolstad, 1973). It is logical to assume that a group will be less affected by observation if they are unaware of the observer's *role-as-observer* than if they are aware. Using the complete participant role will not, however, guarantee a lack of reactivity. The role must be played properly. The complete participant should play as passive a role in the group as possible without raising suspicion. Even when the beginning of an observational study is accomplished without undue reactivity, the researcher must be careful that reactivity does not increase during the course of the project. Such a situation occurred in a study of a group that had predicted the end of the world (Festinger, Riecken, & Schachter, 1956). On numerous occasions, situations arose that forced the observers to become active participants to the point where subsequent group activity was influenced in significant ways by their actions.

ASSESSMENT OF FIELD RESEARCH AND QUALITATIVE METHODS

Advantages

1. The advantage most often claimed for qualitative research is that it provides deeper and more insightful data than those generated by most other methods. Especially with participant observation, researchers immerse themselves in the daily activities of those studied to a greater degree than with other techniques. This places them in a position to gain information that would likely be missed with techniques such as a questionnaire or interview. This is particularly true for the complete participant or those who have become "invisible": as accepted members of the group, they see people behaving freely and naturally. Even techniques such as in-depth interviews and case studies can produce a deep and rich understanding of people's lives. The survey interview, on the other hand, may generate socially acceptable responses and a carefully orchestrated presentation of self. The participant–observer and the in-depth interviewer, how-

ever, are better able to go beyond these public fronts and penetrate the behind-the-scenes regions of human behavior.

2. Unlike surveys, which are limited to dealing with verbal statements, many forms of qualitative research can focus on both verbal and nonverbal behavior. This is an advantage because *actual behavior* is being studied in addition to people's *statements* about how they behave. By dealing with behavior, field research avoids a potential source of error: the gap between what people say they do and what they actually do. The ability of observational techniques to consider both verbal and nonverbal behavior places the researcher in a better position to link the verbal statements with behavior.

3. Much qualitative research is inherently longitudinal in nature and thus enables researchers to make statements concerning changes that occur over the time of the research. In addition, by following activities over time, qualitative researchers will have less trouble establishing the correct causal sequence than would be the case with surveys. As noted in Chapter 4, establishing the causal order with survey data can at times be difficult.

4. Field research is particularly capable of studying groups and behavior that would be closed to other forms of research. Many studies cited in this chapter involve groups that for various reasons would not be open to research by other methods. They may have something to hide, or may view intrusion by a stranger as threatening to their cohesion and values. The ability of the complete participant to conceal his or her identity and conduct research where it could not otherwise be conducted is a major advantage of this observational technique. Even participant-observers and in-depth interviewers, whose status as researcher is known, may, over time, gain access to groups through the development of trust and rapport; these same groups might reject the more brief and superficial entreaties of an interviewer.

5. A frequently overlooked, but nevertheless significant, advantage of much qualitative

research is that the most qualified person is often directly involved in the collection of data because the senior researcher is often one of the observers or interviewers (Denzin, 1989b). This is very different from surveys, for example, in which the project directors may rarely conduct interviews, leaving this to part-time interviewers hired for the job. This places the most knowledgeable person the farthest from the data-collection effort.

Disadvantages

1. A nagging concern with much qualitative research has always been the possible effect of observer bias on the results. Such research does not have the structured tools of other methods that help assess objectivity. If researchers are not careful, personal attitudes and values can distort research findings, rendering them virtually useless for scientific purposes.

2. Closely related to the issue of observer bias is the problem of the observer "going native," or overidentifying with those who are studied (Shupe & Bromley, 1980). As the observer frequently becomes a part of a group for a substantial period of time, this possibility is quite real. Going native and allowing the research to deteriorate into a propaganda piece for the group studied can surely have as disastrous results for the utility of a study as would observer bias.

3. The lack of structure also makes exact replication—an important part of some scientific research—difficult. Field research projects, for example, are often such individualized projects that the possibility for exact replication is slight. Any observer in a natural setting will be forced to record what occurs selectively owing to sheer volume. There is little chance that a replication attempt would select precisely the same aspects of a given setting on which to focus.

4. The nature of the data gathered in some qualitative research makes them very difficult to quantify. I noted that some participant-observers generate field notes that are basically rambling descriptions. Data in this form can be difficult to code or categorize in summary form. This makes traditional hypothesis testing more difficult. As a result, a field study can fall into the trap of not getting beyond a description of the setting observed. Of course, some nonpositivist researchers would consider this to be an advantage of qualitative research.

5. The ethics of some field research have been called into question by some critics, with the complete participant role generating the most controversy. Some social scientists see it as unethical to conceal one's identity for the purpose of conducting research. The strongest statement against concealed observation is that by Erikson (1967), who completely rejects all field studies that do not inform those being studied in advance. Whether disguised observation is ethical is still an open controversy in the social sciences. As noted in Chapter 3, disguised observation is considered a "questionable practice" that requires approval by an institutional review board. So long as the research is not trivial and the identities of participants are not revealed, disguised observation would likely be allowed.

6. As mentioned earlier, field research often affords the researcher little control over the variables in the setting. A great deal may be occurring, and the researcher will not be in a position to control or moderate these influences. This means that it may be difficult to identify the important causal factors in a situation with any certainty.

7. Because of the physical limitations on the observing capabilities of human beings, field research and some other qualitative methods are often limited to relatively small samples of people. Although one could, with a sufficient number of observers, study a large sample, this is rarely done. Observation is more commonly limited to a small group (such as a family or gang) or one setting (such as a bar or playground). This means that it may be difficult to generalize beyond the people or group being observed because it is not clear who they represent.

REVIEW AND CRITICAL THINKING

One of the strengths of qualitative research is its sensitivity to the perspectives of those being studied. It uses procedures that enable researchers to understand meaning and experience as it is constructed and intended by the people themselves. It avoids, as much as possible, the researcher imposing structure or meaning on people's lives from the outside. These are important lessons that can be utilized effectively in our everyday lives. When you are confronted with information about people or their way of life, consider the following.

1. To what extent is the information purely descriptive or is there an effort to place it in a framework that gives it some meaning?

2. Who provides the meaning framework? Does it flow from the words and actions of the people themselves? Or are there other "authorities" who, in some fashion, tell you what it means?

3. In the information available, are there any efforts to deal with problems relating to sampling, reactivity, validity, or reliability in observations?

Computers and the Internet

Not too many years ago, the use of computers in social science research was limited to the analysis stage with quantitative data. Computers had little use in qualitative research, especially when it was qualitative and nonpositivist in nature. This has changed dramatically. Field researchers, for example, now enter their field notes directly into small and portable notebook or laptop computers using word processing or other kinds of software. If the field observations are more quantitative in nature, spreadsheet or other software can be used to enter codes indicating which behaviors were observed. One simple way to do this would be to have the columns in the spreadsheet represent different behaviors and the rows identify time sequence. An entry into a column indicates that that behavior was observed. The row in which an entry is placed indicates the time sequence of the observations. In this way, an observer can enter observations of fairly complex sequences of behavior, certainly much better than can be done by hand. Some software can also utilize the computer's internal clock to record the time that each entry is made, which makes possible precise measurements of time rather than just the sequencing.

Another major development in the use of computers in qualitative research is the emergence of software to assist in the analysis of qualitative data in the form of field notes, in-depth interviews, or the transcripts from focus groups. Called computer-assisted qualitative data analysis software (CAQDAS), it is discussed in Chapter 15.

Interest in qualitative research has exploded in the past two decades in many disciplines, and this upsurge in interest is reflected on the Internet. Looking for Web sites related to qualitative research produces a plethora of hits. As you look online, pay attention to the number of different fields in the social sciences and human services that show an interest in this topic. A good way to begin is to search using terms such as "qualitative methods," "participant observation," or "field research," although the last will produce many sites relating to the natural sciences since it also conducts something called "field" research. However, this offers an important learning exercise: Can you identify the ways in which the research or science discussed at the natural science Web sites are similar to or differ from that done by social scientists, as discussed in this book? For example, do the natural scientists appear concerned with the positivist versus nonpositivist debate?

I found a number of Web sites that contained useful information and/or links to other valuable Web sites. For example, the School of Social and Systemic Studies at Nova Southeastern University maintains a Web site titled "The Qualitative Report Homepage" (**http://www.nova.edu/ssss/QR/**). That same site also has a comprehensive section on ethics in qualitative research (**http://www.nova.edu/ssss/QR/nhmrc.html**). Two other comprehensive sites are Resources for Qualitative Research (**http://wwwedu.oulu.fi/sos/kvaltutk.htm**) and QualPage: Resources for Qualitative Researchers (**http://www.ualberta.ca/~jrnorris/qual.html**). Another good Web site is for the CAQDAS Networking Project: **http://www.soc.surrey.ac.uk/caqdas/**. This site includes an extensive bibliography as well as opportunities to review qualitative data-analysis software.

Main Points

1. Qualitative research is research in which the data comes in the form of words, pictures, narratives, and descriptions rather than in numerical form. Field research is one type of qualitative research. Two important characteristics of qualitative research are: it is contextual in nature and uses a grounded theory approach.

2. One type of field research is participant observation, which is similar to the anthropologist's ethnographic research. Many field researchers basing their work on the interpretivist approach often use the method of *verstehen*. Knowledge in participant observation is gained through both participation (empathic understanding) and observation (deep appreciation of context).

3. Major decisions for the participant-observer are whether his or her status as observer will be revealed to those studied and the extent to which the researcher's role will stress participation or observation.

4. Unobtrusive observation, including both hidden observation and disguised observation, is designed to minimize reactivity. Unobtrusive observation raises the issue of informed consent.

5. Another form of qualitative research is in-depth interviewing. It is an informal and unstructured interview that can explore a wide range of topics and may last for a long time, even days or weeks. It is quite different from the interviews done in survey research. Qualitative research also takes the form of narratives, life stories, or case studies, which all involve a detailed descriptive account of part or all of a particular individual's life or with some case studies, an organization, or an event. The goal is to gain understanding through the depth and richness of detail achieved with this method in comparison to quantitative methods. Focus groups are also a form of qualitative research.

6. Steps involved in conducting field research include: problem formulation, selecting a field setting, entering the field, developing rapport with people in the field, becoming "invisible," observing and recording, and exiting the field.

7. When variables cannot be quantified easily or when using the *verstehen* strategy, data in field research is collected in the form of field notes, which are detailed, descriptive accounts of the observations made during a given period. When variables can be measured quantitatively in field research, coding schemes are used to measure and record observations. Coding sheets contain the categories of the coding scheme and are designed to facilitate the recording process.

8. Time sampling is often used in observational research to reduce the volume of

observations that have to be made. Validity in observational research means that the observations reflect reality correctly and accurately. Some observational techniques, especially unobtrusive observation, minimize reactivity and are thus good for studying sensitive topics such as racism and sexism.

Important Terms for Review

coding
ethnographic interviews
ethnography
field notes
field research
grounded theory
in-depth interviews
nonreactive observation
participant observation
reactivity
time sampling
unobtrusive observation
verstehen

For Further Reading

Anderson, Elijah. (1990). *Streetwise: Race, class, and change in an urban community*. Chicago: University of Chicago Press. This is an excellent example of participant observation research in a community setting. In this case, changes in community life are described as the racial and social class composition of a neighborhood changes.

Berg, Bruce L. (1998). *Qualitative research methods for the social sciences* (3rd ed.). Boston: Allyn & Bacon. Berg provides more detail on the qualitative research methods discussed in this chapter, as well as some others, such as ethnographies, that I have not included.

Ferrell, Jeff, & Hamm, Mark S. (Eds.). (1998). *Ethnography at the edge: Crime, deviance, and field research*. Boston: Northeastern University Press. This remarkable book takes the pioneering, and not necessarily popular, stance that *verstehen* in field research requires that researchers gain experiential immersion in the criminal and deviant activities of those being studied. It is a realistic view of what happens in field research and raises many key methodological and ethical issues.

Liebow, Elliot. (1993). *Tell them who I am*. New York: Free Press. This book is an observational research account in the genre of Talley's Corner. The author carefully documents the patterns and routines of homeless women, showing how they meet their needs and struggle to keep hope and humanity alive.

Marshall, Catherine, & Rossman, Gretchen B. (1999). *Designing qualitative research* (3rd ed.). Newbury Park, CA: Altamira Press. The authors provide a good introduction to qualitative research methods in applied research and policy analysis. Although it emphasizes educational research, the book is relevant to applied social science research concerns.

Silverman, David. (1993). *Interpreting qualitative data*. Newbury Park, CA: Sage. This book discusses the theoretical issues involved in collecting and analyzing data from qualitative research as well as describing some of the particular data-collection techniques.

Whyte, William Foote. (1984). *Learning from the field: A guide from experience*. Beverly Hills, CA: Sage. A delightful book by one of the premier field researchers in the social sciences. Whyte has used his 50 years of experience in the field to produce a practical and accessible volume on the gamut of issues related to field research.

Williams, Constance C. (1991). *Black teenage mothers: Pregnancy and child rearing from their perspective*. New York: Lexington Books. This is a fine example of applying the ethnographic approach to the study of a human problem that is of great interest to social scientists as well as policymakers. It is enlightening for its coverage of a social issue as well as being an example of a research approach.

Chapter 12

EVALUATION RESEARCH

This final chapter in Part III, on commonly used research designs, is actually more about a reason for doing research rather than about a new technique for doing research. **Evaluation research** is the use of scientific research methods to plan and assess social policies, social interventions, and social programs. **Program evaluation** is one kind of evaluation research whose goal is to assess how well social programs or social interventions operate and whether they achieve their goals. Some use the terms "evaluation research" and "program evaluation" interchangeably, but I will use the term "program evaluation" to refer to one type of evaluation research, although by far the most common type. As I will show later in the chapter, some types of evaluation research do not focus on particular programs as does program evaluation.

Evaluation research is a form of applied research (discussed in Chapter 1) because the primary purpose of the research is to produce information that can be useful to create, assess, or change social interventions. Although evaluation research has been around for many years, it has risen to considerable prominence over the past two decades as the amount of public and private funds channeled into social interventions has grown. As funding has increased, those providing the funds have sought valid and reliable evidence regarding whether programs achieve their goals, how efficiently they do so, and whether they produce any unintended consequences. Evaluation research focuses on gathering evidence about these issues and has become an integral part of most social interventions.

In this chapter, I first discuss in more detail what evaluation research is, distinguishing it from basic research. Then I discuss program evaluation, showing how it is conducted, including an assessment of the various research designs commonly used. This is followed by a discussion of other less commonly used—but still important—types of evaluation research. The chapter concludes with a discussion of the problems that often arise when it comes time to utilize the conclusions of evaluation research.

WHAT IS EVALUATION RESEARCH?

Evaluation research is a means of supplying valid and reliable evidence regarding the operation of social policies, social programs, or other types of social interventions—how they are planned, how well they operate, and how effectively they achieve their goals. Evaluation research can take many different forms. In some cases, it is conducted to assess broad social policies, such as welfare policy. Over the decades, for example, research has been conducted to determine whether welfare programs achieve their goals such as helping clients develop job skills and find employment. In other cases, evaluation research is conducted to assist particular organizations to accomplish their goals. For example, it may be helpful for a medical care facility to know if there is a community need for a day care program for the elderly. Or criminal justice authorities may wish to know which criminal offenders are good candidates for parole or probation. Or school administrators may want to know whether a new teaching strategy enhances students' ability to learn material. In the private sector, corporations use social science research methods to assess public opinion about a company or a product or to determine how much demand there will be for a new product.

In the past, these issues were often resolved through subjective impressions, intuition, or anecdotal evidence, but everyone today recognizes that these sources of evidence provide, at best, only a very weak foundation from which to make important and expensive decisions about these social interventions. Today, in these situations and others like them, evaluation research uses all of the research techniques discussed in other chapters to provide evidence about the workings of programs and interventions in many realms of life. A typical evaluation effort might involve some combination of interviews, questionnaires, observation, available data, and an experimental design. However, some new issues and techniques arise in doing evaluation research, and these are the topics of this chapter.

Why Evaluate?

Evaluation research is conducted for three major reasons (Rossi, Freeman, & Lipsey, 1999). First, it can be conducted for *administrative purposes,* such as to fulfill an evaluation requirement demanded by a funding source, to improve service to clients, or to increase the efficiency of program delivery. Evaluations done for administrative reasons tend to focus on assessing the daily operations of a program rather than its overall impact, and the goal is typically to find the most efficient means of running a program or agency. A second reason for conducting evaluation research is *impact assessment*: A program is assessed to see what effects, if any, it is producing. Typically, the goals of the program are identified, and the program is measured in terms of how well it achieves those goals. The results of the impact assessment are then used to make policy decisions regarding whether to expand, change, or curtail a program. Third, evaluation research is often conducted for purposes of *social policy development*, or to develop, evaluate, or change overall strategies of social intervention. Research results may be used, for example, as part of a justification for passing state or federal legislation that establishes a new welfare policy or modifies an existing one. This involves far more fundamental changes than might arise out of other kinds of evaluation research.

Evaluation Research and Basic Research

As a form of applied research, evaluation research involves a special application of the general research techniques used in basic research. Because of this, there are similarities

between the two, but I want to focus first on some of the important differences between evaluation research and basic research (De-Martini, 1982; Weiss, 1998). One difference is that the results of evaluation research have immediate practical use in assessing operating programs and policies. Basic research, on the other hand, is oriented toward more general information gathering and hypothesis testing.

A second difference is that in evaluation research, the needs of the decision makers sponsoring the study are paramount in shaping the form and content of the research, whereas the basic researcher has more control over which issues are to be investigated and how they are to be investigated. In some evaluations, there may be some latitude for the evaluator to expand beyond the issues of direct interest to the decision makers, but control over the content of the research is shared with them. This may become a source of conflict between researchers and sponsors of the evaluation.

A third difference is that evaluation, by its very nature, has a judgmental quality about it that is often not a part of basic research. The evaluation may deem a program or policy as a "success" or "failure" on the basis of how well it achieved its purposes, and this judgmental dimension can be a source of tension between an evaluator and the sponsors of the evaluation. Understandably, the sponsors are concerned that a negative evaluation could have dire consequences for the existence of the program and for their own livelihoods and careers.

A fourth difference between basic and evaluation research relates to the issues given priority in the research process. In basic research, quite naturally, the requisites for producing a scientifically sound study are given strong weight. Evaluation research, on the other hand, takes place in a context in which an ongoing program is in operation, and the demands of the program may conflict with the demands for using the most rigorous scientific practices. When this occurs, the program administrators may give higher priority to the program than to the evaluation. The scientific demands of an evaluation, for example, might call for the random assignment of nurses from a home health care agency to each new client of the agency. The agency administrator, on the other hand, may prefer to assign nurses on the basis of his or her assessment of their competencies and "fit" with the client. In a conflict such as this, the agency administrator may override the requests of the evaluators, forcing the latter to modify their scientific procedures. If the evaluation of a program is required, such as by funding legislation, then the evaluation may proceed using compromised procedures, whereas a basic researcher might decide not to continue under such conditions.

A final difference between basic and evaluation research relates to making the results of the research public. One of the canons of science is that research results be made public for others to see and criticize in order to reduce the likelihood that errors or personal bias might find their way into scientific research (see Chapter 1). In the past, wide dissemination of evaluation results was uncommon. The major journals in sociology, for example, such as the *American Sociological Review* and the *American Journal of Sociology*, rarely published the results of evaluation research. It was not until 1976 that a journal specifically devoted to evaluation reports, *Evaluation Quarterly*, came into existence. Today, the *Journal of Applied Sociology* as well as journals in fields such as criminal justice and social work publish articles on evaluation research. There are now about a dozen journals devoted to this topic, and a number of professional organizations serve as outlets for the dissemination of the results of evaluation research. However, it is still the case that the results of most evaluations are neither published nor reviewed by anyone other than those sponsoring the research. This has the effect of thwarting the accumulation of information so necessary for progress. Even if evaluation results show no effect, those findings are valuable in preventing ineffective programs from proliferating.

Despite the differences between evaluation research and basic research, there are important similarities. First, both may choose from the entire array of data-gathering techniques those that best fit their needs, such as surveys, experiments, or available data. In fact, as noted, evaluation projects often involve a synthesis of data gathered in a variety of ways. Second, both forms of research can focus on determining cause-and-effect relationships. In basic research, researchers seek cause-and-effect relationships between variables of their own choosing, whereas in evaluation research the investigation is focused on variables that are a part of the program being assessed. Finally, both basic and evaluation research must address the variety of methodological problems discussed in earlier chapters, such as measurement and sampling.

It should be apparent that the differences between evaluation and basic research are mostly of a practical nature, deriving mainly from the context in which research is conducted. Although these differences are important, the actual process of inquiry is very much the same in both types of research. In fact, evaluation research illustrates the dynamic and flexible quality of basic social research methods in that the core methods and techniques can be expanded and changed to confront new problems and issues. Evaluation research is a novel and challenging application of methods that have been used in many other contexts. As such, the distinction between basic and evaluation research is really one of degree rather than kind.

The Politics of Evaluation Research

Decades ago, many evaluation researchers took a fairly straightforward positivist view of their activities. By this I mean that they viewed the social programs they were evaluating as interventions that address clearly identifiable problems in the world; these problems were viewed as obvious social ills that needed rectifying; it was assumed that almost everyone agreed on what the betterment of those social ills would entail; and evaluators' methods were seen as objective and scientific tools that would assist in achieving that social betterment. You may have guessed where I am going with this: The arguments of the nonpositivist paradigms discussed in Chapter 2—the interpretivist and critical approaches—have led many evaluation researchers to recognize that reality is not as simple as the straightforward positivist view would suggest (Greene, 1994; Patton, 1987; Rule, 1978). In particular, these approaches have led to a recognition that evaluation research, and the programs they evaluate, are inherently political in nature. Politics has to do with power: Who has it, who exercises it, and who controls resources. Social interventions have to do with controlling and distributing resources.

Let me illustrate this with one example: programs to reduce teenage pregnancies. The problem of teenage pregnancy can be approached in a number of different ways, such as programs to encourage sexual abstinence, to make available safe and reliable contraceptives, or to provide abortion services to those who become pregnant. Which of these approaches a particular social program takes will be determined by the values of those who fund and control the program and thus who have the power to see that their values prevail. It is they who will shape what the program is like, which services will be provided, and how they will be provided. In doing so, these policymakers and program managers will have great influence in shaping people's definitions of what the problem of teenage pregnancy is and what acceptable solutions to the problem are. So social programs are the product of political decisions involving the establishment of priorities and the allocation of resources.

In recent decades, evaluation researchers have come to recognize the importance of **stakeholders** in social programs, all those who have an interest in whether a social program operates or how well it does so. Evaluation researchers also recognize that any program has

a large variety of stakeholders, including program funders, program administrators and personnel, and the program's clients or beneficiaries. In most cases, these stakeholders will have varied and competing interests. The teenagers receiving services from a pregnancy prevention program, for example, may benefit from different kinds of interventions than the bureaucrats who run the program or the policymakers who fund it may want to provide. As one instance of this conflict, the teenagers might wish to have abortion services available to them whereas those who fund the program may ignore such services because of their moral opposition to abortion. However, evaluations are typically sponsored by only one or a few stakeholders, such as the government agency that funds the evaluation or the program managers who run the program. These sponsors may want to see the evaluation come out a certain way and may pay for an evaluation that addresses some questions but not others. If the sponsors want to promote sexual abstinence, for instance, they may include in the evaluation measures that assess how well abstinence works. But they may not even ask the teenagers if they would use contraceptives or abortion referral services if these were available. Thus, it is possible that an evaluation could be directed toward some conclusions and away from others because the evaluation is being funded, and in part directed, by only some stakeholders in the program.

So this new approach recognizes that social ills and their betterment are not objective conditions about which there is social consensus. Rather, social ills and social betterment are political issues based on differing social definitions of reality about which people disagree and come into conflict. In designing and conducting evaluation research, then, researchers need to recognize these competing viewpoints and interests and consider the possibility that some research methodologies may be biased in the direction of certain conclusions and the interests of certain stakeholders. When I discuss research designs appropriate to evaluation research later in this chapter, I will return to this issue.

Formative and Summative Evaluation Research

As mentioned, much of what is called evaluation research is program evaluation, or the assessment of a particular social program. Program evaluations can achieve two basic goals (Scriven, 1991). **Formative program evaluation** focuses on providing information to guide the planning, development, and implementation of a specific program. It is primarily concerned with ensuring a smooth-running, well-integrated program rather than with the ultimate worth or impact of the program. **Summative program evaluation** is concerned with the program's effects. Here the purpose centers on assessing the effectiveness and efficiency of programs and the extent to which the outcomes of the project are generalizable to other settings and populations. Formative program evaluation has traditionally received less attention in the evaluation literature than has summative. However, the two forms are closely linked and may be likened to a foundation and a building. The formative component of evaluation may not be especially glamorous or attract much attention but, as you will see, unless it is carefully prepared and well done, the summative type of study may be difficult to carry out and obtain high-quality results. Each type of program evaluation will be discussed separately.

FORMATIVE PROGRAM EVALUATION

Formative research involves applying research to the types of questions that arise in the planning, implementation, and operation of programs.

Program Planning and Implementation

For initiating a program, certain basic information is essential. First, it is necessary to gather data on the target population and their characteristics. The nature of the problems they have, the number of potential program users, their location in the community, and other demographic information would also be essential to planning a good program. Second, it is important to be aware of existing services that the program under development might duplicate or on which it may rely for referrals or auxiliary services. Third, the program planners need to be knowledgeable about the specific intervention strategies that might be applied to the problems. Fourth, the program operators must be able to specify the skills that staff must possess in order to deliver the program successfully. Fifth, it must be determined if it is feasible to offer and monitor the program as it has been conceptualized.

To provide answers to some of these questions, formative research might take the form of a **needs assessment**: collecting data to determine how many people in a community will need particular services and to assess what level of services or personnel already exist to fill that need (McKillip, 1987). In doing needs assessments, researchers commonly utilize a sampling strategy and then survey members of the target population. Thus survey research methods form the backbone of such needs assessments. However, one could also rely on direct observation or utilize available data to gain a profile of the population and identify needs. Essentially, any techniques used for descriptive research may be employed to answer this question. Knowledge of existing and related services might be gained through interviewing potential clients and representatives of existing agencies. Knowledge of possible intervention strategies and program components often comes from conducting a thorough literature search, a fundamental step in any research project. (Appendix A, on the use of the library, should assist you in this task.)

Formative evaluations also sometimes utilize *focus groups*, the group interview technique discussed at length in Chapter 9. Focus groups are especially useful when the formative evaluation is exploratory in nature or when planners need to learn about the very personal and subjective meanings and experiences of people. Focus groups can be a useful strategy for drawing such information from people.

Formative evaluation research can also take the form of a *pretest* or *trial run*, in which all the procedures to be used in a program are tested before the full program is implemented. An example of this can be found in a program to provide financial aid to newly released offenders (Rossi, Berk, & Lenihan, 1980). The researchers began with a very modest project involving six released inmates who received six weekly payments of $60, and 20 controls who received no payment. Relying on a review of the literature and existing statistics on the problems of released offenders, this modest project helped prepare the way for a large-scale program in several ways. It helped determine that it was logistically possible to make the payments and to interview and keep track of ex-inmates in the community. Had this pilot project shown that these operational elements could not be accomplished effectively, then it would have been foolish to continue with a larger project even though one might be convinced of the overall effectiveness of the program.

Because the questions addressed in formative research are often modest in scope, the research is frequently conducted by the staff of the agency administering the program. However, such research may also be done on a national scale in conjunction with large programs. Programs such as welfare, food stamps, and vocational rehabilitation involve large sums of money, and initiation of such programs involves estimating the number of potential recipients, which can vary considerably depending on which definition of *poverty* is applied. Before initiating such programs, formative research based on needs assessment is essential. Many programs fail because there

are really no potential users of the programs or the design of the program precludes clients from using it.

Program Monitoring

Besides serving as a tool for the planning of an intervention program, formative research may also be used to monitor the implementation of new programs and the ongoing operation of existing ones. Experience with evaluations for program effectiveness has shown that a major factor in program failure is often the fact that the program as planned was never really implemented (Berman & Pauly, 1975). Several basic issues in program monitoring are counterparts to program planning: (1) Is the target population in fact being served? (2) Are the services that are supposed to be delivered actually being delivered? (3) Is the quality of the service adequate?

As a means of supplying answers to the first question, it is common to use a census of program users and compare their characteristics to the characteristics of the population for whom services were intended. Any discrepancy suggests that some members of the target population are not receiving services intended for them. Service delivery may be assessed in several ways, including questionnaires and direct observation. Time samples may be used to determine if behaviors associated with service delivery are occurring as planned. Quality control techniques not unlike those used in industry are employed to monitor the delivery of services in many settings. Major financial assistance services such as welfare are routinely monitored by state departments of social services. Typically, this process involves selecting a random sample of recipients and examining the most recent action on their files to determine if the action resulted in a correct payment, an overpayment, or an underpayment. States are expected to keep their error rate within certain specific limits. Other settings may use follow-up questionnaires to service users to determine if expected services were delivered.

Finally, agencies commonly use a time-reporting system to be sure that staff are spending the expected amount of time on specified aspects of the program. Such program monitoring does not address the question of whether or not the program is actually doing any good. Rather, they address the narrower issue of the extent to which the services actually being delivered are true to the intended plan.

Social science researchers are often employed to gather data used in many of these formative evaluation tasks. However, programs are often developed and implemented without using such research and may suffer as a result. Social scientists are also employed in doing summative evaluations on programs after they have been implemented, and it is less likely today that a social intervention will be allowed to operate for long without a summative evaluation being conducted.

SUMMATIVE PROGRAM EVALUATION

Summative program evaluation involves assessing the impact of a program, although good summative evaluations usually investigate other matters concerning the program's operation. The results of summative evaluations are intended to be used for policy-making decisions—whether to continue, expand, or cancel a program, and whether to generalize the findings of this project to other settings and populations. Summative evaluations are typically large-scale projects involving considerable time, personnel, and resources. For this reason, as well as to avoid the biasing effect of personal interest, summative evaluations are often conducted by outside consultants rather than by the staff of the agency running the program.

Although the steps in the research process are much the same for all types of research, including program evaluations, some special planning is called for to take into consideration the unique problems of evaluation research.

Evaluability Assessment

During the planning of a program evaluation, an **evaluability assessment** is often conducted to enable investigators to decide whether a program has the necessary preconditions to be evaluated. An evaluability assessment involves four steps (Rutman, 1984; M. F. Smith, 1989). First, the purpose of the evaluation—from the standpoint of the eventual users of the results—is determined. This identifies which aspects of the program are to be assessed for impact. To be evaluated, a program needs clearly specified goals. Having clear goals is necessary because these are the major criteria of a program's success. Goals are the dependent variables that are supposed to be influenced by the program. Unfortunately, the goals of many programs are either vague or so global (improving family functioning, for example) that they are unusable for evaluation purposes. As we shall see, it may be necessary for an evaluator to become actively involved, together with the staff, in developing a set of clear goals before a program is evaluable.

The second step in an evaluability assessment is to study and gather information about the program. What are the inputs (independent variables), the expected results or goals (dependent variables), and the linkage between the two? This linkage is essentially the theory on which the program is based (Weiss, 1997). It involves answering the question: Why, given these program inputs, should certain outcomes be expected? Figure 12.1 diagrams the linkage between the inputs and the goals, presenting first a model for input-goal linkage and then a hypothetical illustration of a teenage pregnancy program. Note that some inputs might impact on more than one goal. (Also note that if some input did not impact on any goal, one would have to question why a program expended any resources on that input.) The theoretical linkage between inputs and goals may

FIGURE 12.1   Linkage Between Inputs and Goals in Social Programs

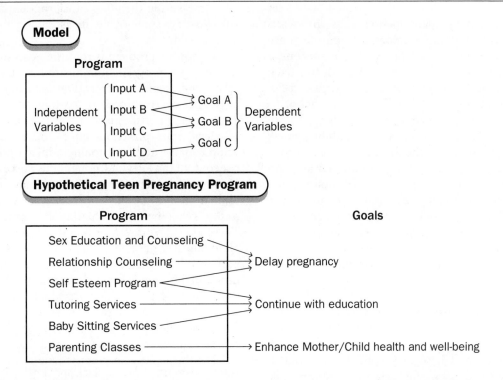

be explicitly recognized as the basis of the program, or it might be implicit but derivable from the program's operation. The purpose of identifying this rationale is that it is important in understanding the success or failure of the program. For example, some programs may identify certain outcomes as goals but not direct any effort toward realizing them. (Quite obviously, if evaluated against these goals, the program would likely turn out to be a failure.) If we do not specify the linkage, the reason for failure may be unclear and, in the case of a success, we would be at a loss to explain why. Knowledge of the linkage reveals the source of the problem as the failure to direct input resources toward those particular goals. During this step, it is important to monitor the program *as implemented* because gaps sometimes develop between the stated program and the program as operated.

The third step in an evaluability assessment is to use information gathered during the second stage to develop a *flow model* of the program. This model traces program inputs, clients, and interventions as they affect the expected results, specifying any assumed causal linkages along the way. Modeling of programs is extremely useful in program evaluation because it provides a clear picture of the structure and operation of the program. It can also help explain a program's successes or failures. For example, as a part of an evaluation of a group counseling program in a correctional system, a careful model of the program was developed (Kassebaum, Ward, & Wilner 1971). When the results of the evaluation showed that inmates in the counseling program did not have lower recidivism rates than other inmates, the evaluators were hardly surprised. Their model revealed that counselors were poorly trained and unmotivated, counseling sessions tended to be unfocused "bull" sessions, inmate participation was motivated largely by a desire to impress the parole board, and the inmates did not view the sessions as likely to be helpful to them. Against this backdrop, the failure of counseling to reduce recidivism is understandable. The flow model of the program helped identify these elements.

The final stage of the evaluability assessment is to review the program model to identify those aspects of the program that are sufficiently unambiguous in terms of inputs, goals, and linkages that evaluation of them appears feasible. The result of the assessment may be that the entire program can be evaluated, none of the program can be evaluated, or, most commonly, only certain parts of the program are amenable to evaluation.

Specification of Variables

As in any research, an important part of program evaluation research is the specification of variables and how they will be measured. In some evaluations, the variables of interest take the form of independent and dependent variables (see Figure 12.1). The inputs to a social program, for example, might constitute the independent variables. In some cases, the independent variable takes the form of a dichotomous variable: participation or nonparticipation in a program. In other cases, the independent variable might be the degree or duration of participation in a program, such as the frequency of contact with a counselor, the level of financial aid received, or the length of time a service is provided. Commonly, the dependent variables in program evaluation are the goals of a program: precisely, what it is supposed to accomplish. Some program evaluation research, of course, especially the formative type, would not involve independent and dependent variables. Whatever form the variables take, a central issue in evaluation research is that the variables be clearly and properly specified, and numerous problems can

arise along these lines. This is especially true when measuring the goals of a program, and I will illustrate some of the measuring problems in this realm.

A frequent problem is that the goals, as articulated by program administrators, do not easily lend themselves to evaluation. They may be vague, overly broad, or so long term that evaluation is not feasible. For example, a goal of Head Start preschool education is to develop capable and functioning adults who can rise out of poverty. Although the success of Head Start in achieving this goal may be its ultimate test, it would be necessary to wait 20 to 30 years before evaluating the program. These are called *distal*, or long-term goals and, although they may be a laudable and essential part of the program, funding agencies are understandably reluctant to expend funds for that length of time with no evaluation. Thus, such program evaluations normally include what are called *proximate goals*, or goals that can be realized in the short run and that are related to the achievement of the long-term goals (Weiss, 1998). In the case of Head Start, academic achievement would be a reasonable proximate goal because performance in school is associated with social and occupational success in adulthood.

There are of course programs in which it is possible and desirable to assess the long-term impact. For example, there was a 30-year follow-up of the Cambridge–Sommerville Youth Study, a five-year experiment in delinquency prevention begun in Boston in 1939 (McCord, 1978). This field experiment focused on 506 young boys, half receiving counseling and other assistance while the remainder served as a control group. The experimental condition consisted of counseling sessions every two weeks, tutoring, medical and psychiatric assistance, summer camps, and organized youth activities. Thirty years later, 95% of the participants in the experiment were located. Many of the comparisons showed no differences between the groups. But what differences were found suggested that the experimental variables had the *opposite* effect from what was expected! The experimental group committed more crimes as adults, had higher rates of alcoholism, poorer mental and physical health, and less occupational success. The only positive result for the program appeared to be the participants' own subjective evaluations. Two-thirds of the experimental group indicated that they thought the program had been helpful to them even though their objective situation was worse than that of people in the control group. The program may have produced harmful effects because it raised participants' expectations to an unrealistically high level. When those expectations were not realized, participants suffered frustration and added stress, which resulted in a greater tendency toward criminality or alcoholism as well as the deleterious effects on their mental and physical health (McCord, 1978). Despite the desirability of early program evaluations based on proximate goals, long-term evaluation, accomplished so well in this study, is still essential in assessing many programs.

When goals are vague or overly broad, they must be clarified or reduced in scope so that they are amenable to evaluation. In this regard, it is helpful if program administrators consult with evaluators in developing the goals for a program. During discussions with program administrators, it is important that evaluators not simply accept as all-inclusive the goals articulated by the administrators. In fact, one of the major reasons for finding that a program does not have the intended result is that program goals are too limited (Chen & Rossi, 1980). Administrators too often state program goals from the standpoint of what they desire, resulting in goals that are unattainable given program inputs. When evaluated against these goals, the program naturally appears to be a failure. Evaluators should "cast a wide net" in seeking program effects, including not only those suggested by administrators but also others logically expected given the nature of

the program. The model developed during the evaluability assessment and the theoretical basis of the program are productive places to look for possible program effects to be included in the evaluation. This approach promises to provide a better chance of finding nonzero program effects and supplying information on what the program does as well as what it does not do.

Another problem that can arise in specifying the goals and variables in an evaluation is that different stakeholders may have different conceptions of what the program should accomplish. All sides may be very clear about what the goals should be but simply disagree with one another on them. This points again to the political dimensions of evaluation research and the possibility that those who control the resources (usually the program funders or administrators) will impose their version of what the program goals should be. The evaluation researcher needs to ensure that other significant stakeholders have some input into the process of defining and clarifying the goals.

Measuring Variables

The goals specified for many programs tend to be abstract statements of desired outcomes. Before an actual evaluation can proceed, researchers must develop measurable criteria or operational definitions specifying exactly which observations will be made to determine goal achievement. Evaluators thus distinguish between *goals*, which are the desired end states for a program; and *objectives*, which are the measurable criteria for success. The objectives are the operational definitions of whether a program will be considered a "success" or a "failure." These operational definitions of "*success*" should be established before the evaluation is begun. For example, a program goal of a substance abuse prevention program might be to reduce experimentation with alcohol and cigarettes among junior high school students. An objective of the program might be that 85% of program participants be able to correctly list at least five health hazards associated with alcohol use. You can see how the objective is very specific and concrete. Some goals can be readily measured because they have clearly quantifiable outcomes. For example, the academic effects of compensatory education programs can be readily measured by standard achievement tests. Such effects of Head Start as improved self-esteem, better adjustment to the classroom environment, or enhanced parent-child relationships will likely require more inventiveness on the part of the evaluator to measure.

When measures for program effects are being considered, there may be alternative indicators of the same program effect. For example, in measuring the effects of a family planning program, we might use as a measure (1) the proportion of participants adopting contraceptive practices, (2) the average desired number of children, (3) the average number of children actually born, or (4) the attitudes toward large families. All these indicators are logically related to the effects of a family planning program. Multiple indicators—though sometimes prohibited by budget—are more useful than single ones. Multiple indicators are more sensitive and therefore more likely to show an effect if the program produced one (Weiss, 1998). If multiple indicators are impractical, a decision must be made as to which of the alternatives is best. In the case of the four alternatives for the family planning program, the most valid indicator of program success would be a low birthrate among participants. This indicator, however, might be impractical for some purposes because we would have to wait many years before the evaluation could be completed. The two indicators dealing with attitudes would not be the best choices because attitudes can and do change, and sometimes only a weak relationship exists between attitudes and behavior. Indeed, attitudinal measures should be avoided whenever a behavioral alternative is available. Of the four possible measures, then, the proportion of participants who

adopt contraception is probably the best single indicator of the effectiveness of the family planning program. It would not be perfect because contraceptives must be used conscientiously for them to be effective, but for a short-run measure it would be adequate.

As I have recommended on a number of occasions in this book, it is preferable to use existing measures where possible. This avoids the work involved in creating new measures, pretesting them, and establishing their validity and reliability. Further, existing measures contribute to the accumulation of knowledge because they make evaluations of different programs more clearly comparable. The preference for existing measures extends only to the point where good existing measures for the variables of interest can be found. If existing measures are only tangentially related to what one seeks to measure, then it is far better to develop new measures.

Assuming that some new measures must be created, it is important to keep in mind the dual criteria for assessing them: validity and reliability. Measures used in evaluation research must meet the same standards of validity and reliability as those used in basic research, and the methods for assessing validity and reliability discussed in Chapter 5 apply here.

Designs for Program Evaluation

Summative evaluation research is often concerned with cause-and-effect relationships, especially for those using a positivist paradigm. For example, the program being evaluated is presumed to bring about some changes in such factors as people's behavior. A research design developed for such an evaluation needs to be based on an awareness of this cause-and-effect dimension. In Chapter 8, I noted that true experimental designs involving randomization are the better choice for establishing cause-and-effect relationships because they best control the validity threats that can lead to false causal inferences. The ideal approach to determining the effects of a program is therefore the randomized experiment. Any of the true experimental designs discussed in Chapter 8 is appropriate for evaluation purposes. However, as will be shown shortly, quasi-experimental designs are quite common because these are often far more feasible and expedient than true experimental designs (Orr, 1998).

Randomized Experimental Designs

Recall the requirements of a true experiment: two randomized equivalent groups, one that experiences the experimental condition and one that does not and serves as a control group. In summative program evaluation, the experimental condition requires some level of participation in the program under consideration. The crucial feature of the true experiment is that members of the comparison groups are assigned randomly. This is the surest and most reliable way of producing equivalent groups.

Virtually all randomized experiments in evaluation research are field experiments, taking place in the setting where the actual program is administered. Because of this, evaluators may encounter a number of impediments to conducting a randomized experiment. The first impediment centers around the control group and the randomization procedure used to form it. It is necessary, in order to have a control group, to deny some members of the target population access to the program under evaluation, and that denial must be on a random basis. Evaluators may encounter substantial resistance to such denials. For example, the enabling legislation of some programs mandates that all persons who meet the eligibility requirements have a legal right to participate in the program. If such is the case, random denial of service is ruled out.

In other programs, resistance to random denial of services may spring from program administrators and staff who are accustomed to providing services on the basis of need and may be disinclined to use a table of random numbers instead. It seems cold, insensitive,

even immoral, to withhold available services from people who need them, especially if intuition leads us to believe that the services would have beneficial effects. In advocating a randomized experiment, evaluators sometimes find themselves in a "no win" situation. If they discover that the program produces harmful effects, they are blamed for subjecting the experimental group to the harmful program. If the program produces positive results, they are blamed for withholding this valuable service from the control group. Imagine the ethical implications of a randomized experiment for evaluating something like a suicide prevention program where life-and-death issues are at stake.

A second impediment to conducting randomized experiments in evaluation research is that they may be more time consuming and expensive than other designs. Experimental evaluations of programs are typically longitudinal: Sufficient time must pass for programs to have an effect. With many social programs, such as compensatory education or job training, the minimum length of the experiment might be a year. Further, a listing of the target population required for randomization may be difficult or expensive to obtain. These practical considerations mean that randomized experiments are limited to those cases where both money and time are available for an elaborate, rigorous evaluation.

Randomized experiments are clearly the best designs from which to assess causality, and many statistical procedures are based on the assumption of equivalent experimental and control groups. Reports in journals such as *Evaluation Quarterly* document the widespread use of these designs today. With some inventiveness on the part of evaluators, much of the resistance to randomized experiments can be overcome. Indeed, there is a growing consensus about the desirability of randomized experiments and an expanding literature documenting their use (Orr, 1998). This should work to reduce the barriers to future randomized experiments in evaluation.

APPLIED SCENARIO **Evaluating Programs for Reducing Recidivism among Released Prison Inmates**

Even the most reformed and well-intentioned former prisoner faces a host of obstacles in his or her efforts to begin a new life outside prison. And recidivism—repeating another crime after being released from prison—occurs at distressingly high rates in the United States. It is a very costly problem, and policymakers have designed a variety of programs in attempts to reduce it. One such approach is to provide assistance to former prisoners to help them make the transition to civilian life and find a job. Because many crimes are economically motivated, it seems reasonable to hypothesize that such assistance would reduce the motivation for a newly released inmate to turn to crime.

To test this hypothesis, the Baltimore Living Insurance for Ex-prisoners (Baltimore LIFE) Project was developed (Lenihan, 1977; Rossi, Berk, & Lenihan, 1980). This field experiment used a good randomized experimental design. In the Baltimore LIFE Project, high-risk inmates scheduled for release by the Maryland Department of Corrections were randomly divided into four groups, three experimental groups and one control group. One experimental group received financial assistance of $60 per week for 13 weeks after release and was offered job placement services. A second received the financial assistance but no job placement services. The third experimental group received only the job services, whereas the control group received neither financial assistance nor job services. This is a factorial design, as described in Chapter 8, and I have diagrammed it in Figure 12.2. The intent of this phase of the research was to determine if the two independent variables—*financial assistance* and *job placement services*—would have any positive effect on recidivism. For two years, the ex-prisoners were observed and their experience in employment and criminal activity was noted.

The Baltimore LIFE Project illustrates several key features of careful program evaluation. First, the project was preceded by a much smaller exploratory study that enabled the project staff to test and refine some of their procedures. Second, the evaluation utilized a variety of research techniques. In addition to the basic experimental design, the research relied on structured interviews and case studies of several participants to provide insight into the problems,

One common situation in particular contributes to the possibility of a randomized experiment. When the target population is larger than the program's capacity to serve it—in short, when demand for services exceeds supply—services must be denied to some people. Because some members of the target population will not be served anyway, the determination of who will be served may as well be random unless there is some other clear-cut and defensible criterion that could be used, such as severity of need. In fact, a reasonable argument can be made that random allocation of services is the fairest method when resources are inadequate to serve all and no other criteria seem applicable.

The Applied Scenario box describes a program evaluation that used a randomized experimental design.

Alternatives to Randomized Experiments

The barriers to a randomized experiment may be sufficiently formidable that an alternative design must be utilized. It is important to

reactions, and frustrations of these men as they attempted to cope with life in the community. Third, a cost–benefit analysis was conducted to ascertain if the program was worthwhile. Finally, great care was taken in selecting participants, in explaining the program to the participants, and in dispensing the financial aid and job counseling according to the design.

The results of the study indicated that the released inmates who received financial aid had an 8% lower rate of arrest for charges of theft. Arrests for other types of crime were not significantly different. Also, those not receiving financial aid were arrested earlier, were more likely to be convicted, and were more likely to be returned to prison. Job placement services had no apparent impact on recidivism or occupational success.

Based on the modestly optimistic findings of this research, a much larger program, which involved dispensing financial aid through the existing Employment Security Offices in the states of Texas and Georgia, was implemented. The outcome of that project was also evaluated, but the overall findings were not positive. The researchers concluded that the disappearance of positive outcomes may have resulted from the way the larger program was administered, thereby emphasizing the need for careful program implementation as well as identification of appropriate input variables. Thus evaluation research focuses not only on the outcome of programs but also on how they can be best implemented.

Program evaluations are important to do in order to assess the outcomes of programs. They can also be difficult to conduct if constraints are placed on how programs are run. Given what you have learned in this chapter, do the following.

1. Locate some programs on your college campus or in your community that could benefit from a program evaluation. Describe how that evaluation could be designed.

2. Could the evaluation be designed so as to make it a randomized design?

3. If a randomized design is not possible, what could be done to increase our confidence in the results?

remember that anything other than a randomized experiment will produce results in which confidence is reduced because such alternative designs are weaker on internal validity (see Chapter 8). Properly conducted, however, these designs can allow evaluation with a reasonable degree of certitude (Achen, 1986).

One alternative to the randomized experiment is to use a *quasi-experimental design*. When a program is meant to affect behavior about which data are routinely collected, a time-series design may be appropriate. One research group, for example, used a time-series design to evaluate the impact of a Massachusetts gun control law that mandated a one-year minimum prison sentence for anyone convicted of carrying a firearm without a permit (Deutch & Alt, 1977). The law went into effect in April 1975. The researchers traced the monthly occurrences of homicide, assault with a gun, and armed robbery in Boston from 1966 to 1975, which provided a baseline of gun-related offenses for about nine years prior to the introduction of the gun control law. Their analysis revealed that in the first seven months the law was in effect, statistically significant decreases

FIGURE 12.2 The Design of the Baltimore LIFE Field Experiment

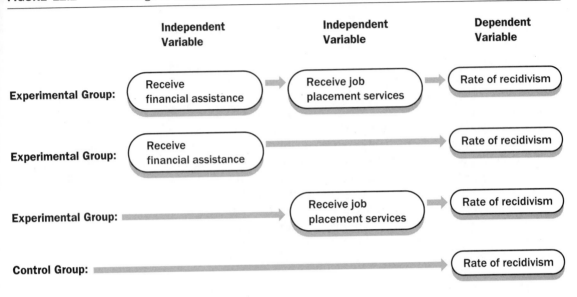

in both armed robbery and assault with a gun occurred. No change was registered for homicide, however. Despite their frequent utility for evaluation, it is important to remember that the chief weakness of time-series designs is the validity threat of history. Unless it is possible to use a control group, as in the multiple time-series design, it is always possible that some extraneous variable can intervene and confound the results.

A second alternative to randomized experiments is *matching*. If randomization is not feasible, it may be possible to match persons in the experimental group with persons having similar characteristics in a control group. Recall, however, that matching can be unreliable because relatively few variables can be used for matching, which leaves uncontrolled variables that might confound the results. An example of an evaluation in which a form of matching was used was a study of the effects of a housing allowance program implemented by the Department of Housing and Urban Develop-

ment (HUD) in 1970 (Jackson & Mohr, 1986). Because of the nature of the program, people could not be assigned to experimental and control groups. Instead, for the experimental group, a random sample of clients enrolled in the program was used. For a control group, a sample was selected from the Annual Housing Survey administered by HUD. The HUD survey contained data similar to that collected from the experimental group. The matching involved selecting clients from the Annual Housing Survey who were eligible for, but not enrolled in, the housing allowance program. This provided a comparison group to assess the impact of enrollment in the housing allowance program on such things as changes in housing quality and the extent of the rent burden.

A third alternative to the randomized experiment, really another form of matching, is the use of *cohort groups*. Cohorts are groups of people who move through an organization or a treatment program at about the same time. For example, the following are cohorts:

students in the same grade in a school, people receiving public assistance at the same time from a particular agency, and people in a drug rehabilitation program at the same time. Cohorts are valuable alternatives to randomized experiments because we may be able to assume that each cohort in an organization or program is similar to the preceding cohorts in terms of the characteristics that might affect a treatment outcome. In other words, each group should be alike in age, sex, socioeconomic status, and other important characteristics. However, there can also be very significant differences, and cohorts should always be assessed to detect any possible systematic variation.

An elaborate cohort study evaluating curriculum revision and televised instruction was conducted in El Salvador from 1969 to 1973 (Mayo, Hornick, & McAnany, 1976). Seventh grade classes in 1969, 1970, and 1971 made up three separate cohorts of students. Within the cohorts, some classes received a new curriculum, some received the new curriculum with televised instruction, and some received the old curriculum. The cohorts from 1969 and 1970 were followed for three years; the cohort from 1971 was followed for two years. Comparisons among the groups produced mixed results. The new curriculum was consistently superior to the old one, and televised instruction was superior during the first year it was experienced. However, the superiority of televised instruction wore off as the students became accustomed to it.

The major weakness of cohorts is, again, the threat to validity from history. Because the measurements are taken at widely spaced time intervals, extraneous variables may intercede and possibly affect the results. The El Salvador example is instructive on this point. It would be unlikely that a similar cohort study could have been reliably conducted 10 years later as the country became unstable owing to guerrilla warfare. Comparing a cohort from a period of peace to one from a period of near civil war would have obvious problems.

A fourth alternative to the randomized experiment is called the *regression discontinuity design*: Persons are selected to receive a treatment based on their score on a test, their eligibility for a program, or some other criterion (Reichardt, Trochim, & Cappelleri, 1995). An example of such a design is a study modeled after the Baltimore LIFE experiment described in the Applied Scenario box. This study assessed the effectiveness of a state-mandated program to provide unemployment benefits to newly released prison inmates as a means of reducing recidivism (Berk, Boruch, Chambers, Rossi, & Witte, 1985). Prison inmates were eligible for the experimental group (receive benefits) if they had worked sufficient hours in prison to be eligible for unemployment benefits when released. Inmates who had worked fewer hours were put in the control group (no benefits). The researchers concluded that the program saved California $2,000 for each inmate involved because fewer inmates were sent back to prison and the state saved those prison costs. However, since the random assignment of inmates to experimental and control groups was not done, the best research design for assessing program impacts—the randomized experiment—was not used. This made the results of the evaluation less certain: Any differences in recidivism rates between the two groups might be due to the fact that those who work more hours in prison are also less likely to commit crimes once released—the validity threat of selection discussed in Chapter 8 or possibly statistical regression. Despite its limitations, the regression discontinuity design is often implemented for evaluation research field experiments when it is not possible to create a randomized control group.

The last major alternative to randomized experiments is the use of *statistical controls*, which are procedures that allow the effects of one or more variables to be removed or held constant so that the effects of other variables can be observed. These procedures allow comparisons to be made between groups that differ from one another on some characteristics thought to be important. The effects of the variables on which the groups differ are removed through statistical manipulation so that they cannot obscure the results. Statistical controls, however, even in their most elaborate application, can only approximate the level of control achievable in randomized experiments. Like matching, only those variables known to the researcher to be potentially important can be controlled statistically, so there is always the possibility of important extraneous variables left uncontrolled. Further, statistical controls tend to underadjust for between-group differences because of the error component in the measurement of the control variables (Berk & Rossi, 1990; Weiss, 1998). The error allows at least some of the effects of the control variables to remain even after the statistical controls have been applied. Because of these limitations, statistical control alone may not be appropriate. However, it is well suited as an adjunct to the physical control obtained through design. For example, in a matched design we might discover, after the fact, that an important variable was left unmatched. Assuming the necessary data were collected, we could correct this error by applying statistical control to that variable.

The combined use of both design control and statistical control is probably the best overall approach for evaluation because statistical control can even be useful in randomized experiments (Rossi, Freeman, & Lipsey, 1999).

Nonpositivist Approaches to Evaluation

The interpretivist and critical approaches make quite different assumptions about what is going on in a summative program evaluation than do positivists. First, they assume that there are multiple stakeholders in, and thus multiple perspectives on, social interventions (Greene, 1994). A given research methodology may assess the program in terms of the interests of some stakeholders but not others. In fact, some critics of positivist approaches to evaluation argue that because such evaluations are typically funded and supported by the sponsors and managers of the programs, they tend to address issues of concern to those stakeholders. Such issues typically include, for example, the economic efficiency of the programs, the numbers of people served, or other issues that can be measured in quantitative, often monetary, ways. These approaches tend to focus on the importance of assessing program outcomes, program efficiency, and accountability. And in fact these issues are the ones that are most readily addressed through the randomized experimental designs that positivists consider to be the ideal evaluation methodology. For evaluators who address different issues or the interests of other stakeholders, the randomized experimental design with control groups may be less important or possibly irrelevant as a research design (Cook, 1985). The Eye on Diversity box explores some ways in which diversity plays a part in considerations such as these.

A second assumption of nonpositivist evaluators, especially those using the interpretive approach, is that interpretation and social meaning are at the core of social interventions (Denzin, 1989a; J. K. Smith, 1989; Weiss, 1998). In this view, social reality does not just exist "out there" but rather is created by people as they interact and exchange meanings (Guba & Lincoln, 1989). An interpretivist program evaluation would focus attention on how a social intervention is experienced by all the stakeholders—the sponsors and managers as well as the recipients of services. From some perspectives, such as that of those receiving the services, economic efficiency and other quantitative matters may not be the key elements of the program as they experience it. The point is that nonpositivist evaluators refuse to define issues and solutions solely from the perspective of the more powerful and dominant stakeholders. These concerns lead

EYE ON DIVERSITY **Valid Considerations of Gender in Program Evaluations**

One area in which evaluation research has played an important role is in assessing the operation and impact of policies and programs affecting minorities. When the women's movement hit full stride in the 1970s, women in large numbers began to enter what had previously been considered exclusively male occupations. Before long, studies were being done assessing the relative performances of men and women working the same jobs. As more and more of this research was amassed, serious questions began to be raised concerning the quality and fairness of these evaluations, particularly toward the women in traditionally male occupations.

Take the case of evaluating male and female police officers. Merry Morash and Jack Greene (1986) reviewed nine major evaluations of women police officers that had been conducted up to that time. Not surprisingly, given the varying methodologies of the several studies, results were inconsistent. With one exception, however, the studies concluded that female police officers were neither better nor worse than their male counterparts, but that they were different. That is, women were better than the male officers at some police activities but were not as good as the male officers at others. If these differences are real, such information could be quite valuable in allocating police personnel to maximize the strengths and minimize the weaknesses of both male and female officers. But what if the differences are merely artifacts of a faulty research design? Because assignments in police departments have implications for career advancement, we had better be sure that the findings of such evaluations are indeed valid before we proceed to use them as a basis for allocating personnel—or for any other reason.

Morash and Greene's (1986) review of the nine studies is not encouraging on the validity issue, but it does point out issues that future researchers should address. Among the problems they found were a tendency for evaluations to emphasize conformity to male stereotypes (such as marksmanship or the frequency of arrests), a failure to evaluate performance on a representative sample of police tasks, an overemphasis on the violent and dangerous aspects of police work, a failure to consider the differences between men's and women's experiences in the workplace (there is greater camaraderie among male officers, for example), a failure to evaluate variations in performance within each sex as well as variation between the sexes, and unclear or unspecified definitions of what constituted good police work. Quite a litany of criticisms!

Morash and Greene's (1986) analysis clearly indicates that conducting performance evaluations of men and women in job settings can be very difficult and fraught with possibilities for invalid conclusions. It also points to the importance of recognizing the different stakeholders in an evaluation. In this case, gender produces different stakeholders, each serving to benefit if the evaluation is done in a particular way. In other cases, it might be variations in race, ethnicity, or sexual orientation that produces different stakeholders. The challenge for evaluation researchers is to ensure that the voices of all significant stakeholders are heard in the design of the evaluation.

nonpositivist evaluators to research methodologies that are more useful for discovering the interpretations and perspectives of the various parties to the programs: participant observation, in-depth interviewing, case studies, and other more qualitative approaches. In these approaches, control groups, random assignment, and random samples are less important than discovering meaning in the social contexts in which people live and allowing participants to frame their own issues and define problems using their own meaningful categories.

A third assumption of nonpositivist evaluators, especially those using a critical approach, is that social interventions typically reflect and reinforce inequitable distributions of power and resources in society. A central goal of the critical program evaluator is to reveal the mechanisms whereby inequities are reinforced and increased and to show how social programs promote the agenda of the powerful while doing relatively little for the less powerful. Critical evaluators might use either quantitative or qualitative approaches, but in either case their approach would be more participatory in nature: Less powerful stakeholders would participate in designing and carrying out the evaluation. Their participation increases the likelihood that the research will be designed to discover facts and relationships that are beneficial to them. They would be consulted about the research all along the way, and one of the goals of the research would be the empowerment of these groups. A part of this empowerment would involve the evaluation serving as a catalyst for social change.

As stressed in other chapters, the distinction between positivist and nonpositivist research does not follow the separation between quantitative and qualitative research rigidly, although there is a relationship. And this is true in the field of evaluation research. While positivists and nonpositivists do have the methodological inclinations described in the preceding three paragraphs, the key distinctions between them has to do with the assumptions stated rather than the particular research methodologies used. By looking at both the positivist and nonpositivist perspectives, we achieve a much broader view of what evaluation research can accomplish and whose issues and values can be attended to in such research.

OTHER TYPES OF EVALUATION RESEARCH

Program evaluation is, without doubt, the most common kind of research that goes under the heading of evaluation research or applied research. Yet there are some other research techniques that fall into this general category, such as needs assessment that was discussed with formative program evaluations. I next discuss a few additional examples of evaluation research to give you a good overview of what is generally called "evaluation research." The first (social impact assessment) is somewhat analogous to but distinct from program evaluation. The other three are often conducted as a part of program evaluations but are also often done as a part of the assessment of general social policies rather than program evaluations.

Social Impact Assessment (SIA)

Social impact assessment refers to procedures for estimating the impact of programs and projects on people, groups, communities, and social institutions (Finsterbusch, 1981). Large-scale social impact assessments typically involve a team of social scientists working together since the impact of a large project might be quite widespread and call for the expertise of sociologists, anthropologists, economists, and other social scientists.

How do program evaluations and social impact assessments differ from one another? They can overlap to an extent, but SIAs focus on projects or programs that have effects on many or most members of some particular community. Most social programs, on the other hand, directly affect only those for whom they are intended. For example, a program to place social workers in elementary schools has direct effects on the children and families who receive services from the social workers. However, that social service program will have little impact on the teachers and other students in the schools or on most members of the community at large. If these other groups see any benefit, it will be only very indirectly. On the other hand, a project to build a hazardous waste incinerator in a community can have a severe impact on the lives of many, possibly all, people living in the area. Property values of all residents in the community might be affected, and all residents might suffer if toxic wastes are released into the environment. Social impact assessment is used to evaluate the broad impact of such projects, and some of the techniques that are used are quite different from the techniques used in program evaluation.

There are two types of SIAs. A *pre-impact assessment* is done during the planning stages of a project and is used to design the project to achieve certain goals (somewhat analogous to formative program evaluation). A *post-impact assessment* (which is analogous to a summative program evaluation) is done after a project is completed and is used to determine what impact, both positive and negative, the project actually has had. Post-impact assessments may lead to changes in the project or provide valuable information that might be used by future projects that are akin to it. Pre-impact assessments are more difficult to do and often more speculative because they attempt to project a future state of affairs. Post-impact assessments can be more precise and quantitative since they measure how many people were actually affected, to what extent, and in what way.

SIAs often begin by looking at existing post-impact assessments conducted on earlier projects that are like the proposed one. A detailed description of the past and present circumstances of the community would be then developed using surveys, field observation, existing records, and other research methods described in Chapters 8 to 11. Then a projection is made regarding how things will change as a result of the program. Projections could be made about changes to individuals, households, groups and organizations, the community, and social institutions such as church or family (Finsterbusch & Motz, 1980). You can see that the thrust of social impact assessments is very wide ranging, often focusing on disparate aspects of society.

In forecasting the future, SIAs vary considerably in exactly what they do (Bowles, 1981). Some merely extrapolate past trends into the future (e.g., if current increases in burglary rates continue, then the burglary rate will double in the next decade). Others are more complete and useful and provide a *forecast*, which presents a range of possible outcomes, tells the conditions under which each will occur, and presents a causal model that explains the predictions. In projections of population growth, for example, organizations such as the U.S. Census Bureau and the United Nations usually make three projections based on low, medium, and high levels of fertility. Based on past trends in birthrates, one could make a case for low fertility in the United States in the future, but a reasonable case could also be made that fertility will be higher.

The final stage in an SIA is to modify the project to alleviate the problems it creates. While decisions at this stage are made by politicians or project managers, the researchers, through their data analysis and conclusions, provide the information and clear articulation of possible outcomes that are used by policymakers in making decisions.

Cost–Benefit Analysis

In an era of increasing accountability for social programs, *cost–benefit analysis* has proven to be of considerable utility, although it must be understood well in order to avoid misusing it. **Cost–benefit analysis** involves calculating the dollar cost of a program, subtracting that from the dollar value of the benefits of the program, and producing either a net gain (benefits exceed costs) or a net loss (costs exceed benefits). Such an approach is very appealing to many policymakers because it seems to clarify complex issues and programs through quantification. They would logically support and perhaps expand programs showing a net gain and curtail those producing a net loss. If only it were that simple. As we shall see, quantifying benefits and costs can be extremely difficult and often involves a number of unproved assumptions and estimates.

Cost–benefit analysis can be applied to a program during its planning stages (called *ex ante* analysis) or after it has been in operation (*ex post* analysis) (Rossi, Freeman, & Lipsey, 1999). The major difference between the two is that an ex ante analysis requires the use of more estimates and assumptions because no hard data may exist on many of the costs or benefits. In an ex post study, there are records of most actual cost outlays, and benefits can be determined empirically through normal evaluation research procedures. The use of estimates in ex ante analyses means that results are far more tentative. This also accounts for why ex ante analyses conducted by different parties sometimes come to widely divergent conclusions: They use different estimates and assumptions regarding costs and benefits. Sorting out whose estimates are most valid has produced some lively debates among policymakers. Ex post analyses require fewer estimates and are therefore more reliable.

plemental unemployment compensation program must assume a certain unemployment rate. If the actual rate differs, the cost of the program could skyrocket or fall dramatically.

Considerably more difficult to estimate than the direct costs of programs are what economists call *opportunity costs.* **Opportunity costs** are the value of forgone opportunities. Suppose, for example, that you are fortunate enough to win $1,000 in a contest. You can invest the money and receive a monetary return or spend it on anything you like. Suppose also that you decide to spend it on a notebook computer. The direct cost of the computer is the $1,000 you spent on it, but there are also opportunity costs. The opportunity costs are what you forgo by buying the computer. You lose the return that you could have received by investing it or the value of other items you might have purchased, such as new clothes or the down payment on an automobile.

A similar situation applies to the funding of social programs. Agencies have limited resources. If they decide to fund a given program, the cost of the program includes the opportunity costs of not funding alternative programs. Normally, the estimated value of benefits of competing programs is used as the basis for computing the opportunity costs of the program being analyzed. Computing the benefits that are lost by not initiating a program can be very complex, which makes it difficult to calculate the opportunity costs of the funded program. Such estimates need to be done, however, to provide an accurate picture of the total costs of a program.

Estimating Costs

The easiest part of cost–benefit analysis—although by no means simple—is determining the **direct costs** of a program. There is either a record of actual expenditures (in ex post analysis) or a proposed budget for the program (in ex ante analysis). A budget proposal, however, is based on certain assumptions that may not be accurate. For example, the budget for a sup-

Estimating and Monetizing Benefits

The really difficult and often unreliable part of cost–benefit analysis comes in determining program benefits and monetizing—attaching a dollar value—to them. This may be either a fairly straightforward or a mystical process, depending on the program. In general, if a program's benefits are related to some economic activity, they are easier to monetize. For example, the value of subsidized day care can be monetized easily. The market price of private day care plus the added income of the parent who otherwise could not work would constitute the major dollar benefits from the program. But what about program benefits less related to economic activity? How can a dollar value be placed on such program benefits as improved mental health, improved self-esteem, reduced domestic violence, or other noneconomic outcomes? Cost–benefit analysis attempts, through complex procedures, to place a dollar value on practically anything. Doing so, however, requires many assumptions and value judgments that are likely to be controversial. Because of this, cost–benefit analysis is of the greatest utility when the relationship between program benefits and a certain dollar value is fairly clear.

Another complicating factor in cost–benefit analysis is that benefits and costs do not accrue at the same time. Most costs are incurred immediately upon the program's implementation, whereas benefits may accrue at some later date, possibly far in the future. In some programs, such as education or job training, at least part of the benefits may indeed be very long term. This temporal gap is a problem because the value both of costs and of benefits changes with time. We are well aware, for example, that a given number of dollars today does not have the same purchasing power it did 10 years ago. To make meaningful comparisons over time, we must adjust costs and benefits so that comparisons are made in constant dollars. This involves the calculation of what is called the *discount rate*. The discount rate is that amount by which future costs and benefits are reduced to make them comparable to the current value of money. Actual determination of the discount rate involves highly complex accounting procedures. There are also several competing approaches to its calculation (Rossi, Freeman, & Lipsey, 1999). As many saddened investors will attest, predicting the future value of money is a risky business. Fur-

ther, the discount rate used has a marked effect on the outcome of the analysis. For all these reasons, it is commonplace to run several analyses using differing discount rates to see how the program fares under different sets of assumptions.

Whose Costs? Whose Benefits?

An important consideration in cost–benefit analysis is that costs and benefits are calculated from particular perspectives. Three different perspectives may be used: program participants, the funding source, or society as a whole. A comprehensive cost–benefit analysis would include all three.

Let's take the example of early childhood intervention programs, which are efforts by governmental or other agencies outside the family to support and improve the quality of life for youngsters from the prenatal period to the school years. These programs focus on providing health, education, and social service interventions. The theory behind such programs is that they will provide broad benefits to society because children who are supported by them will generally be healthier, do better in school, and be less likely to create problems of crime or welfare dependence as adults. Table 12.1 lists some of the potential costs and benefits of such early intervention programs, along with an indication of which perspective might see each as a cost or benefit.

Let's look at some elements in the table. For the mothers (participants) who can work at paying jobs because of the services provided by the programs, the costs are the loss of welfare payments they would receive if they did not work. The mothers' benefits would be the income received by working (together with less tangible benefits such as enhanced social status, job satisfaction, and freedom from child care responsibilities). Participants would also benefit from any reductions in criminal activity the mother or child experience because of participation in the program.

TABLE 12.1 Perspectives on the Costs and Benefits of an Early Childhood Intervention Program

	Perspective		
Benefits and Costs	**Society**	**Participants**	**Funding Source**
Costs of home visits to children	−	+	−
Reduction in emergency room visits by child	+	+	0
Increase in taxes paid by mother because of her increased employment income	+	−	+
Decrease in cost of government welfare payments to mother	0	−	+
Decrease in mother's arrest and jail costs	+	+	0
Decrease in child's arrest costs as an adolescent	+	+	0
Decrease in child's arrest costs as an adult	+	+	0
Income from mother's increased employment	0	+	0
Decrease in welfare payments to mother	0	−	+
Decrease in losses to crime victims	+	0	0

Key: + = an expected benefit from a given perspective
 − = an expected cost from a given perspective
 0 = neither a cost nor a benefit
Adapted from Lynn A. Karoly et al., *Investing in Our Children: What We Know and Don't Know about the Costs and Benefits of Early Childhood Interventions* (Santa Monica, CA: Rand, 1998).

From the perspective of the funding source—in this case we will assume it is the federal government—the costs and benefits are quite different. The costs would be the direct costs of running the program (e.g., costs of home visits) together with the opportunity costs of not using the money for something else. The benefits would be the reduced costs of other public assistance programs and an increase in tax revenues derived from the incomes of working mothers.

The societal perspective is the broadest and frequently the most difficult to calculate. (Further, if the funding source is the government, we should not assume that the government's perspective coincides with the societal perspective. The government represents those groups that happen to control a particular government agency.) The costs to society of the early intervention program would be the increased taxes or federal borrowing necessary to fund the program, plus the opportunity costs involved. Benefits would be increased productivity of the mothers who are now freer to make economic, social, and cultural contributions to society (although we need to remember that performing as a parent or homemaker is also an essential contribution to society). Other less direct benefits might accrue if working and the additional income it provides have positive effects on family relationships, the children's well-being, future aspirations, and the like. And, of course, society as a whole benefits when costs associated with crime are reduced.

Early intervention programs involve more elements—and more costs and benefit—than listed in Table 12.1, but the table helps you understand the idea that the costs and benefits of programs need to be assessed from a variety of perspectives. People might also disagree over whether a particular element listed in Table 12.1 is a cost or a benefit from a particular perspective. The possibility of such disagreements reinforces the point that cost–benefit analysis is difficult, complicated, and often contentious. By the way, most cost–benefit analyses of these early childhood intervention programs conclude that, if well run, the programs do provide substantially more benefits than costs (Karoly, 1998).

Cost Effective Analysis

Because it is difficult to monetize benefits, interest has developed in an alternative approach that does not require that benefits be ascribed a dollar value. **Cost effective analysis** compares program costs measured in dollars to program effects measured in whichever units are appropriate, such as achievement test scores, skill performance levels, coping abilities, or whatever effect the program is supposed to produce. Such analysis is most useful for choosing among competing programs rather than evaluating a single program. For example, a cost effective appraisal of a compensatory education program might reveal that total per-pupil costs of $300 raised reading performance by one point on a standard achievement test. This can then be compared to other programs in terms of the cost to raise reading performance by one point. A cost effective comparison of remedial programs for disadvantaged children conducted by the General Accounting Office (GAO) found programs with wide ranging cost effectiveness (Socolar, 1981). The average per pupil cost of all such programs was $778. Interestingly, the GAO found that six especially effective programs cost only $180 per student. Such cost effective analyses of many competing programs make it possible to select the most efficient approach.

Interpreting a single cost effective analysis, however, is less clear-cut than a cost–benefit analysis because the costs and benefits are not expressed in the same units. For this reason, cost effective analysis is not an interchangeable substitute for cost–benefit analysis. They answer very different questions.

Although cost–benefit analysis is a valuable tool in assessing programs, it is important that it not be oversimplified or overemphasized. Results must be carefully interpreted in light of the data used to produce them. The complex components that go into a cost–benefit analysis need to be considered when interpreting the results. Like all forms of analysis, cost–benefit is only as good as the data, the estimates, and the assumptions on which it is based. All these components should be explicitly discussed in the report, and the users should be encouraged to evaluate the soundness of them. The real risk associated with cost–benefit analysis comes when bottom-line results are accepted blindly and become the overriding factor in decision making. At its current level of development, cost–benefit analysis can be very useful, but must be cautiously interpreted as only one of many factors in the decision-making process concerning social programs.

Social Indicators Research

Social indicators research refers to studies whose focus is the development of quantitative measures of important social phenomena (de Neufville, 1981). It attempts to measure quantitatively the extent of or change in some key behaviors, groups, institutions, or larger social systems. So, for example, the crime rate, as published by the Federal Bureau of Investigation, is often used as a summary measure of how well law enforcement does its job or of how safe society is. Likewise, the divorce rate is one way to measure the change over time in the social institution of the family.

The central issue in social indicators research is that of measurement and operational definitions discussed in Chapter 5 and Chapter 6. At one level, social indicators are a measure of some concrete phenomenon. The divorce rate, for example, is a measure of how many couples divorce during a given year. However, social indicators are also intended tell us something about a more abstract level of reality. The divorce rate, for example, can inform us about the state of and changes in the social institution of the family. It might even tell us something about how well society as a whole is doing. Understanding these abstract linkages, of course, depends on having a theory that posits how the concrete (divorce) ties into the more abstract (the family and society as a whole).

Developing social indicators is much like developing operational definitions. You need to begin with a clear conceptual analysis of what is to be measured. Then you need to develop valid and reliable indicators of the phenomenon. This process is described in Chapters 2, 5, and 6. Once developed, social indicators are often used in developing and changing social policies. They can be used to inform the public and the policy makers about key issues: the well-being of society, the extent of change in social structures, and so on. This information may be used to shape legislation targeting certain social ills. Social indicators can also be used in making decisions about the allocation of resources, such as which groups need which kinds of social services provided.

BARRIERS TO THE USE OF EVALUATION RESEARCH

At the beginning of this chapter, I described the purposes of evaluation research as including such things as evaluating a social program or aiding in the policy-making process. The assumption is, of course, that the research results will be used to produce some change in the status quo. In actuality, this is often not the case, since there are many barriers to the use of the results of evaluation research.

One barrier is the fault of evaluators. Owing to poor design or execution, the evaluation may not produce clear-cut results. It is obviously difficult to overcome resistance to change unless the reasons for change are strong and the direction that change should take is clear. All too often, the basic conclusion of an evaluation is: "The program as currently operated is not achieving its intended goals." What is one to do with such a conclusion? There is no indication as to why the program is failing or any suggestions for improving it. To avoid such results, I suggested earlier that evaluation be conceived broadly so that findings indicate not only what the program does not do, but also what it does do and why. Such detailed findings are of far greater utility for pointing the way for the future and for producing positive program changes (Bedell, Ward, Archer, & Stokes, 1985).

A second barrier to the use of evaluation research results is poor communication on the part of evaluators (L. P. Miller, 1987). Researchers are used to communicating with other researchers who share a common technical language and background. When communicating with one another, researchers assume those commonalities and write their reports accordingly. If this is done in an evaluation report, the results may be quite unclear to practitioners, program administrators, and policymakers who are supposed to use those results. Evaluation reports should be written so that they are understandable to the audience who will use them, and evaluators should work through the report with sponsors, explaining it thoroughly and answering all questions. Those sponsoring an evaluation or using the results should demand such accountability from evaluation researchers.

A third barrier to the use of evaluation research is the failure of evaluation researchers to press for the adoption of their research findings. Such an advocacy role is foreign to many researchers who feel that their job terminates once the data have been analyzed. However, implementation of modifications to a program is often complex, and program staff are faced with competing interests. Without active participation by the researcher, adoption of recommendations may very well not take place.

A fourth barrier is the resistance to change that can be found in many quarters. People become accustomed to established procedures and may be disinclined to change. There are vested interests that are difficult to overcome. One example of this is what happened to a study of group counseling in a correctional system (Kassebaum, Ward, & Wilner, 1971). This was a well-conducted study that not only found few positive effects of counseling but also provided many suggestions as to why these effects did not occur. The reaction to the report by the Department of Corrections was swift, but not what would have been predicted given its contents. The counseling program was not dropped but was expanded to every prison in the system, and the expanded program was not modified to take into account suggestions in the report (Ward & Kassebaum, 1972). The evaluators speculated in a rather discouraged tone that the main effect of their report would be to limit future outside evaluations of prison programs.

One change suggested to improve research utilization is increased dissemination of results (Weiss, 1998). Earlier I expressed concern over the fact that evaluation reports are often not widely circulated. Broader dissemination may bring a report to the attention of someone willing to implement the results. Evaluation research reports are disseminated in a number of ways. Sometimes they are published in social science or human service research journals, such as *Evaluation Review, Journal of Applied Behavioral Science,* or *Journal of Applied Sociology.* In the past 10 years, more outlets for the publication of evaluation research have developed. Evaluation research reports are also published in government documents and reports, especially when the study concerns a governmental program; these can be accessed like other government documents (see Appendix A).

Some evaluation reports are not published, however, and this can make them more difficult, but not impossible, to find. Even though unpublished, the results may be presented at meetings of applied research organizations, such as the Society for Applied Sociology, the NTL Institute for Applied Behavioral Science, and the National Association of Social Workers. In some cases, unpublished research reports can be obtained from the researchers themselves or from the organizations or agencies for whom the research was done. Unpublished reports can be difficult to locate because they are not included in the indexes or abstracts used in information searches during a literature review. This explains why researchers participate in a communications network with others doing research in the same area: These personal contacts are a way to learn what others are doing and obtain unpublished research reports.

Another change that has been shown to increase the use of the results of evaluations is involving the potential users in the evaluation research itself. Users can help design the research or serve as interviewers. When they do, there is better communication between evaluators and users, the users perceive the evaluation as more relevant and credible, and the users are more committed to the evaluation (Dawson & D'Amico, 1985; Greene, 1988).

What is really needed to improve research utilization is for policymakers and program administrators to develop an increased willingness to implement evaluation results. In fact, resistance to research utilization appears to be declining, and there may be a growing awareness that common sense and conventional wisdom are inadequate bases for designing and operating effective social programs. Years of experience with ineffectual programs have made this conclusion quite evident. Such changing perspectives are encouraging for the future of evaluation research and its increased utilization.

REVIEW AND CRITICAL THINKING

Chapter 12 emphasizes the importance of evaluating social policies and programs through systematic observation. If a program or practice has been around for some time, well-intentioned people may just assume that it works, or may use some kind of casual or anecdotal evidence to support their assertion. Or people may have a vested interest in believing that some social practice has positive effects. But this can lead people to miss or ignore information that doesn't support their belief. To avoid these problems, consider the following in assessing social programs or practices.

1. What are the goals, both stated and unstated, of the program or practice?

2. Can you identify what might be called independent and dependent variables? How are these variables measured? Are they valid measures? Are there other ways to measure variables? Would these other measures change the conclusions?

3. If a systematic evaluation is done, what kind of research design was used? A randomized experimental design? If not, what procedures were used to enhance confidence in the conclusions?

Computers and the Internet

As the chapter suggests, social science researchers conduct evaluation research in many different fields, such as criminal justice, education, social services, and substance abuse, to name just a few. One way to broaden your understanding of evaluation research is to review

examples of such evaluation studies in a variety of these fields, and the Internet is an excellent place to do this. If the field of substance abuse interests you, for example, you could contact the National Clearinghouse for Alcohol and Drug Information (NCADI) prevention database at **http://www.health.org/dbarea/index.htm**. NCADI offers substance abuse information via searchable databases. Click on "Prevention Materials database," and you can access each database by clicking on the button for each database at this URL and then using the search mode attached to each database. You can research these databases for bibliographic abstracts pertaining to such subjects as alcohol, tobacco, marijuana, cocaine, and other drugs.

The Justice Information Center, operated by the National Criminal Justice Reference Service and located at **http://www.ncjrs.org/homepage.htm**, provides access not only to bibliographic references but also to the full text versions of a wide range of evaluation projects in criminal justice. By selecting "Research and Evaluation" from among the menu options, you can review the studies and learn about the actual methodology used to evaluate the projects. One of the pages associated with this site is devoted to evaluation grants (**http://www.ncjrs.org/fedgrant.htm**). Here you can review grant application instructions. I point out in the chapter that evaluation research differs from basic research in the role of the decision makers in program evaluation. I suggest that you review the specifications for one of the program evaluation projects with an eye toward how the sponsoring organization's needs affect the design of the research project. The Bureau of Justice Assistance also maintains a Web site with many resources and Web links related to evaluating criminal justice programs: **http://www.bja.evaluationwebsite.org/**.

An excellent source of program evaluation related resources in the field of education is the ERIC Clearinghouse on Assessment and Evaluation (**http://www.ericae.net**). In addition to directing you to various evaluation studies, this site provides links to assessment tools and professional organizations involved in evaluation. One such site worth mentioning in its own right is the American Evaluation Association (AEA), whose home page is located at **http://www.eval.org/**. The AEA is an international professional association of evaluators devoted to the application and exploration of program evaluation, personnel evaluation, technology, and many other forms of evaluation. This organization maintains a current listing of Internet sites of interest to program evaluators, so whether you are interested in health, education, or any other human service area, this site will help you to locate additional resources.

Main Points

1. Evaluation research refers to the use of scientific research methods to plan and assess social policies, social interventions, and social programs. The most common form of evaluation research is program evaluation. Evaluation research has some similarities to basic research, but there are also some clear differences between the two.

2. Social interventions have to do with politics, or the control and distribution of societal resources. This means that there are a variety of stakeholders in social programs. Evaluation researchers recognize that particular research studies might support the perspective of some stakeholders but not others.

3. Formative program evaluation focuses on using social science research to plan, develop, and implement social programs. Needs assessments are often used to do this. Summative program evaluation assesses the effectiveness and efficiency of programs and the extent to which program effects may be generalized to other settings and populations.

4. Prior to beginning a program evaluation, an evaluability assessment is conducted to gain knowledge about the program as operated and to identify those aspects of it that can be evaluated. This involves determining the program inputs and goals and finding appropriate measures for these. Proximate goals are short-run goals logically related to the achievement of long-term distal goals that may take many years to evaluate.

5. Some evaluation researchers, especially positivists, support randomized experiments as the strongest, most desirable designs for assessing program impact. When randomized experiments are not feasible, evaluators turn to such alternatives as quasi-experimental designs, matching, cohort groups, regression discontinuity designs, and statistical controls.

6. Nonpositivist approaches to evaluation, recognizing that there are multiple stakeholders in social programs and that interpretation and meaning are at the core of social interventions, argue that qualitative research is often more useful in program evaluations than quantitative, randomized experiments.

7. Other evaluation research techniques include social impact assessments, cost-benefit analysis, cost effective analysis, and social indicators research.

8. A number of barriers can arise that deflect people away from actually using the results of evaluation research. However, steps have been and are being taken to reduce those barriers.

Important Terms for Review

cost–benefit analysis
cost effective analysis
direct costs
evaluability assessment
evaluation research
formative program evaluation
needs assessment
opportunity costs
program evaluation
social impact assessment
social indicators research
stakeholders
summative program evaluation

For Further Reading

Boruch, Robert. (1996). *Randomized experiments for planning and evaluation.* Thousand Oaks, CA: Sage. The author stresses the point made in this chapter that good program evaluations should be based on randomized experimental designs. The book is packed with useful examples of how to do this, often in situations where it might seem impossible.

Campbell, Donald T., & Russo, M. Jean. (1999). *Social experimentation.* Thousand Oaks, CA: Sage. Donald Campbell, now deceased, is one of the giants in the fields of experimentation and evaluation research. This book presents his approaches to how social experiments should be designed and how they can improve society.

Carley, Michael. (1983). *Social measurement and social indicators: Issues of policy and theory.* Boston: Allen & Unwin. This book presents an in-depth look at various social indicators that may be monitored to assess aspects of social change. These indicators may be useful in measuring the impact of social programs for evaluation purposes.

Guba, Egon G., & Lincoln, Yvonna S. (1989). *Fourth generation evaluation.* Newbury Park, CA: Sage. This book focuses mostly on program evaluation in educational settings. It challenges conventional ways of thinking in that it takes a nonpositivist approach, viewing all knowledge as a social construction. Even scientific knowledge is seen as socially constructed, as only one version of reality.

Krause, Daniel R. (1995). *Effective program evaluation: An introduction.* Chicago: Nelson-Hall. A very readable textbook that covers all the basics of conducting a program evaluation, this book also includes some selected readings that help facilitate the research process.

Nas, Tevfik F. (1996). *Cost–benefit analysis: Theory and application.* Thousand Oaks, CA: Sage. The author shows, in greater detail than could be included in this chapter, how to conduct good cost–benefit analyses. The book covers all the relevant issues.

Patton, Michael Quinn. (1996). *Utilization-focused evaluation* (3rd ed.). Thousand Oaks, CA: Sage. This book considers both theoretical and practical issues in conducting program evaluations. It also uses a case study approach and stresses the perspective that good program evaluations must be designed to be useful to program managers and policymakers.

Shadish, William R. Jr., Cook, Thomas D., & Leviton, Laura C. (1993). *Foundations of program evaluation.* Newbury Park, CA: Sage. This volume looks at the origins of program evaluation and at the accumulated experiences of veteran program evaluators to provide an insightful discussion of the development of the field and of key issues that are relevant today.

Part IV

DATA PREPARATION AND ANALYSIS

CHAPTER 13:
Data Preparation and Presentation

CHAPTER 14:
Descriptive and Inferential Statistics

CHAPTER 15:
Analysis of Qualitative Data

The three chapters in this part of the book focus on data analysis, which refers to deriving some meaning from the observations that were made during the research process. The first two chapters address issues of quantitative data analysis, which means that the observations are in some numerical format and manipulated in terms of their mathematical properties. Chapter 13 focuses on preparing quantitative data for analysis, including entering the data into a computer file. This chapter also addresses how to present data to show conclusions, especially in terms of various kinds of graphs, charts, and tables. Chapter 14 deals with more sophisticated data analysis, where various kinds of statistics are used to summarize data or to draw conclusions about populations from data collected on samples drawn from those populations. Chapter 15 addresses the analysis of qualitative data, showing how meaning can be extracted from data in the form of narratives and descriptions without necessarily quantifying the data.

Chapter 13

DATA PREPARATION AND PRESENTATION

All research involves some form of **data analysis**, which refers to deriving some meaning from the observations that have been made during a research project. Data analysis can take many forms. In some cases, it is qualitative, such as a summary description of an investigator's field notes from a participant-observer study. Our focus in this chapter, however, is with quantitative data analysis, in which observations are put into numerical form and manipulated in some way based on their arithmetical properties. The analysis of quantitative data typically involves the use of **statistics**, which are a set of mathematical procedures for assembling, classifying, analyzing, and summarizing numerical data so that some meaning or information is obtained.

This chapter begins our coverage of common data-analysis methods, which is continued in Chapter 14 and Chapter 15. Primarily designed for those who have not taken a course in statistics, these chapters may also serve as a refresher for those who have. Actually, to learn how to do statistical analysis would require at least a full course devoted solely to that topic. However, my goal is more basic: to acquaint you with the issues of preparing a data set for analysis, to help you select the most appropriate statistics to accomplish a particular task, and to provide guidance in the interpretation of statistical results. This will prepare you to better understand various statistics when you encounter them in research reports and the popular media and also to assess whether the statistics are being used properly. This chapter explores some of the fundamentals of preparing and managing data. After discussing how observations are put in a form that makes it possible to perform statistical procedures on them, I examine initial steps in describing and displaying the data and in exploring relationships between variables through the use of graphs, charts, and tables. Chapter 14 follows on with a dis-

cussion of statistical procedures for detecting and estimating relationships between variables.

PREPARATION FOR DATA ANALYSIS

Imagine that you have completed a survey of sample size 400. The completed questionnaires are neatly stacked on your desk. The questionnaires contain the data for assessing the hypotheses you set out to test with the survey. But as long as the data are on the questionnaires, or in any other raw form for that matter, they are useless. In such a state, the data are in a highly inconvenient form from the standpoint of deriving any usable meaning from them. Indeed, even if we tediously read through the 400 questionnaires, we would have little idea of the overall contents. The data were collected in an organized way that made sense in terms of the collection process, but this format is inconvenient for drawing conclusions. The collected data must be reorganized before we can apply the necessary statistics or derive some other meaning from them.

A common first step is to enter the data into a computer file or into what is called a *data file* or *data set*. Recall from Chapter 1 that data analysis is part of the overall research process and that many questions regarding data analysis should be resolved before any data are actually collected. As emphasized in Chapters 2, 4, and 5, theoretical and conceptual considerations are important in determining the nature of the data to be collected, and the nature of the data determines the kinds of statistics that can be applied. Further, as discussed in Chapter 9 in relation to survey data collection, procedures such as computer-assisted telephone interviewing (CATI) essentially merge the steps of data collection, data coding, and data set creation into one process because responses of participants are entered directly into a computerized data file. So even though the actual data analysis occurs toward the end of a research project, many of the issues discussed in this chapter will have been addressed, or at least envisioned, *before* any data are actually collected.

Coding Schemes

As explained in Chapter 10 and Chapter 11, **coding** refers to categorizing a variable into a limited number of categories. In the current context, coding refers to the process by which a researcher transforms raw data into a format suitable for entry into a computerized data file, and this requires that each observation be translated into a numerical value or set of letters. The numbers or letters assigned to an observation of a variable are called a *code*. The raw data set may have resulted from an observational study, a survey, an experiment, or an analysis of available data and can be in one of several forms, including completed written questionnaires, survey interview schedules, observation notes, or organizational records. In the following discussion, I use survey data to illustrate the coding process, but the basic principles apply to other research methods as well. The top part of Figure 13.1 shows what part of a completed survey form might look like.

Coding data requires adapting the data as collected to the constraints of the computer software used for statistical analysis. On the one hand, the researcher must consider the data source, such as a completed questionnaire, and determine how the data can be translated into a coded form. On the other hand, the researcher must also be cognizant of the capabilities and restrictions of the computer software that will be used to analyze the data. Although I present these perspectives somewhat independently, in practice the researcher must give simultaneous attention to both. The plan by which the researcher organizes responses to a variable or an item, together with how the variable is defined for

computerization, constitutes the **coding scheme** for a variable. The initial purpose of the coding scheme is to provide the rules and directions for converting the observations into code. Referring to our illustration of the 400 completed questionnaires stacked on your desk, the first task is to convert the responses into numbers and letters for computer entry.

Sometimes the coding of a variable is built into the way a question is asked and answered (see Chapter 6). This is the case with the following question format.

In which of these groups did your earnings from (OCCUPATION IN PREVIOUS QUESTION) for last year—1999—fall? That is, before taxes or other deductions. Just circle the number.

1. Under $5,000
2. $5,000 to 9,999
3. $10,000 to 14,999
4. $15,000 to 19,999
5. $20,000 to 29,999
6. $30,000 to 39,999
7. $40,000 to 49,999
8. $50,000 or over
9. Don't know

In this case, the circled number would be the code for how a particular respondent answered a question. This example also illustrates one reason why coding and data analysis issues need to be considered before data collection is initiated. The number of options in the question predetermines the maximum number of response categories that can be used later in looking at possible relationships between *income* and other variables. Because we did not provide an option for $50,000 to 69,999, we could not use this as a category in our data analysis.

With some variables, the actual value of the response is a number and can be used to code the data. *Family size, income,* and *number of arrests* are variables of this type (see Figure 13.1). However, when the data take the form

FIGURE 13.1 Data from a Survey Form Transferred to a Data Entry Screen of the Kind Commonly Used in Windows Versions of Statistical Software

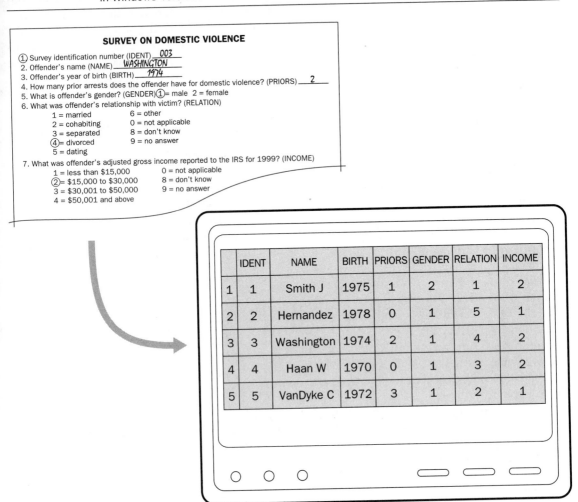

SURVEY ON DOMESTIC VIOLENCE

① Survey identification number (IDENT) __003__
2. Offender's name (NAME) __WASHINGTON__
3. Offender's year of birth (BIRTH) __1974__
4. How many prior arrests does the offender have for domestic violence? (PRIORS) __2__
5. What is offender's gender? (GENDER)①= male 2 = female
6. What was offender's relationship with victim? (RELATION)

    1 = married   6 = other
    2 = cohabiting   0 = not applicable
    3 = separated   8 = don't know
    ④= divorced   9 = no answer
    5 = dating

7. What was offender's adjusted gross income reported to the IRS for 1999? (INCOME)

    1 = less than $15,000   0 = not applicable
    ②= $15,000 to $30,000   8 = don't know
    3 = $30,001 to $50,000   9 = no answer
    4 = $50,001 and above

	IDENT	NAME	BIRTH	PRIORS	GENDER	RELATION	INCOME
1	1	Smith J	1975	1	2	1	2
2	2	Hernandez	1978	0	1	5	1
3	3	Washington	1974	2	1	4	2
4	4	Haan W	1970	0	1	3	2
5	5	VanDyke C	1972	3	1	2	1

of responses to open-ended survey questions, field notes, or other nonnumerical entities, for quantitative analysis, the data must be translated into numbers. This process is essential for most data analysis because the substitution of numbers for observations greatly reduces the volume of data that must be stored and facilitates its analysis, especially by computer. (As noted in Chapter 11, qualitative analysis often does not involve the substitution of numbers for the actual observations; qualitative data analysis is discussed in Chapter 15.) Many computer programs permit you to use words or letters to represent categories, but these are generally more cumbersome and only infrequently used. (Recall from Chapter 5, however, that assigning a number to coding categories does not necessarily mean that we can perform on them the various mathematical functions, such as addition and subtraction. Whether we can do this depends on the level of measurement of the variable.)

When coding categories are being established, two general rules should be followed. First, the coding categories should be *mutually exclusive*; that is, a given observation is coded into one and only one category for each variable. The universal practice of categorizing persons by sex as either male or female exemplifies mutually exclusive categories. Second, the coding categories should be *exhaustive*, which means that a coding category exists for every possible observation that was made. For example, it might be tempting not to bother to code the "no opinion" response with Likert-type questions, on the grounds that you will probably not include those responses in the data analysis. You might very well decide not to analyze those responses in the end, but the coding stage is not the time to make such decisions. If you fail to code a response and later decide to include it in the analysis, it would then be necessary to develop a new coding scheme and reenter the data into the computer. A good rule of thumb when coding is to do it in a way such that *all* information is coded. Coding is a difficult and tedious task, but it is critical that the job be done well. Searching out errors after the fact is often time consuming and expensive. Unless caught and rectified, coding errors can ruin the most carefully collected data set by causing distortions in relationships and resulting in meaningless or misleading results.

Computerizing a Dataset

Today, virtually all quantitative, and much qualitative, data analysis is done via computer. This means that not only must the coding scheme specify logically meaningful categories, as discussed thus far, but the coding system must also provide specifications required by the computer program in order to use the data once they are entered into a computer file. These specifications include attributes for each variable that permit the computer program to display the data, to use the data in calculations, and to record the data in various files for storage and retrieval.

In its most basic form, a computerized dataset can be conceptualized as a rectangular table of cells where the columns represent variables and the rows designate individual cases. Figure 13.1 illustrates this with a portion of a simulated data set showing how the data might appear on a survey form and then how that same data would appear on a typical computer program data entry screen. You can see how each item on the survey form represents a variable and the data is located in the corresponding column on the computer screen. The first column of numbers on the left of the screen is simply the numbers identifying each row on the data entry screen and is not actually a part of the data set. The second column, under "ident," consists of a unique identification number that is assigned to each case. Although most social science data analysis is done in the aggregate and thus without a need to identify particular cases, such identification is occasionally needed and so a case identification number is usually used. The third column displays the name of each person from whom data was collected. Most social science data files would not include the name because the data will be analyzed in the aggregate and the identification number will serve to identify individuals without creating a threat to confidentiality. In fact, having both name and identification number in the data set is largely redundant. However, names can be included if that information might be important in the analysis, and I wanted to demonstrate that a computerized data set can include letters as well as numbers. The remaining five columns in Figure 13.1 contain the data for five variables in the data set: *birth year, prior arrests, gender, relationship to victim,* and *annual income.* Note that the top of each column displays the variable name (IDENT, NAME, BIRTH, and so on). So, the third case in the data file is a person named Washington, who was born in 1974, had two prior arrests, and so on. The variable names and other variable attributes, together with information about the position of the variable in the dataset, are also vital elements of the coding scheme for computerized data.

For variable names, computer program data entry screens typically use a default designation, such as C1, C2, C3 (in the case of Minitab) or VAR001, VAR002, VAR003 (in SPSS) to identify variables. Obviously, such designations convey little information about the nature of a particular variable. So researchers normally provide more recognizable variable names. Variable names should be brief in order to minimize entering long lists of variable names when conducting statistical analysis. In addition, most statistical programs limit the number of characters that can be used in a variable name, such as a maximum of eight characters. Although brief, a variable name should indicate the content of that variable. A common practice is to use a mnemonic device when naming variables. Thus BIRTH and PRIORS clue the reader that the variable names represent "when that person was born" and "prior arrests," respectively. Variable names may also be selected to help group together variables. For example, a dataset on domestic violence may include similar variables on both the suspect and the victim, such as date of birth, age, and alcohol use. Beginning each variable related to victims with the letter *v* and each suspect-related variable with an *s* helps the researcher quickly distinguish which party a given variable refers to. When a variable is measured by combining people's responses to a number of separate items, it is often useful to use a number as part of the name to identify the variable's position in the set. The 10-item self-esteem scale in Table 6.6, page 161, for example, would be entered into a data file as a separate variable for each item, with each designated as SE1, SE2, SE3, and so on through SE10. Such a designation can reduce the amount of work in combining the items into a single score. Thus, a total score on self-esteem can be generated by a computer command like "Sum SE1 to SE10" instead of requiring a separate listing of all 10 variable names.

In addition to the variable name, a coding scheme also specifies the *variable format*. For example, the variable format tells whether the variable is numeric or nonnumeric in nature. Figure 13.1 displays several types of variables; some consist of letters and others are only numbers. **Numeric variables** are sometimes called "values" because they have the property of a quantitative value. **Alphanumeric variables,** on the other hand, consist simply of letter characters; they are sometimes also referred to as "string" variables. An alphanumeric variable has no quantitative meaning and cannot be used in mathematical computations. Thus, the variable NAME is an alphanumeric variable because its field (column) consists of text, such as "Smith" and "Hernandez." However, what makes a variable alphanumeric is determined by how the variable is defined in the coding scheme and not simply by the characters displayed on the computer monitor. For example, we could enter Social Security numbers in a dataset and designate these as alphanumeric values. Even though the entire column consists of numerical characters, such as 375426174, a statistical program such as SPSS would not compute any statistics on this variable because it is designated as an alphanumeric or string variable.

Additional elements may be added to a coding scheme to enhance its usefulness. Besides a brief variable name, the researcher may specify a variable label, which is an extended description of a variable that the data analysis program will print in addition to the variable name whenever output is generated for that variable. Thus, in a table using the variable PRIORS, the designation "Number of prior arrests" can be printed to make the table more understandable. Similarly, the coding scheme may include "value labels" that designate in words what a given value represents. For example, the variable RELATION, which is the offender's relationship to the victim in the spouse abuse study, was coded using the following categories: 1 = married, 2 = cohabitating, 3 = separated, 4 = divorced, 5 = dating, and 6 = other. Because this is a nominal variable and thus without an inherent quantitative meaning to the numbers, the value labels are printed on outputs to help make it clear what quality each number represents.

Finally, the coding scheme should also show how to interpret special codes. For example, missing values arise when there is no response recorded for an item for a particular case. So special codes might include 9 if the respondent refused or neglected to answer a question, 8 if he or she didn't know the answer to a question, or 0 if the item does not apply to a particular respondent (see Figure 13.1). For a two-digit variable such as *age*, the corresponding codes would be 99, 98, and 00. The reason data are missing can influence how the data analysis is done, so the coding scheme must be clear on this. In addition, you must not use codes for these special purposes that can be actual values on a variable. With the variable PRIORS, for example, 0 is a meaningful category since a person could have no prior arrests; so this value should not be used to indicate "not applicable."

Preparing and Using a Codebook

Researchers commonly develop a **codebook** for the data set, which is an inventory of all the individual items in the data collection instrument together with the variable names, variable labels, special codes, and all the other information necessary to understand the coding scheme. Table 13.1 shows what a codebook for the data file in Figure 13.1 might look like.

TABLE 13.1   A Codebook for a Study of Domestic Violence

Item	Variable Name	Variable Label	Value Labels	Position
1. Identification number	IDENT	identification number		1
2. Offender's name	NAME	offender's name		2
3. Offender's year of birth	BIRTH	year of birth	(actual year)	3
4. How many prior arrests for domestic violence?	PRIORS	number of prior arrests	(actual number)	4
5. What is offender's gender?	GENDER	offender's gender	1 = male 2 = female	5
6. What is offender's relationship with victim?	RELATION	relation to victim	1 = married 2 = cohabiting 3 = separated 4 = divorced 5 = dating 6 = other 0 = not applicable 8 = don't know 9 = no answer	6
7. What was offender's adjusted gross income reported to the IRS for 2000?	INCOME	offender's 2000 income	1 = less than $15,000 2 = $15,000 to $30,000 3 = $30,001 to $50,000 4 = $50,001 and above 0 = not applicable 8 = don't know 9 = no answer	7

Most statistical computer packages today have a built-in codebook that contains all the information and can be accessed for each variable. Normally a hard copy of the codebook is also available for use. I stated that the initial use of the coding scheme is to guide the process of converting responses into codes. But the need for a codebook does not end once the data are entered into a computer file. The codebook provides permanent documentation on how the dataset is constructed. Any time there is a question about how a variable is constructed or how it can be used in analysis, the researcher can turn to the codebook for help. Not only does the researcher who collected the data need the codebook, but without such documentation, it would be impossible for someone else to conduct a follow-up analysis on a previously collected dataset or to explore new hypotheses with existing data.

The preliminary codebook is normally prepared before data are entered into the computer file, and is used as a guide for data entry. However, the codebook is not a static document. As I will describe shortly, researchers often recode original variables and compute new ones. It is critical that these manipulations of the original data also be documented in the codebook so that any researcher can determine how these new variables were derived, what they mean, and where they are located in the data set.

DATA ENTRY

The goal of the data entry process is to produce a complete error-free data set, that the data analysis software program can access and process. Data entry is a critical part of the research process, and the researcher needs to take steps to ensure that it proceeds with accuracy and efficiency. Significant strides have been made in recent years both to minimize data entry error and to increase the speed of transforming raw data into a data set that is ready for analysis.

Raw Data Entry

When the raw data must be extracted from existing documents such as court records, or printed questionnaires or survey schedules, the data must be entered into the computer manually. Several options exist for this process. Sometimes the data are taken from the records or the questionnaires and coded onto pages known as "transfer sheets," consisting of numbered rows and columns that look much like the computer screen in Figure 13.1. In other cases, questionnaires have space along the border for writing the code, referred to as "edge coding."

Whether one uses a transfer sheet, edge coding, or simply the raw questionnaires themselves, a common way to enter a data set is to use the data entry facility of the statistical software that will eventually be used to analyze the data. Modern statistical software packages for the personal computer, such as SPSS and Minitab, include full-featured data entry facilities as part of the package. The data entry facilities look and work much like common spreadsheet programs such as Excel, Lotus, and Quattro. Conventions and individual features vary from program to program, but the basic process is similar. The data entry screen is laid out in columns and rows, as shown in Figure 13.1. Each column represents one variable, and each row represents one case. Assuming that the variable names, labels, and other specifications have already been entered during the preparation of the codebook, the value for each variable is simply typed into the respective cell in the row corresponding to the individual case. Data entry can then proceed by entering all the data for one variable (column) at a time or all the data for each case (row) at a time. Errors can be corrected by moving the cursor with mouse or arrow keys to the cell in question and reentering the correct code.

Another way to enter a data set into a computer file is with spreadsheet or database programs. You might do this if you do not have immediate access to the data analysis package itself. Or an organization may have collected the data and wants it entered into a spreadsheet or a database for its own use. Such spreadsheets as Lotus, Excel, and Quattro have made popular the row-by-column cell display now used by statistical packages discussed earlier. Major statistical packages can read data files stored in these spreadsheet programs. So even though you may not have the latest version of SPSS or Minitab installed on your own personal computer, you can basically prepare a dataset on your spreadsheet program, save it on disk, download it into the statistical package, and begin analysis.

Database programs such as Access and Paradox have some special advantages as data entry vehicles. For example, some of them have form design features that make data entry speedy and more accurate. Basically, the researcher prepares a data entry form that looks very similar to the actual questionnaire. For complex items, the form may also provide coding instructions for the person entering the data. Instead of staring at an endless sea of rows and columns, the data entry person views a form with blank fields in which data may be entered. The actual data are stored in a standard file structure, but the form helps guide the data entry process. The form can even be programmed to follow the skip pattern of the questionnaire (see Chapter 9), eliminating the need to make decisions about where to enter the next response.

Instead of making data entry a distinct step in the research process, it may be combined with data collection. One way of doing this, as described earlier, is to use a data entry form prepared with a database as the actual questionnaire. Either the interviewer or the respondent views the items on the computer monitor and enters the responses. For studies consisting primarily of closed-ended questions, recording the responses directly into the computer program's data set can reduce error and increase the speed of data entry.

Another innovation in data entry is the use of optical scanning technology. A data collection instrument is designed using a special program. Respondents may either mark a circle corresponding to their choice or, in more sophisticated programs, write in a response. The questionnaire is then read by a scanning instrument or faxed into the computer and the coded values are entered directly into the database for analysis. Whether one uses a specialized data entry such as this or relies on manual entry into the data analysis program depends on several factors. The specialized equipment is cost effective for large projects where a staff of data entry personnel would otherwise need to be employed. On the other hand, designing, setting up, and testing an automated data entry program would be prohibitively expensive for a small research project. Researchers today have many options to choose from, and the most important issue is to select the process that will yield an accurate, usable data set within financial constraints.

Data Cleaning

No matter how much care one takes during the data-entry process, errors can be expected. Errors include such things as skipping variables for certain cases, entering the wrong value, entering the value in the wrong column, or entering an alphanumeric character in a numeric variable column. If uncorrected, some errors might cause the analysis program to abort a statistical procedure, or at least may seriously distort the findings of the analysis. So, before analysis begins, it is highly recommended that the researcher examine the data set carefully and make any necessary corrections. Although no system is foolproof, researchers have developed a variety of techniques for locating errors.

If the dataset is small, you may be able to detect some errors simply by scanning the data using the data editor screen. Scrolling through the rows and columns of data on the screen can turn up some obvious errors, such as blank cells, unusually large numbers, or stray alphanumeric characters. However, it is best to rely on a more systematic approach. One simple technique with a statistical package, spreadsheet, or database is to use a sort procedure. Sorting rearranges the order of all cases in the data set on the basis of the values of the variable selected as the sort key. Alphanumeric variables are sorted alphabetically, and numerical variables are sorted in numerical order. By scanning the sorted data on the monitor, one can detect misspellings and missing or out-of-range values. For example, if the variable used in the sort procedure is a scale item where expected values are whole numbers from 1 to 5, blank entries and entries of 0, 6 or above, or anything with a decimal place must be an error. At the top of the list would be 0s, decimals such as 1.5 would show up between 1 and 2, and values of 6 or more would be at the end of the list. This can be a cumbersome process with large data sets, but if there are only 50 or so variables, sorting the data works well. With a statistical package such as SPSS, another option is to use a "Frequencies" procedure on all variables. This generates a list of every value that actually occurs in the data set for each variable in ascending order and the number of cases for each respective value. This aids in detecting variables with out-of-range values or values that should not be present.

Having identified a variable with one or more suspect cases, the researcher can use a conditional selection process to find which case or cases has the wrong values assigned. The exact command differs from program to program but, basically, you enter a command to select those cases for which the value for the variable in question meets a certain condition. For example, if no one in the data set should have an age greater than 18, you execute a command to select all cases where the variable AGE is greater than 18. You then include a second command to list the identification numbers or the last and first names of all cases meeting that condition. (This data cleaning is one of the reasons why a researcher might have to identify individual cases and why I recommended earlier that all data files include an identification number for each case.) Armed with this information, you can scroll through

the data set to the identified cases and correct them. You may edit the entry by replacing it with the correct value, entering a missing value code if the correct value is not available, or, in really serious cases, eliminating entirely the whole case.

The techniques covered thus far will help locate data entries that are too large or too small, but sometimes the data are within the acceptable range, yet wrong for that case. Some of these errors can be detected by looking for logical inconsistencies. For example, if the value in a month variable is April, June, September, or November, then the value in a day variable associated with that month should be no greater than 30. If the variable entry for the number of adults in the household is 2, and the number of children is 4, then the entry for total household size should be 6. Depending on the software program, it is possible to build queries that detect cases where such inconsistencies are present. Another rather expensive approach to data cleaning is to have the data input twice and compare the two sets of entries for differences. This approach is based on the assumption that an error is unlikely to be made in the identical way twice. Any differences between the two data files must reflect an error in data entry.

Data Transformation

When all the data have been entered and checked for error, one more task remains before proceeding to data analysis. The variables as recorded may not be in the final form for the desired analysis procedures, and the researcher may need to do what is called **data transformation** by modifying existing variables or generating new ones. Such data manipulation may sound unusual but, as you will see, it is a legitimate and a necessary part of the research process. Here is one example of how and why data transformation is done. Recall from Chapter 6 that, when using multiple-item scales to measure variables, some scale items are stated in positive terms and other items are stated in negative terms to avoid problems such as response set. With a 10-item Likert scale of self-esteem (see Table 6.6, page 161), for instance, with choices ranging from 1 to 4, half of the items would be stated such that a positive response ("strongly agree" or "1") would indicate high self-esteem. The other half of the items would be stated such that a negative response ("strongly disagree" or "4") would indicate high self-esteem. To conduct statistical analysis on these items, some need to be reverse scored so that a score of 1 on all items signifies high self-esteem and a score of 4 signifies low self-esteem. This can be accomplished by use of a *recode* procedure in statistical programs. A recode procedure specifies that the existing values in a given variable are to be changed to some specific new values. In our example, for the five negative items, we would change the values as follows.

Old Item Value		New Item Value
1	$\longrightarrow$	4
2	$\longrightarrow$	3
3	$\longrightarrow$	2
4	$\longrightarrow$	1

Having recoded the negative items, we now have 10 consistent items where low values indicate high self-esteem and high values indicate low self-esteem. However, what we really need is a total scale score on *self-esteem*. This can be obtained by using another data transformation procedure in the data analysis program. A *compute* procedure creates a new variable by performing mathematical computations using one or more existing variables. Depending on the conventions of the particular software, we would enter a command like:

Let SELFESTEEM = sum(SE1 to SE10)

A new variable called SELFESTEEM, which is the total scale score, is created. Its value is computed for each case by summing all 10 individual items that make up the self-esteem scale (SE1 through SE10). The new variable is added to the data file. As with the existing variables, the researcher can attach a variable label and value labels to such newly created variables. Whenever the researcher modifies the data set, whether by recoding or computing new variables, the steps used in the data manipulation must be recorded and added to the codebook. Documenting the transformations this way assures that any researcher using the data file will be able to confirm how new variables were generated. As we shall see, recoding and computing new variables is an essential step in the process of unlocking the findings contained in the data set. There are many other types of data transformation that are done, but this illustration of recoding and computing shows generally how and why the procedure is done.

DATA DISTRIBUTIONS

Once the data have been stored in a computer file, systematically inspected for error, cleaned, and possibly revised through recoding and variable creation, we are ready to begin actual data analysis. *Data analysis* is the process of seeking out patterns within individual variables and in relationships between variables. The term *univariate analysis* refers to the process of describing individual variables. **Univariate statistics** are statistics that describe the distribution of a single variable. Even though our ultimate goal in a research project may be to determine how two or more variables are related, the process of describing the data by specifying the characteristics of individual variables is critical to a research project. Learning how individual variables are distributed can help determine which variables to use in studying relationships and which data analysis procedures we should use.

Frequency Distributions

One of the first steps usually taken with a data set is to look at the range of values for each variable. To accomplish this, some type of *frequency distribution* is constructed. When data are collected, they are in the form of a *raw data distribution*, which means that the distribution contains all the different values of

TABLE 13.2 Two Types of Data Distributions of Patients' Ages

A. Raw Data Distribution

98	82	78	70	68	60
96	81	76	70	68	60
95	80	76	70	68	59
92	80	75	70	68	59
91	79	74	70	68	59
89	79	73	70	68	58
87	79	72	70	67	58
85	79	72	70	66	56
85	79	71	70	66	54
84	79	71	69	66	52
82	79	71	69	64	51
82	79	71	69	63	50

Total *N* = 71

B. Simple Frequency Distribution

X	f	X	f	X	f
98	1	76	2	59	3
96	1	75	1	58	1
95	1	74	1	56	1
92	1	73	1	54	1
91	1	72	2	52	1
89	1	71	4	51	1
87	1	70	9	50	1
85	2	69	3		
84	1	68	5		
82	3	67	1		
81	1	66	3		
80	2	64	1		
79	8	63	2		
78	1	60	2		

all the cases that were observed on a variable. Table 13.2A displays a raw data distribution of the ages of a sample of 71 residents in a veterans' facility.

Simple Frequency Distributions

A first step in data analysis might be to construct a *simple frequency distribution*, in which each value of a variable is listed only once, along with the number of cases that have that value. Table 13.2B shows a simple frequency distribution of the ages of the people in the veterans' facility. The X column refers to the values or categories of the variable, while the *f* column indicates the frequency, or number of cases, that have each value.

Simple frequency distributions can be constructed on variables at any level of measurement. In Table 13.2 note how much easier it is to gain some sense of the age distribution of the residents from the simple frequency distribution in comparison to the raw data distribution. Of particular interest is the *shape* of a frequency distribution. A distribution's shape derives from the pattern the frequencies produce among the various categories of the variable. A number of terms are used to describe the shapes of distributions. First, distributions may be symmetrical or asymmetrical (see Figure 13.2). *Symmetrical distributions* are balanced, with one-half of the distribution being a mirror image of the other half. In reality, most distributions only approach perfect symmetry. *Asymmetrical distributions* have cases bunched toward one end of the scale, with a long "tail" caused by a small number of extreme cases trailing off in the other direction. Asymmetrical distributions are said to be *skewed*, with positively skewed distributions having long tails extending in the direction of the higher values and negatively skewed distributions having tails going in the direction of lower values. (Note that while frequency distributions can be constructed on variables at any level of measurement, the concepts of positive and negative skewness apply only to distributions of data of ordinal level or higher. Because the categories of nominal data have no inherent order, the shape of a distribution of nominal data is purely an arbitrary matter of how one chooses to arrange the categories.) Determining the

FIGURE 13.2   Symmetrical and Asymmetrical Distribution

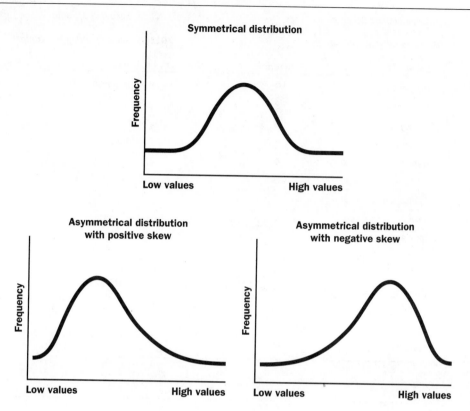

amount of skewness is an important preliminary step for later analysis. Certain statistics, for example, are based on the assumption that the variables are normally distributed, and a normal distribution is symmetrical. Other statistics give a distorted impression if they are calculated on a highly skewed distribution. Our analysis at this stage may therefore be critical for helping to decide later which statistical procedures are appropriate. These issues are addressed in greater detail in Chapter 14 when specific statistical procedures are discussed.

Simple frequency distributions, while useful in some cases, can be too cumbersome in other cases. In Table 13.2B, for example, there are still a lot of categories to look at, and many categories have no cases or only a few cases in them. Searching for patterns in data like this is akin to the old adage of not being able to see the forest for the trees. The large number of observations, the small variations between some observations, and large gaps between others combine to hide the pattern inherent in the data. In such a situation, it is often preferable to collapse a distribution further by creating a grouped frequency distribution.

Grouped Frequency Distributions

Grouping is the process of combining a large number of individual variable categories into a smaller number of larger categories. It permits us to step back from the data and see the big picture by ignoring minor variations between cases and focusing attention on larger patterns. One reason for grouping is to summarize the data into a manageable number of categories;

TABLE 13.3   Illustration of Grouping a Nominal Variable

Original Code Scheme			Recoding Scheme		
RELIGION religious affiliation			RELIG2 religious affiliation groups		
Value	Label	Frequency	Value	Label	Frequency
1	Roman Catholic	10	1	Roman Catholic	10
2	Methodist	3	2	Protestant	20
3	Presbyterian	2	3	Jewish	10
4	Lutheran	5	4	Other	5
5	Baptist	4			
6	Unitarian	3			
7	Episcopalian	3			
8	Jewish	10			
9	Moslem	1			
10	Eastern Orthodox	0			
11	Hindu	1			
12	Buddhist	1			
13	Shintoist	0			
14	Atheist	2			
15	Other	0			

Total $N = 45$

a table or graph may be unreadable if it contains too many categories. Another purpose for grouping would be to eliminate categories that have no or very few cases in them. In addition to making it easier to see patterns in the data, grouping is often necessary for certain data analysis procedures. For example, a statistical computation may require that there be enough cases so that each cell in a table could theoretically have at least five cases in it. Grouping may help meet this requirement because reducing the number of categories increases the number of cases falling into the newer and larger categories.

The manner in which grouping is done depends in large part on the level of measurement of the data. Table 13.3 shows the hypothetical results of a small survey that includes the variable *religious affiliation*. This frequency distribution clearly is in need of grouping: Some categories have no cases at all, others have few cases, and there are a large number of categories in the coding scheme. When grouping nominal data such as these, one must rely on logic or some theoretical considerations to group categories together that in some way fit each other. For example, it might make sense, both logically and theoretically, to group all the Protestant denominations into a new category called "Protestant," because they all share some core beliefs and practices. On the other hand, those categories with few or no frequencies could reasonably be grouped into a category called "Other" simply because there is an insufficient number of cases in any one of them. The difference in reasoning

TABLE 13.4   Grouping Ordinal Data

Social Class	f	Social Class	f
upper-upper	8		
middle-upper	14	upper	57
lower-upper	35		
upper-middle	56		
middle-middle	92	middle	212
lower-middle	64		
upper-lower	44		
middle-lower	32	lower	87
lower-lower	11		
Total (N) = 356		Total (N) = 356	

behind forming each group is important to note because it may be significant for later analysis. For example, if we are evaluating how religious affiliation relates to attitudes toward volunteering for community service, "Catholic," "Jewish," and "Protestant" are meaningful categories where the members of each share something in common theologically. The "Other" category, on the other hand, contains a potentially wide variety of religions that we really can't describe in terms of a common theology, and it may be more difficult to draw solid conclusions regarding this group.

Grouping ordinal data can be a more straightforward process of merging adjacent categories. For example, if it were decided that the nine *social class* categories in Table 13.4 were too many, it would be easy to collapse them down to the conventional three as shown in the table. In this illustration, going from nine to three categories makes logical sense because the nine small categories are gradations of the three larger groupings. However, this is not always the case. A survey may include an item that asks for *years of education* but not have a variable for *highest degree attained*. The desired variable could be approximated by grouping the existing variable. Although *years of education* could con-

ceivably be construed of as a ratio-level variable, if it is applied as an indicator of preparation for employment, then it might be better treated as ordinal. In terms of getting a job, having a high school diploma, an associate degree, or a bachelor's degree may be the critical determinants. Persons who dropped out without completing the twelfth grade are all dropouts, whether they completed 9, 10, or 11 years of education. Persons who have 13–15 years of education probably have some college, but not a bachelor's degree. Rather than using equal multiples as in the social class example, a researcher might approximate employment preparation by grouping years of education into the following categories:

> *1 = <9 years (no high school)*
> *2 = 9–11 years (some high school)*
> *3 = 12 years (high school grad)*
> *4 = 13–15 years (some college)*
> *5 = 16+ (college grad)*

The scheme would be imperfect because not everyone with 12 years of schooling necessarily graduated from high school, but it would approximate the desired variable. As this example illustrates, the fact that the original categories are ordered is a starting point for grouping a variable, but the researcher must still rely on logic and knowledge of the application to which the grouped variable will be put to decide the boundaries of the intervals.

Grouping interval-level data is a fairly direct process. Unlike ordinal data, the distance between units of interval-level data are by definition equal, so one general principle is to use equal-width intervals when grouping data. Nevertheless, common sense and knowledge of the use to which the grouped data will be put are essential to planning the group intervals. Intervals of $10 might make perfect sense if grouping data on weekly earnings of high school students, but completely inappropriate if the data concern annual income of single-parent households. The size of the grouped intervals is called the *interval width*.

TABLE 13.5 Patient Age Data Grouped by Intervals of 5 Years

Interval Width	Value	Frequency	Percent	Valid Percent	Cum Percent
50–54	52	4	5.6	5.6	5.6
55–59	57	5	7.0	7.0	12.7
60–64	62	5	7.0	7.0	19.7
65–69	67	12	16.9	16.9	36.6
70–74	72	17	23.9	23.9	60.6
75–79	77	12	16.9	16.9	77.5
80–84	82	7	9.9	9.9	87.3
85–89	87	4	5.6	5.6	93.0
90–94	92	2	2.8	2.8	95.8
95–99	97	3	4.2	4.2	100.0
	Total	71	100.0	100.0	

Valid cases = 71; Missing cases = 0

Consider the simple frequency distribution of ages for residents in a veterans' residential facility in Table 13.2B. This distribution is clearly in need of grouping as it is so spread out that any pattern is difficult to see. I illustrate the grouping of interval data by grouping this distribution into 10 new intervals. On the one hand, we want enough intervals so that we don't obscure significant variation in the data; on the other hand, we want to reduce the clutter of too many categories. Once the decision on the number of intervals is made, one must determine the number of measurement units that will go into each new interval; in other words, we need to determine the interval width. In our particular example, the youngest person is age 50 and the oldest is age 98, with a range of 48 years on the *age* variable. With the number of grouped intervals set at 10, it is fairly easy to determine that the interval width should be 5. In other situations, however, the interval width may not be so easy to determine, so here is a formula that one can apply to find the interval width:

$$\text{Interval Width} = \frac{H_s - L_s}{N_i}$$

where:

H_s = *Highest actual score in the distribution*
L_s = *Lowest actual score in the distribution*
N_i = *Number of desired intervals*

If we applied this equation to the data in Table 13.2B, we would have:

$$\text{Interval Width} = \frac{98 - 50}{10} = \frac{48}{10} = 4.8$$

If the result of this formula is a decimal, as in this case, one merely rounds up to the nearest whole number, which would be 5. (If you round down, the 10 intervals may not reach all the way from the lowest value in the distribution to the highest.) The resulting frequency distribution is displayed in Table 13.5.

If possible, it is advantageous to select interval widths that contain odd numbers of measurement units, such as 3, 5, or 7. This should be done because intervals with an odd-numbered width have a whole number for their midpoints. For example, the interval 50–54 in Table 13.5 has five measurement units and a midpoint of 52. If instead the interval were

50–53, then it would have a midpoint of 51.5. When working with grouped interval data, the midpoints have a number of uses (such as category labels—the "Value" column in Table 13.5), and it is far more convenient to work with midpoints that do not have decimals.

Now compare the grouped data in Table 13.5 with the raw data and simple frequency distributions in Table 13.2. Do you notice anything in the grouped distribution that you did not see when you first looked at those other distributions? The grouped distribution clearly shows the characteristic accumulation of people in the middle of a distribution with the frequencies tapering off toward both tails of the distribution. Such a distribution is symmetrical and is commonly referred to as a "normal curve"; this is an important characteristic of data distributions that is used in choosing some statistics in Chapter 14. It is more difficult to detect that pattern or other patterns when the distributions have too many categories.

GRAPHICAL DISPLAY OF DATA DISTRIBUTIONS

In addition to describing variables by means of a frequency distribution, another common procedure is to use a graph (Henry, 1994). The visual impact of a graph can help identify and summarize patterns in data that might not be detected as readily by perusing frequency distribution tables. Today, most statistical and spreadsheet software provide a diverse array of full-color graphing options, either as a special set of tools or as a part of other statistical analysis procedures. For example, by simply checking a box in the SPSS Frequencies Menu, the program provides a histogram with each frequency distribution table. The easy access to graphical presentation of data, however, is a two-edged sword. Used correctly, graphs are a powerful tool for communicating information about data; used incorrectly, they can confuse and even mislead viewers on the meaning of a distribution. One situation in which graphs are

popular occurs when the results of a frequency distribution must be presented to an audience that is unfamiliar with reading tables.

I next present a few of the more common ways of graphing individual variables. Basically, these graphs are visual representations of a frequency distribution. Earlier, I discussed ways of grouping data for presentation in a frequency distribution table. Many of the same issues apply to graphing data. For example, a graph is more effective if there are a manageable number of categories. If several categories have a very low frequency, they will be hard to see on a graph. An effective graph, then, begins with organizing the data appropriately as one would for a frequency distribution table.

Bar Graphs

One of the most commonly used types of graphs is the **bar graph**. A distinguishing feature of the bar graph is the space between the bars. These spaces illustrate that the categories of the variable being represented are separate, or discrete. In Chapter 5, I pointed out that only certain categories are theoretically possible in a discrete variable, such as race, religious affiliation, and household size. Nominal and many ordinal variables are by definition discrete variables, so bar graphs are especially useful for these levels of data. Figure 13.3 illustrates a typical bar graph, with the height of the bars representing the frequencies in each category of the variable. As can be seen in Figure 13.3, a bar graph makes it easy to note any trends, such as the upward trend over time.

When constructing a bar graph (and most other types of graphs, for that matter), care should be taken when establishing the dimensions of the graph. The vertical axis, which represents the frequencies, and the horizontal axis, which represents the categories of the variable, should be about equal with reasonably equal spacing of the categories on both axes. Again, Figure 13.3 illustrates this. This

FIGURE 13.3   Example of a Bar Graph

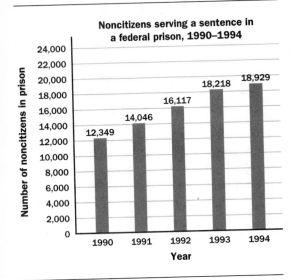

Noncitizens serving a sentence in a federal prison, 1990–1994

Source   Based on Bureau of Justice Statistics, *Nonciti- zens in the Federal Criminal Justice System (1984–1994)* (Washington, D.C.: U.S. Department of Justice, 1996), p. 9.

FIGURE 13.4   A Histogram and Frequency Polygon of Patient Ages from Table 13.5

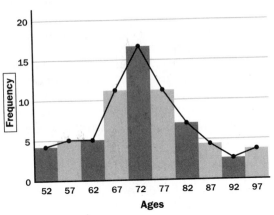

concern for roughly equal dimensions is im- portant because it is possible to accidentally or purposefully construct a graph that gives a false impression. For example, by expanding the dimensions of the vertical axis, a graph can be constructed that makes small differences among categories appear large, at least to the casual observer. On the other hand, expanding the horizontal dimension has the opposite ef- fect. A graph can be constructed that appears to minimize differences among the categories. Purposeful manipulation of graph dimensions in an attempt to deceive viewers is considered unethical. It should be noted, however, that misleading graphs are produced all the time, so always carefully inspect any graph you are reading to be sure that you are not being mis- led by its initial appearance.

Histograms and Frequency Polygons

Although some variables, such as *household size* or a frequency count of a *behavior*, are

discrete variables, at the interval- or ratio- level of measurement, variables are conceived of as being continuous, meaning that, unlike the discrete variables, there are no gaps or spaces between the categories. Thus *age* is a continuous variable. This continuous nature of interval and ratio data should be reflected in graphs used to present them. One wishing to graph interval or ratio data has a choice of two popular methods. The first is called a *histogram* and bears a considerable resem- blance to the preceding bar graph. Once again, as can be seen in Figure 13.4, bars of varying lengths are used to represent the magnitude of the frequencies from a fre- quency distribution. The only difference be- tween a bar graph and a **histogram** is that the bars in a histogram touch, signifying the con- tinuous nature of the data.

Figure 13.4 also illustrates the alternative technique of graphing interval or ratio data, the *frequency polygon*. A **frequency polygon** is a line graph that connects the midpoints of each category of the variable. The choice of ei- ther a histogram or frequency polygon is purely a matter of personal preference, as they are interchangeable.

FIGURE 13.5   An Example of a Pie Chart

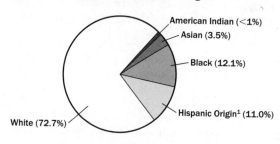

Population of the United States, by Race and Hispanic Origin

American Indian (<1%)

Asian (3.5%)

Black (12.1%)

Hispanic Origin[1] (11.0%)

White (72.7%)

[1]Person of Hispanic origin may be of any race.

Source   U.S. Bureau of the Census, *Statistical Abstract of the United States (1998)* (Washington, D.C.: U.S. Government Printing Office, 1998), p. 19.

Pie Charts

Pie charts are another type of commonly used graph that are particularly good for showing how some whole amount is divided. Figure 13.5 is a pie chart showing the distribution of the population of the United States in terms of race and Hispanic origin.

The **pie chart**, quite obviously, gets its name from its resemblance to the slices of a pie. The size of each "slice" gives a visual depiction of the percentage of the whole that each category represents. Preparing a pie chart via a computer program now involves simply choosing a variable and the pie chart option from the graphing menu. However, the process that the graphing program uses involves determining how many of the 360° of an entire circle are to be allocated to each of the categories. For example, Figure 13.5 indicates that 11% of the U.S. population consists of people of Hispanic origin. To determine how large a slice to allocate to this group, the data analysis software multiplies 360° by 11%, obtaining a value of 39.6°. So, the slice devoted to people of Hispanic origin is drawn such that it occupies 39.6° of the 360° of the circle.

The Eye on Ethics box discusses some ethical issues that are important in graphical displays of data as well as all other types of data analysis.

CONTINGENCY TABLES

Thus far I have been discussing univariate statistics: ways of describing the distribution of cases on a single variable and ways of presenting single variable data in the form of frequency distributions and graphs. Now we turn our attention to ways of describing and exploring how two or more variables are distributed together. That is, given the values in a data set on one variable, how are values distributed on one or more other variables? Classifying or organizing the data on one variable according to values on a second variable is the basis for contingency table analysis.

Bivariate Relationships

Most data analysis in research involves dealing with two or more variables simultaneously. Statistical procedures used to describe the relationship between two variables are called **bivariate statistics**, whereas **multivariate statistics** deal with three or more variables. When two or more variables are analyzed with descriptive statistics, the major feature of interest is the relationship between the variables, especially the extent to which they covary, or vary together. If two variables are related, a change in one of the variables is associated with a change in the other.

A convenient way of investigating bivariate relationships is to cross-tabulate the data in the form of a table. A *contingency table* contains raw frequencies, or percentages, or both. I discuss the construction of these tables for only two variables. However, you should be aware that contingency table analysis can be applied to three or more variables.

Table 13.6 illustrates the general form of a contingency table. I utilize two conventions to standardize the construction of contingency tables. First, when there is an independent variable

⚖️ **EYE ON ETHICS**   **Thou Shalt Not Lie (or Even Mislead) with Statistics!**

Many people mistrust statistics. Mark Twain is alleged to have said: "There are three kinds of lies: lies, damned lies, and statistics." Another sentiment regarding statistics was expressed by Mrs. Robert A. Taft:

> I always find that statistics are hard to swallow and impossible to digest. The only one I can remember is that if all the people who go to sleep in church were laid end to end they would be a lot more comfortable (Patton, 1982, p. 243).

Sentiments such as these suggest that you can't trust statistics because they can be manipulated or that one can reach any conclusions one wants to based on statistics. These sentiments represent a real concern because it is possible to manipulate statistics to support misleading, or in some cases totally inaccurate, conclusions. That, however, is not the fault of statistics but rather of the person who uses them unethically or of the consumer who is uninformed about statistical analysis. The more people know about statistics, the harder it is to deceive or mislead them. In Chapter 1, I discussed the scientific community and how people who become researchers are socialized during their educational training to accept certain norms as a part of their ethical framework. A fundamental part of that ethical framework is to always give a thoroughly accurate representation of research data.

It's important to recognize that researchers rarely ever present all of the data they have to audiences. After all, that would mean presenting the raw data, which is typically not very informative. Whether the data are to be presented to other professionals in a research journal, to policymakers in a public presentation, or to the general public, the researcher presents some sort of summary or distillation that purports to be an accurate characterization of the data. Researchers use a variety of data analysis techniques, (discussed in this and the two following

and a dependent variable, the vertical *columns* are used to represent categories of the independent variable, with the horizontal *rows* representing categories of the dependent variable. In Table 13.6, the variable *Income* is the independent variable and *Hours of Television Watched* is the dependent variable. Second, when ordinal level or higher data are cross-tabulated, the categories should be ordered as illustrated in Table 13.6, with columns running from lowest on the left to highest on the right and rows from lowest at the bottom to highest at the top. Be advised that these conventions are not applied universally, and you may encounter tables constructed differently. Following the conventions, however, contributes to consistency and ease of interpretation. Further, the computational routines for

some statistics assume that tables are constructed according to these conventions and must be modified to produce correct results with tables structured differently.

Several labels are used to refer to the various parts of tables. The squares of the table are called *cells*, with the frequencies within the cells labeled *cell frequencies*. Values in the Totals column or row are called *marginals* (only the column marginal is shown in Table 13.6). Tables are often identified according to the number of rows and columns they contain. A table with two rows and two columns becomes a 2×2 (read "2 by 2") table. A table such as Table 13.6 is a 3×3 table. In addition, because the number of rows is always designated first, a 2×3 table is not the same as a 3×2.

chapters) to figure out what trends, patterns, or themes can be found in the data. Then statistics and other techniques are used to present to an audience an accurate picture of those trends or themes. It is the ethical obligation of the researcher to be sure that the picture presented is accurate.

While many ways exist to misrepresent data, I give one example to illustrate how this can be done with graphical representations. Figure 13.6 presents data on the infant mortality rate in the United States for each year from 1990 to 1996. The data are presented in two separate graphs, although it is the exact same data in each. The actual data on infant mortality for that period suggests a modest decline: 20% fewer infant deaths in 1996 than in 1990. Now if we look at the visual impression given by the two figures, Figure 13.6A seems to imply a very sizable decline; it seems to suggest a much lower infant mortality rate at the end of the period. The reason is that the vertical axis is designed so that each measurement unit (each additional infant death) takes a lot of space on the axis. The vertical axis also doesn't show the whole range of measurement units. When we lengthen that axis and add in all the measurement values, as in Figure 13.6B, the visual impression given by the graph is of a far more modest decline, which is a more accurate representation of the trend found in the actual data.

A number of good reasons can be found to shorten one or another axis on a graph, such as saving space, but it must not be done in a way that produces a misleading visual impression of the patterns actually found in the data. And it certainly must never be done to deliberately distort the data. What is said regarding this illustration is true of all data analysis in the social sciences: Data analysis can distort results, but social science researchers consider it highly unethical to do so. The prime responsibility of the researcher is to give a complete and accurate representation of the data and conclusions.

Contingency tables almost always include *cell percentages* because, by themselves, cell frequencies are difficult to interpret if the number of cases vary in each column and row. Converting frequencies to percentages makes interpretation far easier, and this can be done by dividing each cell frequency by the appropriate marginal total and multiplying the result by 100. The column marginals would be used if we were interested in seeing how the dependent variable is distributed across categories of the independent variable. Percentages in the first column were obtained by dividing the cell frequencies 49, 164, and 42 by the column marginal 255, with the result multiplied by 100. Below each column, the percentages are totaled and indicated as equaling 100 percent.

This informs the reader that the column marginals were used to compute the percentages. Sometimes the cells in a contingency table contain only percentages and no frequencies. When this is done, the marginal frequencies should always be included. They supply valuable information regarding the number of cases on which the percentages are based and enable the reader to compute the cell frequencies if needed.

Reading percentage tables is a straightforward process, similar to constructing them. Whereas we computed the percentages down the columns, we read percentage tables by comparing percentages along the rows. Of particular interest is the *percentage difference* (%d) between any two categories within a

FIGURE 13.6   Two Graphs Showing the Rate of Infant Mortality, 1990–1996

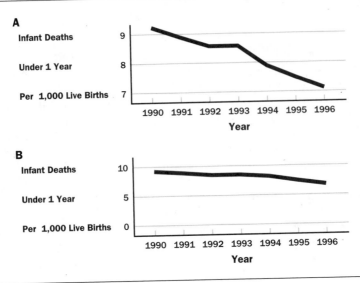

Source   U.S. Bureau of the Census, *Statistical Abstract of the United States (1998)* (Washington, D.C.: U.S. Government Printing Office, 1998), p. 76.

given row. For example, in Table 13.6, the difference between the Low Income column and the High Income column in the top row is 12.5%, meaning that substantially more low income people watch high levels of television. The %*d* suggests that high levels of television watching is less common among high income groups. Note that with a 2 × 2 table, a single

TABLE 13.6   A Contingency Table with Frequencies and Percentages

dependent variable		*independent variable*		
		INCOME		
		Low	Medium	High
NUMBER OF HOURS OF TELEVISION WATCHED	High	49 (19.2%)	19 (8.6%)	10 (6.7%)
	Medium	164 (64.3%)	136 (61.5%)	86 (57.7%)
	Low	42 (16.5 %)	66 (29.9%)	53 (35.6%)
	Totals	255 (100%)	221 (100%)	149 (100%)

cell frequency   *cell percentage*   *column marginals*

Source   James A. Davis and Tom W. Smith. *General Social Surveys (1972–1994)* (Chicago: National Opinion Research Center; Storrs, CT: The Roper Center for Public Opinion Research, University of Connecticut).

%*d* summarizes the complete table. As the number of rows and columns in a table increases, the number of %*d*s that can be calculated increases rapidly.

The magnitude of %*d*s is a crude indicator of the strength of the relationship between two variables. It should not be surprising that small differences of 1% or 2% indicate very weak and possibly meaningless relationships. On the other hand, %*d*s of 15% or more usually indicate substantial relationships. Unfortunately, no hard-and-fast rules concerning evaluating the magnitude of %*d*s can be offered because of the complicating factor of sample size. For example, if we are dealing with employment data for the entire nation, a difference of 1% or less could represent a million more workers with or without jobs. With very large samples, smaller %*d*s are more important. Alternatively, with small samples, %*d*s must be large before they indicate a substantial relationship. (Other statistics for assessing the strength of relationships in contingency tables are discussed in Chapter 14.)

Multivariate Analysis

As noted in Chapter 4, analyzing an independent and a dependent variable and finding a bivariate relationship between them does not prove that the independent variable actually *causes* variation in the dependent variable. To infer causality, the possible effects of other variables must be investigated. Although research projects may focus primary attention on two variables, they will typically consider others to assess the full complexity of social phenomena. A set of procedures for conducting this kind of multivariate analysis is called **table elaboration** or **contingency control**. Table elaboration involves examining a relationship between an independent and a dependent variable while holding a third variable (also called a test factor) constant (Rosenberg, 1968). Three variable tables are constructed such that the relationship between the independent and dependent variable can be examined within each category of that third variable. The general format of contingency control is illustrated in Figure 13.7.

FIGURE 13.7   General Format for Partialing Tables

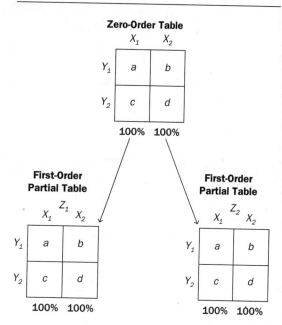

$X =$ categories of the independent variable.
$Y =$ categories of the dependent variable.
$Z =$ categories of the test variable.

The original table that you begin the analysis with is called the *zero-order table*, and the relationship contained within it is known as a *zero-order relationship*. The "zero" in these labels indicates that no control or test variables are being used. As we introduce test variables, the tables produced are called *partial tables* and the relationships within them are known as *partial relationships*. A partial relationship is a relationship between an independent and a dependent variable within one category of the test variable.

Partial tables are referred to by the number of test variables being controlled at one time. If we are controlling one test variable in our partial tables, they are referred to as *first-order partials*, indicating that the number of variables controlled is one. If more variables than one are simultaneously controlled, the tables are called second-order partials, third-order partials, and so on. Oftentimes, research does

FIGURE 13.8 Table Elaboration Showing "No Effect" of the Test Variable (Replication)

Zero-Order Table

Is Life Exciting

	Gender	
	Male	Female
Exciting	213 (50.1%)	221 (39.8%)
Routine/ Dull	212 (49.9%)	334 (60.2%)
Totals	425 (100%)	555 (100%)

First-Order Partial Tables

Single

Is Life Exciting

	Gender	
	Male	Female
Exciting	69 (54.3%)	48 (46.2%)
Routine/ Dull	58 (45.6%)	56 (53.8%)
Totals	127 (100%)	104 (100%)

Married

Is Life Exciting

	Gender	
	Male	Female
Exciting	114 (47.7%)	107 (40.4%)
Routine/ Dull	125 (52.3%)	158 (59.7%)
Totals	239 (100%)	265 (100%)

Source James A. Davies and Tom W. Smith. *General Social Surveys (1972–1994)* (Chicago: National Opinion Research Center; Storrs, CT: The Roper Center for Public Opinion Research, University of Connecticut).

not go beyond the first order with contingency control. One reason is that as we add more test variables, the number of partial tables that are generated increases rapidly, and interpreting all of these tables may be difficult. The other reason table elaboration is often not taken beyond the first order is that the sample gets divided quickly, to the point that individual cell frequencies may become too small and reduce our confidence in the results. We literally run out of cases. This effect can be seen in Figure 13.7, where in the zero-order case the sample is divided among the four cells of the Zero-Order Table; but when we move to the partials, the same number of cases is spread among eight cells, thus reducing the magnitude of the cell frequencies. For analyzing multivariate relationships beyond the first order, techniques other than table elaboration are normally used.

I want next to describe some of the outcomes that can occur when a test variable is introduced and partial tables created and explain what they mean. This will give you an idea of the basic logic behind multivariate analysis. First, the partial tables may show *no effect*. This is also called *replication* because the results in the partial tables replicate or repeat the results in the Zero-Order Table: The relationship in the partial tables is approximately the same in terms of strength and direction as it was in the Zero-Order Table. In this case, we have tested a variable that is unrelated to either the independent or the dependent variable; it could not therefore affect the relationship between those variables. This is illustrated in Figure 13.8 where the Zero-Order Table shows the relationship between *gender* and whether people believe life to be *exciting, routine,* or *dull*. In Figure 13.8 you can see in the Zero-Order Table that about 10% more of the males believe that life is exciting. It is possible, however, that marital status might influence

FIGURE 13.9 Relationship Between Reading Smoking Reports and Smoking Cessation, with Level of Education as a Control Variable

Zero-Order Table

		Person Reads Report	
		Yes	No
Person Quits Smoking	Yes	200 (50%)	135 (27%)
	No	200 (50%)	365 (73%)
	Totals	400 (100%)	500 (100%)

First-Order Partial Tables

Less Than High School Education

		Person Reads Report		
		Yes	No	Totals
Person Quits Smoking	Yes	20 (20%)	75 (19%)	95 (19%)
	No	80 (80%)	325 (81%)	405 (81%)
	Totals	100 (100%)	400 (100%)	500

High School Education or More

		Person Reads Report		
		Yes	No	Totals
Person Quits Smoking	Yes	180 (60%)	60 (60%)	240 (60%)
	No	120 (40%)	40 (40%)	160 (40%)
	Totals	300 (100%)	100 (100%)	400

whether men and women perceive life to be exciting. So, in the partial tables, I look at the relationship between the two variables separately for married and single people (I have left out the other possible marital statuses to simplify the example). You can see that this third variable has almost no effect on the original relationship. In each of the categories of marital status, about 7%–8% more men see life as exciting in comparison to the women.

A second possibility when doing table elaboration is that the strength of the relationship found in the Zero-Order Table may be substantially reduced or even disappear entirely in the partials. I described this kind of situation in the discussion of causality in Chapter 4 without calling it by these names; that data is reproduced in Figure 13.9. The Zero-Order Table in this figure shows how reading reports about the health consequences of smoking impacted on whether people quit smoking. The Zero-Order Table shows that people who read such reports are substantially more likely to quit smoking (%d = 23%). However, when we control for *education*, the original relationship disappears. In each educational grouping, quitting smoking is as common among those who have not read the report as among those who have.

This result can be difficult to interpret because there are two possibilities and we need more information (which may or may not be available) to choose between the two. One possibility is that in terms of temporal order, the test variable intervenes between the independent and the dependent variable (see Figure 13.10). When such an intervening variable is controlled, it is called *interpretation*; the independent variable is changing the test variable, which in turn is changing the dependent variable. In other words, there is a causal sequencing being specified. When there is such a causal chain, the relationship between the independent variable and the dependent variable is effectively blocked in the partial tables. In

FIGURE 13.10 A Comparison of Intervening and Antecedent Variable Relationships

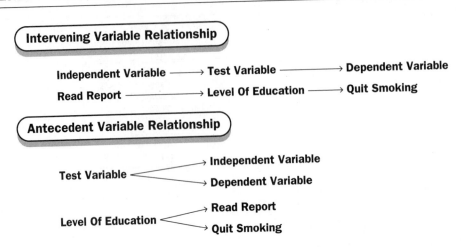

our example, an intervening variable relationship would mean that people's reading of the report is what produces their level of education, and level of education in turn causes them to quit smoking. However, this is an inappropriate assessment of these variables because there is no logical or theoretical reason to assume that reading a health report would change people's level of education. So, interpretation is probably not what is going on in these partial tables.

This brings us to the other possibility; namely, that the zero-order relationship is either all or partly *spurious*. In this case, which is called *explanation*, the test variable is temporally located before either the independent or dependent variable and is related directly to both. This is called an "antecedent variable relationship." When the effects of the test variable are controlled, the apparent relationship between the independent variable and the dependent variable is reduced or eliminated because the operation of the test variable was to inflate the actual relationship between the other two variables in the Zero-Order Table. In other words, the relationship shown in the Zero-Order Table was a false reading. In Figure 13.9, we know that people's educational

levels are antecedent: People's educational levels were established before they had a chance to read the report or quit smoking. In addition, there are theoretical reasons to conclude that *level of education* can influence both whether people read the report and whether they quit smoking. This tells us that the relationship in the Zero-Order Table is spurious.

So, when a relationship disappears in a partial table, we need to establish the appropriate temporal location of the test variable or use some theoretical or logical considerations to determine whether *interpretation* or *explanation* is the proper assessment of what is going on. If that cannot be done, as might occur with survey data, we cannot be sure whether the partials are indicating the presence of an intervening variable or that of a spurious relationship. As noted in Chapter 4, determining temporal order is crucial in sorting out causal relationships among variables.

Yet another possibility can be discovered in table elaboration. *Specification* refers to a situation where the relationship between the independent and dependent variables is found in the Zero-Order Table, but the partial tables show different relationships from one table to another. One partial table may show no

FIGURE 13.11   Table Elaboration Showing Specification: Relationship Between Social Class and Self-Esteem Controlling for Gender

Zero-Order Table

Self Esteem	Social Class		
	Lower	Middle	Upper
High	38%	46%	51%
Medium	26%	25%	23%
Low	35%	29%	26%
Total	340 (100%)	2686 (100%)	195 (100%)

First-Order Partial Tables

Males

Self Esteem	Social Class		
	Lower	Middle	Upper
High	36%	47%	55%
Medium	26%	25%	17%
Low	39%	28%	28%
Total	168 (100%)	1383 (100%)	89 (100%)

Females

Self Esteem	Social Class		
	Lower	Middle	Upper
High	41%	46%	47%
Medium	27%	25%	28%
Low	32%	29%	24%
Total	172 (100%)	1311 (100%)	106 (100%)

Source   Morris Rosenberg, *Society and the Adolescent Self-Image* (Princeton, NJ: Princeton University Press, 1965), pp. 40–41.

relationship or a weak relationship, while the other shows a strong relationship. In other words, the table elaboration specifies the categories of the test variable in which the relationship does or does not occur. Figure 13.11 shows one example of what specification can look like. The figure shows the relationship found in one study between social class and self-esteem among adolescents. The Zero-Order Table in Figure 13.11 shows a fairly strong relationship, with adolescents in higher social classes being more likely to have high self-esteem ($\%d = 13\%$ in the high self-esteem category). When gender is introduced as a test factor, the original relationship gets dramatically specified: Upper-class boys are substantially more likely to have high self-esteem than are the lower-class boys ($\%d = 19\%$ in the high self-esteem category). However, the relationship is much weaker, although still in the same

direction among girls ($\%d = 6\%$ in the high self-esteem category). So the introduction of the test factor enables us to specify that social class is a fairly powerful factor in shaping self-esteem among boys but much less so among girls.

One final possibility that can be found in table elaboration must be mentioned here. The partial tables may show an even stronger relationship between the independent and dependent variables than was indicated by the Zero-Order Table. This outcome occurs because the action of the uncontrolled test variable in the Zero-Order Table is to suppress the relationship between the independent and the dependent variables. In fact, variables that produce this result are called *suppressor variables*, and the result itself is known as *suppression*. This occurs when the test variable and the independent variable are exerting forces on the dependent variable that are, at least in part,

Social scientists who conduct basic research most commonly communicate the results of their research to other social scientists, usually by presenting a paper at a professional conference or publishing an article in a professional journal. These papers and articles are often filled with complex statistical jargon that is alien to most people. However, this is not a problem because the social scientists who are the primary consumers of these papers and articles will likely possess sufficient statistical understanding to grasp the meaning. Applied social science researchers, on the other hand, confront a distinct challenge because the audiences interested in the results of their research will typically include people who are not social scientists and who may possess little if any statistical expertise.

Applied researchers can find themselves presenting their research to very diverse audiences: policymakers, governmental bureaucrats, congressional representatives, state legislators, city commissioners, school board members, or ordinary citizens interested in some issue. This might be done through written reports or public presentations to legislators, city commissions, school boards, and other groups. These audiences are decisionmakers and policymakers who will use the research results to help them make a decision about a social policy or a social intervention. As you can well imagine, the statistical expertise of people in these diverse groups is highly variable, and applied researchers have to be careful that their presentations do not leave people confused and unenlightened.

In presenting statistical results, applied researchers follow a few guidelines. One is to always make complete and accurate representations of the results. This is an important ethical responsibility because other interested parties, including those sponsoring the research, may not themselves possess the expertise to figure out what the data relate. This means that they are dependent on the researcher to get accurate results from the data.

offsetting. That is, as values of the independent variable increase, they influence the dependent variable to increase, but as the test variable increases, it is influencing the dependent variable to decrease, thus masking part of the relationship between the independent and the dependent variables. When the influence of the test variable is controlled, this masking effect is removed and the relationship between the independent variable and the dependent variable is now revealed to be stronger than in the Zero-Order Table.

There are a number of other possibilities that can occur in table elaboration, but these examples illustrate the importance and the difficulty of this kind of multivariate analysis. It is important because it enables us to discover the complexity of social reality where dependent variables are influenced by a number of factors at the same time. It is also a difficult kind of analysis because it requires time, patience, and creativity to tease out what can sometimes be hard to discover relationships. The Applied Scenario box discusses some issues that researchers address when presenting data and statistics to audiences who are not well trained in these subjects.

A second guideline is to present statistics that are comprehensible to the audience. Some of the most sophisticated statistical procedures that researchers use to uncover meaning in the data may not be comprehended by most lay people. Presenting such statistics to the audience may confuse or even discourage them from trying to understand the results. This doesn't mean that applied researchers shouldn't use such statistics but rather that they should be careful in how they present their information to an audience. In many cases, this means presenting only frequency and percentage distributions, graphical presentations, or contingency tables.

A third guideline, which really stems from the second, is to keep the statistical analysis as simple as possible while still providing a complete and accurate representation of the research findings. This provides decision makers and policy makers with the information they need to accomplish their tasks. Any more complex analysis may make it difficult for them to use the information in coming to a decision.

A fourth guideline is to use the presentation of research results as an opportunity to educate the audience. This education involves not only the substance of the research results but also the logic and rationale of social science research. This is done by not just presenting the conclusions but also explaining how those conclusions were arrived at. What is it about the data that leads to one conclusion rather than to another? What about the analysis produces a certain level of confidence in the results? This might call for being sure that people grasp the patterns displayed in a frequency distribution or a contingency table. Or it might call for educating the audience on some of the meanings of more sophisticated statistical procedures. The goal is to leave the audience better informed than they were before the presentation. But don't overwhelm them! Most audiences will not become master statisticians after a 30-minute presentation.

REVIEW AND CRITICAL THINKING

Chapter 13 is about how to understand numerical data, and we are surrounded by, some would say drowning in, numerical information in our daily lives. Modern computer and communications technologies have vastly amplified this problem. This makes it especially important to develop some critical thinking skills that will enable you to evaluate this deluge of information and to separate what is legitimate and useful from that which is meaningless and deceptive. The material presented in this chapter can serve as a guideline for this critical thinking. When confronted with statistical information, ask the following questions.

1. Was some kind of coding or other data-reduction strategies used in preparing the

data for analysis? If so, does this introduce any distortions or oversimplifications into the data?

2. Is the raw data presented? Or does the analysis involve some summarization of the data into a simple frequency distribution, a grouped frequency distribution, or a contingency table? If so, does this introduce into the analysis anything distorting or misleading?

3. Are the data displayed in any graphical form? If so, is the chosen form appropriate? Is the graph designed in a way that the visual impression communicates a different conclusion than the numerical data seem to suggest?

Computers and the Internet

The Internet can provide a number of tools for doing research, including data entry software. Although one can enter data directly into a statistical software program, such as SPSS or Minitab, it is not always convenient to do so; neither are there always safeguards against entering wrong values. A solution is a freeware (so-called because there is no charge for obtaining it) program, called QDATA that is available over the Internet. QDATA is an easy-to-use program, specifically for designing datasets, entering the data, and cleaning the data. One can begin by defining variables to be included in the data set—including variable names, extended labels, the range of acceptable values, and labels for the values. Once the variables are defined, you can enter your dataset. After the data are entered, the program guides you through several data cleaning steps. Once the data are entered and cleaned, the data file can be saved in a statistical package format or in ASCII (a generic computer format that most statistical packages can read). By downloading the QDATA program and its manual, you can either use the program with actual data that you collect or simply experiment by making up

a hypothetical dataset to gain a sense of the process involved.

Graphical aids to data analysis and presentation have exploded with the advent of the personal computer, and the Internet has an abundance of resources available to enrich your understanding of graphical data analysis and to provide tools to aid you in generating graphs of your data. One of the most inclusive sites is Michael Friendly's site, Statistics and Statistical Graphics, which can be accessed at URL **http://www.math.yorku.ca/SCS/StatResource.html**. This page provides a topic-based collection of available resources for statistics, statistical graphics, and computation related to research, data analysis, and teaching; it now contains over 350 links. An interesting feature of this Internet site is a page devoted to illustrating some of the best and worst examples of using graphs to depict data. Examining these graphical depictions of data will show you the value of using graphs and some of the pitfalls to avoid as well. The Gallery of Data Visualization can be accessed at **http://www.math.yorku.ca/SCS/Gallery/**.

Another example of an Internet site focusing on the analysis of graphical data analysis is an Internet course offered by Jeff Banfield (**http://math.montana.edu/~umsfJban/STAT438/Stat438.html#Graphics**). A perusal of his site will provide leads to texts and other Internet resources, as well as show you a variety of graphing procedures.

Main Points

1. Data analysis refers to deriving some meaning from the observations that have been made as a part of a research project; quantitative data analysis involves putting observations into numerical form and manipulating them with statistics.

2. Data analysis begins by developing a coding scheme and a codebook. Since most statistical

analysis is done with computers, the raw data must be entered into a computer file. After this is done, data must be carefully checked for coding or input errors.

3. New or revised variables may be created through recoding existing variables or by computing new variables by combining existing variables.

4. Data analysis begins by searching for patterns in the distributions of single variables (univariate analysis) and in the relationships between two or more variables. Frequency distributions reveal the pattern the frequencies produce among the various categories of the variable and may be symmetrical or asymmetrical.

5. Frequency distributions can take the form of simple frequency distributions or grouped frequency distributions. Frequency distributions may be effectively communicated by using graphical displays of data distributions, such as bar graphs, histograms, frequency polygons, and pie charts.

6. Contingency tables are particularly useful for analyzing bivariate and multivariate relationships. In such tables, the percentage difference (%*d*) can be used as a crude measure of the magnitude of the effect of one variable on another.

7. One way to conduct multivariate analysis is through table elaboration. Researchers construct partial tables and analyze how a zero-order relationship changes in a partial table. A number of results might occur in the partial tables: replication, explanation, interpretation, specification, or suppression.

Important Terms for Review

alphanumeric variable
bar graph
bivariate statistics
codebook
coding
coding scheme
contingency control
data analysis
data transformation
frequency polygon
histogram
multivariate statistics
numeric variables
pie chart
statistics
table elaboration
univariate statistics

For Further Reading

Evans, James D. (1996). *Straightforward statistics for the behavioral sciences*. Pacific Grove, CA: Brooks/Cole. This is a standard but well-written introductory textbook in statistics. It covers all the topics discussed here and in Chapter 14, going into more detail and including additional topics.

Foster, Jeremy. (1999). *Data analysis using SPSS for Windows: A beginner's guide*. Thousand Oaks, CA: Sage. This book provides an introduction to the use of the most popular statistical package in use today: SPSS. This handbook shows you how to do all of the things discussed in this and the next chapter.

Fox, William. (1992). *Social statistics using Microcase*. Chicago: Nelson-Hall. Written in a light and accessible style, this book teaches statistics through the use of the increasingly popular and easy-to-use Microcase software, a copy of which is included with each book.

Healey, Joseph F. (1999). *Statistics: A tool for social research* (5th ed.). Belmont, CA: Wadsworth. This is another standard but well-written introductory textbook on statistics. Along with the Evans book, it can help you understand all the material in this and the next chapter.

Jones, Gerald F. (1995). *How to lie with charts*. Alameda, CA: Sybex. An ideal guide for all who want to understand how presenters can use charts to deceive an audience. Readers learn how to make their presentations more effective.

Kachigan, S. (1991). *Multivariate statistical analysis* (2nd ed.). New York: Radius Press. An advanced textbook covering a variety of more sophisticated statistics, with primary emphasis on multivariate techniques.

Wallgren, Anders, Wallgren, Britt, Persson, Rolf, & Jorner, Ulf. (1996). *Graphing statistics and data: Creating better charts*. Walnut Creek, CA: Altamira Press. This book suggests how to create graphs and charts that make viewers aware of the qualities of data. It introduces the elements of charts, such as axes, scales, and patterns; describes steps to make charts clearer, using real data for examples; and walks through the entire process from data to finished chart.

Zeisel, Hans. (1984). *Say it with figures* (6th ed.). New York: Harper & Row. A classic analysis of how to present data with statistics. It is very useful for policymakers who wish to improve their understanding of the statistics presented to them.

Chapter 14

DESCRIPTIVE AND INFERENTIAL STATISTICS

This chapter is a continuation of the introduction to data analysis that began in Chapter 13. You will recall from that chapter that *data analysis* is the process of deriving some meaning from the observations that have been made during the research process. Thus far, the focus has been on describing data using relatively simple procedures, such as frequency distributions, graphs, and contingency tables. Although these means of summarizing and displaying data are important, researchers also rely on more sophisticated statistics in order to describe data precisely and to examine relationships between variables.

Statistics have become common to the lexicon of modern living. We often hear people casually referring to something being "above the norm," "below average," or "correlated" with something else. Tune in to the evening news and you might assume that death from cancer is imminent because research has shown a "statistically significant correlation" between your favorite food and the dreaded disease. Or you might hear a report stating that the relationship between underage drinking and traffic-related injuries to adolescents is "statistically significant." Because statistics are ubiquitous in our modern world, people may use statistical terms, such as "correlation" or "statistical significance," without fully comprehending what they mean. It is imperative to have a basic understanding of statistical concepts in order to be an informed consumer of both the popular media and the professional literature in the social sciences.

Chapter 14 begins with some guides to help you identify which statistical procedures might be appropriate in a given situation. Next, I introduce the common statistics used for describing and summarizing the characteristics of a data distribution. Finally, I examine inferential statistics, which are used to draw conclusions about populations from samples drawn from those populations.

CONSIDERATIONS IN CHOOSING STATISTICS

Each statistical procedure is designed to be used with data possessing particular characteristics. One of the major errors that can occur in data analysis is selecting a statistic that is inappropriate for the kind of data gathered in a research project. Although many factors need to be taken into account in choosing statistics appropriately, five major considerations are especially important.

Level of Measurement

One consideration is the level of measurement of the data collected. Chapter 5 addresses the four levels of measurement: nominal, ordinal, interval, and ratio. Each level of measurement involves different rules of permissible mathematical operations that can be performed on the numbers produced while measuring variables at that level. As you will recall, the nominal level involves merely classifying observations into categories; the categories have no order, and the numbers associated with a category only serve as a label for the category—the numbers have no mathematical value and performing any arithmetical operations on them is inappropriate. On the other hand, a ratio level of measurement has a true zero point, making all mathematical operations permissible.

Each statistical procedure involves mathematical operations that are appropriate at one of the levels of measurement. For example, a *nominal statistic* assumes that the data have only mutually exclusive and exhaustive categories and none of the mathematical properties of the other levels of measurement; likewise, a *ratio statistic* assumes that the data have a true zero point and require mathematical operations appropriate to that. The basic rule is: *a statistic designed for a given level of measurement can be used to analyze variables measured at that or higher levels of measurement.* So, for example, as Table 14.1

TABLE 14.1   Relationship Between Statistics and Level of Measurement

Level of Measurement Statistic Is Designed For	Level of Measurement of the Data			
	Nominal	**Ordinal**	**Interval**	**Ratio**
Ratio	No	No	No	Yes
Interval	No	No	Yes	Yes
Ordinal	No	Yes	Yes	Yes
Nominal	Yes	Yes	Yes	Yes

No = Inappropriate to use the statistic with this data
Yes = Appropriate to use the statistic with this data

illustrates, a statistic designed for ordinal data can be used with a variable that is ordinal, interval, or ratio level. The reason for this is that the ordinal statistic assumes that the categories of the variable are ordered, and this is true for ordinal-, interval-, and ratio-level variables. This is *not* true for nominal variables, however, so using an ordinal statistic on them would be inappropriate.

However, even though data measured at higher levels can be analyzed with lower-level statistics, there is a trade off in terms of the kind of information and relationships that can be discovered about the data based on the statistics. With annual income, for example, if we analyze it as ratio data, we might say that the people in the upper 10% of the income distribution earn *eight times more money* than people in the bottom 50%. If we analyze income as an ordinal variable, on the other hand, then we may only be able to say that one group has a *higher income* than another—not nearly as precise a conclusion. Therefore, it is generally preferred to use the highest level of statistic that is appropriate for a particular variable. In addition, it is absolutely inappropriate to use a statistic designed for a higher level of measurement than the level of measurement of the data being analyzed.

Thus far I have been discussing the role of levels of measurement in selecting a statistic on the basis of individual variables. However, many research questions are bivariate or multi-variate in nature, examining relationships between two or more variables simultaneously, and these may involve different levels of measurement. For example, we might study the relationship between *gender* (nominal) and *income* (ratio) or between *religious affiliation* (nominal) and *educational degrees attained* (ordinal). Some bivariate and multivariate statistics are designed such that all variables included in the analysis must be at least at a certain level of measurement. Some statistics, for example, are designed with the assumption that all variables are at the interval-ratio level. In this case, all the variables must be at the appropriate level of measurement or a higher level of measurement. In other cases, statistics are designed to be used with variables at two different levels of measurement, such as one nominal variable and one interval-ratio variable. For any data analysis problem, then, the researcher must be able to classify each variable in terms of its level of measurement and then choose a statistic that is designed for the appropriate levels of measurement.

Goals of the Data Analysis

The second consideration in assessing the appropriateness of statistics is to determine which goals the statistic is to accomplish. Each statistical technique performs a particular function, revealing certain information about the data. A clear conception of the analytical goals of the data analysis is a prerequisite for selecting the best statistics for achieving those goals. Statistical techniques have one of two general goals: *description* or *inference*. **Descriptive statistics** are procedures that assist in organizing, summarizing, and interpreting data. The data might be from a sample or from a whole population; in either case, descriptive statistics organize and summarize the body of data. **Inferential statistics** are procedures that allow us to make generalizations from sample data to the populations from which the samples were drawn. Recall from Chapter 7 that the ultimate reason for making observations on samples is to draw conclusions regarding the populations from which those samples were drawn. We make observations on a sample of employees in an organization because it is too expensive, time consuming, or impractical to observe all the employees; yet we want to draw conclusions about all the employees. Inferential statistics are based on probability theory and basically tell us the probability of being wrong if we extend the results found in a sample to the population from which the sample was taken.

Statistics must be chosen that are appropriate to the goals of the data analysis. In some cases, both descriptive and inferential statistics are used because the analysis is intended to accomplish both goals. However, if the goal is to summarize and describe a set of data, then descriptive statistics would be the choice; if you want to generalize sample results to a population, then inferential statistics are appropriate.

Number of Variables

A third consideration in choosing an appropriate statistic is the number of variables to be analyzed. **Univariate statistics** analyze only one variable; **bivariate statistics** analyze two variables; **multivariate statistics** analyze three or more variables. An example of a univariate statistical problem would be an investigation of the average income paid to correctional officers in a particular correctional facility. The only

variable in the problem is *amount of income*; "correctional facility" is not a variable but a constant—it has only one value or category. This problem could be changed into a bivariate problem by introducing a second variable: How does gender affect the average income paid to the correctional officers in the facility? Now the problem has two variables: *gender* and *amount of income*. A multivariate problem could be hypothesized by adding additional variables: How do gender, race, and seniority influence levels of income in the facility?

Each statistic is designed to be used on either a univariate, bivariate, or multivariate problem. However, a point of clarification is warranted: Some univariate statistics are also calculated as a step in some bivariate or multivariate statistics. For example, among the statistics to be discussed in this chapter, the mean is a univariate statistic, but it is also calculated as a part of some bivariate statistics (such as a difference of means test) and some multivariate statistics (such as analysis of variance). In these cases, however, the univariate statistic is only one step in the more complex and lengthy calculation of the bivariate and multivariate statistics.

Special Properties of the Data

In addition to the mathematical properties of the data already discussed in relation to levels of measurement, other properties of the data must be considered in choosing appropriate statistics. I illustrate a few of these. One such property is whether the data distribution involves *partially ordered data* or *fully ordered data*. A variable is partially ordered when many cases tie, or have the same value, on the variable. For instance, if we classified *social class* into three groups (lower, middle, and upper), then it would be a partially ordered variable since all the people classified as "lower social class" would be tied on that variable. A fully ordered distribution is one in which practically every case has its own rank and there are few ties on the variable. As an example, if we ranked individuals according to the order in which they finished a cross-country ski race, there might be a few ties, but most persons would cross alone and thus have their own rank. Some statistics are designed for use with partially ordered data and others with fully ordered data.

Another property of the data is whether the cases selected for various groups to be compared are independent of one another or whether they are related. For example, if we were to compare the incomes of men and women in a company, we could select a random sample of men and a separate random sample of women. In this case, exactly which people are selected for one group is independent of who are selected for the other since all the selections are random. A different research problem might call for comparing the incomes of husbands and wives. In this case, if a husband is selected for the sample, then his wife would also be included. Now the samples are not independent, but rather are dependent, or matched. Another example of matched comparisons is when you compare individuals' performances on a test at two points in time. The two groups are identical each time, so the samples are again dependent or matched. The point is that some statistics are designed for comparing independent groups and others for matched groups.

Another consideration in selecting some statistics, which is discussed in the next section, is the shape of the distribution of observations on a variable. Some statistics require that this distribution be normal or symmetrical. If the distribution is not normal, then some statistics may supply a misleading result.

As a part of the process of selecting an appropriate statistic, then, researchers review these and other assumptions about the nature of the data they are analyzing and assess the risk of using a statistic that requires assumptions that are not met, or not fully met, by the data. The more the data violate these assumptions, the greater the risk of producing a misleading result.

Audience

If all the preceding considerations are weighed and more than one statistic could be appropriately used, then consideration is given to the audience for whom the data analysis is intended. If that audience has limited statistical expertise, then a relatively simple statistic that can be understood by the audience would be preferred over a more complex statistic that might confuse. Among the simpler statistics would be the visual forms of presenting data discussed in Chapter 13 and the univariate statistics described in the next section. Some bivariate statistics, and especially the multivariate statistics, may well be beyond the comprehension of audiences without at least an elementary introduction to statistics.

With these considerations in mind, the chapter now turns to the descriptive statistics that are most commonly used in social science research.

DESCRIPTIVE STATISTICS

Descriptive statistics provide quantitative indicators of what is common or typical about a variable, how much diversity or difference there is in the variable, and how values on one variable are associated with values on one or more other variables. In the following, I describe the univariate descriptive statistics known as *measures of central tendency* and *measures of dispersion* and the bivariate descriptive statistics known as *measures of association*.

Measures of Central Tendency

Once a frequency distribution has been created (see Chapter 13), it may be desirable to further summarize or reduce the data. One way to do this is with **location measures**, which are statistics that identify some point, or location, in a distribution. One type of commonly used location measure is **measures of central tendency**, more commonly known as *averages*, which summarize distributions by identifying the "typical" or "average" value, or the center of the distribution. The three most widely used measures of central tendency are the *mode, median,* and *mean,* each designed for use with a particular level of measurement and having unique qualities.

The *mode* is the category in a frequency distribution that contains the largest number of cases. The mode for the grouped age distribution in Table 13.5 (on page 404) is the 70–74 (or the midpoint of 72) because more people fall into this age category than into any other. Although the mode can be determined for data of any level of measurement (such as the age distribution, which is ratio), it is designed for use with nominal variables and usually used with data at the nominal level, for two reasons. First, the mode is the least stable of the three measures; that is, its value can be changed substantially by rather minor additions, deletions, or alterations in the values making up the distribution. With the other measures of central tendency, adding more cases produces less dramatic shifts in their values. Second, if two or more categories are tied with the largest number of cases, we could have two, three, or more modes, none of which would necessarily be very "typical" of the distribution. The presence of several modes also undermines its utility as a summary statistic to describe the average case.

With ordinal data, the *median* is the appropriate measure of central tendency. The median is the point in a distribution below which 50% of the observations occur. For example, a distribution of six scores—10, 14, 15, 17, 18, 25—results in a median of 16. Note that when the distribution contains an even number of cases, an observed score does not fall on the median. The median is then the value halfway between the two central scores. If we add another score greater than 16 to the preceding distribution so that it contains seven scores, the median is 17, an observed score. Because the median does not take into account the actual value of all the scores, only the number of observations, whether the seventh score we add to that distri-

bution is 19, 100, or 1,000 makes no difference—the median is still 17. This makes the median very stable because the presence of extremely high or low scores in a distribution has little effect on the value of the median.

The *mean,* the measure of central tendency most people think of when they hear the word *average,* is calculated by summing all the values in a distribution and dividing by the number of cases. The calculation of the mean of the six scores in the previous paragraph would look like this:

$$\overline{X} = \frac{\sum X}{N}$$

$$= \frac{10 + 14 + 15 + 17 + 18 + 25}{6} = 16.5$$

Where

$\overline{X}$ = the mean score

$\sum X$ = sum up all the values

N = the number of cases in the distribution

The mean, however, is only suitable for interval- or ratio-level data, where there is equal spacing along a scale and various mathematical functions can be performed (see Chapter 5). (There is one exception to this: dichotomous-nominal or ordinal-level variables that have only two values. Many interval level statistics, such as the mean, can be meaningfully computed on such variables. This is called "dummy variable analysis" and is beyond what I wish to introduce in this chapter.) Because the mean takes into account the actual value of all scores in a distribution, it is less stable than the median. The presence of a few extreme scores will "pull" the mean in that direction. Because of this, the median is often the preferred average when summarizing skewed distributions, even with interval-level data, as it more accurately reflects the central value. For example, although the mean could be used

FIGURE 14.1   Hypothetical Distribution of Incomes for Ten Employees of a Firm, by Gender

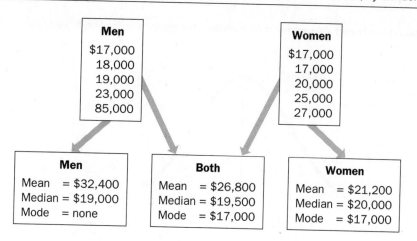

to summarize income data, the U.S. Bureau of the Census typically reports median income because the presence of a relatively few people with very high earnings tends to pull the mean to such a high level that it overstates the average family income. This and other issues relating measures of central tendency are illustrated in Figure 14.1. You can see in that figure that the group of men contains one case with an income that is much higher than the incomes of all the other nine men and women in the illustration. Figure 14.1 also shows the measures of central tendency calculated for three different groups (men alone, women alone, men and women together). The median is almost the same in all three groups, illustrating the stability of that measure of central tendency whether or not a distribution is skewed. The mean, on the other hand, fluctuates considerably from one group to another because of the extreme value for that one very high income. Figure 14.2 shows the effect of a skewed distribution on the mean and median. Basically, the mean is more strongly affected by skew, being pulled more toward the skewed end than is the median. On the other hand, the mode can't be calculated for the men in Figure 14.1 because there is no category with more cases than any other category.

Selecting the most appropriate measure of central tendency for a given set of data is not difficult. Level of measurement and skewness are the primary considerations. With relatively symmetrical distributions, the only factor is level of measurement.

Measures of Dispersion

Like measures of central tendency, *measures of dispersion* are used to describe and summarize distributions. Whereas central tendency indicators describe the middle or average of the distribution, **measures of dispersion** indicate how dispersed or spread out the values are in a distribution. Measures of dispersion add valuable information about distributions. On the basis of central tendency measures alone, we might assume that two distributions with similar averages are basically alike. Such an assumption would, however, be erroneous if the spread of the distributions were different. As illustrated in Figure 14.3, distributions *A* and *B* have the same mean, but distribution *A* is more dispersed, with values deviating more widely from the central value. The values in distribution *B* are more tightly clustered near the average. For avoiding possible erroneous assumptions about the spread of distributions, it

FIGURE 14.2   The Effect of a Skewed Distribution on the Mean and Median

A. An Unskewed, Symmetrical Distribution (The mean and median are equal)

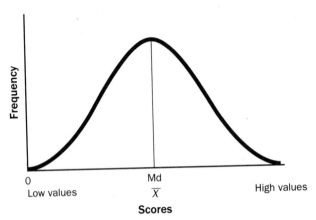

B. A Positively Skewed Distribution (The mean is greater in value than the median)

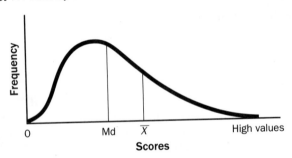

C. A Negatively Skewed Distribution (The mean is less than the median)

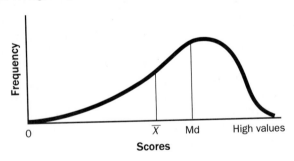

FIGURE 14.3   Two Distributions with the Same Average but Different Dispersions

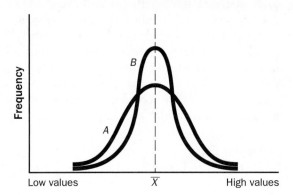

is desirable to report a measure of dispersion along with a measure of central tendency.

Three commonly used measures of dispersion are the *range, semi-interquartile range*, and the *standard deviation*. The *range* is the simplest of these, referring merely to the difference between the highest and lowest scores in the distribution. As such, the range indicates the total spread of a distribution. Knowing the end points of a distribution, however, tells us nothing about how the remaining values are dispersed within the distribution. Further, because the range is based on only two values, it is unstable. Adding or deleting extreme scores causes the range to vary widely, whereas the bulk of the distribution may change little. Moreover, despite its simplicity, the range is suitable only for interval data. The operation of subtraction used to obtain the range assumes that the values of the scores in a distribution have meaning, and that is true only for interval- and ratio-level data. Because of these limitations, the range is usually used as an adjunct to other measures of dispersion and not reported alone.

The *semi-interquartile range* (sometimes called the "quartile deviation") is conventionally symbolized by the letter Q. The semi-interquartile range is obtained by first dividing a frequency distribution into fourths, or *quar-*

tiles. The first quartile is the score below which 25% of the scores occur, and the third quartile is the score below which 75% of the scores occur. Q is calculated by subtracting the first quartile from the third quartile and dividing by 2. The semi-interquartile range is closely related to the median and is the measure of dispersion usually reported with it. In fact, the median is actually the second quartile, the point below which 50% of the scores occur. The larger the value of the semi-interquartile range, the more dispersed the scores are from the median.

The *standard deviation*, symbolized by the letter s, indicates the average (or mean) spread of the scores from the mean and is therefore the measure of dispersion usually reported along with the mean. The larger the value of the standard deviation, the more dispersed the scores are from the mean. Actual calculation of the standard deviation is relatively complex and beyond the scope of this chapter. In addition to its use as a descriptive statistic, the standard deviation has important applications in inferential statistics.

One final note on measures of dispersion. Although their overall purpose is to assist in the description of distributions, a single dispersion value from a single distribution is not particularly revealing. The major utility of dispersion indicators is in comparing several distributions because they enable us to tell at a glance which has more or less spread.

Measures of Association

Measures of association describe the nature of relationships between two or more variables, particularly looking at the *strength* and *direction* of the relationship between the variables. The *strength* of the relationship refers to how closely related the variables are. The strongest relationship is a *perfect* one in which a given change in one variable is associated with a given change in the other variable in every case. Perfect relationships are never found in social science research. A less-than-perfect relationship indicates only a tendency for the variables to vary together. Most measures of association describe the strength of a relationship through a coefficient that varies from 0.00 to 1.00, with 0.00 indicating no relationship and 1.00 indicating a perfect relationship. The closer the value of the measure is to 1.00, the stronger the relationship. So a coefficient of 0.72 indicates a stronger relationship than that of 0.52.

With ordinal- or higher-level data, measures of association also assess the *direction* of relationships between variables, which can be positive, negative, or curvilinear. A *positive relationship* is one in which the change in value of the variables is in the same direction; that is, both increase or both decrease. For example, there is generally a positive relationship between income and level of education: the more education a person possesses, the higher his or her income tends to be. This is true for most people, although not all. A *negative relationship* is one in which one variable increases while the other decreases. An example of this would be the relationship between social class and family size: People in higher social-class positions tend to have fewer children, and thus smaller families, than do people in lower-class positions. So, as social class goes up, family size tends to go down.

Finally, a *curvilinear relationship* is one in which the direction of change in one variable is not consistent with changes in the other. For example, in a U-shaped curvilinear relationship, both low and high values of the independent variable may be associated with high values of the dependent variable. In this case, a negative relationship exists between the two variables for low values on the independent variable and a positive relationship for high values on the independent variable.

The direction of the relationship between variables is indicated by a plus (+) or minus (−) sign. A coefficient of +0.72 would indicate a strong positive relationship, whereas

−0.09 would indicate a weak negative relationship. Keep in mind that the number indicates the strength of the relationship, so a coefficient of −0.79 indicates a stronger relationship than does a value of +0.56.

The overriding determinant in selecting a measure of association is the level of measurement of the data at hand; I next consider measures of association according to the level of measurement for which each was designed.

Nominal Data

Some data are dichotomous in form. That is, the variables have only two values, such as yes or no, male or female. A useful measure of association for two dichotomous variables is the *phi* (pronounced *phee*) coefficient (Φ). The data are cast into a 2 × 2 table as illustrated in Table 14.2 (this same table was used in Chapter 13 as a part of Figure 13.8 to illustrate a zero-order partial table). *Phi* indicates the strength of the relationship between variables by yielding a value between −1.00 and 1.00. Although *phi* may at times yield negative values, the negative sign is simply ignored because it has no meaning in the case of nominal

TABLE 14.2 Nominal Data Suitable for Phi

Is Life Exciting	Gender	
	Male	*Female*
Exciting	213 (50.1%)	221 (39.8%)
Routine/Dull	212 (49.9%)	334 (60.2%)
Totals	425 (100%)	555 (100%)
		Phi = 0.11

Source James A. Davis and Tom W. Smith. *General Social Survey (1972–1994)* (Chicago: National Opinion Research Center; Storrs, CT: The Roper Center for Public Opinion Research, University of Connecticut).

TABLE 14.3   Nominal Data Suitable for Lambda and Goodman and Kruskal's Tau

		Region			
		Northeast	Midwest	South	West
Religious Preference	Protestant	54 (41.2%)	140 (64.5%)	206 (84.8%)	80 (53.3%)
	Catholic	55 (42.0%)	56 (25.8%)	28 (11.5%)	43 (28.7%)
	Jewish	10 (7.6%)	1 (0.5%)	1 (0.4%)	3 (2.0%)
	None	12 (9.2%)	20 (9.2%)	8 (3.3%)	24 (16.0%)
	Totals	131 (100%)	217 (100%)	243 (100%)	150 (100%)

Lambda = .07
Goodman and Kruska's tau = .07

Source   James A. Davis and Tom W. Smith. *General Social Survey (1972–1994)* (Chicago: National Opinion Research Center; Storrs, CT: The Roper Center for Public Opinion Research, University of Connecticut).

data. In Table 14.2, the *phi* coefficient of .11 indicates a fairly weak relationship between gender and whether people find life exciting.

Phi is considered a good measure of association for three reasons. First, it is quite easy to compute. Second, it is mathematically related to measures of association suitable for other levels of measurement, which makes possible comparing the strength of different relationships. Measures of association that are not mathematically related have different operating characteristics and produce values that are not comparable, precluding meaningful comparisons. Third, *phi* is a member of a group of measures of association that can be given what is called a **proportional reduction in error** (PRE) interpretation. The PRE interpretation means that the measure shows how much the independent variable helps to reduce error in predicting values of the dependent variable. To interpret *phi* in this way, it is first necessary to square it (Φ^2). For example, $\Phi = .11$ is squared to become $\Phi^2 = 0.01$. This latter value is treated as a percentage and is interpreted to mean that the independent variable of *gender* reduced the error in predicting values of the dependent variable by only 1%. All PRE statistics are interpreted this way.

Because *phi* is suitable only for two dichotomous variables, a different measure of association must be used for nominal data with more categories. The most generally useful measure for data of this type is *lambda* (λ). As with *phi*, the data are cast into a contingency table, as illustrated in Table 14.3. *Lambda* is always positive because it is a direct reading PRE statistic. That is, *lambda* indicates the proportional reduction in error as calculated and need not be squared as *phi* is. This is an important point to remember as the values *lambda* produces usually look rather small. The reason is not that *lambda* understates relationships, but that the *lambda* values are "presquared," which makes them appear small.

TABLE 14.4   Ordinal Data Suitable for Gamma, Somer's D, or Kendall's Tau

		Income		
		Low	Medium	High
Number of Hours of Television Watched	High	49 (19.2%)	19 (8.6%)	10 (6.7%)
	Medium	164 (64.3%)	136 (61.5%)	86 (57.7%)
	Low	42 (16.5%)	66 (29.9%)	53 (35.6%)
	Totals	255 (100%)	221 (100%)	149 (100%)

gamma = − .33
Somer's D = − .18
Kendall's tau-b = − .20

Source   James A. Davis and Tom W. Smith. *General Social Survey (1972–1994)* (Chicago: National Opinion Research Center; Storrs, CT: The Roper Center for Public Opinion Research, University of Connecticut).

Although *phi* and *lambda* are the most common nominal measures of association, several others are available. For nominal data with more categories, Goodman and Kruskal's *tau* (τ) is often used. *Tau* can be given the PRE interpretation. *Tau* is sometimes preferred to *lambda* because the former uses data from all cells in the table, whereas the latter only includes in its computation data from some of the cells. Also there are some cases where *lambda* will produce a coefficient of 0.00 even though there is a relationship between the two variables. However, *tau* also has its weaknesses; in particular, it does not always vary from 0.00 to 1.00, especially when the dependent variable has more categories than the independent variable.

Ordinal Data

There are many commonly used measures of association for ordinal data. A major consideration in selecting one has to do with whether the data are partially ordered or fully ordered.

When there are only a few ordered categories and many cases to place into them, the data will contain many ties. Data of this type are called partially ordered and are usually put into a contingency table, as illustrated in Table 14.4. Several alternative measures of association are available for data of this type. *Gamma* (γ), Somer's *D*, and Kendall's *tau-b* (τ-*b*) are all suitable and vary between −1.00 and +1.00. They all can be given the PRE interpretation. There are, however, some differences among them that make one or another more appropriate in a given situation.

Gamma is the easiest of the three to compute but, unfortunately, it does not take into account any of the tied scores and tends to overstate the strength of the relationship. Indeed, on the basis of a positive *gamma* alone, we cannot safely assume that as X increases, Y also increases, which is a normal assumption of a positive relationship. Because of its failure to consider ties, all a positive *gamma* allows us to conclude is that as X increases, Y does not decrease, which, of course, is a much weaker conclusion.

Somer's D is used when we are only interested in our ability to predict a dependent variable from an independent variable. When predicting Y from X, Somer's D takes into account ties on the dependent variable. This has the effect of reducing the value of D in comparison to *gamma* when the two are computed on the same data. Somer's D, however, gives a more accurate indication of how much the independent variable reduces error in predicting the dependent variable.

Finally, Kendall's *tau* takes into account all of the tied scores. It indicates the degree to which the independent variable reduces error in predicting the dependent variable *and* how much the latter reduces error in predicting the former. Because it considers the relationship both ways, *tau* is particularly appropriate when we do not have a clearly identifiable independent and dependent variable and merely wish to determine if two variables are related. By including all the ties, a positive *tau* allows us to conclude correctly that as X increases, Y also increases. As this is the type of statement we expect to make on the basis of a positive result, Kendall's *tau* is more generally useful than *gamma*.

Interval Data

The most used measure of association for interval data is the *correlation coefficient,* or Pearson's r. As with other measures of association, Pearson's r varies between -1.00 and $+1.00$ and may be squared (r^2) and given the PRE interpretation. When squared, this value is called the *coefficient of determination.*

The correlation coefficient has a unique characteristic that is important to remember when applying or interpreting it. Pearson's r indicates the degree to which the relationship between two interval-level variables can be described by a straight line when plotted on a scattergram, as in Figure 14.4. The formula for r mathematically determines the best fitting straight line and then considers the amount the

scores deviate from perfect linearity. This feature of Pearson's r means that if the relationship between the two variables is somewhat curvilinear, as the two lower scattergrams in Figure 14.4 describe, r will understate the actual strength of the relationship. Therefore, it is advisable to plot, or have the computer plot, a scattergram. A *scattergram* is a technique for visually plotting the relationship between two variables. In a scattergram, the values of one variable (usually the independent variable) are arrayed along the horizontal axis and the values of the other (dependent) variable along the vertical axis. In the scattergram, each dot represents a case and its placement is determined by its value on the two variables. The scattergram is useful for uncovering curvilinearity that might produce a misleading correlation coefficient. Figure 14.4 illustrates three typical scattergrams. In those instances in which curvilinearity is discovered, a measure of association other than Pearson's r should be selected. Curvilinear measures of association do exist, but are beyond the scope of our discussion. They may be found in more advanced textbooks about Statistics (see the For Further Reading section in this chapter).

Another commonly used statistical procedure for two or more interval-ratio variables is *regression analysis.* Multiple regression is used for a variety of purposes, but a typical application involves estimating the effect of multiple independent variables on a dependent variable. For example, we might be interested in determining if a person's income is influenced by *gender, race,* and *age*—plus, how much each of these variables affects *income.* (Note: While regression calls for all variables to be interval-ratio, nominal variables such as *gender* and *race* can be used if they are dichotomous and used as the dummy variables that I mentioned earlier in this chapter.) Correlation will indicate whether two variables are associated, but regression analysis permits one to estimate how much change in the dependent variable is produced by a given change in an independent

FIGURE 14.4   Three Typical Scattergrams

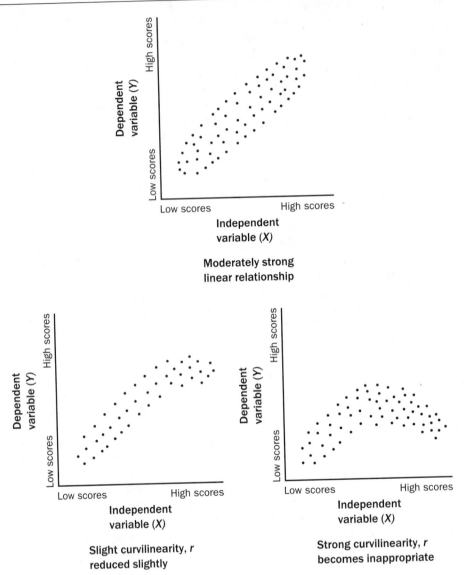

Moderately strong
linear relationship

Slight curvilinearity, *r*
reduced slightly

Strong curvilinearity, *r*
becomes inappropriate

variable. Multiple regression is especially useful in that it can handle a large number of independent variables simultaneously, thus permitting researchers to estimate the effects of one variable while controlling for others. We know, for example, that *income* is correlated with *race, education, age,* and *occupation,*

among many other variables. If we collect these data from a number of respondents, regression analysis permits us to estimate the contribution of each independent variable to *income.* Multiple regression produces coefficients that indicate the direction and amount of change in the dependent variable to be

FIGURE 14.5 Descriptive Statistics Categorized by Level of Measurement

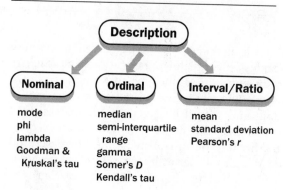

Nominal	Ordinal	Interval/Ratio
mode	median	mean
phi	semi-interquartile	standard deviation
lambda	range	Pearson's *r*
Goodman &	gamma	
Kruskal's tau	Somer's *D*	
	Kendall's tau	

expected from a unit change in each independent variable. Thus, if the dependent variable is *dollars of income*, a regression coefficient of +580 for the variable *years of education* indicates that one more year of education is "worth" an additional $580 in income, assuming the other independent variables are held constant. In our illustration, regression would produce an equation as follows:

$$Y'(\text{income}) = a + b_1(\text{race}) + b_2(\text{education})$$
$$+ b_3(\text{age}) + b_4(\text{gender})$$

where *a* is the value of *Y* before other factors' effects are considered, and each *b* is an estimate of the effect of its variable on income. The Applied Scenario box suggests some of the ways in which regression analysis is used in the development of social policy.

Before we leave measures of association, one important matter requires emphasis. Association or correlation does not imply causality! Just because one variable is labeled independent and another dependent and a relationship is found between them does not prove that changes in one variable caused changes in the other. As I emphasized in Chapter 4, correlation is only one step toward inferring causality. In addition, it is necessary to affirm the appropriate temporal sequence and rule out rival causal variables. Figure 14.5 provides a sum-

mary of the descriptive statistics discussed and links them with level of measurement.

The Normal Distribution

Chapter 13 discusses frequency distributions as descriptive statistical procedures that transform a raw data distribution into a more meaningful form. There is another way to transform raw data to derive additional meaning from it, and this transformation also serves as our transition from descriptive to inferential statistics. A *transformation* is a set of arithmetical operations that are executed on a variable to obtain a new variable. One common transformation is known as *standardizing*. Raw scores are transformed into *standard scores*, also known as *z-scores*. Symbolically, a standardized score, *z*, is given by:

$$z = \frac{x - \bar{x}}{s}$$

Standard scores are obtained by following these two steps:

1. Subtract the mean, $\bar{x}$, from a raw score, *x*.
2. Divide the $x - \bar{x}$ difference by the standard deviation, *s*, of the distribution.

If all the raw scores in a distribution are transformed to *z*-scores, a new distribution is obtained that will always have $\bar{x} = 0$ and $s = 1$. This is called a *standard normal distribution* and, as it turns out, this distribution has some convenient properties.

One way to think about the *z*-transformation is that it expresses raw scores in a distribution in terms of standard deviation units rather than the original unit of measurement. A *z*-score of 1.50 indicates a score point that is 1.50 standard deviations greater than the mean, whereas a *z*-score of −1.50 indicates a score point 1.50 standard deviations below the mean. In other words, a standard score *z* is a number indicating the distance that a raw score deviates from the mean as measured in

Regression analysis is often used as a critical tool in making social policy decisions because the predictions that are made with regression can assist people or organizations assess where their group, organization, or city should be on some factor. Let me give an example of this. I have seen a number of organizations conduct pay equity studies using regression analyses to help them decide whether women are underpaid and should be given "catch-up" raises. I will keep the example simple by discussing just one independent variable that affects *income*; namely, *seniority* in an organization. Generally, because of cost-of-living pay increases and promotions, a person's income rises with seniority in a job. Or at least the impact of *seniority* on *income* should be the same for both men and women. So, the organization can conduct a regression analysis with *seniority* as the independent variable and *income* as the dependent variable, but only for the men. This regression analysis enables the organization to essentially predict, for any level of seniority, what the average income is for men. Then they make the assumption that the relationship between *seniority* and *income* should be the same for women: A woman with a given level of seniority should earn an income that is at least the same, or not far below, the average for men at that seniority level. The income of each woman in the organization can be compared to the regression prediction of income for her seniority level. If she falls significantly below that level, then she would be a candidate for a "catch-up" pay increase.

Regression analysis has also been used to assist city managers and social service personnel in determining whether they were adequately detecting child abuse in their communities (Ards, 1989). One of the problems in this realm is that much child abuse is difficult to detect. If a community has low abuse rates, it may be because much abuse is hidden or because the rates

standard deviation units. Further, the sign of a z-score indicates whether the score point is above $(+)$ or below $(-)$ the mean of the distribution.

On first encounter, it may seem strange to express raw scores in terms of standard deviation units, but doing so allows one to make comparisons between distributions that would otherwise be difficult to make because their original units of measurement differ. Standardizing equalizes the units so that a meaningful comparison can be made. For example, the comparison of an individual's raw scores on two different IQ tests would not be legitimate unless the IQ tests have a common unit of measurement and operate on the same scale. Therefore, mental testers commonly employ the z-transformation as a way of making different IQ tests comparable.

Consider the data in Table 14.5. On both tests, someone received a score of 99, but are both of those scores really equal? In terms of raw scores they certainly appear to be, but in relative terms—that is, compared to the other scores in the distribution—the 99 on Test 2 is far more remarkable and a z-score transformation will reveal this. On Test 1, with a relatively high mean and small standard deviation, the 99 transforms to a z-score of 1.46, which means it is 1.46 standard deviations above the mean of 94.6. The 99 on Test 2, with a low mean and high standard deviation, transforms to a z-score of 2.47, which means it is 2.47 standard deviations above the mean of 49.6. The comparison of z-scores shows that the two scores of 99 are in fact far from equal compared to the other scores in their respective distributions.

are actually low. And many cities do not have the resources to aggressively pursue hidden cases that may in fact not be there. How do you determine whether there may be hidden cases in your community? The Department of Health and Human Services (DHHS) has done extensive research in a number of communities to uncover undetected child abuse. The results of this research serve as a database to determine what levels of abuse to expect in communities with various characteristics. The DHHS study found, for example, that levels of child abuse were related to per capita income, unemployment levels, population density, and other factors. This produced a regression formula with these factors as the independent variables that are used to predict levels of abuse in a community. Any community could plug into the regression formula their per capita income, unemployment rate, and so on, and predict what the child abuse rate in their community should be. If their abuse rate falls significantly below that level, it may be due to the existence of undetected cases, and this may warrant the expenditure of resources to locate those cases.

So you can see how regression predictions are based on the known relationship between a set of independent variables and a dependent variable in some sample (the men in the organizations in the first example, and the group of cities studied by DHHS in the second). Regression analysis for policy decisions is virtually always multiple regression because we know that something like *income levels* or *child abuse rates* are affected by many factors. Making good predictions requires that we consider as many of those factors as possible. A final point to make is that regression analysis would typically be only one of a number of factors that shape policy decisions. Many nonscientific considerations might also be important, but regression can provide an important bit of information that can assist policy makers.

TABLE 14.5 Two Sets of Hypothetical Test Scores

	Test 1	Test 2	
	99	99	
	98	45	
$\bar{x} = 94.6$	96	44	$\bar{x} = 49.6$
$s = 3$	95	44	$s = 20$
	95	43	
	93	41	
	91	41	
	90	40	

Allowing for these types of comparisons to be made is one important use of z-scores.

Another important use of z-scores has to do with what is called the *normal distribution*. The normal distribution is a continuous, bell-shaped distribution, as shown in Figure 14.6. The horizontal axis in Figure 14.6 represents values or categories of a variable, and the height of the curved line indicates the number of cases in the distribution that fall at various points on the variable. The Figure also shows the location of the mean and various standard deviation points. You can see that most of the cases fall near the mean, and then the number of cases drops off as you move away from the mean in either direction. By three standard deviation units from the mean, relatively few cases are found. The **normal distribution** is a symmetrical, unimodal distribution in which the mode, median, and mean are identical.

FIGURE 14.6 Proportions of the Area Under the Normal Curve Between Selected Points

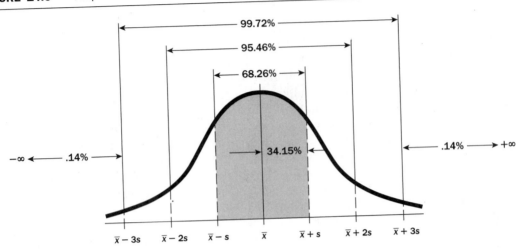

Actually, there is no single normal distribution, but many. Normal distributions may have different means and different standard deviations, for example. What makes them all normal is that they are symmetrical and unimodal, and have the three measures of central tendency at the same point. In addition, all normal distributions have the same proportion of cases between the same two ordinates. This statement means that between, say, +1 and −1 standard deviations from the mean for any normal distribution, the proportion of the total cases in the distribution bounded by those two points will be the same.

In Figure 14.6, the shaded area is the proportion of the total area under the curve bounded by points that are 1 standard deviation unit on either side of the mean. For any normal distribution, no matter what the mean and standard deviation happens to be, the area bounded by $x - 1s$ and $x + 1s$ always contains 68.26% of the cases in the distribution, whereas 95.46% of the cases in the distribution fall between $x - 2s$ and $x + 2s$. In place of $x - s$ and $x + s$, one may substitute any two points one wishes and carve out a proportion of the total area that will be the same for all normal distributions. Figure 14.6 shows

selected points and the proportions of the total area that fall between them for any and all normal distributions.

This points to another use of z-transformations: They are used to assess the relative position of a score in a distribution of scores. For example, suppose you took a national merit exam and received a score of 580. How well did you do compared to the others who took it? If you know that the scores on this exam are normally distributed, and you know the mean score and standard deviation, you can then calculate your position relative to the others who took the exam. Suppose the mean was 490 and the standard deviation was 72. You could then use the z formula to transform your exam score of 580 into a z-score (standard deviation unit) as follows:

$$z = \frac{x - \overline{x}}{s} = \frac{580 - 490}{72} = 1.25$$

So your score is +1.25 standard deviations above the mean for all the students who took the test. By looking at Figure 14.6, you can determine that you scored better than at least 84% of the students who took the exam. You can figure this out because 50% of the students scored below the mean and another 34.15% scored between the mean and $+1.00s$ above the mean. To figure out what additional percentage is included between $+1.00s$ and $+1.25s$, you can refer to tables that tell which proportion of cases fall between any two points in a normal distribution. (These tables can be found in any introductory statistics text.) From these tables, it can be determined that 89.44% of all cases in a normal distribution fall below $+1.25$ standard deviations above the mean. So you did better than over 89% of all the people who took that examination! Transforming your test score into a z-score enables you to make a very precise statement about your place in the distribution relative to others in the distribution.

The standard normal distribution is useful for comparing distributions with different means and standard deviations, and helps us make precise statements about relative position in a distribution. A third important use of the standard normal distribution is that it serves as the foundation for inferential statistics, which will be the next focus of our attention.

INFERENTIAL STATISTICS

As noted in Chapter 7, most social research is conducted on relatively small samples drawn from much larger populations, and all the statistical procedures discussed thus far are designed to assist in describing, summarizing, and interpreting data from samples. Findings from sample data, however, would be of little scientific value if they could not be generalized beyond the members of the sample to the larger population from which the samples were drawn.

Probability Theory

How do we know when research findings are generalizable? Generalization is always an uncertain business, but inferential statistics can reduce the uncertainty to the point where rea-

sonably safe generalizations can be made and the probability of a given amount of error estimated. Inferential statistics are based on probability theory, the same probability theory that works to make probability samples representative. Because of this, inferential statistics can be meaningfully applied only to data based on probability samples or experiments in which random assignment has been implemented.

Probability theory allows the mathematical calculation of the likelihood or probability that random events or outcomes will occur. For example, if 10,000 raffle tickets are sold and the winning ticket is chosen in a way such that each ticket has an equal chance of being selected, then each person who purchases one ticket has 1 in 10,000 chances of winning. However, if you buy 10 tickets, then you have 10 in 10,000, or 1 in 1,000, chances of winning. You have increased the probability of your winning even though the mechanism for selecting the winner is unchanged. This does not mean, of course, that you will win. Less likely events do occur, although with lower frequency than more likely ones, and a person who purchased one ticket could certainly win the raffle. In other words, probability theory tells us the *likelihood* of something's occurring; it does not tell us what *will* occur. For this reason, inferential statistics can tell us whether the odds are on our side that a particular generalization is accurate, but cannot tell us whether the generalization is in fact true.

As noted in Chapter 7, probability samples are supposed to represent the populations from which they were drawn, but we expect differences between a sample and its population owing to chance alone. Because these differences are due to the random process used in the selection of samples, their probability can be readily calculated. For example, suppose a researcher draws a probability sample of 50 delinquent boys and applies a special intervention designed to reduce future delinquency involvement. At the end of one year the success rate of the sample is 75%, whereas for the population of untreated delinquents the rate is 60%. Can we conclude that the experimental intervention produced the higher success rate? At the level of the sample alone, it is clear that the treated boys had a better success rate, but the differences could be due all or in part to sampling error. Even with random sampling, the researcher could have obtained a sample of boys who were on average better risks than those in the general population of delinquent boys. The question that inferential statistics answers is whether the difference between the sample results and the population results is too great to be due to chance alone. In the example, running the appropriate inferential statistic would tell us what the chances are of obtaining, through random error, a 15% difference between a population and a probability sample drawn from that population. If such a difference were highly probable, we would conclude that the seeming effect of the experimental treatment was due to chance differences between sample and population. On the other hand, if the result indicated that there was only a small likelihood that a 15% differential was due to chance, we would conclude that the treatment did produce a generalizable effect.

Sampling Distributions

Our ability to determine the likelihood that a sample difference is due to chance derives from the properties of the normal distribution (discussed earlier in this chapter) and what are called *sampling distributions* (see Chapter 7). A **sampling distribution** is a distribution of sample statistics. For instance, in the example in the preceding section, a sample of 50 delinquent boys was selected; it had a success rate of 75%. We could have selected more samples than this; we could have selected 10 separate samples of 50 boys each and computed a success rate on each sample. The success rate found in each sample would probably differ slightly because each sample consists of a different set of 50 boys. For the 10 samples, we

might have found the following success rates: 72%, 71%, 73%, 69%, 66%, 70%, 73%, 69%, 71%, 67%. These 10 sample results are called a sampling distribution. If we knew the success rate among all delinquent boys in the population from which these samples were taken, this would be referred to as the *population parameter*. Some of the samples will show the same success rate as in the population; other samples will show a success rate higher than the success rate in the population, whereas other samples will show a lower success rate. In other words, the sample success rates will be distributed around the actual success rate in the population from which the samples are taken. Or, put another way, a sampling distribution involves sample statistics that are distributed around a population parameter. (A sampling distribution is another form of data distribution, discussed in Chapter 13; it is a simple frequency distribution in which each case is a sample statistic.)

Sampling distributions have certain properties that are critical in inferential statistics. In particular, the **central limit theorem** from mathematics tells us that when a large number of random samples are selected from a population, and when each of those samples is itself large in size, the resulting distribution of sample statistics has two key properties: (1) the sampling distribution will approximate a normal distribution, and (2) the mean of the sampling distribution will be equal to the population parameter. In the example in the preceding paragraph, this means that the distribution of all sample success rates will be normal, and the success rate in the population—which is what we are trying to discover or infer—will be equal to the average of all those sample success rates. So, we can determine the probability that a given sample result will fall between any two points in the sampling distribution, just as we did with z-scores and the normal distribution. This means that we can determine the probability that a sample result will fall far from the population parameter. Recall from the earlier discussion that approximately 68% of all cases in a normal distribution fall within 1 standard deviation unit above and below the mean of the distribution, and approximately 95% of cases fall within 2 standard deviations (see Figure 14.6). Likewise, in a distribution of samples, 95% of the sample results will fall within 2 standard deviations of the mean of the sampling distribution, which is also the population parameter. Only 5% of samples will fall more than 2 standard deviations away from the population parameter.

Going back to our previous example, we can choose one of two alternatives:

- the treatment didn't work, and we have selected a sample of boys that, by chance, is very different from the boys in the population (i.e., sampling error), or

- the treatment worked, and this is what makes the treatment boys very different from the nontreated boys in the population.

However, if the success rate in the population is 60% and the sample success rate of 75% is more than 2 standard deviations away from 60%, then the *z*-transformation tells us that 75% is far out in one of the tails of the normal distribution and obtaining such a sample by chance would be highly unusual. Therefore, we would be on safer grounds, just in terms of probabilities, to conclude that the 75% is probably not due to sampling error, but rather is showing that the treatment had an effect on this group of boys.

Although many different kinds of inferential statistics are used, they all follow the basic logic involving sampling distributions and normal distributions just described. I next describe two different kinds of inferential analyses that are done: *statistical estimation* and *statistical hypothesis testing*.

Statistical Estimation

One set of inferential statistics used by social scientists is called **statistical estimation**, and in-volves using a sample statistic as an estimate of a population parameter. This is often done in public opinion polling, for example, where the opinions of a sample are used as an estimate for the opinions of a population. If a national public opinion poll, for example, shows that 58% of the sample believe President Clinton is doing a "good job," then we want to use that 58% as our estimate for what the whole public believes of Clinton. However, we realize that we could have obtained a sample proportion of 58% even if the actual population proportion is slightly different. So, what we do instead is to draw an interval around that proportion, called a *confidence interval*, in which we are fairly sure is contained the actual proportion of people in the population who believe Clinton is doing a good job. How big that interval is depends on, among other things, how much error we would be willing to accept. I won't show you the calculations here, but let's suppose that we were willing to accept 5% error rate, and this produced a confidence interval of ±3%. This means that we are 95% sure that the proportion of people in the population who think Clinton is doing a good job is between 55% (58%– 3%) and 61% (58% + 3%). More technically what this means is: If we draw a large number of samples from the same population, calculate the proportion on each and draw a confidence interval around each proportion with a 5% error rate, then 95% of those confidence intervals will include the population proportion and 5% won't include the population proportion. That 95% is called the **confidence level**: It is how often we will be correct, in the long run, when we say that the population value falls in the intervals that we create.

If we wanted to be more certain of our estimate, then we would drop the error rate to, say, 1%; the consequence of that would be to widen the confidence interval. However, there is a trade-off here. While we gain confidence, we lose some precision: A wider interval is less precise than a narrower interval. Using the above example, if we wanted to be 99% sure of our conclusions, we might have to tolerate a ±5% confidence interval. Then we would be saying that the proportion of people in the population who believe Clinton is doing a good job is between 53% and 63%. We are more confident that this conclusion is true, but the conclusion is also considerably less precise (in other words, we are less clear about exactly where the population value is).

The statistical estimation just described using proportions can be done with nominal or higher levels of measurement because you can calculate proportions with nominal level variables. We could do the same kind of analysis with interval–ratio-level variables, but we would be doing it by calculating means rather than proportions. For example, if we wanted to know the mean income in a community, we could select a sample and calculate the sample mean income and then draw a confidence interval around that mean to serve as the estimate of the population mean.

Statistical Hypothesis Testing

Statistical estimation is one type of inferential statistic. The other type is a special form of hypothesis testing in which two hypotheses are developed that are precisely opposite of each other. The outcome of a statistical test is then used to determine which hypothesis is most likely correct. The first hypothesis, called the **null hypothesis**, states that no relationship exists between two variables in the population or that there is no difference between a sample statistic and a population parameter. "There is no relationship between income and gender in the population" or "There is no difference between the success rate of the sample and the success rate of the population" would be two examples of possible null hypotheses. Alternatives to null hypotheses are called **research hypotheses**. Suitable research hypotheses to go with the two preceding null hypotheses would be: "There is a relationship

between income and gender at the level of the population" and "There is a difference between the sample success rate and the population success rate." In each application of a statistical test, we assess the research hypothesis by determining whether the opposite null hypothesis is probable or improbable.

It is not uncommon, upon first exposure to statistical hypothesis testing, to question the utility of null hypotheses. If we believe the research hypothesis is true, why not just test that hypothesis? The need for null hypotheses stems from the fact that it is not possible with inferential statistics to prove research hypotheses directly. However, by determining that it is very likely that a null hypothesis is false, we indirectly provide evidence that its opposite, the research hypothesis, is probably true. To illustrate this point, I give an example where there is a clear criterion to use in testing a hypothesis. Imagine attempting to determine whether a die is unbiased. If it is, each side should appear approximately an equal number of times—about one-sixth of the times the die is thrown. If the die is biased, one or more outcomes will occur a disproportionate number of times—one or more numbers occurring substantially more than one-sixth of the time. Suppose the die is tossed 12 times, with each outcome occurring twice, or one-sixth of the time. Would this prove the die unbiased? The answer is no. The bias, if it exists, might be too slight to appear after only 12 trials. What if the die is tossed 144 times and still each outcome occurs one-sixth of the time? Such a result would still not make it certain that the die is unbiased, for a slight bias might be revealed by even more trials. Indeed, no number of trials would provide absolute assurance that the die is unbiased. However, a large number of trials with no evidence of bias would make the hypothesis of a biased die so unlikely that it is reasonable to reject it. By finding no evidence of support for one hypothesis, we indirectly obtain support for its alternative. In all applications of statistical hypothesis testing, if the evidence

TABLE 14.6   Illustration of Type I and Type II Errors

Decision	Condition in Population	
	Null Hypothesis Is True	**Null Hypothesis is False**
Reject null hypothesis	Type I (alpha) error prob. = alpha	Correct decision (power of test) prob. = 1 − beta
Fail to reject null hypothesis	Correct decision prob. = 1 − alpha	Type II (beta) error prob. = beta

fails to support the null hypothesis, then the opposing research hypothesis is accepted as probably true.

Although their computational routines vary, all inferential statistical tests use the properties of sampling distributions and normal curves to yield a result indicating the probability that the null hypothesis is true. In choosing between the competing hypotheses, we ask: How unlikely must it be for the null hypothesis to be true before we are willing to reject it as false and accept the research hypothesis? This is determined by the **alpha level**, or the probability at which the null hypothesis will be rejected (see Table 14.6). Researchers have some discretion in setting alpha levels but must guard against the two types of inferential errors that can be made. **Type I error**, or **alpha error**, is the probability of rejecting a null hypothesis that is actually true. The alpha level selected determines the amount of alpha error we are willing to tolerate, so the researcher directly controls Type I error by setting alpha. Alpha levels are usually written as $p < .05$ and read as "a probability of less than 5%." Alpha is the opposite of the confidence level discussed earlier. An alpha of .01 is the equivalent of a confidence level of .99. The former states the probability of being wrong, whereas the latter states the opposite: the probability of being correct.

Alpha levels can be understood in the following way. Suppose the null hypothesis is "No relationship exists between variable X and variable Y in the population." We draw a sample from that population to test the hypothesis and indeed find a relationship in the sample data. An alpha level of .05 means that if the null hypothesis is actually true (i.e., there is actually no relationship between the variables in the population), then we would find a relationship as large as we did by chance in 5% of the samples we draw. We are now faced with a difficult dilemma: Is the null hypothesis true and our sample an unusual one, or is the null hypothesis false? Inferential statistics can't answer that question. It can tell us that, in the long run and by testing many hypotheses, we would be correct 95% of the time by rejecting null hypotheses when differences of a given size have an alpha level of .05.

If we set an alpha level higher, say .20, this makes it easier to reject the null hypothesis and conclude that the research hypothesis has merit. However, it also makes it likely that we will quite frequently reject a null hypothesis that is actually true. To avoid so much Type I error, we might establish a very stringent alpha, such as .001, meaning that we would reject the null hypothesis only if its odds of being true were less than 1 in 1,000. If the null hypothesis can be rejected at such an alpha level, we would be quite confident that it was false and that we were not committing a Type I error.

But setting extremely rigorous alpha levels raises the probability of making the other possible inferential error. **Type II error**, or **beta error**, is the probability of failing to reject null hypotheses that are actually false. Alpha levels such as .001 make it so difficult to reject null hypotheses that many false ones, which should be rejected, are not. Thus, in selecting suitable alpha levels, we face a dilemma. Guarding against one type of error increases the chances of committing the other type. Fortunately, conventions regarding appropriate alpha levels have been developed. In general, social researchers use alpha levels of .05 or .01 as the point at which they will reject null hypotheses. Although these levels are the most common you will see in research reports, nothing about them is sacred. The researcher must consider the purpose of the research and the alternative risks in selecting a given alpha level. For example, medical research into the safety and effectiveness of new drugs or treatments typically operate with very stringent alpha levels because life-and-death matters are at stake. If the null hypothesis in a research project is "Drug X is not safe," then we would want to reject it only if we were very sure that it is false. If it is wrongly rejected, people would be exposed to an unsafe product. On the other hand, a program evaluation with a null hypothesis of "Reminder phone calls do not reduce the number of no-shows for clinic appointments" might select an alpha level as low as .10. If the procedure shows any chance of working, the clinic may want to try it. In this case, the consequences of rejecting a true null hypothesis would not be very severe.

A common way of expressing that the null hypothesis was rejected is to indicate that a given result is "statistically significant" at some specified alpha level. Unfortunately, the use of the word *significant* can cause some confusion. In popular usage, *significant* means important or notable. Its meaning in statistics, however, does not have the same connotation. Whether a set of research findings is important or notable depends on the topic, theory, sample, and quality of measurement as well as on statistically rejecting null hypotheses. Research on trivial matters or research that is flawed procedurally

cannot produce important results no matter how many "statistically significant" findings it might contain. Be sure to remember that the meaning of *significant* in statistics is simply that a null hypothesis has been successfully rejected.

Hypothesis-Testing Statistical Procedures

The five considerations in choosing statistics discussed in the beginning of this chapter are applicable to choosing inferential as well as descriptive statistics. I organize this review of hypothesis-testing inferential statistics by level of measurement, but you will see that it gets more complicated here because some statistics are especially designed for cases where there is a nominal independent variable and an interval-ratio dependent variable.

Nominal and Ordinal Data

A widely used bivariate, inferential statistic suitable for nominal data is chi-square (χ^2), which is applied to data in contingency table format. Chi-square is also routinely used on contingency tables where one or both variables is ordinal level. (Review Table 14.1 to see why this is acceptable.) Chi-square operates by comparing the number of cases actually found in each cell of a contingency table to what would be expected in those cells if there were no relationship between the variables and the cases were thus randomly distributed in the table. For example, Table 14.7 shows the relationship between region and religious preference in a sample of 741 persons in the United States. The null hypothesis for this table would be: "There is no relationship between the two variables." Chi-square compares the actual values in a table such as this to what would be expected if the variables were unrelated. Table 14.7 also contains these expected values in parentheses in each cell. Expected values can be calculated from the marginals in the table. The expected proportion of cases in each cell is what would be found in the population if

TABLE 14.7   Nominal Data from Table 14.3 Suitable for Chi Square, Showing Observed and Expected Values

		Region				
		Northeast	Midwest	South	West	Totals
Religious Preference	Protestant	54 (exp = 85)	140 (exp = 141)	206 (exp = 157)	80 (exp = 97)	480
	Catholic	55 (exp = 32)	56 (exp = 53)	28 (exp = 60)	43 (exp = 37)	182
	Jewish	10 (exp = 3)	1 (exp = 4)	1 (exp = 5)	3 (exp = 3)	15
	None	12 (exp = 11)	20 (exp = 19)	8 (exp = 21)	24 (exp = 13)	64
	Totals	131	217	243	150	N = 741

Chi Square = 107, $p < .001$

Source   James A. Davis and Tom W. Smith. *General Social Survey (1972–1994)* (Chicago: National Opinion Research Center; Storrs, CT: The Roper Center for Public Opinion Research, University of Connecticut).

there is no relationship between the two variables. For example, because people in the Northeast make up 17.6% of the entire sample, we would expect 17.6% of all Protestants in the sample (85 out of 480) to be from the Northeast if the variables of religious preference and region are unrelated. Due to random variation, one would expect that the actual values in the sample will not be exactly the same as the expected values even when there is no relationship. However, the more the actual cell frequencies in a sample diverge from the expected frequencies, the more likely it is that the null hypothesis is false and that an association does exist between the two variables in the population.

Small values of chi-square indicate little or no association, whereas large values indicate that an association is likely. With the data in Table 14.7, chi-square is sufficiently large that we would be able to reject the null hypothesis that there is no relationship between the two variables and conclude that there is a relationship between religious preference and region in

the population that this sample was selected from. Furthermore, the alpha level indicated in Table 14.7 is .001, which means that we are very confident in rejecting the null hypothesis. With chi-square, the value of the statistic is influenced by sample size and by the number of categories on each variable. Computer data analysis packages automatically take this into account in reporting significance level; however, when doing hand calculations, one must refer the statistical value to a special table that states the probability of obtaining an χ^2 of that magnitude by chance, given the sample size and number of variable categories. Chi-square does not indicate the strength of the association but whether an association exists at the level of the population.

Interval and Ratio Data
With two interval–ratio variables, Pearson's correlation coefficient, discussed as a measure of association, also has an inferential component. In this case, the null hypothesis would

FIGURE 14.7 Inferential Statistics Categorized by Level of Measurement

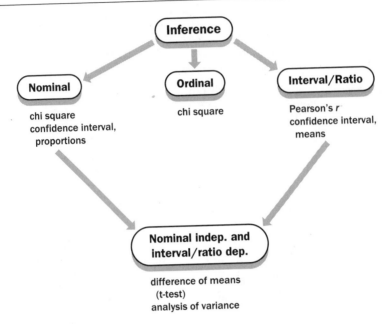

state that $r = 0.00$ in the population whereas the research hypothesis would say that there is a correlation (although it doesn't say how strong the relationship is). If the correlation coefficient obtained in the sample is sufficiently large (how large depends in part on the alpha level set), then we would decide to reject the null and accept the research hypothesis.

Some inferential statistics are designed for situations where the dependent variable is interval but the independent variable is at a lower level of measurement, often nominal. For example, if we wanted to test whether the males employed in a corporation earn higher incomes than the females, we would have one nominal variable (*gender*) and one ratio variable (*income*). We would compare the mean incomes of the males and females. The null hypothesis would be that there is no income difference between the two groups. We would then perform what is called a difference-of-means test (sometimes called a t-test) to decide whether to reject the null hypothesis. If the in-

dependent variable were not *gender* but *race* (with categories of "Anglo," "African American," "Asian," and "Other"), then we would do the same basic analysis by comparing the means of the four groups. However, for technical reasons that I don't have space to explore here, we don't compare the means directly but rather compare the variances, or dispersions, in the four groups. This is called *analysis of variance* (or ANOVA).

Difference-of-means tests are often used in experiments where one is comparing a control group and an experimental group. If the experiment involves more than two groups on the independent variable, then analysis of variance could be used. For example, if four treatment groups each receive differing amounts of tutorial help, we might be interested in the effect each level of help has on the number of problems the members of the different groups can solve. For this situation, involving an experimental design and a ratio-level dependent variable (number of problems solved), ANOVA would be appropriate. We would not expect all members of a treatment group to do equally well. In fact, even if one level of treatment works better than another, some members of a lower group might do better than some members receiving a better tutorial program. ANOVA permits the researcher to estimate how much of the variance in performance between groups is due to the treatment.

Figure 14.7 provides a listing of the inferential statistics discussed, categorized by level of measurement. In this chapter, I have presented a basic introduction to the statistical analysis of data. For the consumer of research, this introduction provides some basic guidelines for interpreting statistical analyses that you will encounter when reading research. Those who will personally conduct research and engage in statistical analysis will need to go beyond the materials presented in this chapter. For such people, I have suggested a few valuable books to consult in the For Further Reading section.

REVIEW AND CRITICAL THINKING

Statistical analysis became more sophisticated and complex in this chapter. Information in professional journals use much of the technical jargon used and discussed in this chapter, such as alpha level, statistical significance, or chi-square. Information from other sources, however, will not likely discuss these matters even though the conclusions they arrive at may relate to them. You can still ask some general questions of the information that you are presented.

1. What is the level of measurement of the variables being considered? Do the conclusions drawn have to do with goals of description or inference? In other words, are conclusions being drawn about a population on the basis of sample data?

2. If any statistics are presented, are they properly used and interpreted?

3. If the conclusions involve an inferential goal, is there any statement about a probability level (that is, an alpha level)? Is there any discussion of "statistical significance" or "margins of error"? Is enough information presented to use this information properly?

Computers and the Internet

Although computing statistics by hand or by calculator is possible, virtually all data analysis today is done via computer. Personal computers today are sufficiently powerful that they can run sophisticated statistical packages such as SPSS, and this has placed state-of-the-art statistical capabilities at the disposal of virtually everyone, and at a relatively low price. At the same time, the bewildering array of software has made software selection a complicated choice.

Because researchers now use the computer for so many different tasks, we can no longer evaluate statistical software solely on the basis of the number of statistical procedures it can run; we must also consider the other applications of the software. Fortunately, several excellent sources are available for assistance in selecting statistical software. Many popular periodicals about computers, such as *Byte* and *InfoWorld*, publish articles evaluating software. In addition, a number of professional journals publish articles on statistical software. One of the better sources for social research applications is *Social Science Computer Review*.

As for selecting the correct statistic to use for a set of data, computer programs are available to help the researcher identify key issues and systematically put that knowledge to use. One example is the program "Methodologist's Toolchest" (Brent & Thompson, 1996). This program does not analyze data; rather, it presents options for data analysis from more than 200 statistical procedures and a description of how well each option fits the stated objectives and assumptions of the research. The program is an example of an Expert System in that the program contains information for decision making. The software leads the user through an interview by presenting on-screen questions. The answer to one set of questions determines the next set of questions to appear. By answering these questions, the researcher tells the program whether the data analysis is descriptive or inferential, how many variables are involved, and so on—basically the kind of information discussed in this chapter. It should be emphasized that the program does not select the statistic for the user; rather, it guides the user through the steps necessary to choosing a proper statistic. The program prepares a report that includes as many as four statistical procedures, rated by how well they match the assumptions and objectives of the researcher as well as the expectations of the intended audience. The report also includes a comprehensive description of each selected test, complete with references and a list of statistical software packages that include the test.

Using a computer program such as this does not ensure that a researcher will always select the appropriate statistic. Neither can one expect it to replace a knowledge of research methods and data analysis. But the program

systematizes the selection process, makes a large amount of technical information immediately accessible, and encourages the researcher to consider design options that might otherwise have been overlooked.

I have mentioned several different data-analysis software programs in the course of this chapter. You can learn more about data-analysis software by contacting software firms at their Internet addresses. One way to locate such sites is to simply enter the name of the software, for example, "SPSS," in a search engine. Many social science research related sites also include links to these sites. For example, at the Internet site for ERIC (located at **http://ericae.net/**), click on "resources," "assessment and evlauation on the Internet," and "statistics" to find links to software sites.

Main Points

1. The major considerations in choosing statistics properly are the level of measurement of the variables, the goals of the research, the number of variables involved, the properties of the data, and the audience.

2. Descriptive statistics are procedures that assist in organizing, summarizing, and interpreting sample data.

3. Measures of central tendency, or averages, summarize distributions by locating the central value of frequency distributions.

4. Measures of dispersion indicate how dispersed or spread out the values are in a distribution, with most indicators revealing the average spread of the scores around the central value.

5. Measures of association indicate the strength of relationships and, with ordinal or higher-level data, the direction of relationships.

6. Data are sometimes transformed into *z*-scores or standard normal distributions. These distributions enable us to compare

distributions to one another, assess the relative position of cases in a distribution, and conduct inferential statistics. The properties of the normal distribution are highly important in this regard.

7. Inferential statistics are procedures that allow generalizations to be made from sample data to populations from which the samples were drawn. Inferential statistics derive from the properties of the normal distribution, sampling distributions, and the central limit theorem. Two different kinds of inferential statistics are statistical estimation and statistical hypothesis testing.

8. In statistical estimation, we use a sample statistic as an estimate of a population parameter by creating a confidence interval.

9. In statistical hypothesis testing, a research hypothesis is paired with an opposite null hypothesis and the results of a statistical test are used to decide which is most likely correct. Attention is paid to Type I and Type II errors, and particular inferential statistics are linked to level of measurement. Among the inferential statistics are difference of means tests, analysis of variance, Pearson's *r*, and chi-square.

Important Terms for Review

alpha error
alpha level
beta error
bivariate statistics
central limit theorem
confidence level
descriptive statistics
inferential statistics
location measures
measures of association
measures of central tendency
measures of dispersion

multivariate statistics
normal distribution
null hypothesis
proportional reduction in error (PRE)
research hypothesis
sampling distribution
statistical estimation
Type I and Type II errors
univariate statistics

For Further Reading

Aron, Arthur, & Aron, Elaine N. (1997). *Statistics for the behavioral and social sciences.* Upper Saddle River, NJ: Prentice-Hall. This very readable introduction to statistics explains the issues clearly and minimizes the use of formulas.

Cuzzort, R. P., & Vrettos James S. (1996). *Elementary forms of statistical reasoning.* New York: St. Martin's. This fairly brief textbook introduces statistical analysis to the student. It focuses more on the logic of statistics and how to reason through problems rather than on computational issues.

Evans, James D. (1996). *Straightforward statistics for the behavioral sciences.* Pacific Grove, CA: Brooks/Cole. This is a standard but well-written introductory textbook on statistics. It covers all the topics discussed in this and the preceding chapter, going into more detail and including additional topics.

Gravetter, Frederick J., & Wallnau, Larry B. (1996). *Statistics for the behavioral sciences* (4th ed.). Minneapolis: West. This text provides a fairly comprehensive introduction both to descriptive and to inferential statistics. It can serve as an important resource and reference work as you learn statistics.

Healey, Joseph F. (1999). *Statistics: A tool for social research* (5th ed.). Belmont, CA: Wadsworth. This is another standard but well-written introductory textbook on statistics. Along with the Evans book, it can help you understand all the material in this and the preceding chapter.

Chapter 15

ANALYSIS OF QUALITATIVE DATA

This chapter focuses on data that is fundamentally qualitative in nature, primarily in the form of the field notes and in-depth interviews discussed in Chapter 11. However, qualitative data also comes in the form of diaries, narratives, video recordings, and other sorts of texts that are nonquantitative. In qualitative data analysis, the researcher is attempting to transform this raw data and extract some meaning from it, mostly without quantifying the data. I pointed out in Chapter 11 that field research can produce some quantitative data; when it does, these data are analyzed using the procedures discussed in Chapter 13 and Chapter 14.

GOALS OF QUALITATIVE DATA ANALYSIS

The goals of qualitative data analysis are both similar to and different from the goals of quantitative data analysis. While the specific strategies used in qualitative data analysis are different from those used in quantitative data analysis, it is, notwithstanding, still data analysis: extracting meaning from observations that have been made. The goals of a qualitative research project might be the same as those discussed in Chapter 1, especially description, explanation, and evaluation. In addition, qualitative research often strives for understanding by generalizing beyond the data to more abstract and general concepts or theories. The ultimate end may, in some cases, be efforts to generalize the results to people, groups, or organizations beyond those observed. So, qualitative data analysis creates meaning, in part, by using the raw data to learn something more abstract and general. In this respect, qualitative and quantitative data analysis are similar.

Beyond the similarities, however, qualitative and quantitative approaches have clear differences from one another. One difference is that qualitative research recognizes that abstraction and generalization are matters of degree and that they may be of less importance in some studies. A second difference is that qualitative research devotes more attention to the effort to contextualize; that is, to understand people, groups, and organizations within the full context or situation in which they act. In fact, some qualitative data analysis strategies devote more effort to contextualizing than to abstracting or generalizing. This is based on the position that scientific knowledge is not found only in abstracting and generalizing; such knowledge can also derive from a deep and full description of a context. In other words, some qualitative research focuses on idiographic rather than nomothetic explanations (go back and reread pages 42–44 on different types of explanations in Chapter 2). A third difference is that qualitative research tends to place more emphasis on inductive reasoning than on deductive reasoning. As we shall see, qualitative researchers stress the value of allowing concepts and abstract ideas to emerge from the data rather than using the data to provide evidence for preexisting concepts and theories.

As a part of the goal of stressing the contextual, qualitative research maintains a close, interactive link between data collection and data analysis. In Chapter 1, I presented the stages in the research process as a sequence in which one stage is mostly completed before the next is begun. In particular, I suggested that the data collection stage is completed before the data analysis stage begins and that data analysis is largely finished before conclusions are drawn (see Figure 1.2 on page 22). This is the way many quantitative and positivist researchers describe the process. One major reason for this sequencing is to ensure that the data collection procedures used are the same over time; if the measuring procedures change from the beginning to the end of the data collection phase, then you may be measuring different variables (see Chapter 5 and Chapter 6). This is an important consideration in research where there are clearly stated and quantifiable

FIGURE 15.1   The Stages of Social Research as Conceptualized by Some Qualitative Researchers

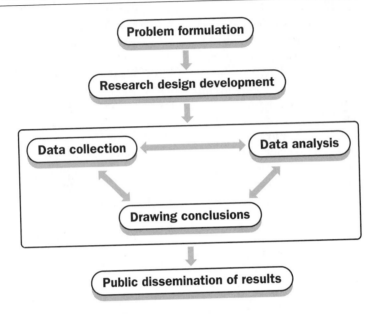

variables and hypotheses and quantitative measuring procedures are used.

For many qualitative researchers, on the other hand, the process looks more like that outlined in Figure 15.1: The stages of data collection, data analysis, and drawing conclusions are more simultaneous and interactive (Maxwell, 1996; Seidman, 1991). The researcher begins doing analysis and drawing conclusions almost as soon as data collection begins, and the analysis and the conclusions provide direction for what additional data collection needs to occur. Unlike the positivists, for whom comparability over time is essential, many qualitative researchers see that aspect of data collection as problematic in the sense that it is a decision to be weighed given the particulars of a research project. In many cases, the advantages to be gained by adjusting and refocusing the data collection efforts as they proceed outweigh the disadvantages of changing the manner in which data are collected.

One of the advantages gained when data collection and analysis overlap is **theoretical sensitivity**: having data collection and analysis closely guided by emerging theoretical issues (Glaser, 1978; Strauss, 1987). Data collection and analysis are guided by theoretical issues in all research, but in a different way. In most research, theoretical issues are used to create measuring devices before the data are collected, and what occurs during the process of data collection does not lead to changes in the measuring devices. However, since many qualitative researchers do not see a rigid separation between data collection and data analysis, this stance opens the door for the possibility of theoretical issues arising *during data collection* to change which kind of data is collected or from whom it is collected. So theoretical sensitivity involves a constant interaction between theory and data collection. Data collected in one interview, for example, may raise some theoretical issues for the researcher such that later interviews are modified to collect data addressing those issues.

Following Miles and Huberman (1994), I find it helpful to organize qualitative data analysis into three categories of things that are done: data reduction and analysis, data displays, and drawing conclusions and verifying theories. Keep in mind that these are not sequential steps but rather overlapping activities that support one another. As Miles and Huberman state: "The three types of analysis activity [data reduction, data displays, and drawing conclusions] and the activity of data collection itself form an interactive, cyclical process" (p. 12). And all three activities occur, at least in part, during the process of data collection and may influence and change data collection.

DATA REDUCTION AND ANALYSIS

As different kinds of qualitative data analysis are described in this chapter, it may seem at times as if I am really talking about the process of data collection in the field (the topic of Chapter 11). That confusion can arise because in qualitative research, data collection and data analysis often occur simultaneously. In qualitative field research, for example, the data collection phase involves collecting field notes and other materials. Data analysis refers to the application of coding schemes and other procedures to those field notes that I describe in this section. So the analysis is sometimes occurring as the data are being collected. The data analysis strategies discussed in this section tend to be of two types. **Categorizing strategies** attempt to generalize and abstract by generating concepts and even theories from the raw data. **Contextualizing strategies** attempt to treat the data as a coherent whole and retain as much of the raw data as possible in order to capture the whole context (Maxwell, 1996). Actually,

many specific qualitative data analyses might involve elements of both. I will first review categorizing strategies in the form of coding, reflective remarks, and memos. Then the contextualizing strategies will be discussed.

Codes and Coding

I have talked about the process of *coding* at a number of points in this book. It is discussed as a part of doing content analysis of available data in Chapter 10, as a part of making structured field observations in Chapter 11, and as a part of quantitative data analysis in Chapter 13. Coding is also used as a form of data analysis with qualitative data and has some similarities to coding done in other contexts. **Coding** refers to the classifying of observations into a limited number of categories. However, coding in most qualitative data analysis is distinct from these other forms of coding in at least three important ways. First, it is not an effort to quantify the data or create a set of numerical categories, as is often the case with other types of coding. Coding in qualitative analysis reduces and simplifies the data, but it does so by retaining words and their essential meanings. Second, codes and coding schemes in qualitative analysis are created (at least in part) from the data itself during the process of data collection; in other words, the data creates the codes. The quantitative coding schemes are more typically derived from some preexisting theoretical stance and then an effort is made to see if the data "fit" the coding scheme. Third, the purpose of coding in qualitative analysis is different. In quantitative research, coding is usually part of the process of measurement: The coding categories constitute the operational definition of the underlying variable. While this is, to an extent, true in coding in qualitative research, qualitative coding goes beyond measurement; it is also an integral part of conceptual development and theory building. As you will see, the process of coding in qualitative research spans both the realms of measurement and of the

TABLE 15.1   A Section of Transcribed In-Depth Interview, with Codes

Codes	Interview Statements
	A 29-year old man with renal failure was discussing his high school years, and events that occurred long before he was diagnosed.
Self-perception Awareness of difference Identifying self through ill health Comparing health to others'	. . . I knew I was different. I caught colds very easily and my resistance was very low, and so I knew that generally speaking my health wasn't as good as everybody else's, but I tried to do all the things that everybody else was doing.
	A 29-year old woman with colitis was recounting her first episode of illness.

Normalizing the context of illness
Self-esteem:
    feelings of failure
    failure of self
Reality contradicts idealized experience
	. . . I was under a great deal of stress as a result of all this bouncing around and trying to get a job and trying not to have to go home to my parents and admit that I had failed. [I] failed at life. I had left college, and left there saying, "Gee, I can do it on my own," so I was trying this exciting existence I read about and there was something wrong; I had all this pain. I didn't know what to do about it.

	A 54-year old woman who had had cancer and currently had a crippling collagen disease was explaining her view on why she had had a recurrence of cancer.
Self in retrospect Self-esteem Outcome of timed struggle Improving self-esteem as treatment goal	. . . When I look back on my second bout of cancer, I was not feeling good about myself and the whole struggle of the last three years put me into X (a cancer institute) to try and get me to feel better about myself.

Source   Reprinted by permission of Waveland Press, Inc. from Kathy Charmaz, The Ground Theory Method: An Explication and Interpretation. In Robert M. Emerson, ed., *Contemporary Field Research: A Collection of Readings.* (Prospect Heights, IL: Waveland Press, Inc. (1983) [reissued 1988]). All rights reserved.

more abstract process of developing concepts and theories.

So, in qualitative research, the data would be in the form of field notes, a narrative produced by someone, or possibly written documents or archival material. Coding is the process of categorizing sections of that data, which might mean a phrase, a sentence, or a paragraph. It is a way of seeing which parts of the data are connected to one another in terms of some issue, concept, theme, or hypothesis. Some researchers use different terms, such as *thematic analysis*, for data analysis strategies that basically involve categorizing the data (Seidman, 1998). Table 15.1 gives an example of this. It shows a few sections of the transcript of in-depth interviews on the right-hand side and the codes that the researcher applied to each section on the left. These interviews were conducted as a part of a study of how people

with chronic illnesses experience time and personal identity. You can see the conceptual codes that the researcher identified, many of them having to do with self-esteem and feelings about oneself. These are the concepts that this researcher identified as important in each part of what was said in the interviews.

Approaches to Coding

There are three approaches to developing coding schemes that are used by qualitative researchers. One is to create a fairly complete coding scheme prior to going out into the field to collect observations (Miles & Huberman, 1994). This would be based on theoretical considerations regarding what will be observed in the field, and which are the important variables, social mechanisms, and causal processes. This approach might rely on prior research or on preexisting coding schemes that had been developed by others and used in research on similar topics. The list of coding categories should be fairly complete in terms of what the observer expects to see in the field, and may be quite detailed and lengthy. However, one thing that distinguishes this approach to coding from what is done by most quantitative researchers is that the qualitative researcher still expects to change and adapt the coding scheme as observations are made in the field. With almost any coding scheme, some categories will prove useful, others less so. Some categories might not be used at all, whereas others are used too much—many observations fall into a category that needs to be divided into subcategories based on differences among the various observations. The coding scheme therefore continues to develop as the data are collected.

The second approach to developing coding schemes is the reverse of the first: Observers enter the field with no pre-established coding scheme (Strauss, 1987). Coding categories are developed as observations are made in the field context. One example of this is the grounded theory approach described in Chapter 11. This is a very data-driven, contextualized approach to data analysis. Without the restrictions of a preexisting coding scheme, the observer describes what happens, tries to identify relevant variables, and searches for explanations of what is observed. Beginning with these concrete observations, the researcher develops a coding scheme that points toward more abstract concepts, propositions, and theoretical explanations that would be plausible given those observations. In this way, preexisting theory does not limit the kind of data collection that occurs. However, even this open approach to coding is given some structure by some proponents of grounded theory, such as Anselm Strauss (1987), who argue that coding should address four general categories of phenomena:

- conditions or causes
- interaction among people
- strategies and tactics
- consequences

A third approach to developing coding schemes falls in between the first two: A general coding scheme is developed that identifies domains of observation rather than referring to specific content within those domains. Then coding schemes would be inductively developed within those domains (Bogdan & Biklen, 1992; Lofland & Lofland, 1995). These domains are more specific and detailed than the four above-mentioned general categories that are used by some grounded theory researchers. A possible list of such domains might be as follows:

- actions/events
- activities (actions of some duration)
- meaning (what people say and do to define a situation)
- perspectives (ways of thinking or orientations)
- relationships
- setting/context
- social structure (statuses, roles, and their relationships)

Most or all of the above domains are likely to be relevant in a field setting, yet the categories are so general that the context can still play a strong part in shaping the observations made and the specific coding scheme that emerges.

To begin to show more concretely how coding is done, we can use an example from a field study of gang life that relied on ethnographic interviews as a source of data (Decker & Van Winkle, 1996). The study began with some theoretical ideas regarding what attracted young men to join gangs, especially that certain instrumental benefits of joining a gang, such as the ability to be protected by others or to join in the sale of drugs, were a major attraction. Then the researcher goes

TABLE 15.2 Excerpts from Ethnographic Interviews of Gang Members, with Suggested Coding

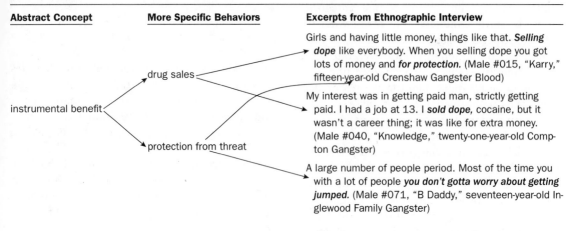

Abstract Concept	More Specific Behaviors	Excerpts from Ethnographic Interview
instrumental benefit	drug sales	Girls and having little money, things like that. *Selling dope* like everybody. When you selling dope you got lots of money and *for protection.* (Male #015, "Karry," fifteen-year-old Crenshaw Gangster Blood)
		My interest was in getting paid man, strictly getting paid. I had a job at 13. I *sold dope,* cocaine, but it wasn't a career thing; it was like for extra money. (Male #040, "Knowledge," twenty-one-year-old Compton Gangster)
	protection from threat	A large number of people period. Most of the time you with a lot of people *you don't gotta worry about getting jumped.* (Male #071, "B Daddy," seventeen-year-old Inglewood Family Gangster)

Adapted from Scott H. Decker and Barrick Van Winkle, *Life in the Gang: Family, Friends, and Violence.* (Cambridge, UK: Cambridge University Press, 1996), pp. 63, 153. Reprinted with the permission of Cambridge University Press.

through the transcribed interviews to find words or phrases that seem to represent these behaviors as reasons for joining gangs. Table 15.2 presents some excerpts from the interviews, with the key words that indicate a particular coding category boldfaced. The individuals in these interviews were responding to questions about why they got involved in gangs. The researchers then use these codes to assess the extent to which, throughout all of the coded interviews, there is support for their theoretical ideas about the role of instrumental benefits. There are a couple of things to notice in the coding in both Table 15.1 and Table 15.2. First, the coding does not focus on counts of how often things happen but rather on descriptions of what is applicable to a particular individual or context. The data analysis produces words, phrases, and descriptions as meaning rather than using numbers to extract meaning. Second, there is a close link between the codes and the data: When a code is applied, it is linked to a particular section (word, phrase, sentence, or paragraph) of the raw data. This is important in terms of assessing validity, which will be discussed later in the chapter.

Descriptive versus Abstract Coding

Coding schemes vary in terms of how general or abstract the coding categories are. Different practitioners of qualitative research have developed a number of ways to describe this. Miles and Huberman (1994), for example, suggest three different types of coding. *Descriptive codes* are coding categories that involve fairly directly observed behaviors or events; the coder need not do a great deal of interpretation in order to place a statement in field notes into these categories. So field notes that say "Jane left the room" or "Jim shook hands with the nurse" could be fairly directly coded as "departure" and "greeting," respectively. *Interpretive codes* are coding categories that require field researchers to use some of their deep understanding of the social context in order to place a section of field notes into a category; in other words, they must interpret the meaning of a particular entry in the field notes. Interpretive codes might involve assessing people's motivations or moods or the meanings that people attach to things. So if the field notes say that "Jane left the room angrily" or "Jim gave the nurse a cool greeting," the words *angry* and *cool* are interpretive in nature—they involve a judgment or interpretation on the part of the researcher. Placing these parts of the field notes into categories of "angry" and "distant" is more interpretive and subjective than are the descriptive codes.

The third type of coding involves *pattern codes*: "explanatory or inferential codes, ones that identify an emergent theme, configuration, or explanation. They pull a good deal of material into more meaningful and parsimonious units of analysis" (Miles & Huberman, 1994, p. 69). Pattern codes reduce the data into more manageable amounts, and help focus later data collection on that which would assist in better understanding the patterns or themes. A section of field notes is identified as representing some abstract or theoretical theme or pattern in the events that are occurring. An explanatory code may refer to some abstract social process. So "Jane left the room angrily" might be coded as exemplifying the social process of "rejection of deviant," whereas "Jim gave the nurse a cool greeting" might be coded as the social process of "role distance." Pattern codes usually focus on one of four general categories of phenomenon: themes, causes or explanations, relationships among people, and theoretical constructs. The danger with pattern coding is that a researcher might prematurely impose a certain meaning on the data and then try to fit everything else into that established pattern. To prevent that from occuring, Miles and Huberman (1994) recommend that the researcher remain tentative and flexible.

> The trick here is to work with loosely held chunks of meaning, to be ready to unfreeze and reconfigure them as the data shape up otherwise, to subject the most compelling themes to merciless cross-checking, and to lay aside the more tenuous ones until other informants and observations give them better empirical grounding (p. 70).

Keep in mind that a single section of a text or field notes could be given more than one type of code; in fact, it might be given a descriptive, an interpretive, and a pattern code.

Going back to Table 15.2, note that it is fairly straightforward to give a descriptive code to some segments, such as "sells drugs"; however, it is more difficult—requires more interpretation and judgment—to assign interpretive codes. For example, it requires some judgment to say that "you don't gotta worry about getting jumped" is an instance of the code "protection from threat." It is even more inferential to assign a pattern code, such as "selling drugs as instrumental benefit that motivates joining a gang." Such abstract coding might be based on things said in other parts of the interview or on the researchers' deep knowledge of the field setting in which the interviewees live and the interviews occurred. This is why Miles and Huberman caution against settling on pattern codes too quickly.

Open Versus Focused Coding

Another approach to coding is to begin by looking for any types of codes that might emerge naturally from the data and then focus on a limited number of codes to see how well they fit various parts of the data (Charmaz, 1983; Glaser, 1978). The initial coding of field notes, interviews, or other documents is called *open coding*, and involves unrestricted coding to produce concepts and dimensions that seem to fit the data fairly well. These codes can be linked to the various dimensions mentioned above (conditions or causes, etc.), and the coding categories are provisional and tentative. The point of open coding is to open up possibilities, rather than to finalize anything. Anything that is wrong or unclear at this stage will be rectified and clarified at later stages of coding. As Strauss (1987, p. 29) puts it, at this stage of the coding, researchers "play the game of believing everything and believing nothing." Whereas in quantitative research the goal of coding is to produce counts of things (how many cases fall into each category), the point of open coding, according to Strauss

(1987, p. 29), is to "fracture, break the data apart analytically": to create new categories and to split apart and rearrange existing categories. There is nothing fixed about the category system, at least at this stage; it is emerging out of the data and is thus in flux. Especially early in the process of open coding, the coding is likely to be more of the descriptive variety, but it may also include some interpretive and pattern codes.

After open coding has proceeded for a while, the field researcher can turn to *focused coding*, or what Strauss calls *axial coding*: intense analysis around one or a few of the coding categories. (Each of the general coding categories can be called an "axis," hence the name axial coding.) Focused coding is more selective and conceptual; it involves applying a limited set of codes to a large quantity of data. It enables the researcher to assess how extensively a set of codes applies and to discover the various forms in which the categories appear. This enables the researcher to explore one of these dimensions of social reality throughout the field notes and thus expose relationships that may not have been obvious with open coding. Focused coding cannot be done until open coding has already provided a plethora of potential concepts, but after a period of open coding, focused coding can be done periodically. In open coding, the center of attention is the raw data; with focused coding, the emerging coding categories are the focus of attention. Focused coding begins to expose the core concepts and categories and the relationships among them that will emerge from the data. However, focused coding is still strongly linked to the data since the researcher must constantly go back to the original texts to validate decisions about what are core concepts and what appears to be relationships among them. Further, focused coding may lead to revisions in the coding done during open coding as new concepts emerge from the data. What is occurring at this stage is essentially conceptual development (as discussed in Chapter 4), but

here conceptual development grows out of the process of data analysis rather than occurring before research design development.

Selective coding is a term used by Strauss (1987) to describe coding that focuses on the core concepts and categories that emerged during focused coding. It is, in a sense, a more intense focused coding. All elements of the text are coded in terms of how they relate, or don't relate, to the core concepts and theories that are emerging from the data. This is similar to Miles and Huberman's (1994) pattern coding. Selective coding builds on the open and focused coding and can begin only after some of those forms of coding have been accomplished. As data analysis proceeds, selective coding comes to predominate the process. This selective coding also serves as the guide for the theoretical sampling discussed in Chapter 7: As core concepts and theories emerge from selective coding, it points to which additional data needs to be collected to provide further tests and comparisons for the theory.

Operational Definitions and Reliability

However a coding scheme is developed, each category in it must have a good operational definition to ensure that all observers uniformly use the coding scheme properly and consistently. The operational definition would be a verbal statement of the kind of observations that should be placed in a particular category. This operational definition might emerge from some existing theoretical framework or out of the observations, as in grounded theory. Nevertheless, it is critical that the definition be clearly specified and understood by all observers. Otherwise, observations that are placed in one category by one observer might be placed in a different category by another observer or possibly ignored altogether. One way to clarify these operational definitions is through *double coding*: Two observers code the same set of field notes, and then cases where they disagree on the coding can be evaluated. Discussion about their disagreements usually produces a sharper and clearer operational definition or a revision in the coding scheme that takes the difficulty into account.

Reliability of coding schemes can be assessed in a number of different ways. As discussed in Chapter 10, one way is double coding, and then the degree of agreement between the two coders can be assessed. One simple way to do this is to calculate the percent of judgments on which coders agree out of the total number of judgments that they must make:

$$\text{percent of agreement} = \frac{2 \times \text{number of agreements}}{\substack{\text{total number of observations} \\ \text{recorded by both observers}}}$$

There is some disagreement about what level of agreement is acceptable, but many researchers argue that the final coding scheme, after revisions and adjustments have been made, should achieve at least 85%–90% agreement. Another way to check the reliability of a coding scheme is to have each observer code the same set of field notes twice, separated by a period of at least a few days. Again, the ultimate code–recode reliability should achieve at least 85%–90%.

Reflective Remarks and Memos

In doing qualitative data analysis, researchers need some mechanisms for moving away from the immediate raw data and toward the more general and abstract. Pattern coding and selective coding are ways to achieve this, but field researchers have also identified other ways. One way to do this is through **reflective remarks** on field notes: reflections, interpretations, connections, or other thoughts that occur to the researcher while transcribing field notes or coding the data. This could take many different forms: questioning the original interpretation of some event in the field notes, a recollection of something about the relationship between two people that didn't get put into the field notes, an elaboration on some-

thing that was only sketchily described in the field notes, or a new hypothesis to explain something that happened in the field. All of these might be useful in understanding and coding the field notes and should be committed to writing.

If the field notes are being transcribed when the reflective thoughts occur, then they can be incorporated directly into the field notes. However, it is a good idea to keep the data recorded in the field separate from these reflective remarks because the latter could be influenced by selective recall or retrospective interpretation. One way to keep the field notes separate from the reflective remarks is to set the remarks apart from the notes with some device—a different typeface, double parentheses, brackets, or any other device that will not be used anywhere else in the field notes—that clearly identifies what it is. If the reflective thoughts occur during coding, they could be written into a margin of the field notes, but not the margin where codes are placed.

Another significant step away from the raw data are what are variously called memos, analytical memos, or theoretical memos. **Memos** refer to attempts at theorizing: the researcher writes down ideas about the meanings of the codes that are emerging from the data and the relationships between the various codes (Glaser, 1978; Strauss, 1987). Memos do not just describe the data; rather, they are more conceptual and abstract in nature, showing that a particular part of the data is an instance of a particular concept or social process. Or a memo might link together various pieces of data as all sharing some abstract property in common. In whatever form they take, they represent a move beyond the raw data and toward more abstract theorizing. This memoing can be done as data collection and data analysis proceed. In fact, memoing can begin as soon as these abstract ideas begin to occur to the researcher, which may be fairly early in the data collection process. This is quite different from quantitative research where data collection and analysis must be almost complete before the theoretical implications begin to emerge.

Memos can be short or long, a few sentences to a few pages. In addition, most researchers identify on the memo exactly which parts of the field notes the memo refers to. This is theoretically important for approaches such as grounded theory because of their stance about the close link between raw data and abstract theory: All abstract ideas should be tied to specific parts of the data. In these approaches to qualitative data analysis, it is not adequate for researchers to report on "general impressions" from the data; this would be considered to have low validity. Instead, memos should identify the lines in the field notes or other memos that it relates to so that there is a clear link between bits of data and more abstract conclusions. The memos should also identify to which codes or concepts they refer and be dated so that it is clear when the thoughts occurred in the research process. Some of the memos, or at least parts of them, may later be incorporated into the final research report.

Beyond saying that memos are abstract and general, it is difficult to specify their content because that is highly variable. Some memos clarify an idea or an existing coding category, suggest some new coding categories or subcategories, or link data from different parts of the field notes or possibly even from other research projects. A memo might propose a hypothesis or a new pattern code, or it could identify something puzzling that does not fit with the conceptual framework emerging in other memos. As the memoing process continues, what begins to emerge, either gradually or in some cases quickly, is a more formalized and coherent set of propositions or explanations of what is found in the data. I discussed the development of propositions and theories at length in Chapter 2. You can see here how the propositions in qualitative data analysis emerge out of the data in a very inductive process. In fact, some qualitative researchers eventually force themselves to state some tentative propositions that can then be reviewed in light of all the memos, codes, and data to see if the propositions have any validity. In some cases, propositions may emerge early in the process of data analysis. Some propositions will eventually be discarded, as they are shown to be inconsistent with much of the data. But, eventually, a set of propositions emerges that appears to be consistent with all or most of the observations. Propositions can take a number of different forms, such as the following:

When A happens, B is also found.

A causes B to occur.

A is found when B, C, and D exist.

A is found when B and C but not D occur.

In whatever form they take (discussed in Chapter 2), the propositions are abstract explanations of phenomena. In qualitative research, these explanations are viewed as tentative; another sweep through the data by the same or different researchers could produce different coding, memoing, and proposition development. Additionally, field observations of different groups or different locations could produce different outcomes. Figure 15.2 diagrams the categorizing strategies that are discussed in this section, and the double arrows illustrate once again how interactive this process is. The researcher constantly goes back and forth among the memos, reflective remarks, and coding as analysis proceeds; in addition, the researcher goes back and forth between these three elements and the data and conclusions.

Contextualizing Strategies

Generally, qualitative researchers stress the importance of context and of viewing and analyzing data with an appreciation of the complete context in which the data were produced. However, the categorizing strategies just discussed tend, to a greater or lesser degree, to move the researcher and the analysis away

FIGURE 15.2   The Data Analysis Process in Some Types of Qualitative Research

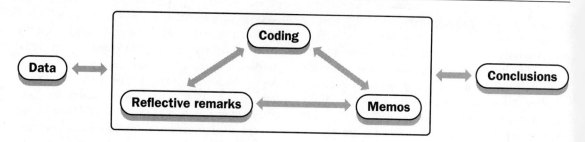

from the concrete data and toward more abstract categories and propositions. Some qualitative researchers argue that something is lost using this strategy (Maxwell, 1996; Seidman, 1998). They argue for *contextualizing strategies*: approaches to data analysis that treat the data as a coherent whole and retain as much of the raw data as possible in order to capture the whole context. Contextualizing strategies are less concerned with abstracting from one set of data in order to generalize to other people or circumstances, and more interested in a deep and rich appreciation of the individuals or situations from whom the data were collected. In other words, these contextualizing strategies focus more on idiographic than nomothetic explanations (see Chapter 2).

One type of contextualizing strategy is the use of *profiles*, promoted by Irving Seidman (1998). **Profiles** refer to vignettes of a person's experience, usually taken from in-depth interviews, that are stated largely in the person's own words, with relatively little interpretation or analysis by the researcher. Profiles are, fundamentally, narratives, or stories. Telling stories is a key way for people to make sense of themselves and their world and hearing others' stories is a valuable way to learn about those people's lives. Profiles are constructed by transcribing in-depth interviews and then identifying parts of the interview that seem to the researcher to be especially important in telling this person's story. Then the parts of the interview are put together into a preliminary pro-

file. This profile is then reviewed with a critical eye to see if any passages are redundant or irrelevant and whether the passages are sequenced properly to tell the most effective story. This review process continues until the researcher is satisfied with the profile. The final profile would be mostly in the person's own words, although the researcher might include some comment to clarify or provide transitions. Any comments by the researcher should be clearly identified as such.

The researcher has to exercise considerable judgment in what to include in the narrative and how to sequence passages, and the possibility exists that some bias might enter into this selection process. Yet Seidman (1998) argues that the researcher is in the best position to make these judgments. It is, after all, the researcher who has "done the interviewing, studied the transcripts, and read the related literature; . . . mentally lived with and wrestled with the data" (p. 110). Certainly, the researcher might consult with those interviewed to see if they concur with the researcher's judgment, but ultimately the researcher needs to be confident that his or her intuitive and professional judgment about how to shape the best profile is correct. The final step in developing a profile is analogous to the conceptualizing or theorizing of other researchers: The researcher spells out what he or she has learned out of the whole process of collecting data, reviewing data, and developing the profile. There would undoubtedly be some abstracting and conceptualizing done in this conclusion, but it is probably grandiose to call it theory building.

An approach similar to profiling is **narrative analysis**, in which interviews, autobiographies, life histories, letters, and other personal materials are used to form a detailed account of a person's life or circumstances (Clandinin & Connelly, 2000; Merriam, 1998). As with the profile, this narrative would rely heavily on the person's own words, as taken from their letters, autobiography, or other sources. However, narrative analysis does not rely so heavily on interviews; it uses a broader range of data sources than do profiles.

A similar contextualizing strategy, but one that doesn't rely so heavily on the subject's own words, is the **case study**: a detailed, descriptive account of one individual, situation, organization, group, or other entity (Merriam, 1998; Patton, 1990). As with profiles, data analysis in case studies focuses on description and narrative rather than categorizing strategies. Further, the description in case studies is a detailed and what is sometimes called "thick" description: a complete and literal accounting of the person or setting under study. While some quantitative data might be presented as a part of a case study, the emphasis is on telling a story in prose, although not completely in the words of the people being studied, as in the profile or narrative. But the idea is the same: The researcher describes people's lives and experiences in great detail. In addition to the researcher's description, the case study might use quotations from those being studied, photos, videos, artifacts, and any other material that would provide an in-depth description of the subject of the case study.

Case studies can be based on direct observation, interviews, document analysis, organizational records, or some combination of these. Basically, any data that would contribute to a description of the case under study could be used. Historical research often focuses on a case study, and might use a wide variety of materials as data: historical accounts, published newspapers and magazines, biographies, letters, governmental statistics and records, and so on.

The primary goal of the case study and the other contextualizing strategies is an idiographic explanation that focuses on an in-depth understanding of this particular case. Such an understanding might enhance one's comprehension of other cases and situations, but the primary focus is description, not generalization. The advantage of profiles, narratives, and case studies is the richly detailed descriptions they provide of people's lives, experiences, and circumstances. One disadvantage of these contextualizing strategies is that they depend heavily on the subjective and intuitive judgments of the researchers closest to the data. Another researcher producing a case study or profile of the same person or situation might come up with quite a different story. Another disadvantage is the limited ability to generalize beyond the individual case.

DISPLAYING DATA

A **data display** is an organized presentation of data that enables researchers and their audiences to draw some conclusions from the data and to move on to the next stage of the research (Miles & Huberman, 1994; Strauss, 1987). In any data analysis, the researcher must display the data so that a convincing argument is made that supports the conclusions reached in the research. In science, we don't take the researcher's word in regard to conclusions; we need to be convinced by a display of the data. In quantitative data analysis, an important part of the display of data is the numbers in the form of contingency tables, charts, or descriptive and inferential statistics (see Chapter 13 and Chapter 14). Some narrative is required to explain how the numbers were arrived at and to clearly explain their interpretation, but the core of the argument is the display of the numbers. In qualitative data analysis, the core of the argument is not numbers; while there may be some numbers to show, different types of data displays are used. Creating data displays is very much a part of data analysis; creating the displays assists the researcher in identifying and clarifying the concepts and categories that are emerging from the coding and other strategies being used on the data.

Narrative Text

One key type of data display in qualitative research is the description, narrative, or verbal argument made by the researcher. In the contextualizing strategies, the narrative provides as full a description as possible in order to give an in-depth picture of the center of attention: the field, interview, or person. The narrative attempts to be "true" to the meaning of the original experience, as interpreted by the researcher. The anthropologist Bourgois (1995), for example, in his study of street drug dealers, points to "the impossibility of rendering into print the performance dimension of street speech. Without the complex, stylized punctuation provided by body language, facial expression, and intonation, many of the transcribed narratives of crack dealers appear flat, and sometimes even inarticulate" (p. 341). The challenge for Bourgois was to produce a narrative that conveyed the richly detailed meanings that he perceived in the field setting, based on all the linguistic and nonlinguistic cues available to him.

In the categorizing strategies, the researcher's narrative describes the concepts and propositions that emerged from the research, along with excerpts from the field notes or other data sources that illustrate the concepts and propositions and corroborate the conclusions drawn (see Table 15.1 and Table 15.2). In neither quantitative nor qualitative research do researchers report *all* the data, which would mean the raw data. A research report is always a summarization of the data, and it is ethically incumbent on the researcher to give an accurate summarization of the data. For the qualitative researcher, this means that the illustrations presented to corroborate the conclusions

are representative illustrations and not selective or distorted.

Visual Displays

Qualitative researchers also use a variety of visual formats to display data, just as quantitative researchers use contingency tables, graphs, and charts in their data analysis (Miles & Huberman, 1994; Patton, 1981; Strauss, 1987). These visual data displays are considered a part of the data analysis process because the process of their development serves as an assist in conceptualization and theory development. By summarizing what is found in the data, these displays can help clarify conceptual categories and give further insight into relationships between categories. They also enable the researcher to see weaknesses in the analysis and thus suggest areas where additional data analysis is called for. In addition, of course, visual data displays serve as an effective adjunct to the narrative text in communicating results to a variety of audiences. While these visual displays can take a variety of formats, there are generally two types. One is a figure that serves as a visual mapping of some physical or conceptual terrain. The second is a table, chart, or matrix into which some text, phrases, or other materials are placed. Although the specific form that these visual displays can take is highly variable and very dependent upon the specific data and concepts, I shall provide a few illustrations to offer a sense of their nature, function, and development.

One type of visual display is a **context map**, which describes in graphic form the physical or social setting where the observations were made. Qualitative researchers stress the centrality of the context in understanding people's behavior; they are fond of pointing out that everything must be contextualized or situated. Positivist, quantitative research often collects and analyzes data without regard to the context. A survey researcher, for example, will collect data on a questionnaire without regard to whether those attitudes might be important, different, or possibly irrelevant *depending on the context in which the person is*

FIGURE 15.3   A Sociogram as an Example of a Context Map

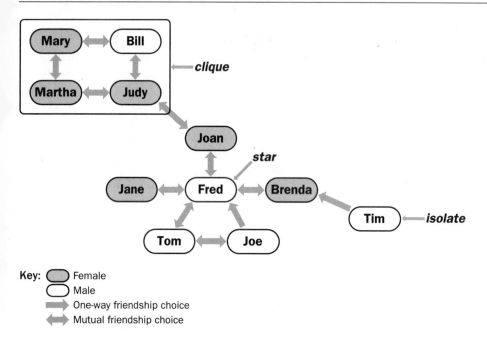

behaving. Many nonpositivists refer to this as "decontextualized" data collection and analysis and argue that those attitudes only take on a life (i.e., become important) in a particular context. Just as a percentage table informs the audience about some of the meaning in data, a context map ascribes meaning to the data by describing the context in which the behaviors occurred.

In some cases, the context map is a drawing of the physical or geographical setting in which observations are made. In fact, I suggested in Chapter 11 that each session of field observation might begin by drawing a map of the setting. The map might be a drawing of a room, building, street, or neighborhood; it might include physical objects, entrances and exits, the location of people or groups, and whatever else is relevant to understanding the social interaction described in the narrative. In another case, the context map might describe the social setting: the individuals, statuses, roles, groups, or organizations that are relevant to the observa-

tions; it would also show connections and relationships between these elements by the use of lines and arrows. Some forms of this are called **sociograms** or **network analysis**. Figure 15.3 gives an example of what this might look like, showing the friendship choices made among a group of people. This visual representation helps us to better see certain social phenomena, such as certain patterns of relationships. The figure provides evidence for certain social phenomena, such as the clique (a group of people who all make mutual friendship choices) or the star (a person who is selected as a friend by a large number of people). Figure 15.3 is based on friendship or attraction choices, but it could map other aspects of relationships, such as showing deference, submission, or dominance, or giving and receiving help. As with other context maps, the specifics would depend on the specific observational settings and the concepts that emerge from it.

Another type of visual display in qualitative research is called a **matrix display**. A matrix

TABLE 15.3   Example of a Matrix Display

Presence of Supporting Conditions

Condition	For Users	For Administrators
Commitment	*Strong*—"wanted to make it work."	*Weak* at building level. Prime movers in central office committed; others not.
Understanding	*"Basic"*—("felt I could do it, but I just wasn't sure how.") for teacher. *Absent* for aide ("didn't understand how we were going to get all this.")	*Absent* at building level and among staff. *Basic* for 2 prime movers ("got all the help we needed from developer.") *Absent* for other central office staff.
Materials	*Inadequate*: ordered late, puzzling ("different from anything I ever used"), discarded	N.A.
Front-end training	*"Sketchy"* for teacher ("it all happened so quickly"); no demo class. *None* for aide: ("totally unprepared. I had to learn along with the children.")	Prime movers in central office had training at developer site; none for others.
Skills	*Weak-adequate* for teacher. *"None"* for aide.	One prime mover (Robeson) skilled in substance; others unskilled.

Source   Matthew B. Miles and A. Michael Huberman, *Qualitative Data Analysis: An Expanded Sourcebook*, 2d ed., p. 95, copyright © 1994 by Sage Publications, Inc. Reprinted by permission of Sage Publications, Inc.

looks very much like a contingency table with its rows and columns. The difference is that, whereas the cells of a contingency table contain numbers or percentages, the cells of a matrix contain some text, quotations, phrases, or symbols that derive from field notes or other sources of qualitative data. The rows in the matrix could represent the categories of a variable (as in contingency tables) or they could represent different concepts that emerged from the field notes or different points in time. The columns could also represent such a range of possibilities. Table 15.3 is a matrix used to analyze data in a study of the implementation of a new reading program in a grade school. The rows represent different conditions that affected the acceptance and implementation of the reading program, whereas the columns represent different kinds of people (users of the reading program, such as teachers versus educational administrators). Note that the material in the cells is not an attempt to say how many people felt or acted a certain way (to

quantify) but rather to give an overall assessment or judgment of each group's reaction to the innovation and then include some quotes from the field notes to support that judgment. In this study, for example, the researchers' analysis of the field notes suggested that the users of the reading program were strongly committed to it, whereas the administrators were only weakly committed. Sometimes in matrix analysis, the rows might be each individual who was observed whereas the columns are different categories from the data.

So, matrix displays are a good way to summarize data from the field notes. It is possible to summarize as many as 20 pages of field notes into one matrix, which makes it much easier and more effective to communicate the results to an audience. Matrix displays are also used as an assist in the data analysis because the researcher can review the matrix in an effort to detect patterns, themes, or trends that may be difficult or impossible to detect by reading and rereading the field notes or other texts. This is analogous to what a quantitative data analyst would do in looking at a contingency table rather than the raw numerical data. Beyond what I have said here, there are few fixed rules about creating matrices. It is a creative process, and the exact nature of the matrix is very much driven by the data. However, if you begin qualitative data analysis thinking in terms of matrices, then it may be easier to discover in the data how a variety of matrices could be created.

CONCLUSIONS AND VERIFICATION

Data analysis involves discovering meaning in the data. In qualitative research, this means identifying themes, patterns, regularities, and in some cases stating propositions, causal connections, and developing theories. In this regard, qualitative research is identical to quantitative research. However, the differences between the two are important to emphasize. One difference is that drawing conclusions in qualitative research occurs, at least in part, during the process of data collection and analysis, whereas in quantitative research this stage is usually defined as occurring after the data have been collected and analyzed (compare Figure 1.2 on page 22 and Figure 15.1). Second, most qualitative researchers hold their conclusions more tentatively or "lightly," especially in early stages of the analysis, because they recognize that additional analysis could lead to reinterpretations or reconceptualizations of the data. As analysis proceeds, the conclusions in qualitative research become more complete and certain. Third, the conclusions in qualitative research are more "grounded" in the data, which means that the conclusions are more clearly allowed to flow out of the data rather than be imposed on the data, such as by measuring devices that are created before the data are collected. In a grounded approach, the data analysis and con-

clusions are allowed to adapt to what is discovered in the data. Finally, in qualitative research more so than in quantitative research, the primary data-gathering instrument is the researcher herself or himself, and this raises concerns about the many ways in which the instrument-observer may bias the observations. This is also a problem in quantitative research, and we discussed ways in which a researcher might build a bias into, for example, a question or a multiple-item scale that is used in survey research. However, when the observer is the data-gathering instrument, then any personal values or biases of the observer can seriously compromise the collection of data and must be taken into account in assessing conclusions.

Qualitative researchers not only draw conclusions but also make efforts to verify those conclusions, or find evidence of their truth and validity. Are the conclusions reasonable and plausible? Without verification, one researcher's conclusions are simply his or her account or story of what is going on. This verification is a part of the overall argument that a researcher makes to convince both himself or herself, as well as any relevant audience, that the conclusions are reasonable and justified. In Chapter 5, I discussed some of the procedures used to assess validity in quantitative research, such as criterion and construct validity. In Chapter 11, I discussed ways to assess the validity of observations made in qualitative research. Here, I am discussing the validity of the data analysis and conclusions, but some of the points made about observations also apply to conclusions. There are also some additional points that can be made.

1. *Assess whether the conclusions are based on a thorough description and interpretation of the situation, including competing interpretations and explanations.* As soon as something resembling a pattern, generalization, or conclusion seems to be emerging from the data, researchers need to ask what rival explanations or conclusions are possible. This

can serve as a check against bias, misinterpretation, or just laziness. This will likely require additional passes through the data to see which evidence supports the rival explanations or conclusions. It can also be helpful to enlist other researchers in this process since the researcher conducting the research may have a bias that prevents him or her from recognizing and clearly evaluating alternatives. If the conclusions are justifiable, then the rival explanations will find considerably less support in the data, and this process provides more verification for the conclusions.

2. *Consider negative evidence and deviant cases.* While much of data analysis is a search for evidence to support a theme or conclusion, it is equally important to search for evidence that does not support it. Especially in qualitative research, it can be easy to miss negative evidence because the researcher is so focused on developing evidence to support a conclusion. (As one example of how this can occur, recall the Big Man bias discussed in the Eye on Diversity box in Chapter 11 on page 350.) If possible, it is helpful to have other researchers go back to the raw data—the field notes—and see if they can find any negative evidence. If they don't, this is fairly powerful support for the conclusions. A part of this is to search for *deviant cases* or *outliers*; that is, cases that don't follow the themes or patterns that are being discovered in the data. Deviant cases may be difficult to find, but the search for them is important and can be assisted by using some of the data displays discussed in this chapter. What is important is not only discovering the deviant cases but also explaining them. Can their deviation be explained and understood within the context of the overall themes and propositions of the research? Explaining them may call for modification in the concepts and propositions or, in extreme cases, may call for totally revamping the theoretical approach.

3. *Carefully assess the desires, values, and expectations of the person analyzing the data to see if these might bias the conclusions drawn.* People's perceptions are drastically shaped by their expectations or lack of them. If we expect something to occur, we are much more likely to observe it—whether or not it actually occurs. If we expect welfare recipients to be lazy, then we will be acutely aware of all those entries in field notes that might be interpreted as laziness. Thus, the validity of conclusions will be reduced to the extent that expectations—whether or not recognized—mold interpretations of data. On the other side of the coin, an absence of expectations may lead us to miss something of importance in the data.

4. *Have other researchers analyze the data to see if they come to the same conclusion.* There are problems with this because those other researchers may not have been involved in collecting the data in the field, and that experience of being in the field provides all kinds of information and insights that may not appear in the raw data (the field notes). Nonetheless, if these other researchers do arrive at the same conclusions, this will provide validation for the conclusions.

5. *Compare the conclusions reached through field observations to conclusions reached via other research methodologies,* whether it be observational research in other settings or survey, available data, or experimental research. This is a variation on criterion validity discussed in Chapter 5. It is also sometimes called *triangulation*, or approaching a problem from several different directions. If the various methodologies yield the same conclusions, then we have greater confidence that the field observations have validity. Field research whose conclusions are at wide variance with the results of other research should be accepted with caution, especially if none of these other checks on validity are available.

6. *Consider how the condition of the observer might influence observations and conclusions.* Hunger, fatigue, stress, or personal problems can lead to very distorted perceptions and interpretations. Likewise, physical

characteristics, such as the lighting in an establishment, may lead to invalid observations. (This is another good reason for keeping complete field notes—field conditions affecting validity can be assessed at a later point.) If a number of these conditions exist, it may be judicious to terminate observation and resume when conditions are more favorable. Or, at the data analysis stage, it might be wise to be very cautious about drawing conclusions from field notes produced under such conditions.

7. *Look for behavior that is illegal, stigmatizing, potentially embarrassing, or risks punishment.* If people engage in these types of behaviors, especially when they know they are being observed, then they are probably acting naturally and not putting on a performance for the benefit of the observer. For example, in his study of a sexually transmitted disease clinic, Sheley (1976) argued that the validity of his data was quite strong because "staff members dropped their professional masks and displayed quite unprofessional behavior and ideas in the company of the researcher" (p. 116). Under such conditions, you can assume that people are reacting to environmental stimuli that normally guide their behavior rather than shaping a performance for the benefit of the investigator. When these kinds of behaviors constitute the evidence for conclusions, then we can have added confidence in the conclusions.

8. *If possible, make a video or audio recording of the scene.* While such recordings have their weaknesses as records of what occurred, they do provide another way for observers to check and validate their observations and conclusions. These recordings can also be reviewed by others, which offers yet further checks on possible bias or misinterpretation.

9. *Assess the representativeness of the individuals, groups, informants, and observational sites.* Qualitative researchers do not typically use probability samples, but this does not absolve them from addressing the issue of how representative their observations are and how valid their generalizations are. I addressed this issue in Chapter 7 in the discussion on some types of nonprobability samples. To generalize beyond the individuals, groups, or sites that were observed, one needs to make a case that those individuals, groups, or sites are representative of other individuals, groups, or sites. This can be done by showing that the characteristics of those observed are similar to the characteristics of other individuals, groups, or sites that were not observed. If no claims of representativeness can be made, this does not completely invalidate the research, but it does throw into question any conclusions that claim to generalize. Of course, with some qualitative methodologies, especially the contextualizing strategies, generalization may be of secondary importance in the research.

10. *Evaluate whether the subjects of the research agree with or support the conclusions.* Who knows the field better than those whose field it is? An important clarifying and validating step in qualitative research can be to assess the reactions to the conclusions by those who are the subjects of the research. Do they agree with the themes, patterns, and propositions that form the conclusions? Depending on who is being observed, they may or may not understand the more abstract and theoretical parts of the conclusions, but many of the patterns and themes will be comprehensible to them, possibly with a little explanation. If they generally agree with the conclusions, then this provides additional evidence of validity. If they disagree with some or all of the conclusions, then this can have two positive effects. One effect is to stimulate the researcher to consider changing or clarifying the conclusions to take into account the objections. Of course, it is possible that the researcher will conclude that no change or clarification is necessary. This can lead to the second positive effect: explaining why the people disagree with the conclusions. In other words, their objections become further observations to be described and explained by the research. This may, in fact, provide more support for the conclusions.

As many of these guidelines as possible should be followed in verifying the conclusions drawn in qualitative data analysis. No one of these, of course, is the key to having confidence in the conclusions; the key is that the more that are incorporated, the more confidence we have in the conclusions.

USING COMPUTERS IN QUALITATIVE RESEARCH

Most people are at least somewhat familiar with what computers do with quantitative data: They "crunch the numbers" by performing various statistical procedures on the numerical data. However, when the data take a more qualitative form, such as field notes or the text of a magazine article, the manner in which computers would be used to analyze the data is less widely understood. Yet, as early as the 1960s, computer programs existed that could perform content analysis on such texts. Since then, advances in computer hardware and software, artificial intelligence, and optical scanning technologies have expanded dramatically the kinds of tasks that computers can do in qualitative research (Bainbridge et al., 1994; Evans, 1996; Weitzman & Miles, 1995). I won't discuss specific software programs, but rather the general capabilities found in some or all versions of what is now called **computer-assisted qualitative data analysis** (CAQDA). CAQDA software provides a valuable assist to qualitative researchers; however, it hasn't substantially changed what is done in qualitative data analysis. Computers simply assist the researcher in doing many of the tasks that I have already described in this chapter. CAQDA software definitely does not do the researchers' work for them. Qualitative data analysis is still difficult and time consuming and calls for substantial creativity on the analyst's part. It is analogous to writing: Even with word processors, writing is still a difficult job that requires a creative mind. But computers definitely make writing, as well as qualitative data analysis, easier and more effective than they would otherwise be.

To illustrate what computers currently do, let's look at a qualitative study of eating disorders among people referred to an addiction treatment center. The data in this study might consist of open-ended interviews with the clients, client social histories and treatment plans, and chart logs taken from client records. There might also be videotapes of meetings where the staff discusses cases. Many of these documents would be in narrative form, and the first step would be to either type or scan them into a computer text file (an ASCII file, in most cases).

Coding and Retrieving Data

At a very elementary level, computer software can provide counts of how frequently particular words or phrases appear in a text or field notes, and of which words and phrases tend to appear together or near one another in the text. However, most qualitative data analysis would go far beyond this. A more important use for CAQDA software is to assist the researcher in creating and using a coding scheme and coding the qualitative data. To do this, the researcher displays the text on the screen and selects a portion of interest. In some software, this is done by blocking the section of text on the screen much the way you highlight a section of text with a word processing program in order to copy it or cut it; other software automatically divides the text into sections based on lines, sentences, or paragraphs and then you place a marker at the beginning or end of the section to which you want to assign a code. For example, a section of text describing the way a mother expressed her disapproval of her daughter's behavior could be selected and marked with a code for "Mother disapproval of daughter behavior." CAQDA software can then search through the text identifying and counting the number of times various codes appear in field notes. Advanced software can also analyze whether certain words, phrases, or codes tend to be used together, how far apart from one another they tend to be used, whether words or codes used in a text tend to be positively or negatively evaluated, and whether certain words or codes tend to be associated with certain other words or codes (Evans, 1996; Kelle, 1995; Weitzman & Miles, 1995). Some software assists in the coding with search procedures and what are called "wild cards." So if you request a search for "moth*," the program will go through the field notes and identify all words that begin with "moth" and display them. Such a search might produce words like *mother, motherly, motherhood,* and *mothballs,* and you can decide which code to assign to each.

To begin analyzing the data, the researcher uses the search function of the program. In this hypothetical study, one might begin by locating all cases that contain the codes "Mother pressured" or "Father pressured." This process is similar to conducting a computerized literature search by entering combinations of key words. By using these codes joined by the key term "OR," a subset of all cases where parental pressure was indicated can be identified. This group of files could then be saved for further analysis. Having selected a subset of cases based on one or more codes, the program can produce a report with whatever information the researcher wishes: the individual's name, the name given to various codes, the frequency of use of various codes, and which source material the data were taken from.

Most CAQDA software programs store the codes for a given case in a special file, sometimes called an "index card." When done coding the material for one case, the index card computer file contains all the codes that the researcher used for all the materials pertaining to that individual. Thus the computer program follows a logical procedure that is much like the traditional practice of manually preparing notecards for cross-referencing cases. To help with coding, the program provides a code list, which contains all the codes used thus far for all cases in the study. Some programs can also "autocode," which means to assign the same code to all identical words or phrases in a document. You can also store information about each case or individual in the study, such as their age, gender, or whatever else is available in the data. In addition, the software allows the user to give more than one code to a given segment of text. You can also revise the coding scheme in several ways. For example, you can create different levels of codes. If a number of codes all share something in common, they can be grouped under a higher-level or more abstract code. The program retains all levels of coding, and the data can be retrieved using any level of code. Or you can go the opposite way: a single coding category can be elaborated into a series of more specific or concrete categories.

Data Analysis and Hypothesis Testing

Some content analysis software can test hypotheses by performing searches for all cases containing particular combinations of codes. Suppose our hypothesis was that "mothers who are critical of their daughter's body image will have a negative effect on their daughter's self-image." The software could search for all cases that include the codes "Mother critical of daughter's body image," "Mother–daughter relationship strained," and "Daughter experiencing weight loss." To each of those cases, it then adds the code "Mother negative influence on daughter self-image," indicating that these cases confirm the hypothesis. Only cases meeting all three conditions would have the new code added to their files. In a similar way, codes can be constructed for other code combinations of interest.

Some content analysis software can perform statistical analyses of the frequency of occurrence of various codes and display the data in graphical form, using some of the graphs and matrices discussed in this chapter. Some software can also import data from and export data to statistical programs and spreadsheets. In addition, beyond merely automating content analysis procedures, some programs provide important advances in the validation, reliability, and generalizability of qualitative data analysis. For example, an independent researcher can code the same data with the codes already developed and stored and thus determine interrater reliability. To assess validity, an independent researcher might code the same material blindly—that is, without benefit of the existing coding scheme—to determine if the second coder develops the same or similar meanings as the first.

Conceptualization and Theory Building

Some CAQDA software is designed to assist in the tasks of conceptualization and theory building (Coffey & Atkinson, 1996; Dohan & Sánchez-Jankowski, 1998). Of course, the software doesn't actually create coding schemes or build theories, but it can provide the researcher with some assistance in these tasks by showing links in the data between concepts, coding categories, and observations that the researcher might not have observed without such assistance. In fact, this CAQDA software has been used heavily by some proponents of grounded theory methodology because the software assists in going back and forth from data to concepts, which is a key feature of grounded theory.

One way the software supports theory building is by providing opportunities for the process of "memoing" discussed earlier in this chapter: You can input (keyboard in) your own comments and the software will link them to sections of identified text. These memos can be stored separately from the field notes, or they can be inserted directly as annotations into the field notes or other text. Annotations to field notes would be identified as such by the software, and the annotations themselves can be coded. As explained earlier in the chapter, these memos and annotations are expressions of conceptual and theoretical ideas, and the software assists in this process by enabling the researcher to gradually build a conceptual scheme and a theory that provides a more abstract understanding of the data while it is also still closely linked to the data. Also, by being able to code these memos, the researcher is helped to notice linkages that might otherwise have escaped his or her attention. Some software programs also provide graphical maps that present a visual representation of links between concepts and other concepts and between concepts and data. This visual picture can be a substantial assist in theory building:

> Maps may help the analyst picture the project's theoretical shape, the concepts in use, the relationships between those concepts, and the ethnographic data that have been collected regarding each of those concepts and links. Theory-building software facilitates experiments with different concepts and links within the research process (Dohan & Sánchez-Jankowski, 1998, p. 490).

All these new technologies do not relieve the researcher from the hard work of qualitative data analysis, which is time consuming even with computer assistance. However, the new technologies do make possible longer and more sophisticated analysis of the meanings embedded in qualitative data than are possible without computer assistance.

REVIEW AND CRITICAL THINKING

Much of the information that we confront in our everyday lives is not quantitative but rather what social scientists call qualitative: narratives and descriptions, either written or verbal. Some of the issues addressed in this chapter can assist you in evaluating that information.

1. What is the context for this information? As presented, has the information been removed from its context and might that change the meaning of the information?

2. Is the presented information the raw data? Or has the information been abstracted or summarized in some fashion (i.e., the equivalent of coding, reflective remarks, or memos)?

3. What abstract or conceptual conclusions are inferred from the information? Are they justified on the basis of the criteria discussed in this chapter?

Computers and the Internet

The use of computers to help analyze qualitative data is discussed extensively in the body of this chapter. The Internet is also a useful resource. As you try various search strategies, pay attention to the number of different fields in the social sciences and human services that show an interest in this topic. A good way to begin, as mentioned in Chapter 11, is by searching under terms such as "qualitative methods," "participant observation," or "field research," although the last will produce many sites relating to the natural sciences since they also conduct something called "field" research. However, this can produce an important learning exercise: Can you identify the ways in which the research or science discussed at the

natural science Web sites is similar to or different from that done by social scientists, as discussed in this book? For example, do the natural scientists appear concerned with the positivist versus nonpositivist debate? Beyond these terms, you can search for various terms discussed in this chapter, such as computer-assisted qualitative data analysis, coding, narrative, case studies, and so on. For this chapter, you are looking for Web sites that deal with qualitative data *analysis*, not how to conduct field research.

I located a number of Web sites that contained useful information and/or links to other valuable Web sites. For example, the School of Social and Systemic Studies at Nova Southeastern University maintains a Web site titled "The Qualitative Report Homepage" (**http://www. nova.edu/ssss/QR/**). That same site also has a comprehensive section on ethics in qualitative research (**http://www.nova.edu/ssss/QR/ nhmrc.html**). Two other comprehensive sites are "Resources for Qualitative Research" (**http:// wwwedu.oulu.fi/sos/kvaltutk.htm**) and "QualPage: Resources for Qualitative Researchers" (**http://www.ualberta.ca/~jrnorris/ qual.html**). Another good Web site is for the "CAQDAS Networking Project": **http://www. soc.surrey.ac.uk/caqdas/.** CAQDAS refers to Computer-Assisted Qualitative Data Analysis Software, and this site includes an extensive bibliography as well as opportunities to review qualitative data analysis software.

Main Points

1. Qualitative data analysis focuses on extracting meaning from nonnumerical data without transforming the data into numbers. While nonpositivist paradigms have been

receptive to qualitative approaches, there is no one-to-one link between a particular paradigm and a tendency to use qualitative rather than quantitative research.

2. In qualitative research, the stages of data collection, data analysis, and drawing conclusions tend to overlap. One of the advantages of this overlap is greater theoretical sensitivity. Qualitative research might pursue goals of description, explanation, and evaluation as well as attempts to abstract and generalize. However, it also seeks the goal of contextualizing the analysis.

3. Qualitative data analysis can pursue a categorizing strategy or a contextualizing strategy. One categorizing strategy is to code the data. One approach to coding is to create a fairly complete and detailed coding scheme before gathering data; other approaches let the coding scheme emerge as the data is collected and analyzed.

4. Specific types of coding can focus on the concrete versus the abstract or on open versus more focused coding. Coding schemes should have clear operational definitions and be tested for reliability.

5. Reflective remarks and memos can assist in coding, conceptualization, and theory building. Contextualizing strategies include profiles, narrative analysis, and case studies. Their primary goal is idiographic: to understand a particular case rather than to generalize to other cases.

6. Data displays are important in helping the researcher develop coding schemes and conceptual categories as well as in communicating conclusions to audiences. Data displays come in the form of narrative texts and visual displays, such as context maps and matrix displays.

7. Qualitative data analysis produces conclusions that must be verified. This verification can follow a variety of guidelines such that confidence in the conclusions increases as more of these guidelines are satisfied.

8. Qualitative data analysis is often accomplished via computer. Computers can assist in coding and retrieving data, analyzing the data, testing hypotheses, and engaging in memoing and other strategies that help in conceptualization and theory building.

Important Terms for Review

case study
categorizing strategies
coding
computer-assisted qualitative data analysis
context map
contextualizing strategies
data display
matrix display
memos
narrative analysis
network analysis
profiles
reflective remarks
sociograms
theoretical sensitivity

For Further Reading

Coffey, Amanda, & Atkinson, Paul. (1996). *Making sense of qualitative data: Complementary research strategies*. Thousand Oaks, CA: Sage. This book focuses on the analysis stage and gives a good overview of the various analysis strategies used, including how computers are used in qualitative data analysis.

Flick, Uwe.(1998). *An introduction to qualitative research*. Thousand Oaks, CA: Sage. This book covers all aspects of qualitative research, including how to analyze data and how to visually present the data.

Gahan, Celia, & Hannibal, Mike. (1998). *Doing qualitative research using QSR NUD.IST*. Thousand Oaks, CA: Sage. These authors describe in clear prose how to use the QSR

NUD.IST software for analyzing qualitative data. It both explains this software and shows clearly how qualitative analysis is done.

Merriam, Sharan B. (1998). *Qualitative research and case study applications in education.* San Francisco: Jossey-Bass. Qualitative research has been popular in the field of education for decades, and this book gives a good introduction on how to collect and analyze qualitative data in educational settings.

Richards, Lyn. (1999). *Using NVivo in qualitative research.* Thousand Oaks, CA: Sage. This book provides a good overview of how computer-assisted qualitative data analysis is done, as well as focusing on how to use one fairly popular CAQDA software package.

Rossman, Gretchen B., & Rallis, Sharon F. (1998). *Learning in the field: An introduction to qualitative research.* Thousand Oaks, CA: Sage. These authors describe the complexity of doing field research while still making the basic data-analysis tools accessible to the student.

Strauss, Anselm, & Corbin, Juliet. (1998). *Basics of qualitative research: Techniques and procedures for developing grounded theory* (2nd ed.). Thousand Oaks, CA: Sage. This book offers a complete introduction to the data analysis techniques developed to support the grounded theory approach. It is both readable and practical.

Part V

PUTTING RESEARCH FINDINGS TO USE

CHAPTER 16:
Research Grant Proposals and Report Writing

CHAPTER 17:
Applications of Social Research

This final part of this book deals with issues relating to how research findings are used. One step in the process of putting research to use is communicating the results of the research to a variety of audiences. One such audience is other social scientists, and Chapter 16 describes the peer review process in the social sciences. Another audience consists of the agencies, organizations, and foundations that provided funds to conduct the research. A central fact of life in scientific research is that much research is funded by grants. So, a key issue in Chapter 16 is how one acquires the grants and then what is the outcome of research that is supported by some funding source. Chapter 16 also focuses on how to properly prepare a research report to communicate research findings to any of these audiences. Chapter 17 describes many of the specific applications of social science research. These are areas in which both basic and applied research are often put to use by both researchers and others in an effort to advance knowledge or improve society.

Chapter 16

RESEARCH GRANT PROPOSALS AND REPORT WRITING

In Chapter 4, I discussed in detail how to formulate a research problem, but what wasn't talked about there were some critical issues: How does a researcher obtain funds to support the research? How are research results communicated to wider audiences? This chapter focuses on these elements of the research process: the role of grants in conducting research, the preparation of research proposals, and the preparation of research reports.

THE SOCIAL WORLD OF RESEARCH

Funds to Support Research

Scientific research occurs in the context of a complex social organization that supports it. I talked about this briefly in Chapter 1 in the discussion of the scientific community. It is necessary to explore it more here in order for you to fully understand the process of research, especially the realms of grant funding and report writing. Scientists conduct research in a variety of contexts, and those contexts influence, to an extent, how they do their research. For example, many scientists are on the faculty of universities and conduct research as a part of their faculty responsibilities. In most cases, they have considerable discretion in terms of the topics of their research, but often the university does not supply the funds to conduct the research, beyond paying the salary of the faculty member and possibly providing some clerical support. In a few cases, the costs of the research may be minimal and can be supported by what the university provides, but in most cases this is not true. So, the scientist must search for funds, which are most often called *grants*.

Grants refer to the provision of money or other resources to be used for either research or other purposes, and are an important funding source for social science research. In some cases, the scientist's university may fund some research through a competition for a limited amount of grant money provided by the university. So the scientist must prepare a research proposal in a quest for those funds. If a scientist's university doesn't provide research funds, or the funds are inadequate, or the scientist's proposal fails to secure some of those funds, then the search turns to sources of funding from organizations outside the university. The amount of grant monies awarded each year is truly staggering. For example, at the federal level, such agencies as the National Science Foundation (NSF) and the National Institutes of Health spend many billions of dollars every year on research. Private foundations award grants worth almost $200 million each year to the social sciences (U.S. Bureau of the Census, 1998, p. 397). Indeed, vast sums are dispersed through this system of grants. Given the large role grants play in funding research, it is highly likely that almost any researcher will be drawn into the grant-funding process at some level. I will talk about some of the sources of those grants later in this chapter.

Many social science researchers are not on the faculty of universities but rather are employed in other settings to conduct research. Some are employed by a governmental agency, often doing program evaluations or other kinds of applied research. Other social scientists work for private research firms whose business is to conduct research for whoever has a research need, which might be a governmental agency, a nonprofit organization, or a private corporation. There are numerous survey research centers—at universities and private firms—that conduct public opinion polling. Corporations also employ social scientists to conduct marketing research and public opinion polling.

The point is that scientists are not independent agents. Rather, their work has to be supported by many organizations, and scientists are to a degree beholden to those organizations. They are beholden in the sense that they must convince the organizations to provide funds to support their research; this means that they must prepare a grant or research proposal to convince one of these organizations to support their research. The scientists are also beholden in that they must report back to those funding organizations the results of their research.

The Peer Review Process

I discussed the scientific peer review process in Chapter 1, and it is worth returning to it as we address this topic of writing grant proposals and research reports. I have stressed throughout this book that many opportunities exist for error and misinterpretation to enter scientific work. Scientists are neither perfect nor are they without personal biases and foibles. One way to guard against these problems is with a process of **peer review**, in which scientists with appropriate expertise review and make judgments about the grant proposals or research reports prepared by other scientists. We saw in Chapter 1 that the norms of the scientific community are promoted through a process of peer review and evaluation that is central to scientific work. When scientists conduct research, they begin by preparing a preliminary research report that contains their research design. Oftentimes, they will seek a grant in order to fund the research. When a request is submitted to a granting agency, that agency typically sends it to other scientists who possess the appropriate expertise to evaluate it. Even when researchers don't seek grants, they still often show the research design to colleagues to get feedback on the adequacy and completeness of the proposed research. This review of research designs is one point of peer review that scientists undergo.

Once the research is done, a report or paper is prepared which describes how the research was done and summarizes the data analysis and conclusions. This report is then (1) presented at a professional conference to other scientists knowledgeable about the topic of the research, (2) submitted for publication in a scientific journal in the field, and/or (3) submitted to the organization that provided the grant for the research or the sponsor who paid for the research. All three of these outlets serve as further points of peer review and evaluation of the research. The professional conferences and scientific journals, for example, only accept papers after they have been reviewed by scientific peers—researchers who are experts on a particular topic—and judged to be acceptable in terms of the research methods used, the data analysis techniques applied, and the conclusions drawn. Likewise, granting agencies and sponsors have expert reviewers to evaluate the research, and a researcher who does not produce quality research will find it difficult to be awarded further grants. Since prestige, jobs, and income for scientists depend on their success in obtaining grants, presenting papers, and publishing research, scientists are highly motivated to see that their research is evaluated positively in such forums. However, the only way to accomplish that is to produce research that stands up to the intense scrutiny of one's scientific peers. If a scientist publishes his or her research in a journal or magazine that does not use peer review, it is then considered a low quality publication and will carry less weight (or possibly no weight) in decisions about employment, promotion, or tenure. So, the final stage in the research process—dissemination of results—is a very important one, both in terms of advancing scientific knowledge and advancing the careers of scientists.

Just as other cultures sanction their members through rewards and punishment in order to encourage acceptable behavior, so, too, does the scientific community. When laypeople learn of this peer review process, they are sometimes inclined to view it as a system of censorship where those who control the journals determine what is scientific knowledge at a given moment. And I would be naive to say that something like this never happens. Yet scientists view this not as censorship but rather as a demand for appropriate standards. Scientists agree that there are certain procedures—the topic of this book—that are more likely to produce accurate and valid knowledge. Peer-reviewed journals in science demand that people use these procedures in order to get published in the journals because this will result in a more accurate body of scientific knowledge.

THE GRANT FUNDING PROCESS

I next describe how researchers seek funds to support their research. However, not all researchers fund their research through grants. Some researchers, for example, are employees of governmental agencies or private research firms, and their employers provide the research money. Even when research is not funded by grants, researchers still have to go through many of the steps described in this section because all researchers need to develop a research proposal in order to justify the expenditure of money on the proposed research.

Funding Sources

Sources of grant money fall into three general categories: governmental agencies, private foundations, and corporations (Bauer, 1999; Smith & McLean, 1988).

Governmental Funding Sources

Governmental agencies are by far the largest source of grant monies. They are also about the only sources of large grants greater than a few thousand dollars. Even in the case of smaller grants, it often pays to start first with governmental sources because many private foundations refuse funding automatically unless you can prove that efforts to obtain government funding have failed.

Among governmental agencies, the federal government is the largest source for grant money. The federal government provides three different types of grants. *Project* or *categorical*

grants (the terms are synonymous) are narrowly focused on some specific need or problem as defined by some agency of the federal government. Grant seekers design approaches, within specified guidelines provided by the governmental agency, to meet the research needs specified by that agency. The agencies make their requests for research opportunities known through what is called a *Request for Proposals* (RFPs), which is a formal request for people, agencies, or organizations to submit proposals on how they would conduct some research. Agencies or researchers then submit proposals in a competition to obtain the grant money.

Formula grants and *block grants* are part of large-scale, nationwide federal programs and serve as mechanisms for allocating the funds for the programs. For example, at a time of high unemployment, the federal government may sponsor various "job-creating" services through formula or block grants. Because unemployment is experienced in varying levels of severity in various places in the country, the money would be allocated by a "formula" that provides most of the funds to those areas hardest hit by the problem and prorates the remaining funds to other areas as needed. These grants, especially block grants, are often channeled through state and local governments, and the service program may ultimately be delivered by a state or local agency or nonprofit organization. Often, a part of these programs is a research component, such as a program evaluation of whether the overall program is achieving its goals. The organization or agency that ultimately runs the program will seek researchers, possibly using RFPs again, to design and conduct the research.

Beyond grants, the federal government also contracts with researchers or research organizations to conduct some research needed by a federal agency. These federal contracts would normally involve some type of applied research because the federal agency has some particular need to be assessed or program to be evaluated.

Many state governments also provide grant money for research, usually applied research to support the provision of state services. States also, of course, serve as the conduit for some federal grant monies. One way to increase your chances of getting state grant monies is to keep in contact with the state departments and agencies that oversee and fund research and services related to the human services, such as the departments of corrections, mental health, and social services. These state agencies may also publish RFPs as a way of soliciting grant proposals.

Private Funding Sources

A **foundation** is a nonprofit, legally incorporated entity organized for the purpose of dispersing funds to projects that meet the guidelines of its charter. The number of private foundations is staggering, upwards of 35,000 (Bauer, 1999; Krathwohl, 1988). In general, private funding sources are good places to seek grants for applied research because they prefer to fund action programs that produce immediate results rather than research projects that, at best, have some long-term payoff. Also, the complexity of proposal preparation and submission is much less stringent than that encountered with governmental agencies.

There are five distinct types of foundations. *Community foundations* exist to serve their immediate local area. Therefore, if a project is modest in scale and serves some local need, a community foundation may be a good choice. *General purpose foundations* are often large, such as the Ford Foundation, and operate nationwide. If a project is large in scope, with the potential of having an impact broader than the local community, these large foundations may be ideal. They particularly like innovative demonstration projects that may show the way for other communities to solve various problems. *Special purpose foundations* carve out a particular area of interest and award grants only to projects that deal directly with that area of specialization. Successful funding from these sources requires some research into

which foundations fund what kinds of projects. Fortunately, large foundations typically publish annual reports, much as corporations do, outlining recently funded projects.

Family foundations are the most difficult to categorize because there are so many—more than 30,000 according to Bauer (1999)—and they are so different. Some are large and have the resources to award fairly substantial grants whereas others have a cap on grant size of a few thousand dollars. Some are quite general in the projects they fund whereas others have very narrow interests. For example, some may fund only projects that benefit a particular religious or ethnic group while others fund only projects that address a particular problem, such as alcoholism or child abuse.

Corporate foundations are used as a conduit for corporate philanthropy. Other corporations engage in philanthropy but do not use the foundation mechanism. In either case, nonprofit agencies are common recipients of corporate giving. As investor-owned, profit-making enterprises, corporations are giving away the stockholders' money. As such, the directors who make the philanthropic decisions are cautious to fund only those activities that can be justified to the stockholders. This tends to mean that the corporation or its employees must stand to benefit in some way from funded projects. For example, a nonprofit child care facility used by many corporate employees might receive a corporate grant or other corporate support.

Learning about Funding Opportunities

Given all the separate agencies and organizations that disperse grants, how do you find specific funding opportunities? Publications and computer databases are available that can help. The *Catalog of Federal Domestic Assistance (CFDA)*, for example, describes all federal government programs. Figure 16.1 provides an illustration from the catalog of a program run by the National Institutes of

FIGURE 16.1   Illustration from the *Catalog of Federal Domestic Assistance*

93.242 MENTAL HEALTH RESEARCH GRANTS

FEDERAL AGENCY: NATIONAL INSTITUTES OF HEALTH, DEPARTMENT OF HEALTH AND HUMAN SERVICES

AUTHORIZATION: Public Health Service Act, Title III, Section 301, Public Law 78-410, 42 U.S.C. 241, as amended; Small Business Research and Development Enhancement Act of 1992, Public Law 102-564

OBJECTIVES: To increase knowledge of basic biological and behavioral processes that underlie mental and behavioral disorders and of processes that contribute to maintaining mental health; to improve methodologies for research relevant to these disorders; and to conduct research on mental health services. Research supported by the National Institute of Mental Health may employ theoretical, laboratory, clinical, methodological and field studies. Studies may involve individuals with a mental disorder diagnosis, individuals with symptom levels that do not meet diagnostic thresholds, and healthy individuals of all ages. Research also may involve animal, computational and mathematical models appropriate to the system being investigated and the state of the field. Areas eligible for research support are: neurosciences, including molecular genetics; behavioral sciences; epidemiology; clinical assessment; etiological studies; treatment; prevention; services research; and research on HIV/AIDS behavior. The Minority Research Infrastructure Support Program provides awards to increase the capacity of institutions with a substantial enrollment of racial ethnic minority students to conduct mental health research projects. The Small Business Innovation Research (SBIR) program and the Small Business Technology Transfer (STTR) program provide awards to increase small business participation in Federal research and development by means of increasing cooperative research and development between small businesses and research institutions (SBIR); and encouraging participation of socially and economically disadvantaged small business concerns and women-owned small business concerns in technological innovation (STTR).

TYPES OF ASSISTANCE: Project Grants (Cooperative Agreements).

USES AND USE RESTRICTIONS: (1) Research project grants provide support for clearly defined projects or a small group of related research activities, and when appropriate, support of research conferences; (2) Program Project and Center grants support large-scale, broad-based programs of research, usually interdisciplinary consisting of several projects with a common focus; (3) Research Resources Development, Demonstrations, Research special Research Exploratory projects, and Dissertation Support; and (4) Small grants support small-scale exploratory and pilot studies or exploration of an unusual research opportunity. Standard small grants are limited to $50,000 direct costs for a period of 2 years or less, while newer small grants are for less time and funds. SBIR and STTR grants are awarded in two stages:

Phase I grants are awarded to establish the technical merit and feasibility of a proposed research and development effort; only Phase I awardees are eligible to receive Phase II support.

ELIGIBILITY REQUIREMENTS:

Applicant Eligibility: Public, private, profit, or nonprofit agencies (including State and local government agencies), eligible Federal agencies, universities, colleges, hospitals, and academic or research institutions may apply for research grants. SBIR grants can be awarded only to domestic small businesses, and STTR grants can be awarded only to domestic small businesses which "partner" with a research institution in cooperative research and development. For further definitions, requirements, and restrictions see the Omnibus Solicitation of the Public Health Service for Small Business Innovation Research Grant Applications (PHS 97-2) and the Omnibus Solicitation of the National Institutes of Health for Small Business Technology Transfer Grant Applications (PHS 97-3).

Beneficiary Eligibility: Public, private, profit, or nonprofit organizations.

Credentials/Documentation: Costs will be determined in accordance with OMB Circular No. A-87 for State and local governments. For-profit organizations' costs will be determined in accordance with 48 CFR, Subpart 31.2 of the Federal Acquisition Regulations. For all other grantees, costs will be determined in accordance with HHS Regulations 45 CFR, Part 74, Subpart Q. For SBIR and STTR grants, applicant organization (small business concern) must present in a research plan an idea that has potential for commercialization and furnish evidence that scientific competence, experimental methods, facilities, equipment, and funds requested are appropriate to carry out the plan. Grant forms PHS 6246-1 and 6246-2 are used to apply for SBIR Phase I and Phase II awards, respectively; grant forms PHS 6246-3 and PHS 6246-4 are used to apply for STTR Phase I and Phase II awards, respectively.

APPLICATION AND AWARD PROCESS:

Preapplication Coordination: Not applicable. This program is excluded from coverage under E.O. 12372.

Application Procedure: The standard application forms, as furnished by PHS and required by 45 CFR, Part 92 (PHS 5161-1), must be used for applicants that are State and local governments. Application kits, containing the necessary forms (PHS 5161-1 or PHS 398, Rev. May 1995) and instructions, if not available at applicant institution, may be obtained from the Division of Extramural Outreach and Information Resources, Office of Extramural Research, National Institutes of Health, 6701 Rockledge Drive, Bethesda, Maryland 20892-7910. Telephone: (301) 435-0714. Fax: (301) 480-0525. E-mail: asknih@od.nih.gov. Consultation on a proposed project may be obtained from the NIMH branch or office responsible for the research area of interest. Applications

FIGURE 16.1   *(continued)*

are reviewed by principally nonfederal consultants recruited nationwide from the mental health field. The amounts of the award and period of support are determined on the basis of merit of the project and the nature of the grant mechanism. This program is subject to the provisions of 45 CFR, Part 92 for State and local governments, OMB Circular No. A-110 for nonprofit organizations, cost principles of A-21 for educational institutions, and 42 CFR, Part 42. Applications for SBIR and STTR grants may be obtained electronically through the NIH's "Small Business Funding Opportunities" home page at http://www.nih.gov/grants/funding/sbir.htm on the Word Wide Web. A limited number of hard copies of these publications is produced. Subject to availability, they may be obtained by contacting the NIH SBIR support services contractor by telephone (301) 206-9385 or by fax (301) 206-9722; e-mail: a2y@cu.nih.gov. The Solicitations include submission procedures, review considerations, and grant application or contract proposal forms. SBIR and STTR grant applications, upon completion, should be submitted to the National Institutes of Health, Center for Scientific Review, 6701 Rockledge Drive, Room 1040-MSC 7710, Bethesda, MD 20892-7710.

Award Procedure: All applications for research grants, cooperative agreements, SBIR and STTR grants are evaluated for scientific and technical merit by an appropriate scientific peer review panel and by the National Advisory Mental Health Council (excepting Small Grants). All competitive applications compete for available funds on the basis of scientific and technical merit, program relevance, and program balance. All SBIR and STTR applications receiving a priority score compete for set-aside funds on the basis of scientific and technical merit and commercial potential of the proposed research, program relevance, and program balance among the areas of research.

Deadlines: New Grants and Centers Renewals: February 1, June 1, and October 1. Other Renewals: March 1, July 1, and November 1. AIDS Grants: January 2, May 1, and September 1. SBIR: April 15, August 15, and December 15. STTR: December 1, April 1, and August 1. Minority Research Infrastructure Program: June 1. Dissertation Research Grants: April 11, August 10, December 13.

Range of Approval/Disapproval Time: Grants: From 240 to 270 days from submission of application. SBIR/STTR applications: About 7-1/2 months; Mental Health Education Programs and Various/Small Grants: From 5 to 6 months. Review of AIDS-related research is expedited.

Appeals: A principal investigator (P.I.) may question the substantive or procedural aspects of the review of his/her application by communicating with the staff of the Institute. A description of the NIH Peer Review Appeal procedures is available on the NIH homepage www.nih.gov/grants/guide/1997/97.11.21/n2.html.

Renewals: Support is recommended for a specified project period, not in excess of 5 years. Prior to termination of a project period, the grantee may apply for renewal of support for a new project period. An application for renewal is processed as a new competing request. Small grants are for 1-2 years (depending on program) and are not renewable.

ASSISTANCE CONSIDERATIONS:

Formula and Matching Requirements: This program has no statutory formula or matching requirements.

Length and Time Phasing of Assistance: Varies, but a project period is generally limited to 5 years or less. Grantee may apply for renewal of support on a competing basis. Within the project period, continuation applications must be submitted on a non-competing basis for each year of approved support. Small Grant support is limited to 1-2 years and is not renewable. SBIR Phase I awards are generally for 6 months; Phase II awards are for 2 years. STTR Phase I awards are generally for 1 year; Phase II awards normally are for 2 years. Payments will be made either on a Monthly Cash Request System or under an Electronic Transfer System. Necessary instructions for the appropriate type of payment will be issued shortly after an award is made.

POST ASSISTANCE REQUIREMENTS:

Reports: Reports must be submitted as follows: (1) interim progress reports annually as part of a non-competing application for previously recommended support; (2) terminal progress report within 90 days after end of project support; (3) annual financial status report within 90 days after termination of annual grant for some programs. In addition, immediate and full reporting of any inventions is required.

Audits: In accordance with the provisions of OMB Circular No. A-133 (Revised, June 24, 1997), "Audits of States, Local Governments, and Non-Profit Organizations," nonfederal entities that receive financial assistance of $300,000 or more in Federal awards will have a single or a program-specific audit conducted for that year. Nonfederal entities that expend less than $300,000 a year in Federal awards are exempt from Federal audit requirements for that year, except as noted in Circular No. A-133. In addition, grants and cooperative agreements are subject to inspection and audits by DHHS and other Federal government officials.

Records: Records must be retained at least 3 years; records shall be retained beyond the 3-year period if audit findings have not been resolved.

FINANCIAL INFORMATION:

Account Identification: 75-0892-0-1-552.

Obligations: (Grants) FY 97 $480,892,023; FY 98 est $518,770,000; and FY 99 est $560,835,000.

Range and Average of Financial Assistance: The range is from $1,500 to $4,400,930; $296,664.

Health and providing funds for research relating to mental health issues. As you can see, the catalog supplies information valuable to the grant seeker. It explains the objectives of the program, eligibility requirements, the steps in the application process, and so on. The *CFDA* comes on both floppy disk and CD-ROM, as well as printed versions. It also can be accessed at a Web site (**http://www.cfda.gov/**). Also on CD-ROM is the Federal Assistance Award Data System (FAADS), which helps you search the *CFDA* and also provides information on people and organizations that have successfully applied for grants from programs listed in the *CFDA*.

Grant seekers can use their personal computers to search for funding opportunities through the use of the "Grants Database" prepared by Oryx Press. This is available on CD-ROM or online (contact Oryx Press or the Dialog Corporation). Oryx Press also has a hard-copy version available titled *Directory of Research Grants*. It is the most comprehensive source of current information on grants offered by government, corporate, and private funding sources. In addition to its convenience, the database has the advantage of being updated monthly, unlike conventional publications that may become dated. It also contains an extensive and useful list of Web sites that relate to obtaining funding for research. There is also a Web page where you can search through this database (**http://www.higheredconnect.com/grantselect/**).

Another useful publication is the *Federal Register*. This daily, magazine-size volume reports on the activities of the federal government. Although it includes a lot of information of little use to the grant seeker, new programs are first announced in it; so it is a good resource for the ever-changing opportunities in obtaining federal funding. Guidelines for obtaining funding under the new programs are also first provided in the *Federal Register*. Eventually, this information gets into the *CFDA*, but because this is only published annually, many months could pass before a new

program gets listed in its latest edition. The *Federal Register* is also available on CD-ROM and at a Web site maintained by the National Archives and Records Administration (**http://www.access.gpo.gov/nara/**). Many funding agencies publish periodic newsletters or bulletins describing their latest activities and programs. It is easy to get on these mailing lists. They often contain RFPs; searching the RFPs, you may find opportunities for research.

An organization called the Foundation Center (**http://www.fdncenter.org**) produces a huge amount of material to help in a search for grants. One of its primary publications, *The Foundation Directory*, contains information on foundations of all types. Figure 16.2 presents an entry from the directory on one such funding organization. As you can see, the entry supplies a considerable amount of information about an organization—information that is useful in the sorting-out process. For example, the financial information gives some idea of the size of grants that a foundation typically awards. It also provides information on "purpose and activities," "types of support," and "limitations" to further screen potential funders. In these sections, you can learn whether a particular foundation provides grants for research purposes, since some do not. Also, procedures for making applications are described and memberships of boards of directors are provided. For those in the business of seeking grants, *The Foundation Directory* is an essential tool.

Another publication from the Foundation Center, *Foundation Fundamentals: A Guide for Grant Seekers*, is particularly useful for the beginner. It outlines the services of the center along with information on locating foundations and preparing and submitting proposals to private foundations. A special guide to corporate foundations, *Corporate Foundations Profiles*, is also offered by the Foundation Center. This useful volume presents detailed descriptions of the 250 largest corporate foundations and less detailed information on 470 more. This directory, among other things, allows you to determine which corporations have operations in your area and might therefore be likely prospects for funding. The Foundation Center also publishes *The Foundation Center's Guide to Grantseeking on the Web* and has much material on a CD-ROM called "FC Search: The Foundation Center's Database on CD-ROM." Some of the center's materials are also available at its Web page.

FIGURE 16.2 Explanation of a Sample Entry Form from *The Foundation Directory*

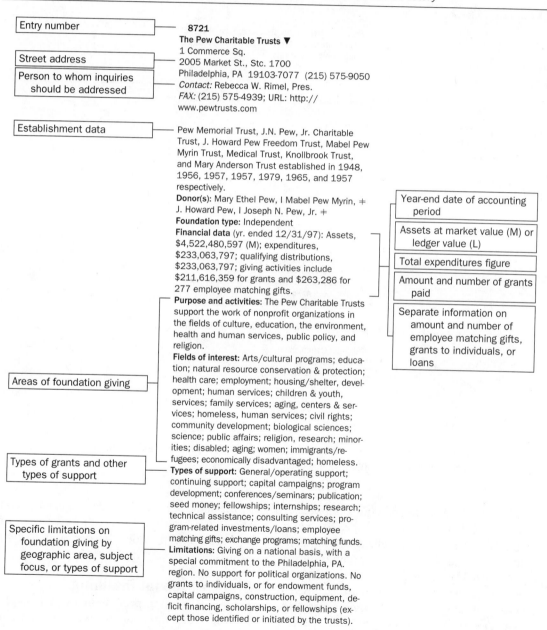

(continued)

FIGURE 16.2   *(continued)*

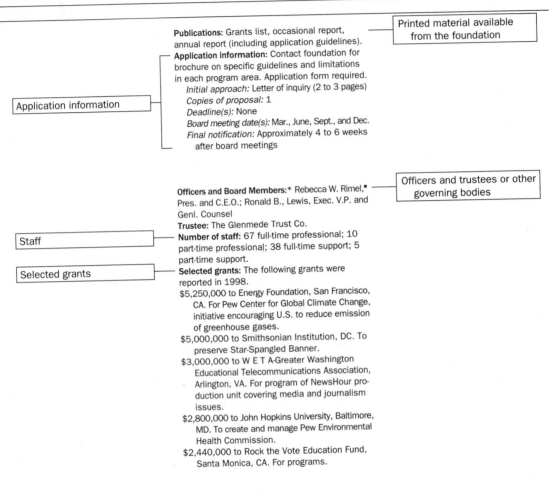

Application information

Publications: Grants list, occasional report, annual report (including application guidelines).
Application information: Contact foundation for brochure on specific guidelines and limitations in each program area. Application form required.
Initial approach: Letter of inquiry (2 to 3 pages)
Copies of proposal: 1
Deadline(s): None
Board meeting date(s): Mar., June, Sept., and Dec.
Final notification: Approximately 4 to 6 weeks after board meetings

Printed material available from the foundation

Officers and Board Members:* Rebecca W. Rimel,■ Pres. and C.E.O.; Ronald B., Lewis, Exec. V.P. and Genl. Counsel
Trustee: The Glenmede Trust Co.

Officers and trustees or other governing bodies

Staff

Number of staff: 67 full-time professional; 10 part-time professional; 38 full-time support; 5 part-time support.

Selected grants

Selected grants: The following grants were reported in 1998.
$5,250,000 to Energy Foundation, San Francisco, CA. For Pew Center for Global Climate Change, initiative encouraging U.S. to reduce emission of greenhouse gases.
$5,000,000 to Smithsonian Institution, DC. To preserve Star-Spangled Banner.
$3,000,000 to W E T A-Greater Washington Educational Telecommunications Association, Arlington, VA. For program of NewsHour production unit covering media and journalism issues.
$2,800,000 to John Hopkins University, Baltimore, MD. To create and manage Pew Environmental Health Commission.
$2,440,000 to Rock the Vote Education Fund, Santa Monica, CA. For programs.

Source   Reprinted with permission from *The Foundation Directory,* 21st ed., copyright © 1999 by the Foundation Center, 79 Fifth Ave, New York, NY 10003 (www.fdncenter.org).

The Annual Register of Grant Support: A Directory of Funding Sources, like *The Foundation Directory,* is an excellent source of information on grant sources. Funding sources are organized according to funding purposes. One category is "Special Populations," which includes subcategories for African Americans, Native Americans, Spanish-speaking people, and women. Additional listings for children and youth, community development, crime prevention, and public health and social welfare are covered under "Urban and Regional Affairs."

Grant Proposal Planning

The grant funding process involves two players: the funding sources, who sift through proposals seeking worthy projects in which to invest, and the organizations or individuals with project ideas that deserve funding. Getting the two together is the heart of the granting enterprise. Having described the funding sources, I now turn to the second part of the process; namely, the development of a fundable proposal (Locke, Spirduso, & Silverman, 2000; Miner & Miner, 1999).

Grant proposal development has many principles in common with the research process introduced in Chapter 1. The first step is, of course, to have an idea that can be developed into a fundable proposal. In seeking funds, the prospective grantee must identify a problem, hone this into a well-defined and manageable topic, develop objectives for the project, search the literature to devise an appropriate research method, and plan for data analysis. Further, just as one research study leads to new questions for study, lessons learned in one grant-funded project lead to new ideas for further projects.

One of the most important components of a successful grant application is establishing the existence of some problem or need that requires amelioration. All funding sources must operate within their annual budgets, so the competition for available funds is fierce. If you can make the case that the problem or need you wish to address is most pressing, you will have greatly increased your chances of being funded. There are a variety of ways of making a case for your proposal, and those successful include information from more than one source. Some of the evidence supporting the existence of need might come from a *needs assessment survey*. With a properly drawn sample, it is possible to make quite accurate estimates concerning the extent of some need within a given population. Additional supporting evidence might come from previous research or *key informants*, persons who are particularly close to and knowledgeable about the problem at issue. *Community forums* can be held to gather testimony about the problem. Examples of individuals suffering from the problem can be used as case studies, which illustrate the problem in more human terms than abstract statistics. Finally, data from *public records* may be used as additional evidence of need.

In some cases, it may be possible to obtain a grant to conduct needs assessment research. Especially with problems we know little about, funding agencies might be willing to underwrite a survey to obtain more information. In other cases, a funding agency may require that a needs assessment survey be included as part of a larger funding proposal, which may include a service delivery program that is intended to alleviate the problem. In any event, needs assessment often plays an important part in the grant-funding process.

The most crucial component of any proposal is the research project itself. Having established a need, it is necessary to translate that need into specific outcomes for the project and to develop a plan by which those objectives can be achieved. In the case of a research grant, the task is fundamentally one of preparing a detailed blueprint of the stages of the research process presented in Chapter 1. Hypotheses must be developed and a method of testing these hypotheses devised. Issues of subject selection, study design, and data collection and analysis must be taken into account. Direct connections between the goals of the project and the program content must be explicated. Even when service delivery is the primary emphasis of the grant proposal, most grants require an evaluation component; so a strategy for monitoring the program and securing data for evaluation purposes must be included.

Before one can organize the final proposal, though, consideration must be given to potential funding sources. As a first step, the most appropriate organizations should be identified from among the myriad governmental agencies and private foundations. However, sending numerous duplicate proposals to whichever funding sources appear most receptive probably won't work. Proposals must be designed specifically for each funding source because

FIGURE 16.3   Elements to be Included in a Letter to a Federal Agency to Obtain Information About Grants

---

Date:

Name
Title
Address

Dear _____:

    Our organization is interested in carrying out a project under your program title _____. The project will deal with meeting the needs in the area of _____.

    Please add me to your mailing list to receive the necessary application forms, program guidelines and any existing priorities statements or information that you feel would be helpful to me. Please include a list of last year's grant recipients under this program.

    If my project is ineligible under your current guidelines or there are no funds available, could you please refer me to a more appropriate agency?

    I have enclosed a self-addressed stamped envelope for your convenience in returning the list of successful grantees. Thank you for your cooperation and assistance in this matter.

                         Sincerely,

                         Name
                         Title
                         Phone Number

---

each has different rules and needs. This requires research into the various funding sources—their rules of submission, project areas they have funded in the past, and their particular political viewpoints. Once you have a list of possible funding agencies, you begin to narrow the list.

Having identified a few agencies or foundations as likely funding sources, contacting them may be the next step. Doing so can sometimes substantially increase the chances of funding (Miner & Miner, 1999; Morth & Johnson, 1999). If the research organization has advocates who are associated with the funding agency or can make contact with the agency, they could make the first contact. Otherwise, contact the funding agency by letter. Figure 16.3 contains some of the elements that such a letter might include: The elements inform the agency of your intent and obtain for you some much needed forms and other information. Assuming that the agency's initial response is not negative, such as "no funds available," it may be worthwhile to arrange for a personal contact. Call the agency and explore your research plans, trying to gather more information about the agency. Some funding agencies review drafts of a grant proposal and offer suggestions for revision and improvement.

In some cases, it may be possible to personally visit the agency. This is especially feasible with local agencies, such as a county social service department or school district office. This visit can accomplish three general purposes. First, it will confirm or reject the selection of this agency as a likely funding source. With large federal agencies that offer funding under numerous special programs, information gained during your visit assures that you are applying to the agency most appropriate for your project. Second, the visit supplies you with additional information on how best to tailor the proposal to the agency's special needs, thus increasing your chances for funding. Third, the visit provides a personal touch for your proposal when it is submitted. Instead of representing a faceless organization, the proposal will be from people who are known personally and who made an impressive presentation of a serious need and their plans to fill it. Whether you call, write, or visit the agency, Table 16.1 suggests some things that are important to learn about an agency.

Note that one element in Figure 16.3 requests a list of last year's grant recipients. Before finalizing the grant proposal, you should

TABLE 16.1 Things to Learn About a Granting Agency

- Does the funding agency have a real commitment to funding in the area of the proposed project as evidenced by previous grants funded?

- Does the amount of funds requested in the proposed grant fit within the granting agency's typical range of funding?

- What proportion of agency grants go to new projects as opposed to the continuation of currently running projects? Will this proportion change in the coming year?

- Does a new grant like mine have a chance for funding, especially when competing against those requesting a continuation of funding?

- To see what has worked in the past, can we review grant proposals that have received funding from this agency in the past?

- What are the most common reasons for rejecting a grant proposal submitted to this agency?

- What is the most common mistake that people make when submitting a grant request to this agency?

- Can we submit a draft of our grant proposal to this agency and receive feedback before submitting the final proposal?

- Are there any "packaging" guidelines that must be followed or the proposal will not be seriously considered (e.g., length, number of copies, binding, format, etc.)?

- What are the deadlines for submitting a grant proposal?

contact one or more of these successful organizations. The range of useful information they can supply is vast. They have experience with the agency, and there is no substitute for that in learning the ins and outs of successful "grantsmanship" with that particular agency. Each agency develops its own particular style of operating as well as a perspective on problems and ways to solve them. You must learn about these bureaucratic idiosyncrasies from the past grant recipients so that you can tailor your proposal to best fit what the agency is looking for.

Writing a Grant Proposal

Having identified a topic, collected supporting documentation that confirms the need for research, formulated a method for addressing the problem, and targeted a funding source, you have completed the groundwork. Actually writing the proposal, with the information gleaned from these earlier steps, should be a relatively straightforward endeavor—but one that requires a great deal of care (Miner & Miner, 1999). In a number of places in this book, I emphasize that a very important part of research is to communicate to others what you plan to do and what you have accomplished. Such communication can occur at a number of different points in the research process, and I emphasize two in this chapter: preparing a grant proposal and writing a research report. If you cannot prepare a comprehensible and convincing grant proposal, you will not gain the financial support needed to complete a research project. If you cannot write a clear and thorough research report, your research findings—no matter how important—will neither add effectively to the body of scientific knowledge nor be translated into policies and practices by others. I do not pretend that I can make you an accomplished writer by reading this chapter, but I hope to offer some useful suggestions and point you in the direction of useful resources.

Appearance and Writing Style

The old adage that you cannot judge a book by its cover may be true, but people routinely make such superficial judgments anyway. Because of this, both outward appearance and style of presentation in a grant proposal are crucial to success. Demand for grant monies is great, therefore funding agencies may look for any excuse to reject proposals in order to reduce the number that have to be given full review. Failure to follow any guideline may be seized on as a reason not to consider your proposal further. So, to begin with, follow the guidelines to the letter, even if they appear senseless to you. In particular, be careful about length restrictions. Submitting an overly long proposal, no matter how worthy, may result in rejection.

Reviewers have a limited amount of time to review many proposals, so you want to make your proposal attractive and capable of being "skimmed" easily. Using uncomplicated sentences and short paragraphs works toward this end. Underlining key phrases or using "bullets" (solid dots used to set off a series of points in a text) for highlighting purposes helps, too. Using different type styles and boldface headings further contribute to overall appearance and readability. Don't forget to include charts and graphs. They enhance visual impact and can sometimes convey much information in a far more abbreviated form than text. For example, a graph depicting the increasing incidence of some problem is far more effective at making the point than a simple reference to that fact in the text. The graph is also less likely to be unintentionally passed over by busy reviewers.

Not too long ago, great effort and a professional printer were required to produce a proposal with these desirable features. The advent of word processing and desktop publishing systems, however, has placed the ability to produce high-quality documents within reach of most researchers. Such systems are strongly recommended for producing your grant proposals. You can bet that at least some of the grant applicants you will be competing against will produce a computer-generated, slick and attractive grant request. If you do not have this kind of technical support available, it may be worth contracting an organization that can assist you in preparing a polished document.

As in all writing, a sense of one's audience is crucial to preparing a successful grant proposal. Of particular importance is the fact that review panels, for both federal agencies and private foundations, may contain at least some nonspecialists (Krathwohl, 1988). This means that you must be careful to communicate your intentions in language that will be clear to someone who is not a professional in your field. More specifically, jargon should be avoided or, if unavoidable, explained. Beyond that, no assumptions can be made regarding such things as prior knowledge of the problem, its importance, previous approaches, measuring devices, or analytical techniques. Everything must be explained in detail and in terms the typical layperson can understand. Doing so and at the same time not boring the specialists requires a difficult balancing act in your writing.

Once a draft of the proposal is completed and before it is submitted, it is a good idea to have several knowledgeable individuals who are not directly involved in the project read and critique it. This step is important because people unfamiliar with the proposal will approach it more as a reviewer would, with no prior knowledge of its content. The proposal will have to stand on its own merits when these people read it, just as it will during review by the funding source. In addition to looking for the usual typographical and grammatical errors, the reviewers should assess the total package for content and presentation. Is all the needed information included? Is the problem adequately documented? Does the proposed project address the problem identified logically? Is the research design appropriate and practical? Is the budget adequately detailed to justify the requested funds? These and many other questions regarding the content of the proposal should be addressed at this final stage.

Components of the Proposal

A typical research grant proposal will contain most or all of the components shown in Table 16.2, probably in the order listed. The precise components may vary somewhat depending on the requirements of the specific agency or organization and on the kind of research being proposed. Generally, grant proposals will have nine elements.

TABLE 16.2   Typical Grant Proposal Contents and Sequence

1. Cover letter	6. Methods
2. Title page	7. Dissemination
3. Summary	8. Budget
4. Problem/Need	9. Attachments
5. Objectives	

1. *The Cover Letter.* The cover letter is probably the last item to be completed, but because it is the first item read by those who will evaluate your proposal, it is crucial. One very important function of the cover letter is to remind agency personnel who you are and that you took the initiative to contact the agency and take into account their suggestions when designing the proposal. The idea is to show that you have done everything right (according to the agency's views) so that, now, your proposal deserves careful consideration.

2. *Title Page.* The title page is often a standard form supplied by the granting organization. Table 16.3 illustrates the common elements of a title page. A good title is one that describes the project and communicates the anticipated results. Thus, the title "Reducing Homicide in Family Disputes" is preferable to

TABLE 16.3   Elements of a Title Page of a Grant Proposal

1. Title of project
2. Program being applied to
3. Grant program contact person
4. Name, position, and institutional affiliation of principal investigator
5. Name of other sources, if any, to which you have applied for funding for this project.
6. Proposed start-up date and anticipated completion date

"Applying Mental Health Crisis Intervention Techniques to Family Violence" because the first title indicates what the project plans to achieve with the funds, whereas the second merely describes a service. Beyond serving as a label for your proposal, the title page may be used to route your application to the various officials who must process it. The page should clearly identify the applicant's name and address as well as the specific program being applied for and the granting organization contact person.

3. *Summary.* Most granting agencies request a brief summary of the proposal so that agency administrators can quickly assess who should receive a copy. This should be no more than a paragraph and should briefly mention all elements of the proposal: research problem, methods used, and anticipated results.

4. *The Problem or Needs Statement.* This is where you really begin to make your case. What should be stated here has already been discussed in the previous section while analyzing the needs assessment. This section should state clearly and coherently what problem is the focus of your research and why your research will help solve it.

5. *Objectives.* State very clearly and precisely in your objective section just exactly what your proposal will achieve—what research will be done. These objectives need to be very concrete and achievable. No funding agency will be impressed if your objective is to discover the "real truth" about spouse abuse. A more concrete objective would be to learn about the role of economic independence on the ability of women to avoid abuse. List all the objectives with no more than a sentence or two devoted to each. They should be presented in the order of their potential importance and contribution, with the most important listed first. In a research proposal, this section should contain theoretical considerations and the development of hypotheses.

6. *Methods.* The proposal should include a complete description of how you plan to conduct the research. This section should con-

tain all the mechanics of carrying out the research: sample size, sampling technique, research design, and procedures to be used in analyzing the data—in short, all the things you have learned about in this book!

7. *Dissemination.* Dissemination refers to spreading the word about your research, the grant, the funding source, and, hopefully, your successes. Inclusion of comments regarding dissemination of results is looked on as an indicator of confidence. Agencies like positive publicity about the "good" that they do and tend to look favorably on opportunities to obtain it. This is a small thing and alone it will not get you a grant, but successful grantsmanship is ultimately a result of doing a lot of little things right and better than the competition.

8. *Budget.* A carefully detailed budget is an important part of any grant proposal because granting agencies are punctilious in their accounting demands of grant recipients. Table 4.5 on page 109 provides an example of a research budget for a survey interview, showing the major elements that would be included. Essentially, every dollar you request must be justified and accounted for. The methods section of your proposal, where you spell out precisely what the research program will do, provides the guide for developing the budget. All costs that will be incurred must be identified and included. Novice grant seekers often underestimate costs or leave out expenses. Your organization may have standard formulas for fixing fringe benefit costs, travel, and overhead, and may require bidding procedures for purchasing equipment. Be realistic. Promising the moon on a shoestring budget will not endear your proposal to an agency. If costs are reasonable and well justified, it may be possible to negotiate reductions if the total amount is too high.

9. *Attachments.* The attachments section provides important supporting evidence for claims made elsewhere in the proposal. Experienced grant seekers suggest the following as appropriate for inclusion in the attachments

section, if they are available: needs assessment and other supporting research, résumés of key personnel, minutes of advisory committee meetings, names of board members, auditor's financial statement, letters of support from advocates, any pictures or diagrams, and copies of any relevant publications (Bauer, 1999; Locke, Spirduso, & Silverman, 2000). Documentation of community support can be very important in some research projects. Some funding sources demand a demonstration that the community is behind a project, but it should be provided even for those who do not require it. Evidence of community support generally comes from two sources: advisory board minutes and newspaper articles.

Submitting the Proposal

Public funding sources have quite rigid guidelines, not to mention firm deadlines, that govern the submission process. You, of course, will have obtained these along with all the other information from the funding agency. Because deadlines are involved, the proposal should be submitted either in person or by registered mail. In either case, it is a good idea to telephone the funding source a bit later to verify that they received the proposal.

With many of the private foundations, generalizing about submission procedures is difficult because each foundation has its own idiosyncratic way of doing things. This places an added burden on the grant seeker. In some cases, private foundations do not require the lengthy and detailed proposals that I have described. They do not have the resources in the form of reviewers to evaluate such complex documents. Instead, they rely on what is called the *letter proposal*. As the name suggests, the letter proposal outlines the need, your plans to meet that need, and your grant request all in a fairly brief letter of no more than a few pages in length. Even something as important as the budget is abbreviated. Sometimes, the estimated total cost is all that is required. Rather than the voluminous detail characteristic of a federal proposal, each important issue in the letter proposal must fit into a paragraph. Brevity and clarity are the watchwords of a letter proposal.

What if your grant proposal is not funded? Understandable disappointment aside, rejection is an opportunity to learn. Contact the funding source and ask what was wrong (and right) about your proposal. Some funding agencies will provide written reviews of your proposal. Learn from them how to do a better job the next time. Oftentimes, a proposal can be rewritten and resubmitted to the same organization during a later funding cycle. And remember, the only grant seekers never turned down are those who never submit a proposal. It is the nature of the game.

WRITING A RESEARCH REPORT

One of the strengths of the scientific method is the public character of scientific results. Publicizing scientific findings accomplishes several important functions. First, unless research findings are made public, they accomplish little social good. How can others learn from the findings of research if those findings are withheld? Clearly, publication of research findings is necessary in order to apply those findings in developing programs and policies.

Second, publication allows the process of replication to ferret out errors, frauds, and falsehoods that inevitably creep into the products of human endeavor. The self-correcting nature of science that has contributed so much to its success depends on the wide dissemination of research results.

Third, publication of research findings makes attempted suppression of those findings more difficult. I noted in Chapter 12 that this can be a problem for evaluation research reports where an agency may be inclined to ignore research results that do not cast a positive light on the agency's programs. A few copies may be supplied to a sponsor who then has control over what is (or is not) done with the results. Broader publicity concerning research findings makes it more likely that they will come to the attention of someone who will use them.

Finally, the publication of research findings, like any written publication, is an effort at persuasion. It is an attempt to influence the readers to accept your ideas or conclusions. Chapter 3 discusses the idea of advocacy in research and shows that social science researchers sometimes advocate some particular use of their research findings. To be effective, advocates must communicate positions to others, and one major mode of communication is through the written word. An interesting, well-written, and smooth presentation is more likely to be persuasive.

Given the central role communication of research results plays in the entire scientific enterprise, the proper preparation of the research report is vital. In this section, I assume that you have learned appropriate English grammar and usage. The complexity of the English language being what it is, I strongly recommend that you consult one or more of the style manuals listed in "For Further Reading." No matter how well you think you write, your writing will benefit from regular usage of a style manual.

Consideration of the Audience

As with writing a grant proposal, an important consideration before beginning any writing assignment is the intended audience. For the writing to be most effective, it must be tailored to its specific audience. In preparing a research report, the report might be intended for one of the following audiences: professional social scientists, sponsors of the research, or the general public. Each of these audiences require some special consideration in preparing a research report.

Social Scientists and Peer Reviewed Journals

Researchers often present their research to other social science professionals at professional conferences or submit their research for publication in professional journals. These conference presentations and journal publications are typically what are called "peer reviewed" or "refereed," which means that a research report must be reviewed and approved by other social science researchers (the "peers" or "referees") before it can be presented or published. This audience will be the most critical one in terms of whether the report is acceptable to the scientific community. A report aimed at this audience contains the most detail in terms of describing the scientific procedures followed and presenting any statistical and other data analysis performed. With this particular audience, certain assumptions can be made, such as familiarity with basic concepts of the discipline and knowledge of common statistical terms. Although these assumptions make writing for other professionals easier, such an audience is likely to be more critical of such things as following a proper format, elements of style, and substantive content.

Because different journals are read by different audiences, requirements of professional periodicals vary from one publication to another. If a research report is being prepared for possible publication, care must be taken to follow the specifications of the particular journal to which the manuscript will be submitted. Readers will find information on the journal's review process, editorial focus, format, and suggested style guide in the most recent copies of the journal.

Sponsors

The sponsors of research are a mixed lot. If the sponsor is the National Science Foundation or some similar large, national organization, then a research report submitted to them will be reviewed by highly knowledgeable social science professionals such as is described in the previous section. On the other hand, if the sponsor of a program evaluation is a local social service agency, criminal justice agency, or school board, then the report will likely be read by persons possessing varying degrees of knowledge about research and statistics. Some of these individuals will probably be professionals with advanced educational degrees, but their training may be in service delivery and educational policy rather than research methods. I have worked with some sponsors who are highly knowledgeable about research issues (that made my job a lot easier!) and other sponsors who didn't have a clue about research and statistics. In addition, with many applied research projects, the sponsors may include the board of a local school system or of a regional social service agency, and many of these people may have no professional training at all.

So, as I said, sponsors are a mixed lot as far as what you can assume about their level of knowledge. It is important to keep these issues in mind when writing reports. Carefully assess who the audience will be. If it is professionals but some of them have limited knowledge of research issues, then the report needs to be modified and simplified in comparison to what would be submitted to a professional social science journal. If the sponsors also include a number of laypeople with no special training, then the report may need to look more like one you would submit to an audience of the public.

The General Public

If the intended audience is the general public, or others less familiar with research and the social sciences, the preceding assumptions cannot be made. In that case, minimize the use of professional jargon, which is likely to be meaningless or possibly misleading to such an audience. Occasionally, jargon cannot be avoided. Social science disciplines do not make up jargon for its own sake but rather to enhance precision or to describe phenomena that our everyday language does not have words for. When used, professional jargon should be carefully explained to the lay audience.

Presentation of data to a lay audience must also be simplified. They probably won't know what a probability coefficient is and may even have difficulty grasping the importance of a percentage table. Use the visual impact of graphs and charts to help get your message across. Also, explain fully what each statistic used accomplishes, and what the result means. Professionals who do not work routinely with statistics and data analysis may also need some of this sort of assistance. If you believe in the importance of your research and want it to be useful, careful attention to one's audience will further that goal immensely.

Organization of the Report

I next present the elements that would normally be found in a research report submitted to an audience of professionals. As just mentioned, some modifications need to be made in what follows for differing audiences.

1. *Title.* The title is an important part of a report and the first thing a reader sees. The major function of the title is to give prospective readers an idea of what the study is about so they can decide whether they are interested in reading further. A good title informs the reader about the major independent and dependent variables and possibly the major findings. A second reason to develop a good title is that the computerized library searches discussed in Appendix A use key words in titles as one means of selecting articles. Therefore, a report with a misleading or poor title may not be selected in searches when it should be. The following examples are titles culled from some recent issues of social science journals:

"The Effect of Parents' Working Conditions and Family Economic Hardships on Parenting Behaviors and Children's Self-Efficacy"

"Social Network Characteristics Associated with Risky Behaviors among Runaway and Homeless Youth"

"The Effect of Short-Term Family Therapy on the Social Functioning of the Chronic Schizophrenic and His Family"

Note that each of these titles provides sufficient information for a reader to decide whether to pursue the article further and for a computerized retrieval system to identify key words.

2. *Abstract.* Most scientific journals and reports contain an abstract (a short summary of the study) that allows the reader to learn enough to decide whether to read the whole thing. Abstracts are also often collected and published in reference volumes, some of which are discussed in Appendix A. In these collections, the abstracts allow the reader to decide whether to locate the complete articles. Because of their importance and brevity (125–175 words), abstracts must be written carefully. The first sentence should be a clear statement of the problem that was investigated by the study. The research methodology and sampling techniques are then indicated. A brief summary of findings and conclusions completes the abstract. Figure 16.4 shows an abstract with its component parts identified.

3. *Introduction and Problem Statement.* The first part of the body of the research report states the research problem and its significance. This should include a literature review of the history of the problem in previous research and theory. This material indicates how the current study flows from that which has gone before. Presentation of the theoretical material sets the stage for presenting the hypotheses that were tested in the study. Of necessity, this section must be kept relatively brief. For example, the literature review typically consists of numerous citations of previous work on the topic area, with only the most relevant aspects of each study commented on. This emphasis on brevity should not be overdone, however. Clarity of presentation in this section is a must because without it, the remainder of the report loses its meaning.

4. *Methods.* The methods section describes the sample that was studied and the research techniques employed. It also explains how concepts are operationalized and what measuring devices, such as scales, were used. This section is very important because it provides the basis on which the validity and generalizability of conclusions will be judged. It is

also the basis for any future replication efforts. As such, this section must be written with sufficient detail so that it can perform both of these functions. Readers must be able to tell precisely what was done in the study and who participated.

5. *Results.* This section is a straightforward presentation of the findings of the study, devoid of any editorializing or comment as to the meaning of the results. That comes later. In quantitative research, the presentation of results involves the use of tables, graphs, and statistics; in qualitative research, there might be some of this but, as we saw in Chapter 15, it could be more in the form of a narrative. As noted previously, you should consider your audience and fashion your presentation so that the data can be readily understood.

6. *Discussion and Conclusions.* In this section, the meaning and importance of the findings are discussed, and conclusions are drawn regarding the implications of the data presented in the results section. Each tested hypothesis should be related to relevant data and a conclusion stated about the degree of support, or nonsupport, the data provide. Beyond that, any broader implications of the findings for either research or social policy should be noted. Any limitations or weaknesses of any of the results should be noted honestly as well. Often, research results raise new questions as they answer others. It is common practice, therefore, to identify those opportunities for future research.

7. *References.* A list of all works cited in the report is presented, usually as the last element of the report. A variety of formats can be used, but the one known as the "Harvard method" is the most common in the social sciences, and in a somewhat modified form it is the format used in this book. As sources are cited in the body of the text, the author's last name and year of publication are placed in parentheses at the pertinent part of the sentence. The references are then listed in alphabetical order, by author's last name and year of

FIGURE 16.4   An Example of an Abstract of a Research Article

Ennet, S. T., S. L. Bailey, and E. B. Federman. Social Network Characteristics Associated with Risky Behaviors Among Runaway and Homeless Youth. *Journal of Health and Social Behavior*, 40 (1999), 63.

1 { Runaway and homeless youth are at high risk for substance abuse and unsafe sexual behavior. Our study describes the personal social networks of these youth and examines network characteristics associated with risky behaviors. In 1995 and 1996, we interviewed

3 { a purposive sample of youth aged 14 through 21 who were living in Washington, DC and were identified on the streets or through shelters or other service agencies (N = 327). } 2

Although we found that most youth reported current social relationships, a significant minority (26%) did not. Youth without a social network were significantly more likely to report current illicit drug use, multiple sex partners, and survival sex than youth with a network. For youth with a network, the networks were small, strong in affective and supportive qualities, comprised primarily of friends, typically included an alcohol or illicit drug user, and usually were not a source of pressure for risky behaviors. Our results indicate that networks had risk-enhancing and risk-decreasing properties in that network characteristics were associated in both positive and negative directions with risky behaviors. } 4

Parts of Abstract: 1. **Statement of problem**
                               2. **Sample selection**
                               3. **Method of study**
                               4. **Results**

Source   Ennet, S. T., S. L. Bailey, and E. B. Federman. Social Network Characteristics Associated with Risky Behaviors Among Runaway and Homeless Youth. *Journal of Health and Social Behavior,* 40 (1999), 63.

publication, at the end of the research report. Any common social science journal, such as the *American Sociological Review, Social Problems,* or the *American Journal of Sociology,* may be consulted for examples of this format.

An alternative format is called the "serial method," in which the citations are indicated numerically in the body of the text. The references are then presented in the reference section in the same order as cited. This method is an older style than the Harvard method and is generally less used by researchers because it is more complicated to prepare and not justified except for long research reports with extensive references. Others, however, prefer it because there is less intrusion in the text. But the serial method is sometimes more difficult to use. For example, if only one additional reference is added after the report is finished, all the references have to be renumbered from the point of the inserted new reference.

Preparing a list of references or a bibliography for a grant or research report is generally a difficult and not very pleasant task (it wasn't my favorite task when working on this book!). All the names of the authors must be spelled correctly, the page numbers and dates must be correct, and the punctuation is not like that used anywhere else. New computer software, however, can remove some of the drudgery of preparing bibliographies. Bibliography-formatting software creates a file of references with all the required information for each reference (still a lot of work). Once the file is complete, the program reformats the information into a variety of popular bibliographic styles. So long as the information in the file is correct, the bibliography will be correct, down to the last punctuation mark. In addition, some programs interface with word processing programs and "search" documents for citations and make up a correct bibliography from the file of references. It also flags any citations that are not in the file and any references in the file that are not cited. Trust me: when revising a long document (such as a book or grant proposal), this feature can save a lot of work and reduce errors.

The Process of Writing

It is not possible, of course, to cover in a brief chapter all the elements involved in writing. This is done in writing courses and through practice at writing. However, a few points are of special importance and so are noted.

First, writing is a process rather than a product. You never finish writing, although you may finish a paper or a report because it has to be submitted by a deadline. However, this doesn't mean that you couldn't write and revise further. Writing aims at the expression and communication of clear thought. It is difficult to write because, in some respects, language is a poor mechanism for communicating the subjective reality of one's thoughts. If you tell someone about the car that almost hit you on the way to school this morning, words like *car* and *hit* seem straightforward and comprehensible. But do these words encompass the reality that you experienced? It was a bright red car, and it was speeding, and the driver seemed not to notice you, and . . . How much detail do you provide to communicate your experience? You tell the person you were frightened by the close call, and then realizing he or she might not understand the wrenching terror that shot through you, you repeat that you were "really" frightened, using qualifiers, emphasis, and inflection to make your point. The words may seem inadequate to communicate your experience, but this is precisely the challenge that any writer confronts: using words to describe a very complex and sometimes confusing reality. Researchers face the same challenge of using words to describe a very complex theoretical and methodological reality to various audiences.

A second point, which really flows from the first, is that rewriting and revision are an inherent part of the writing process. Very few writers are capable of making their first draft the final draft. Most writers, especially professional writers, must rewrite their material several times before it can be considered clear, comprehensible, and smooth. (The word *smooth* in this context simply means the writing contains very few errors or awkward constructions that distract the reader from the content.) One of the keys to revision is to be able to read your prose through the eyes of the intended audience. Would they understand a particular word? phrase? sentence? Can they follow the sequential organization in the report? Do they grasp the transitions that move the reader from one sentence to another, from one paragraph to another? What is perfectly clear from the writer's perspective may be muddled and unclear from the reader's. The writer's talent is to perceive his or her writing from that other perspective.

Rewriting is an essential, if sometimes tedious, task, but it is also creative. As with other creative efforts, the energy and attention needed to create is greater at some times than at others. This means that the best writing is usually not produced in one sitting. Most writers find it useful to approach revision after they have been away from a paper for some time. This enables them to approach it with fresh insight and attention. For this reason, things that are written at the last minute, with a deadline rapidly approaching, may not be the best.

The task of writing and rewriting is, of course, made easier by modern computers and word-processing software. As a writer who has spanned both the pre- and post-word processing eras, I can attest that the latter is far preferable to the former. However, word processors do not write. The writer must still plant himself or herself firmly in a chair for many hours and throw words onto the screen. Writing still takes creativity, perception, persistence, and hard work.

Computers and the Internet

Although the Internet is proving to be a boon to students and researchers for locating research information, it has produced some new problems in terms of how to properly cite sources that one finds on the Internet. One aid in this is the Columbia University Press's "Columbia Guide to Online Style," at **http://www.columbia.edu/cu/cup/cgos/idx_basic.html**. Not only does this site provide information about how to correctly cite Internet sources, it also provides many links to other Internet sites that can offer additional guidance on the topic.

You should search for other Web sites that provide assistance in the writing and referencing of online materials.

In this chapter, I devoted considerable attention to the grant-funding process and writing the grant proposal. You can greatly enrich your knowledge of the grant-writing process by using the Internet to locate sites devoted to this topic. Simply entering such terms as "research grants" or "grant funding" in one of the search engines will generate a large number of potential sites to explore. You might also combine the word *grants* with a key word for a subject of interest to you, such as crime, aging, or education. Some of the major governmental human service sites have extensive material on grants, including abstracts of funded projects and RFPs that you can examine to better understand the issues of writing successful proposals. For example, the Justice Information Center (**http://www.ncjrs.org/**) has extensive information available on grants from the National Institute of Justice, the Office of Juvenile Justice and Delinquency Prevention, the Office for Victims of Crime, the Bureau of Justice Assistance, and other grant opportunities from the Department of Justice. Just enter the word "grants" in their search field.

The Web site of the Catalogue of Federal Domestic Assistance (**http://www.cfda.gov/**) contains a search function where you can search the catalog. By entering terms for a social research issue, you can quickly locate information about related grant opportunities.

A good way to get a better sense of how to write grant reports is to read some completed studies. One source you might try for this is the National Clearinghouse on Alcohol and Drug Information (**http://www.health.org/Dbarea/index.htm**). You can easily search their databases and review abstracts of completed projects. The references indicate where and how to obtain complete reports. Another excellent location is the National Science Foundation (**http://www.nsf.gov**). Click on "Social,

Behavioral Sciences" or on "Grants and Awards."

Main Points

1. Grants have become a very important source of funding for social research. Sources of grant money fall into three categories: governmental agencies, private foundations, and corporations, with government being the largest source. Government funds come in a number of different forms: project (categorical) grants, formula grants, block grants, federal contracts, and state government grants.

2. Although many foundations fund grants, most fund only small projects and some fund research on only limited subject areas. Corporations tend to fund grants that can be justified to the stockholders, such as for services or improvements from which the corporation or its employees will benefit. Many funding sources can be located through online computerized search services.

3. The steps in the grant development process are analogous to the steps in the research process. The first step is to identify a fundable topic; this can be assisted by conducting a needs assessment.

4. Once the project has been identified, you need to target a funding source that will be interested in your project; this might call for contacting potential funding sources to assess their interest.

5. To be successful, grant proposals must be well written, and this requires paying attention to the appearance and to the writing style. Proposals must be neat, interesting to read, and addressed to the audience who will read them.

6. A grant proposal should contain all the components necessary to provide a funding source with adequate information to assess it.

7. Research reports should also be well written because the communication and publication of scientific results accomplish important functions for science. A research report should be written at the level of the audience for whom it is intended. These might be social science researchers conversant with the jargon of research and statistics, other professionals at least partially unfamiliar with such jargon, or the general public.

8. Most research reports include the following elements: a title, an abstract, an introduction and statement of the problem, a description of methods and results, a discussion of the implications of the findings, and a list of references.

9. It is important to recognize that writing is never finished, although a particular project may end. Good writing is based on rewriting and being able to read what you have written from the perspective of the audience.

Important Terms for Review

foundation
grant
peer review

For Further Reading

American Psychological Association's guide to research support (3rd ed.). (1992). Hyattsville, MD: Author. A complete guide for social scientists looking for support for their research. It includes information on funding sources—including names, addresses, and telephone numbers—application procedures, and submission deadlines.

Belcher, Jane C., & Jacobsen, Julia M. (1992). *From idea to funded project: Grant proposals that work* (4th ed.). Phoenix: Oryx Press. This is an excellent guide to the preparation of effective grant proposals. It provides much sound advice and helpful hints to those about to embark on the process.

Chicago manual of style (14th ed.). (1993). Chicago: University of Chicago Press. Prepared by the editorial staff of the University of Chicago Press, the *Chicago Manual* has long been considered the definitive writing reference work. If you take your writing seriously, you should own a copy.

Cuba, Lee. (1997). *A short guide to writing about social science* (3rd ed.). New York: Longman. This book is a good review of the things you need to know about writing research papers and preparing presentations on topics in the social sciences.

Hackwood, S. (Ed.). (1999). *The grants register, 2000* (18th ed.). New York: St. Martin's. This volume is mainly focused on graduate or professional students who are looking for grants to assist them in research and travel that are a part of their educational experience, but it does include some granting organizations that support research by social scientists and human service agencies.

Hall, Donald, & Birkerts, Sven. (1997). *Writing well* (9th ed.). Reading, MA: Addison-Wesley. Hall and Birkerts have produced a book that can help anyone improve his or her writing. Unlike many such books that present the rules in a rather stiff and direct fashion, Hall and Birkerts makes learning to write well interesting.

Johnson Jr., W. A., Rettig, R. P., Scott, G. M., & Garrison, S. M. (1998). *The sociology student writer's manual.* Saddle River, NJ: Prentice-Hall. While this book focuses mostly on undergraduate students writing papers for class assignments, it provides some excellent information for any writer of social science materials on such topics as referencing styles, using the library, and organizing material.

Miner, L. E., Miner, J. T., & Griffith, J. (1998). *Proposal planning and writing* (2nd ed.). Phoenix, AZ: Oryx Press. This is another excellent guide to developing grant proposals for research or other types of grants.

Ries, Joanne B., & Leukefeld, Carl G. *Applying for research funding: Getting started and getting funded.* (1995). Thousand Oaks, CA: Sage. This book is an excellent overview of the steps in preparing a successful grant application for research in the social sciences.

Schumacher, Dorin. (1992). *Get funded! A practical guide for scholars seeking support from business.* Newbury Park, CA: Sage. This is an especially helpful guide for seeking funding from businesses; it points out the differences between the corporate, academic, and human service mentalities and how to incorporate this knowledge into the preparation of grants.

Shertzer, Margaret. (1996). *The elements of grammar.* New York: Macmillan. This little book covers all aspects of correct English usage. It draws upon hundreds of examples from the greatest contemporary authors.

Chapter 17

APPLICATIONS OF SOCIAL RESEARCH

Wen people are introduced to social science research methods, they sometimes ask very pointed questions, such as "Why is it important that I learn this material?" and "Of what use is all this knowledge about how science is done?" These questions are important because most people are not interested in learning something merely for its own sake; they want to know what will be the applications of the knowledge. In addition, from the perspective of society as a whole, it is legitimate to ask what benefit society gains when it expends resources on the training and employment of social scientists.

So this chapter will focus on the applications to which social science research is put. Actually, I have explored this topic throughout the text in the examples of research that I have used to illustrate various points and in the Applied Scenario boxes that are found in many of the chapters. Here, I want to talk more directly and systematically about this topic. This will also give you a clearer idea about the employment settings in which social scientists toil. I address this in Chapter 16 when discussing the different kinds of audiences that social scientists might address in presenting the results of their research, and I will expand on that in this chapter.

This discussion also elaborates on another point made in this book; namely, the flexibility of scientific research methods. In all the applications that I will discuss, researchers use pretty much the same range of research methods, but apply them in a variety of ways. The same methods are used in many different kinds of research (for example, marketing research as opposed to theory building), by many different social science disciplines (such as sociology and political science), and by people in a variety of human service fields (such as social work, criminal justice, education, and nursing). With the training provided by this book, along with some advanced training, you are prepared to conduct research in all of these disparate fields.

KNOWLEDGE PRODUCTION AND THEORY BUILDING

One of the key goals of the scientific endeavor has always been to develop a better and more accurate understanding of how everything works. Human beings are fundamentally curious creatures who have an irresistible urge to probe and test their world, seeing how this works and why that happens. Science is a formal and systematized way of accomplishing this. As scientists do their work over time, a body of knowledge grows and becomes the tentative and provisional accumulated wisdom that science offers us about how the world works. In Chapter 1, I refer to this as basic science whose purpose is knowledge production—advancing our foundation of knowledge about how the world works without any immediate practical benefit in mind. In the scientific community, this knowledge production is generally seen as a good thing; in fact, many scientists believe that this is science's purpose, its obligation. Recall from Chapter 1 the communalism value in the ethos of the scientific community: The duty of science is to produce knowledge for the benefit of the whole community. The underlying belief is that the more accurate knowledge we possess about the world, the better off everyone will be. So, scientists see it as their positive duty to advance knowledge even when we cannot see any immediate practical use to which the knowledge can be put.

At a number of points in this book, but especially in Chapter 2, I talk about the central role of theories in research. The approach to and assessment of theory would differ depending upon which of the three paradigms (positivist, interpretivist, and critical) one is using. However, most social scientists agree that research should produce some abstract and general knowledge about the world that can be generalized beyond the setting being studied. This generalization would be clear and direct for many positivists doing quantitative research, whereas the generalizations might be far more tentative and less extensive when done by qualitative researchers using case studies or narrative analysis, as discussed in Chapter 15. Nevertheless, knowledge production is intended to produce general knowledge (whether you call it theories or conceptual development or hunches) that can be applied to the understanding of new people, groups, and circumstances. This is part of the constant effort to advance the state of knowledge.

Many social scientists, therefore, carry out research whose main goal is knowledge production and theory building. In most cases, these researchers are on the faculty of universities where their employment, promotion, and tenure is contingent upon their completing good research that passes the critical review of their social science colleagues and gets published in peer reviewed journals and books. Many of these researchers support their research with grants from one of the foundations or governmental agencies discussed in Chapter 16.

Having said all that about basic research and the value of advancing knowledge, some readers undoubtedly still have a nagging feeling of incompleteness, still wondering "But why do we do this?" "What's the purpose of advancing our knowledge?" I can give two answers to these questions. The first answer is that for some people, it is remarkably satisfying merely to sate that irrepressible curiosity to know how things work! Knowledge can provide a sense of wonderment and enlightenment, even when the knowledge is not put to any use. Exercising the remarkable intellectual powers that human beings possess is gratifying and irresistible. The second answer is that many scientists believe that abstract, basic knowledge may be of use to human beings at some point in the future. Just because we don't currently comprehend the utility of a piece of knowledge doesn't mean that it has no utility. Others may well be able to put it to use in ways that the basic researcher may not recognize. Once again, the accumulated body of scientific knowledge that researchers produce is a

treasure available to be tapped whenever and wherever people can find a use for it. The more complete and accurate that body of knowledge is, the more useful it will be when someone can apply it.

So this is what motivates scientists in the production of knowledge. And scientists consider it a very serious community responsibility of theirs to always conduct the highest quality research. I find both rationales for conducting basic research compelling. Certainly I value putting knowledge to use, and the research agenda of my career has focused on this. At the same time, my life has been devoted to the very challenging, but also supremely gratifying, endeavor of developing a greater understanding of my world.

SOCIAL POLICY DEVELOPMENT

Social science research is often used by a variety of groups for the betterment of society. This happens in so many ways that it is difficult to describe concisely, but it is sometimes discussed under the general heading of *social policy development*. The term **social policy** refers to all those laws, administrative procedures, court rulings, and other formal and informal social practices that shape the distribution of resources and the treatment of people in society. Social policy is shaped at the national level when, for example, Congress enacts laws establishing welfare regulations, creating environmental restrictions, or modifying benefits for Medicare. Social policy is also shaped at the state level when a state legislature creates new penalties for criminal code violations, or at the local level when a county commission issues a zoning variance to permit a waste treatment plant to be built near a residential community.

Social policy is shaped and changed through the competition of a variety of interest groups, and each group will use whatever available resources it has in the struggle to see its side prevail. In some cases, it is possible to use the tools of social science research to advance one's position in this competition. Often the social science research that enters the fray is basic research and was not conducted to advance any particular agenda. But if the research results support one group's position, that group may bring the results to the attention of Congress, state legislatures, or other decision-making bodies. As an example of this, there have been two national commissions in the United States in recent decades to study the effect of pornography on aggression and violence: a president's commission in 1970 and an attorney general's commission in 1986 (Sullivan, 2000). The task of these commissions was to decide whether pornography affected aggression and violence against women and make recommendations about new laws or other procedures that seemed called for. Many arguments and much evidence were presented to these commissions by a variety of interest groups. Social scientists had been conducting research on the topics of pornography, aggression, and violence since at least the 1950s, so there existed a large body of research results. Much of this research constituted basic research conducted by social scientists who were mainly interested in advancing our knowledge on the topic and certainly didn't envision their research entering into the debate at these national commissions. And, in most cases, it wasn't the scientists who conducted the research who brought the results to the commissions. Rather, one interest group or another would scour the published social science literature for research that supported their position. In some cases, the research was a decade or more old, but the publication practices of science provide a permanent record of research results that can be used by others when the need arises.

Some social science research used for social policy development is applied research rather than basic: It is designed and conducted with the specific intent of influencing the policy process. One type of such applied research is the evaluation research discussed in Chapter 12. This involves program evaluations, needs assessments, and cost-benefit analyses the intent of which is to assess how effectively a particular policy, program, or organization is achieving its goals and what modifications might make it more effective. Numerous examples of this are discussed in Chapter 12. A second type of applied research for policy development is called **policy research** and focuses on broad policies that affect communities or society as a whole. For example, when the Civil Rights Act was passed in 1964, it contained a provision that a study be conducted to determine whether youth in the United States were being denied equal educational opportunities because of racial discrimination. This study was eventually conducted by the Office of Education of the U.S. Department of Health, Education, and Welfare and directed by sociologist James Coleman and six other researchers (Coleman et al., 1966). This was not a study of a particular program or organization; it was a societal-wide assessment of the extent and impact of educational discrimination based on race. Its intent was to help shape the public policy debate over what to do about unequal educational opportunities. And it did that: It became an element in the policy debate that ultimately produced programs of school integration and busing.

The line between evaluation research and policy research cannot always be drawn in an exact way, and some research projects could be classified into both categories. Yet, much applied research can be clearly classified as one or the other, and the distinction between them does illustrate the range of contributions that social science research can make to policy development. The social scientists who conduct research for policy development work in a variety of settings. They might be on the faculty of a college or university, employed by a private research consulting firm, on the staff of a governmental agency, or work as a self-employed research consultant. The research might be funded by grants (discussed in Chapter 16) or paid for by a governmental agency that needs the research done.

A final point to be made is that in the realm of policy development, research clearly becomes a part of the political process. Yet, it is important to remember that this does not mean that the integrity of the research methods are in any sense compromised or that social science research is merely "political." Professional and ethical norms require that researchers conduct the most scientifically sound studies, use the generally accepted procedures described in this book, fully and accurately report the results, and identify any weaknesses in the research. It is not the research results that are political; rather it is the uses to which particular advocates put them that might be political. It is also important to recognize that social science research is only one element that shapes social policy development. Many other interests and values are a part of the policy process and may in some cases play a greater role in shaping the policy outcome.

BUSINESS AND MARKETING RESEARCH

Private businesses and corporations also make use of social science research (DeWeese, 1983; Stevens, Wrenn, Ruddick, & Sherwood, 1997). In the private realm, of course, the research is focused ultimately on assisting the corporation in producing and marketing products and maintaining profitability. Some aspects of business life are, of course, outside of the expertise of social scientists, such as how to produce a more fuel-efficient automobile engine or increase the memory of a computer chip. On the other hand, other aspects of the business world can best be understood by using the theories and methodologies of the social sciences. Businesses function because groups of employees interact together to produce products or services; businesses must also attract customers and persuade them to purchase products. All of this makes the business world an important area for the application of social science

knowledge. For example, issues of whether the public will accept smaller cars as the price of greater fuel efficiency or how to organize a work group to maximize employee satisfaction have to do with theories of social behavior and social organization that can best be comprehended by using the research methods discussed in this book.

Probably the best way to communicate the range of applications of social science research in the business world is to list some common research foci:

- Marketing research (consumer response to a new product; consumer assessment or value of a product compared to other products)

- Customer satisfaction (level of satisfaction of customers with products and services; retention of existing customers)

- Employee satisfaction and morale (attitudes of employees about their job and fellow employees; organization of work setting to increase satisfaction and reduce alienation and employee turnover)

- Advertising effectiveness (the effect of different types of advertisements on various audiences; the penetration of advertisements to targeted audiences)

- Future consumer trends (social and cultural trends that will shape demand for products in the long term in a changing society)

Many of these research topics can be addressed through the use of surveys, and surveys are probably the most common methodology used in business research, as is also the case in social science research in general. Focus groups are also fairly common in business, especially to gather data on consumer responses to products or services. However, experiments, the analysis of available data, secondary analysis, and the other research methods discussed in this book are also found in business research.

Social scientists conducting research in the business world are sometimes employees of a research division of a corporation. In other cases, they might work for a private research firm that contracts with the corporation for specific research needs. There is no need to apply for grants, as described in Chapter 16, because the corporation would fund the research that needs to be done, either as an ongoing part of their business when conducted by an employee, or as a one-time contract with outside consultants. For the most part, social science research in the business world tends to be very interdisciplinary and problem-oriented. The corporation has a very specific problem or need that it wants addressed, and will use researchers who have the skills to solve the problem.

PUBLIC OPINION POLLING

Public opinion polling has become widespread and an important application of social science research methods. Basically, **public opinion polling** involves measuring the attitudes of some identifiable group on some political or social issues of interest. The government, the media, and a variety of nonprofit organizations are the major entities interested in assessing public opinion, but private corporations may also be interested in this. Surveys of public opinion are often used to advance some social or private agenda. Politicians conduct polls in order to assess their own popularity or to determine which issues are popular or unpopular with the voting public. Nonprofit interest groups, such as groups supporting gun control as well as groups opposing gun control, conduct polls in order to demonstrate that their position on issues are prevalent with the public or to assess the kinds of policies the public will support. The media sponsor opinion polls because the results of polls have come to be seen as an essential part of the news that the media reports to the public each day.

As for research methodology, public opinion polling involves doing surveys almost exclusively. Opinion polling is done at the local, state, and national levels and often involves random samples of the whole population of interest. However, at times, a corporation, governmental agency, or television station may want to assess the opinions of a specialized group, such as corrections officers or Catholic priests.

Some social scientists who conduct opinion polling are employed by private polling organizations, such as Gallup, Harris, or Roper. There are also many nonprofit polling organizations, some of which are survey research centers associated with universities. Many private research organizations also do public opinion polling as well as program evaluations or needs assessments. Researchers who conduct opinion polls have formed their own professional organization, the American Association for Public Opinion Research, which publishes a journal, the *Public Opinion Quarterly*. Through the organization and its journal, public opinion pollsters work to develop better survey techniques and serve as a watchdog to ensure that those conducting opinion polling use sound survey techniques.

COLLABORATIVE RESEARCH

Some social scientists conduct research that is called by a variety of terms, such as collaborative research, participatory research, and participatory action research (Nyden, Figert, Shibley, & Burrows, 1997; Stoecker, 1999). I will use the term **collaborative research**: studies in which the sponsors, subjects, and other stakeholders work with the researcher to plan and carry out the research, do the data analysis, prepare the report, and disseminate the results. This approach to research is based on the assumption that the researcher's expert perspective on the research process is only one of a number of legitimate perspectives. It recognizes that research can be strengthened when these other perspectives have input into the design and conduct of research. I discuss collaborative research in this chapter because proponents of this approach argue that it produces research that has more direct and useful applications to a variety of stakeholders. Conventional research controlled by the researcher and possibly sponsors is highly likely to focus on their specific interests and concerns but may ignore the interests of other stakeholders. These other stakeholders may see themselves as benefiting from different kinds of research problems or might have different interpretations of some of the data. After all, the people being studied know themselves and their community from perspectives that researchers or sponsors may not have access to. Collaborative research often provides the most benefit to stakeholders who are relatively powerless, such as the recipients of the services of some social programs. It is the sponsors of research and those in charge of the programs who are typically in a position to see that their perspectives are attended to. The perspectives of other stakeholders are more likely to be ignored. When stakeholders have an input into the research process, it increases the likelihood that the research will address their concerns and produce results that are useful to them.

Collaborative research also recognizes that researchers need to be accountable to the community that is the focus of the research. Accountability means not only involving the community in designing and conducting the research, but also communicating the results to them and having them participate in preparing a final report. Collaborative research is not a distinct kind of research design but rather a distinct orientation to doing research. Collaborative research might take the form of a program evaluation or a needs assessment, but it rejects the positivist model of the objective, expert researcher doing something to passive research subjects. For this reason, collaborative research has gained adherents among interpretivist and critical theorists. However, positivists might conduct collaborative research because the research plan that emerges from the collaborative process could adhere to all of the standards of positivist research. What collaborative research focuses on is the process of developing the plan rather than the content of the plan itself.

Sometimes collaborative research is referred to as **empowerment research**, meaning that one goal of the research is to develop new knowledge, skills, and capabilities among the various stakeholders in the research—in other words, to empower them. The stakeholders might learn more about themselves or their community. By helping to design and conduct the research, these stakeholders might also learn how they can use the research results to advance their own interests. Or they might learn more about the research process that would enable them to participate more fully, and perhaps even direct research projects in the future. In whatever form it takes, empowerment research strives to help the people who are the focus of the research make changes for the better in their lives.

CONSUMING RESEARCH FINDINGS

Many persons who take courses in social science research methods will not continue on to become researchers themselves. However, many will likely find themselves in situations where they will need to read and evaluate the results of social science research or possibly work with social scientists who are conducting research. Practically all of the applications discussed in this chapter could entail researchers having to work with nonresearchers to develop and carry out the research. For example, social service professionals working in an agency that is required to conduct a program evaluation will have to work with the researchers in identifying goals, selecting operational definitions, deciding on an appropriate sampling procedure, and so on. Likewise, workers at a television station or on a political campaign will have to consult with public opinion pollsters in order to get the information they need.

from public opinion polling. As another example, employees of a corporation may find themselves assisting researchers who are doing marketing research or a customer satisfaction survey.

In all of these settings, the nonresearchers can be more effective at their jobs if they possess the skills of an informed consumer of research. That is, if they are sufficiently informed about the research process, they can read and understand research reports, assess the adequacy of the methodology, and evaluate their conclusions. With this background, you may not be able to fully design and carry out a research project yourself, but you could work with researchers and provide assistance to them as they carry out the research. I have worked with human service professionals who were knowledgeable about research and others who were not, and the former make a very substantial contribution to the speedy and efficient completion of a research project. The completion of practically all of the research tasks described in this book is enhanced when those working with the researcher comprehend the key issues.

DEVELOPING CRITICAL THINKING SKILLS

The final application of social science research has to do with everyday life. A lot of the information we confront on a routine basis is not in the form of a research report. Rather, it is in a newspaper or magazine article, on a news report seen on television, on an Internet Web site, or something received by word of mouth. Oftentimes, when trying to understand something that is relevant to us, we confront information from all of these sources. As I have stressed in each chapter, the logic of social science research can serve as a guideline for assessing this information. A thorough comprehension of research methods provides a road map for navigating through what can be an admittedly confusing thicket of information that we confront in the modern world.

The tools of research won't provide you with all of the answers, of course. As with science in general, research methods are good tools for dealing with issues of fact, but are less helpful when dealing with opinion or speculation. In addition, in some cases, despite being equipped with a knowledge of research methods, there just may not be sufficient information to resolve an issue. However, applying the guidelines of research methodology to this information can sometimes help you decide whether there is adequate information to resolve an issue. And you are certainly better off making an assessment of "I don't know" than rushing to judgment and drawing unwarranted conclusions.

So, the final application of the materials learned in this book is to use them as rigorous and disciplined critical thinking guidelines throughout your life. Questions such as "How was that measured?", "What kind of sampling was done?", and "How was the question that produced this information worded?" should become routine parts of how you think about information. I have provided you with a series of these questions at the end of each chapter. In combination, they provide an arsenal of questions to ask when somebody says "Its a fact that . . . " or "Research has shown that . . . " With what you have learned in this book, your arsenal is well stocked!

Computers and the Internet

You can use the Internet to explore the various applications to which social science research is put. A good place to start is the Yahoo search engine. Click on "Social Science," and this can serve as the portal to a variety of social science disciplines. For example, if you select "Sociology" and then "Organizations," you will come to a page with links to a large number of sociology organizations in the United States and around the world. You can explore the Web pages of these various organizations for what you can learn about social science research applications. Do the same for social science disciplines other than sociology.

Using a search engine, enter various terms or combinations of terms discussed in this chapter: "business," "research," "marketing," "collaborative research," and so on. Make note of the various kinds of applications of research that you come across and report the information back to your class. Finally, go to the Web page of the Society for Applied Sociology at **http://www.appliedsoc.org/**. There you will find additional information about the many applications of social science research. Another useful Web page is one maintained by a topical interest group of the American Evaluation Association: **http://www.stanford.edu/~davidf/empowermentevaluation.html**. It focuses on collaborative, participatory, and empowerment evaluation research, and has many links, references, and programs to explore.

Main Points

1. One of the primary applications of social science research is to advance our knowledge of human social behavior and to develop, test, and verify theories of social behavior. This is often done without any immediate, practical use of the knowledge in mind.

2. Social science research is often a part of social policy development, which influences the distribution of resources in society and the treatment of people. This research often takes the form of program evaluations or policy research.

3. Social research has many applications in the business world, such as studies of consumer satisfaction, employee morale, or marketing research. Business research is focused on advancing the goals of the business, which means producing and marketing products in order to maintain profitability.

4. Social research also takes the form of public opinion polling. Opinion polls are conducted by many groups: the government, the media, special interest groups, and so on.

5. Collaborative research refers to research in which the sponsors, subjects, and other stakeholders in the research work with the researcher in planning and carrying out the research, doing the data analysis, and preparing the report and disseminating the results. Collaborative research often produces results that have direct applications in benefiting a wide range of stakeholders.

6. Learning about research methods enables people to read and critically evaluate the results of social science research or possibly work with social scientists who are conducting research. The principles of research methods can also be used as a guide to critical thinking in one's everyday life.

Important Terms for Review

collaborative research
empowerment research
policy research
public opinion polling
social policy

For Further Reading

Finsterbusch, Kurt, & Motz, Annabelle Bender. (1980). *Social research for policy decisions.* Belmont, CA: Wadsworth. An excellent, brief book about research as it might be used by decision-makers attempting to influence public policy.

Freeman, H. E., Dynes, R. R., Rossi, P. H., & Whyte, W. F. (Eds.). (1983). *Applied sociology.* San Francisco: Jossey-Bass. This book covers a wide range of research applications, but focuses more on the locations where research might be conducted, such as business settings, schools, or the criminal justice system.

Olsen, Marvin E., & Micklin, Michael. (Ed.). (1981). *Handbook of applied sociology.* New York: Praeger. This volume describes a wide range of applications of social science research. It focuses on types of research, such as program evaluations.

Patton, Michael Quinn. (1997). *Utilization-focused evaluation: The new century text* (3rd ed.).

Thousand Oaks, CA: Sage. This text on program evaluation provides good insight and instruction on how to do more participatory or collaborative evaluations.

Rossi, Peter H., Freeman, Howard E., & Lipsey, Mark W. (1999). *Evaluation: A systematic approach* (6th ed.). Thousand Oaks, CA: Sage. This book provides an excellent and thorough presentation of the many ways in which evaluation research can be done.

Stringer, Ernest T. (1999). *Action research: A handbook for practitioners* (2nd ed.). Thousand Oaks, CA: Sage. This book describes how to conduct community-based action research that is very collaborative and egalitarian in nature. It devotes much attention to descriptive and qualitative research.

Appendix A

A GUIDE TO THE LIBRARY AND INFORMATION RETRIEVAL

A library is both a repository of information and a gateway for accessing information stored at other sites. Libraries own some of the information products they make available to customers and assist customers in locating and accessing information resources at other libraries or other locations. Some social science research projects can be completed in the library if the necessary information or data can be found there. More commonly, the library and its services are used as means of reviewing existing literature to see what has been done on a particular topic, gaining help in developing a research design, locating existing scales, and the like. However, because of the proliferation of materials accessed through libraries these days, finding what you want may be a daunting task for those unfamiliar with the organization of the library. To make effective use of the library you must first define precisely what it is you need to know; second, determine the best strategy for locating that information; and third, evaluate the information that you do find. This appendix focuses primarily on the second step by providing a basic introduction to the organization of the library and the services provided there. In addition, it points to some specific resources in the library that would be useful to researchers in the social sciences.

ORGANIZATION OF THE LIBRARY

There are two key elements in the organization of a library that you need to understand to use the library effectively. First, the library contains certain *departments* that perform special functions in the search for information. Second, the library contains a number of important mechanisms for *accessing* information.

Library Departments

For students seeking materials, one of the most valuable departments in the library is the *reference department*. In most libraries, the reference department contains the basic tools for searching for information: encyclopedias, indexes, abstracts, dictionaries, handbooks, yearbooks, government documents that summarize statistical data, and an array of other resources. Familiarity with this department is essential for efficient use of the library. In addition to the materials in the reference department, reference librarians are another valuable asset to the researcher. Highly trained and knowledgeable about how to locate materials, librarians have as one of their major duties helping library patrons. By working closely with a reference librarian, you will not only locate a specific item, you can also learn the techniques of thorough and efficient searching for materials. Although you should probably first try to locate materials yourself, do not hesitate to consult the reference librarian if you have difficulties.

Unless you have access to a very large library, such as the New York Public Library or the Library of Congress, you will undoubtedly find that some books or resource materials you need are not available in your library. To overcome this problem, you need to make use of the *interlibrary loan department*. Most libraries participate in a local or regional interlibrary loan program that permits a library to borrow books and other materials from other libraries in the program. Libraries use computers to locate materials in libraries in their region, around the nation or, in some cases, the world. All participating libraries benefit because they can provide a greater range of materials and services to patrons than their own budgets allow. Books are usually sent by mail from one library to another, with the borrower sometimes asked to pay postage and insurance. Many libraries, as a part of their interlibrary loan service, also participate in a periodical reprint service through which you can request a duplicated copy of an article from a periodical or magazine.

The *documents division* of the library houses government publications, magazines, pamphlets, and sometimes collections of special books. Many government documents are now available on microfiche and in electronic formats. In a small library, the documents division might be integrated with the reference department.

The *stacks* are the part of the library where most books and periodicals are stored. At large universities, there may be more than one library, so all the books and journals will not necessarily be in the same building. Some libraries permit anyone to roam the stacks, whereas others are more restrictive, distributing "stack passes" to a limited number of persons such as university faculty, graduate students, or other serious researchers. You should gain as much access to the stacks as possible because browsing in the stacks is a valuable adjunct to the other ways of finding appropriate resource materials. In the stacks, you can review the table of contents and the index of a book, which provide more information about how useful it will be to you than does the brief description of the book found in the public catalog. In addition, because books on similar topics are normally shelved near one another in the stacks, you can browse for books that might be useful once you find the right area in the stacks. Further, if a book is misshelved slightly, you may accidentally find it if the library staff sent to retrieve it were unable to do so.

Accessing Library Materials: Search Strategies

Having a library available with ample materials and services is of little use unless you know how to access the materials you want. Patrons unfamiliar with the library can easily feel overwhelmed because there seem to be so many crannies, both physical and electronic, in which relevant resources can hide. There are, however, four basic strategies for locating materials.

1. Use the *online public access catalog* (OPAC)

2. Use a variety of *abstracts* and *indexes*.

3. Search for government documents, which can be found through the OPAC or the *Monthly Catalog of U.S. Government Publications.*

4. Use the *Internet* or the *World Wide Web*, which has become increasingly important in the search for information; its use is explored in the "Computers and the Internet" sections in each chapter of this book.

These four strategies provide a good start to discovering the information available in and through a library. To do a thorough search that locates the most up-to-date sources, you must be familiar with all four strategies, and I will describe them in this appendix (A).

Accessing Library Materials: Computers in Research

The usefulness of computers in the research process extends into the library. To cope with the proliferation of materials, libraries over the past few decades have made extensive use of computerization to make library searches faster and more effective. For example, most libraries now have a computerized online public access catalog (OPAC) of all the books, periodicals, and other materials in the library. In some cases, the OPAC includes the holdings of all libraries in a region. Library patrons can search for materials by author, title, subject, or by the use of keywords (described in the next section). In addition, many abstracts, indexes, and government documents described later in this appendix are now available on CD-ROM.

A variety of computerized information-retrieval systems exist that can provide a list of citations to books, articles, and government publications on particular topics. In most cases, the user provides a list of keywords that relate to the topic of interest. The keywords might be subject headings, author's names, titles of journals, and so on. The computer then searches through abstracts, indexes, and other materials n its database for citations that are filed under those keywords or contain a keyword or words in the title of the book or article.

Keyword searches work like this: If one were interested in child abuse, then keywords such as "child abuse," "discipline," "cruelty to children," "battered children," and "parent–child relationship" could be entered into the search field. It is also possible to narrow the search by cross-referencing a number of keywords. For example, we could request only materials listed under *all* the following keywords: "battered children—Native American—male—age 1 to 4—Michigan." It is also possible to limit the search to works published between specified dates, sponsored by certain organizations, or qualified by any number of other criteria, depending upon the search service and the database searched. Obviously, the more encompassing terms ("parent–child relationship") will result in the retrieval of a longer list of articles than will narrower terms ("battered children") or terms that are cross-referenced. This is something to take into consideration if there is a fee for this search service based in part on the number of references provided. Also, the longer list will include many sources not directly relevant to your concerns and could waste your valuable time. The librarians operating the computer search service can advise you as to whether or not a computer search is appropriate to your topic (sometimes a manual search is better) and also suggest useful keywords to maximize your search. Many search services provide a dictionary of terms used in the database. One such database, for example, is ERIC, or Educational Resources Information Center. If your library subscribes to this database, then it will also have a hard-copy volume titled *Thesaurus of ERIC Descriptors* that contains all the subject terms used in the ERIC database.

Computers have also made it feasible to provide other services that give patrons greater access to materials: table-of-contents and document-delivery services. *Table-of-contents services* make available the tables of contents of thousands of periodicals, providing author(s) and title as well as other information relevant to accessing the articles. This means that a user can review the contents of many journals rapidly, even journals not owned by the particular library being used. *Document-delivery services*, often a part of table-of-contents services, actually deliver copies of articles to patrons. Even if the library being used doesn't subscribe to the journal, the article will be mailed or faxed to the patron from a library that does. These services have made it possible for libraries to provide materials for patrons even when the library does not physically possess the materials. The electronic and computer era has shifted libraries' focus away from ownership of materials and toward providing access to materials. In some cases, local and regional libraries form a sort of co-op where if one library subscribes to a journal, the others do not, but instead subscribe to a different one. This keeps the availability of journals more widespread through the use of interlibrary loan. Without such co-op agreements, every library might believe it did not have to expend its own funds because it could obtain the material from other libraries; then, no library might purchase any journals.

Many of these computerized library services are operated by the patrons themselves, although some require assistance from library personnel. Some of these services are free or provided at low cost, but some of the services can be expensive. Some of the document-delivery services, for example, can get rather costly. Check with library personnel to determine precisely which computerized or online services are available and what costs you might incur by using them.

BOOKS

The Classification System

Small local libraries might contain a few thousand books. Small- to medium-sized universities have libraries with hundreds of thousands of volumes. Large university and public libraries contain millions of books. Despite the tremendous variations in library size, the systems to classify books in them are designed so flexibly that the system used in the smallest library can also be used in the largest, and, despite the tremendous size of some libraries, a person can easily retrieve a book if he or she knows how the book is classified. Most classification systems arrange books according to *subject matter*. First, books are classified into very general categories, such as history, science, or social sciences. Then, within each general category, they are further arranged into more precise subcategories. The category of social sciences, for example, is further divided into economics, sociology, and so on. In this fashion, the subject matter of a book is further narrowed until it fits into a fairly specific category. Then each book is given a unique *call number*, which contains all the necessary information from the classification system to identify that particular book.

There are primarily two classification systems used in the United States: the *Library of Congress System* and the *Dewey Decimal System*. The Library of Congress System uses a combination of letters and numbers to classify books (see Table A.1). The first letter in the call number is one of 21 letters of the alphabet used to classify books into the most general categories. The letter *H*, for example, indicates the general category of the social sciences. The second letter (if there is a second letter) further narrows the subject matter within the social sciences. The letter *Q*, for example, indicates that the book falls in the more specific social science subject matter of "Family, Marriage, Women," whereas *V* refers to the subject of "Social Pathology." The letters of the call number are followed by a number that further narrows the subject matter. Within the classification *HV*, for example, numbers between 701 and 1420 refer to works on the topic of "protection, assistance, and relief of children,"

TABLE A.1   Library of Congress Classification System

A	General Works	
B	Philosophy and Religion	
C	History (General—Civilization, Genealogy)	
D	History—Old World	
E	American History and General U.S. History	
F	American History (Local) and Latin America	
G	Geography, Anthropology, Folklore, Sports, and Other	
H	Social Sciences	
	HA	Statistics
	HB–HD	Economics
	HF	Commerce
	HG–HJ	Finance
	HM	Sociology
	HQ	Family, Marriage, Women
	HV	Social Pathology
J	Political Science	
K	Law	
L	Education	
M	Music	
N	Fine Arts	
P	Language and Literature	
Q	Science	
R	Medicine	
S	Agriculture, Forestry, Animal Culture, Fish Culture, Hunting	
T	Technology	
U	Military Science	
V	Naval Science	
Z	Bibliography and Library Science	
Example:	HV	
	875	
	F6	

whereas the numbers from 5001 to 5840 refer to books on "alcoholism." This number is followed by a letter that is the first letter of the last name of the author of the book (in cases of

TABLE A.2   Dewey Decimal Classification System

Broad subject areas or classes:

000	Generalities
100	Philosophy and Related Disciplines
200	Religion
300	The Social Sciences
400	Language
500	Pure Sciences
600	Technology (Applied Sciences)
700	The Arts
800	Literature and Rhetoric
900	General Geography and History, and the like

Each class can be subdivided into smaller classes or subclasses:

300	The Social Sciences
310	Statistical Method and Statistics
320	Political Science
330	Economics
340	Law
350	Public Administration
360	Welfare and Association
370	Education
380	Commerce
390	Customs and Folklore
Example:	362.7
	N34

The Dewey decimal system arranges books into 10 general categories based on a three-digit number on the top row of the call number (see Table A.2). Numbers in the 300 range comprise the social science category. The second and third digits of this top row further narrow the classification within the social sciences. For example, the 360s deal with "welfare and association."

The three-digit number is followed by a decimal point and numbers that indicate narrower classifications. In the second row of the Dewey call number is a letter, which is again the first letter of the last name of the author of the book, followed by a number that further identifies the author. This may be followed by a lowercase letter that is the first letter in the first word of the title of the book (excluding *a*, *an*, and *the*). The Dewey decimal system also provides each book with its own unique call number.

The Public Catalog

With knowledge of the classification system in use at your library, you can locate any book. To do so, you need to find the *public catalog*, which is a listing of all the holdings in that library. In most libraries today, the public catalog is computerized as an OPAC of holdings. In some cases, these computer files can tell you not only whether a library owns a particular item but also its status: whether it is checked out, at the bindery, lost, and so on. Some libraries are involved in regional networking systems with other libraries, and their computerized public catalogs list which libraries in the network own a particular book or other holding. Thus, if a book is not available at your library, you can see where it can be found.

In the computerized version of the public catalog, each library holding has a separate entry or record that can be called up and read on the computer screen. To locate a book or other item in the library, find the OPAC record of the book, make note of the location (e.g.,

multiple authors, it is by the first author's last name). This is followed by a number that further identifies the author. These sets of letters and numbers of the call number provide an identification that is unique to this book: No other book has exactly the same call number. Books are placed on the shelves in the order of their call numbers. All the *H*s are placed together, and within the *H*s all the *HM*s go together. Within the *HM*s, books are arranged according to the other letters and numbers.

FIGURE A.1   An Example of a Record from an Online Public Access Catalog

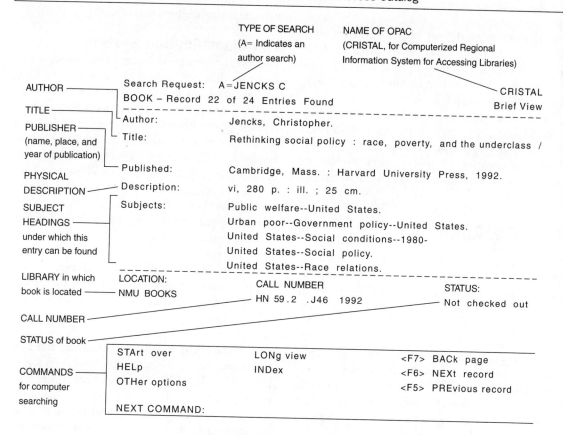

reference, documents, or stacks) and the call number, and find it on the library shelves. (Remember that books with similar call numbers will be about similar topics, so it is a good practice to browse through books in the immediate vicinity of the one whose call number you have located.) Figure A.1 presents a computer record for a book, with the various elements of the record identified. The record shows the author and title of the book, the subject headings under which it can be found in the public catalog, the library in which the book is located, and so on. To find a record in the public catalog, look under the author's name, the title of the work, the subject heading under which it is listed, or by keywords that appear in the record. Author and title searches

are fairly straightforward, but searching by subject can be slightly more problematic because library patrons sometimes have difficulties finding useful subject headings for their topic. To assist in this task, a volume titled *Library of Congress Subject Headings* can be referred to if your library uses the Library of Congress classification system. If your library uses the Dewey decimal system, then the equivalent volume is *Sears List of Subject Headings*. Both these volumes list headings acceptable for use in a public catalog, cross-referencing a number of different topics. These volumes are usually located somewhere near the public catalog itself. I described keyword searching earlier in this Appendix. It is done the same way with the computerized public catalog. If you use the public catalog to locate books, you will have to determine from the title of the book and the brief description of the contents contained in the record whether a book will be useful to you. When you do this, it is better to err on the side of jotting down too many call numbers rather than too few.

Once the organizational key to the library is understood it is easy to locate the information stored in books. There are, however, other repositories of information in the library, one of the most important of which is periodicals.

PERIODICALS

One of the most valuable and heavily used resources of the library, in addition to books, is *periodicals*. Published periodically—weekly, monthly, bimonthly, or semiannually (hence, the name "periodicals")—they contain articles by different authors on a variety of topics. Periodicals come in two basic types, although some are a blend of both. *Magazines* are directed at the general public, written in a light and popular style, and are often commercial ventures. *Journals* are professional periodicals that publish articles primarily intended for members of a particular profession. The *American Sociological Review*, the *American Political Science Review*, and the *Journal of the American Medical Association* are examples of professional journals. I will focus primarily on journals, but both magazines and journals can be very helpful in researching a problem because, first, periodicals tend to be more current than books, especially in reporting research findings. This is so, in part, because the production process for books is longer than that for periodicals. In addition, in many scientific fields, including the social sciences, much research is reported first in journals. In fact, many books contain primarily summaries or assessments of research that has already appeared in journals. Second, the articles in periodicals are brief and specific, enabling you to

gather more information about a particular topic in a shorter period of time.

The Classification System

Periodicals, like books, can be assigned a call number, although not all libraries classify their periodicals according to call number. When periodicals are given a call number, they can be housed in the library in two ways: along with the books by call number, or separately from books by call number. When periodicals are classified by call number, you will have to look in the public catalog or other location to find the call number of the periodical you need. The advantage of classifying periodicals by call number is that periodicals on similar topics will be near one another. Some libraries, however, do not use the periodical call number, arranging periodicals separately from books in alphabetical order by the title of the journal (excluding the first *the* or *a* in the title). Once you understand how the periodicals are stored in your library, any particular periodical can then be found easily because no two periodicals have the same title or call number. Each issue of a journal is further identified by a volume number, the year of its publication, and an issue number indicating its place in the sequence of issues making up a particular volume. Some journals also identify their issues with the month or season (for example, Summer) in which it was published. To locate an issue of a journal, then, you need the following information: the call number or name of the journal, year of issue, volume number, and issue number (or month or season of publication).

There are usually three locations for periodicals in the library. First, most libraries store the *current periodicals* (most recent issues) in a separate place where users can browse through them. Typically, the most recent one or two years are kept there. Second, *bound periodicals* contain several issues that have been bound together in one hardcover volume for extended storage. Each bound volume normally contains one year's worth of issues. Most of a library's periodical holdings are in this form. Third, old volumes are often stored on *microfilm* or *microfiche* or some other space-saving device for long-term storage. These are usually kept in large cabinets located near the machines that magnify them for viewing.

A problem can arise when accessing periodicals that, though minor, is irksome to both patrons and librarians. The problem is that the separate issues that make up a volume of a journal must be bound together. This is done by sending the volume to a bindery, which means that the issues in that volume are temporarily unavailable to library patrons. The period is usually brief, but the periodicals are also usually recent ones because the binding is done within the first few years of publication. Students need to be aware that this problem can occur; it is one reason to begin library research early because some recent journals you are looking for may be available now but sent to the bindery a month from now.

Journals Important in the Social Sciences

Literally hundreds of journals in the library contain materials of use to social science researchers. Although no one can be expected to read all these journals, awareness and occasional perusal of them can uncover much valuable source material. One way to spend free time is to browse through some of these journals in the current periodicals section of the library. Although no list could claim to be complete or comprehensive, here is a list of journals that social science researchers are likely to use:

- *American Journal of Orthopsychiatry*
- *American Journal of Public Health*
- *American Journal of Sociology*
- *American Sociological Review*
- *Child Development*
- *Clinical Sociology Review*

- *Community Mental Health Journal*
- *Crime and Delinquency*
- *Criminology*
- *Current Contents—Social and Behavioral Sciences*
- *Evaluation Review*
- *Families in Society*
- *Gerontologist*
- *Journal of Applied Behavioral Science*
- *Journal of Applied Sociology*
- *Journal of Criminal Law and Criminology*
- *Journal of Drug Issues*
- *Journals of Gerontology* (Series A and B)
- *Journal of Health and Social Behavior*
- *Journal of Marriage and the Family*
- *Journal of Personality and Social Psychology*
- *Journal of Research in Crime and Delinquency*
- *Journal of Studies on Alcohol*
- *Merrill–Palmer Quarterly*
- *Public Opinion Quarterly*
- *Social Forces*
- *Social Policy*
- *Social Problems*
- *Social Psychology Quarterly*
- *Sociological Quarterly*
- *Sociological Practice*

Given this diversity of periodicals, how does one find those articles that relate to a particular topic? Two types of reference publications, called *abstracts* and *indexes*, are the major keys to accessing the information contained in periodicals.

ABSTRACTS AND INDEXES

Some periodicals provide an abstracting service for library patrons. An *abstract* is a brief description, usually no more than a paragraph, of the contents of a book or article. A good abstract provides a complete summary of the work, including the thesis of the author, a description of any data collected, the conclusions drawn, and limitations of the study. Abstracts also frequently provide the list of keywords under which each book or article is catalogued, and this is an important help in searching for other materials on the same topic. Abstracts help you locate relevant research and decide whether it is sufficiently useful to warrant reading the entire work. Abstracting services provide a list of the journals whose articles are abstracted, usually in the first few pages of the volume of abstracts. Likewise, some journals list in each issue the abstracting and indexing services in which they are included.

Using an abstract is relatively simple. I illustrate this with an abstracting service titled *Sociological Abstracts*. Suppose you are interested in the issue of occupational mobility in global service occupations. Abstracts usually contain at least three important elements: a subject index, an author index, and the abstract itself. You can begin by finding a topic heading in the subject index of the abstract that will point you toward relevant articles (see Figure A.2). For your topic of interest, "occupational mobility" or "workers" would be such topic headings and should be consulted. You may have to try a number of headings—some more general, others more specific—in order to locate articles. After locating a specific heading, glance over the subtopics listed under it for one directly related to your interests. When you find the relevant subtopic, there will be one or more numbers following it, each referring to a separate abstract. The abstracts themselves are listed elsewhere in the volume in numerical order. Abstracting services also provide an author index should you wish to search for works by a particular author.

In addition to the summary description, the abstract provides the complete reference so

that you can locate the work. In the illustration, the information presented in the reference, in the order it is presented, is: authors' names, an address where the author can be contacted, title of article, title of journal, year journal was published, volume of journal, issue in volume, and pages of that article in the journal. If an abstract suggests that an article will be useful to you, the complete reference should be noted.

The following are some of the major abstracting services useful for topics in the social sciences:

- *Abstracts in Gerontology: Current Literature on Aging*
- *Child Development Abstracts and Bibliography*
- *Criminal Justice Abstracts* (formerly *Crime and Delinquency Abstracts*)
- *Dissertation Abstracts International*
- *Exceptional Child Education Resources* (*ECER*, formerly *Exceptional Child Education Abstracts*)
- *Human Resources Abstracts* (formerly *Poverty and Human Resources Abstracts*)
- *Psychological Abstracts*
- *Social Work Abstracts*
- *Sociological Abstracts*
- *Wilson Social Science Abstract*
- *Women Studies Abstracts*

In addition to abstracting services, there are other publications that provide indexing services. An *index* is an alphabetical arrangement of materials based on some element of the materials, usually the author's last name, the title of the work, or the subject matter. (This text, for example, has separate name and subject indexes at the end.) Indexing publications usually provide an author index, which lists all the articles published by a particular author, and a subject index listing all articles on a given subject. An index also presents a list of the

FIGURE A.2 Use of an Abstract

Author Index

Abdullah, Firdaus Hj., SA08775
Aber, Lawrence (Ed), a145258
Abilov, A. Zh., SA07684
Abolafia, Mitchel, a145134
Abrahams, Taryn, SA06574
Abrahamson, Peter, a145263, a145335
Ackerman, Peggy T., SA08672
Acosta Silva, Adrián, SA07790
Adair, John, SA06628
Adam, Erik, SA07776
Adam, Heribert, SA07399
Adamets, Sergey, SA08093
Adamjee, Riaz, SA07279
Adams, Guy B., SA07013
Adams, Jacqueline, SA06868
Adams, Rebecca G., SA07306
Adar, Gila, SA08744
Adelantado, José, SA08776
Adelson, Leslie A., SA06812
Adelson, Naomi, SA08374
Adendorff, Ralph, SA07704
Adler, Lomnitz, Larissa, SA07400
Adler, Patricia A., SA07617
Adler, Peter, SA07617
Afkhami, Mahnaz (Ed) a145135
Afonso, Almerindo Janeta, SA07791
Agadjanian, Victor, SA08343
Agata, Kenji, SA07618
Agersnap, Torben, SA06292
Ághová, L'ubica, SA08304
Agnew, Robert, SA08673
Agouba, Mukahtar, SA08638
Aguilera, Ruth V., SA06894
Aguilera-Vaqués, Ruth, a145801
Agustoni, Alfredo, SA08360
Ahlers, Ingolf, SA07365
Ahmed, Akbar S,. a145136
Ahn, Mi Na, SA06556
Ahola, Sakari, a146044
Ahrne, Göran, a145368, a145420
Ahuja, Ram, a145137
Aitchison, Cara, SA07298
Aitov, N. A., SA08800

ABSTRACT

SA07617

Adler, Patricia A. & Adler, Peter (Dept Sociology U Colorado, Boulder 80309 (e-mail: adler@spot.colorado.edu)), **Translence and the Postmodern Self: The Geographic Mobility of Resort Workers,** *The* Sociological Quarterly, 1999, 40, 1, winter, 31-58.

¶ The occupational community of resort workers offers a glimpse into the global postmodern workforce: individuals who relocate around the world, impelled by their career aspirations or their search for the intense experience of the beauty, exotic nature, & extreme recreation in various international destinations. These people have abandoned the conventional lifestyle anchored in security, continuity, & tradition & embarked on a

lifestyle of transience. Drawing on 4 years of participant observation with 400+ workers (including 50 in-depth interviews) in a luxury HI resort & across the mainland US, four dimensions of these workers' lives that primarily engage their transience are identified: work, ideology, family, & friendships. From these patterns, conclusions are drawn about the nature of the global postmodern workforce & community, & about the social psychological nature of the postmodern self. 96 References. Adapted from the source document.

Subject Index

Occupational Achievement
Occupational Achievement; Asian Americans; Whites; Racial Differences; Males; Race; SA06769
Occupational Achievement; Employment Opportunities; Hiring Practices; Job Search; Theoretical Problems; Social Networks; SA07626
Occupational Qualifications; Occupational Achievement; Income; Occupational Status; Social Status; Sex Differences; SA07618
Occupational Choice
Sex Stereotypes; Children; Occupational Choice; Sex Differences; SA08993
Occupational Classifications
Great Britain; Heart Diseases; Mortality Rates; Occupational Classifications; Social Inequality; Methodological Problems; SA08047
Occupational Mobility
Labor Supply; Supply and Demand; Occupational Mobility; Netherlands; Human Capital; Labor Market Segmentation; SA06919
Czech Republic; Slovak Republic; Social Mobility; Occupational Mobility; Demographic Change; Social Structure; SA07610
Postmodernism; Self Concept; Leisure; Tourism; Workers; Occupational Mobility; SA07617
Sweden; Work Skills; Employment Changes; White Collar Workers; Blue Collar Workers; Occupational Mobility; SA06951
Social Mobility; Occupational Mobility; Unemployment; Social Status; Occupational Status; Northern Ireland; SA07593
India; Economic Policy; Human Capital; Labor Market Segmentation; Occupational Status; Occupational Mobility; SA07606
Physicians; Professional Women; Mexico; Career Patterns; Occupational Mobility; Migration Patterns; SA07634
Occupational Status; Occupational Mobility; Undocumented Immigrants; Sex Differences; Social Integration; United States of America; SA08085
Melbourne, Australia; Occupational Mobility; Asian Cultural Groups; Human Capital; Social Background; Refugees; SA06916

Research and Development; Scientists; Social Networks; Promotion (Occupational); Social Mobility; Occupational Mobility; SA07630
Discipline; Surveillance; Railroads; Careers; Occupational Mobility; Personhood; SA07057
Occupational Qualifications
Evaluation; Management; Managers; Peer Review; Self Evaluation; Workers; SA07058
Occupational Qualifications; Occupational Achievement; Income; Occupational Status; Social Status; Sex differences; SA07618
Occupational Roles
Clergy; Females; Pastors; Self Concept; Occupational Roles; United Kingdom; SA07943
Europe; Rural Urban Differences; Health Care Services; Physicians; Comparative Analysis; Occupational Roles; SA08381

journals it indexes. Indexes have both advantages and disadvantages in comparison to abstracts. One advantage is that, because they take less time to prepare, they tend to be more current. Some abstracts are not published until more than a year after the articles covered have appeared. One disadvantage of indexes is that the user must rely on the title of the article to determine whether it is sufficiently relevant to spend the time seeking out the actual work itself. Anyone who has used indexes will attest to the many wild-goose chases where a seemingly useful article turned out to be irrelevant to a particular topic. Likewise, there are undoubtedly useful articles that are ignored because their titles do not seem sufficiently relevant.

Numerous indexes can be of value to social scientists.

- *Criminal Justice Periodical Index*
- *Cumulative Index to Nursing and Allied Health Literature (CINAHL)*
- *New York Times Index*
- *Nursing and Allied Health Index*
- *PAIS International*
- *Readers' Guide to Periodical Literature*
- *Social Sciences Citation Index*
- *Social Sciences Index*
- *United States Government Periodicals Index*
- *Women's Studies Index*

Although these are some of the more important abstracts and indexes in the social sciences, there are others. Consult your reference librarian and become familiar with all the abstracts and indexes in your library relevant to the social sciences. If you compile a list of them, you will have a quick reference for any research you need to do while in college or on the job. Some abstracts and indexes are also available on CD-ROM or online. *Sociological Abstracts'* CD-ROM version, for example, is called "Sociofile"; *Psychological Abstracts'* is called "PsychLit." A widely used online search

service that produces abstracts or indexes is called "FirstSearch." Once again, you should determine which of these is available to you.

REFERENCE BOOKS

One group of highly useful books that are unfortunately often overlooked are general reference books. Included in this rather broad category are encyclopedias, directories, and bibliographies. Although such works do not substitute for a thorough library search, they can save you time, as well as efficiently answer many routine questions. The following is only a partial listing of reference works useful in the social sciences.

- *Companion Encyclopedia of Anthropology*
- *Encyclopedia of Adolescence*
- *Encyclopedia of Aging*
- *Encyclopedia of Alcoholism*
- *Encyclopedia of Associations*
- *Encyclopedia of Drug Abuse*
- *Encyclopedia of Psychology*
- *Encyclopedia of Sociology*
- *Guide to Reference Books*
- *National Directory of State Agencies*
- *Public Welfare Directory*
- *The Social Sciences Encyclopedia*
- *State Executive Directory Annual*

GOVERNMENT DOCUMENTS

The U.S. government is one of the world's largest publishing houses, pouring forth mountains of books, bulletins, circulars, reports, and the like. Researchers in the social sciences find government documents especially important because much of the research done in this area is sponsored by the government. Demonstration projects, program evaluations, and needs assessments, for example, are commonly funded by the government, and the resulting research reports are published by the government. Many of these documents can now be accessed online and downloaded to your computer, as described in several of the "Computers and the Internet" sections in this book.

The Monthly Catalog

Locating useful government publications may at first seem a bewildering endeavor. With the aid of a few basic tools, however, the task can be done quickly, thoroughly, and with relatively little pain. The first thing to learn is whether your library is a *depository library*, which is a designation made by the superintendent of documents at the U.S. Government Printing Office (GPO). A *regional depository library*, of which there can be no more than two in any state, receives everything published by the GPO. *Selective depositories* receive only some government publications, a listing of which can be found in *List of Classes of U.S.*

Government Publications Available for Selection by Depository Libraries. Nondepository libraries will have some government documents, depending upon which they choose to purchase.

The next thing you need to determine is which government documents are relevant to your particular topic of interest. Many libraries today list government documents along with other books in their online public access catalog. The documents can be located via name, title, subject, or keyword searches. Another major source for information on government documents is the *Monthly Catalog of U.S. Government Publications*, which has been published since 1895 and is now available on CD-ROM. This publication includes a listing of most government publications along with the information necessary to locate them in the library or to purchase them from the office of the superintendent of documents.

FIGURE A.3   A Sample Entry in the *Monthly Catalog of U.S. Government Publications*

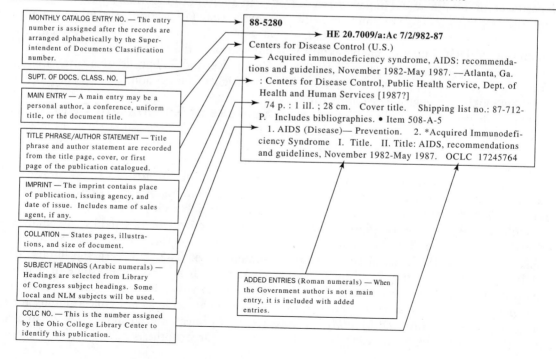

Every government document submitted to the office of the superintendent of documents is given a *Monthly Catalog* entry number (see Figure A.3), and documents are listed in the *Monthly Catalog* in order by this number. This number has two components. The first two digits indicate the year the document was published (1988 in the illustration). The second group of digits locates the record in the *Monthly Catalog*. This second number is derived from sequencing the publications alphanumerically according to the classification number of the superintendent of documents, or SUDOCS number. The SUDOCS number is like the call number in the public catalog for locating books because the government documents are filed in the library according to this number. However, although the Dewey decimal and Library of Congress systems classify entries according to subject matter, the SUDOCS system classifies materials according to the agency issuing the materials. In Figure A.3, for example, the *HE* in the SUDOCS number means that this document was issued by the Department of Health and Human Services. Thus, the *Monthly Catalog* entry number helps you locate the document in the *Monthly Catalog*, and the SUDOCS number enables you to locate where the document is shelved in the library.

Simply looking through the *Monthly Catalog* is an inefficient way of finding publications on the exact topic you want. Several aids can make your search both quicker and more fruitful. These are the subject index, the author index, the title index, and the title keyword index (see Figure A.4). Thus, if you actually know the author or title of a publication, you can look for these in the appropriate index. The index will give the *Monthly Catalog* entry number (and, since 1987, the SUDOCS number). If you know neither the exact title nor the author, the subject index provides a listing of documents according to subject matter and will again provide the title and *Monthly Catalog* entry number (and, since 1987, the SU-

DOCS number). If you know a keyword associated with the title, the title keyword index similarly directs you to the *Monthly Catalog* number. Armed with this number, you simply go to the correct volume of the *Monthly Catalog* according to the entry number. The *Monthly Catalog* provides a variety of details about the publications (see Figure A.3). Generally, in addition to the SUDOCS number that allows you to locate the document, the entry gives the author, title, publication information, price, and subject headings. This is useful for determining whether that particular document is of further interest to you.

To summarize, then, the steps followed in using the *Monthly Catalog* to find government documents are as follows. First, consult an appropriate index and obtain *Monthly Catalog* entry numbers and SUDOCS numbers; second, consult the *Monthly Catalog* for more information about the document, a complete reference, and the classification number of the superintendent of documents; third, locate the document itself.

In addition to the *Monthly Catalog of U.S. Government Publications*, the following agencies publish separate catalogs of their own publications: the Bureau of the Census, the Commerce and Labor departments, and the Civil Rights Commission.

Sources of Government Data

In Chapter 10, I discussed research using available data—data collected by someone else but available to others for analysis. Many of the government documents discussed previously report research findings, describe programs, or analyze social policy. These are not, strictly speaking, sources of raw data to be analyzed but rather interpretations, assessments, or summaries of data. However, the library also contains such available data, some of it made available in government publications. Some of this data can serve as a useful beginning for a research project, and in some cases it can also be used for hypothesis testing and evaluation

FIGURE A.4 Four Indexes Useful in Locating Documents in the *Monthly Catalog of U.S. Government Publications and Identifying the Entry in Figure A.3*

Author Index

Center of Military History.
The Annapolis Convention., 88-5726
Armies, corps, divisions, and separate brigades/, 88-4957
The Army historian: a publication of the United States Army Center of Military History., 88-664
The Army Medical Department, 1818-1865/, 88-3731
Army museum newsletter., 88-663
David Brearly., 88-3727
Dissertation year fellowships/, 88-7892
The German campaigns in the Balkans (spring, 1941)., 88-1990
German tank maintenance in World War II., 88-7894
Jonathan Dayton., 88-3728
The medics' war/, 88-1987
Moscow to Stalingrad : decision in the east/, 88-6529
The profession of arms : the 1962 Lees Knowles lectures given at Trinity College, Cambridge/, 88-1988
The staff ride/, 88-1989
Thomas Fitzsimons., 88-3729
U.S. Army mobilization and logistics in the Korean War : a research approach /, 88-7893
Warfare in the far north., 88-7895
William Jackson., 88-3730

Centers for Disease Control (U.S.).
Acquired immunodeficiency syndrome, AIDS : recommendations and guidelines, November 1982-May 1987 (88-5280)

The agents of non-A, non-B viral hepatitis/, 88-2364
AIDS in Africa : an epidemiologic paradigm/, 88-2361

Title Index

Acid rain. (1 19.114:R 13), 88-5751
Acoustic emission/flaw relationship for in-service monitoring of nuclear pressure vessels : progress report / Hutton, P. H. (Y 3.N 88:25/4300/v.4, no. 1), 88-6126
Acoustic emission monitoring of fracture development / Anderson, Sterling J. (1 28.23:9077), 88-8443
Acoustic-televiewer and acoustic-waveform logs used to characterize deeply buried basalt flows, Hanford Site, Benton County, Washington [microform] / Paillet, F. L. (1 19.76:85-419), 88-9555
Acquiescence rulings United States. Social Security Administration (HE 3.44:(date)), 88-844
Acquired Immune Deficiency Syndrome (AIDS) and the Veterans Administration : hearing before the Subcommittee on Hospitals and Health Care of the Committee on Veterans' Affairs, House of Representatives. One hundredth Congress, first session, June 17, 1987. United States. Congress. House. Committee on Veterans' Affairs. Subcommittee on Hospitals and Health Care. (Y 4.V 64/3:100-19), 88-6304
Acquired immunodeficiency syndrome (AIDS) : fifteenth update, July 1987 through September 1987. 845 citations / Abrams, Estelle J. (HE 20.3614/2:87-13), 88-5261
Acquired immunodeficiency syndrome AIDS : recommendations and guidelines, November 1982-May 1987. Centers for Disease Control (U.S.) (HE 20.7002/2:87-7/2/982-87) (88-5280)
Acquired immunodeficiency syndrome (AIDS) : sixteenth update, October 1987 through December 1987. 1028 citations.

Keyword Index

January — June 1988

 "   to families with dependent children (AFDC) pro 88-2233
 "   to students., Financial 88-2219
AIDS /, Foreign policy implications of 88-9914
 "   / Progress on the treatment of 88-5246
 "   : facts about the disease; how to protect yo 88-4098
 "   : need for immediate OSHA regulations to pro 88-3409
 "   : Opportunities for international scientific 88-10204
 "   : recommendations and guidelines, November 19 (88-5280)
 & hepatitis B., Worker exposure to 88-9857
 "   and teenagers ; 88-8795
 "   and the education of our children : a guide f 88-3754
 "   and the education of our children : a guide f 88-7933
 "   and the law enforcement officer/ 88-9805
 "   and the Veterans' Administration : 88-10232
 "   crisis as related to the federal budget ; 88-7526
 "   drug development program at NCI 88-121
 "   for health services research /, Selected bibl 88-9485
 "   for policymakers /, An Annotated bibliography 88-6830
 in Africa : an epidemiologic paradigm /, ... 88-2361
 "   knowledge and attitudes for September 1987: 88-6869
 "   knowledge and attitudes : provisional data fr 88-6868
 "   patients /, New drug therapy developed for pa 88-6852
 "   prevention program operations, Guidelines fo 88-267
 "   published in the Morbidity and Morality Week 88-4101
 research : 88-4776
 "   research and medical care within the Veterans 88-4794
 "   research and medical care within the Veterans 88-8971
 "   research program /, NCI's 88-5245
 "   to navigation bulletin 88-1451
 vaccine /, Research on development of an 88-5244
 "   , a public health challenge : state issues, po 88-5227
 "   , CDC reports on acquired immunodeficiency s 88-2344
 "   , fifteenth update, July 1987 through Septem 88-5261
 "   informe del jefe del Servicio de Salud Pub 88-6826
 "   , Public Health Service guidelines for counse 88-5283
 "   , sixteenth update, October 1987 through Dec 88-9474
 "   , Straight facts about 88-2265
 "   , Tips on avoiding 88-6861
 "   , twelfth update, October 1986 through Decem 88-2298
 "   , Viruses in cancer and 88-8125

Subject Index

Agriculture — United States.
Country-of-origin labeling on imported perishable agricultural commodities : hearing before the subcommittee on Domestic Marketing, Consumer Relations, and Nutrition of the Committee on Agriculture, House of Representatives. One hundredth Congress, first session, on H.R. 692, H.R. 1176, and H.R. 1246, March 30, 1987. United States. Congress. House. Committee on Agriculture. Subcommittee on Domestic Marketing, Consumer Relations, and Nutrition. (Y 4.Ag 8/1:100-14), 88-6150

Agriculture — United States — Periodicals.
Semiannual report to Congress. United States. Dept. of Agriculture. Office of the Inspector General. (A 1.1/3:3:987), 88-480]

AIDS (Disease) — Bibliography.
Acquired immunodeficiency syndrome (AIDS) : fifteenth update, July 1987 through September 1987. 845 citations / Abrams, Estelle J. (HE 20.3614/2:87. 13) 88-5261

AIDS (Disease)— Prevention.
Acquired immunodeficiency syndrome. AIDS : recommendations and guidelines, November 1982-May 1987 (Centers for Disease Control (U.S.). (HE 20.7009/a:Ac 7/2/982-87) (88-5280)
Human T-lymphotropic virus type III/ lymphadenopathy-associated virus : agent summary statement (HE 20.7009/a:H 88/6), 88-5282
Public Health Service guidelines for counseling and antibody testing to prevent HIV infection and AIDS, (HE 20.7009/a:H 88/7), 88-5283
Recommendations for prevention of HIV transmission in health-care settings. (HE 20.7009/a:H 88/8), 88-5284
Summary : recommendations for preventing transmission of infection with human T-lymphotropic virus type III/ lymphadenopathy-associated virus in the workplace. (HE 20.7009/a:H 88/2), 88-5281

of programs. Three government agencies that produce such publications are the Bureau of the Census, the Department of Labor, and the Department of Justice.

The U.S. Bureau of the Census publishes a vast array of statistical data of use to social scientists:

- *Census of Population.* The nationwide census is conducted once every decade and provides data on employment, income, race, occupation, poverty, and much more.

- *County and City Data Book.* This is a supplement to the *Statistical Abstract*, reporting data on a local or regional basis. This book is published every five years, so the data it contains may be outdated.

- *Current Population Reports.* These are a series of annual publications that report on many of the same issues that the census does but are based on a probability sample of the population. There are series on income, household and family characteristics, marital status, living arrangements, and geographic mobility. These are useful publications for keeping current on changing social trends and problems.

- *Historical Statistics of the United States: Colonial Times to 1970.* This volume provides historical data to supplement the Statistical Abstract, which often includes data from only the past decade or less.

- *Statistical Abstract of the United States.* This is an annual summary of statistical information about the United States. It contains data on population, birth and death rates, marriage and divorce rates, crime, health, education, and social services. This is probably the single most useful statistical summary.

The Bureau of Labor Statistics of the Department of Labor publishes many volumes relating to the labor force, employment, and earnings in the United States:

- *Employment and Earnings and Monthly Report on the Labor Force.* Published monthly.

- *Employment and Earnings, States and Areas.* An annual summary of state and regional trends.

- *Employment and Earnings Statistics for the United States.* An annual summary.

- *Monthly Labor Review.* A journal that presents both statistical and analytical articles relating to work and the labor force.

The Women's Bureau of the Department of Labor also publishes a number of periodicals focusing specifically on women workers, their earnings, their educational attainment, and legislation that affects them.

The Department of Justice publishes the *Bulletin of Criminal Justice Statistics* which contains valuable data on prisons, prison populations, and the criminal justice system in general.

Increasingly, these government sources of data are available on CD-ROM or on the Internet. The Census Bureau, for example, makes some data available on the Internet, and a program called GPO Access also links into online data sources. Information about Internet access to data is found in several of the "Computers and the Internet" sections in this book; in addition, you should check with the reference librarian at your library about CD-ROM and online sources of government data.

INFORMATION LITERACY AND CRITICAL THINKING SKILLS

A library is a veritable cornucopia of information, and I have tried to provide you with basic information on how to access that information. However, the best way to learn about a library is to use it. Roam around the stacks, browse through the public catalog, delve into government documents, explore the Internet. I have emphasized the ways in which you can systematically access information in the library. However, you can also find much useful material in random and casual meanderings through the library. Above all, consult with the reference librarians who are there primarily to assist you. These professionals are trained to help students and faculty devise programs and strategies for finding, analyzing, synthesizing, and evaluating information. Such information literacy is increasingly essential in the modern world, whether such information is found in a book, on a CD-ROM, or online. Information literacy is an important part of critical thinking, or the ability to judge the authenticity, accuracy, and worth of information.

This guide to the library has emphasized its utility to social science researchers. However, a library is not meant solely to help you write a term paper or complete a research project. A library is a repository of cultural knowledge. It houses research reports on alcoholics along with the epic myths of Homer and the ancient and sacred literature of Hinduism, the Vedas. Literature, philosophy, and theology accompany engineering and celestial mechanics. And you can read the daily newspaper there! The point is that a library is, in a sense, the information storehouse of a culture, and I hope that you use it to its fullest extent throughout your life as a source of enrichment and fulfillment. The search programs and strategies devised with the help of reference librarians in your college years can be used after college to help you find and analyze information about social issues or personal problems and to help you keep current in your chosen field of endeavor.

Computers and the Internet

All libraries today serve as gateways to the Internet, although one can access the Internet in other ways. The Internet has become an essential adjunct to the library in the search for information. For example, some of the table-of-contents and document-delivery services mentioned in this chapter can be accessed via the Internet, either through a library or your own personal computer. One such service is the UnCover system, whose Web site is http://uncweb.carl.org. At this site, you can view the tables of contents of many thousands of periodicals and can arrange to have copies of articles mailed or faxed to you. This Web site also tells you what the fees are for these various services.

There are also Web sites that enable you to review the holdings of a variety of libraries across the country and around the world. This is especially useful if the library you have access to is small, with limited holdings. By exploring these other libraries, you can find resources, especially books and journals, that may not be available in a smaller library. The Web site for the Library of Congress, for example, is at http://lcweb.loc.gov/homepage/lchp.html. Another Web site that provides access to many libraries is http://library.usask.ca/hywebcat/. Maintained by the University of Saskatchewan, this Web site will link you to the OPACs of many libraries around the world. You can search by geographical area or by the type of library desired. Under the latter, there are many options, such as medical, religious, or public libraries as well as the libraries of colleges and universities. This Web site also has links to many publishing companies.

Another way to gain library access is to use a search engine. Searching for "library access" will produce lists that enable you to locate many libraries in the United States and abroad, as well as many specialized libraries in areas such as art and law.

For Further Reading

Bolner, Myrtle S. (1995). *Library research skills handbook* (2nd ed.). Dubuque, IA: Kendall-

Hunt. This book provides a comprehensive overview of how to do research in the library.

Cuba, Lee. (1997). *A short guide to writing about social science* (3rd ed.). New York: Longman. This book is mostly about writing term papers and research reports, but it also includes a significant amount of material on using the library and the Internet.

Gates, Jean K. (1993). *Guide to the use of libraries and information sources* (7th ed.). New York: McGraw-Hill. This book is designed as a text for basic courses in library resources, so it can be an excellent introduction to the library that extends the topics in this chapter considerably.

Hahn, Harley. (2000). *Harley Hahn's Internet and Web yellow pages*. Berkeley, CA: Osborne Mc-Graw-Hill. As the name implies, this book is an excellent resource for finding materials on the Internet.

McLaren, Bruce J. (1997). *Understanding and using the Internet 1997* (2nd ed.). Minneapolis/St. Paul: West. This book is an excellent introduction to the Internet, providing all the information you need to utilize that online source of information.

Reed, Jeffrey G., & Baxter, Pam M. (1992). *Library use: A handbook for psychology* (2nd ed.). Washington, DC: American Psychological Association. Focused as it is on psychology, this book is nonetheless a valuable guide to the library for someone preparing to conduct a social research project. Among other things, it covers using abstract services, doing computer searches, and locating various scales and measuring devices.

The Sociology Writing Group. (1998). *A guide to writing sociology papers* (4th ed.). New York: St. Martin's. This book provides useful guidance in terms of doing library research in the social sciences, organizing time and materials, and preparing and writing analyses based on both quantitative and qualitative data.

Appendix B

GENERATING RANDOM NUMBERS

Random numbers are used for many purposes in social research. They are often used, for example, in constructing a probability sample in which elements are selected from a population and placed in the sample on a random basis. In some cases, this is done with a table of random numbers. Tables of random numbers can be found in many places, such as textbooks on statistics. A few volumes contain nothing but random numbers. Table B.1 in this appendix is a brief table of random numbers. In Chapter 7, when discussing simple random samples, I described in detail how to use such a table.

Spreadsheet software, such as Microsoft Excel, Lotus 1-2-3, or Corel Quattro, make generating random numbers easy. Although the specific procedures vary from software to software, the basic approach is the same. I will illustrate it using Quattro. In Quattro, the following command would be entered into the cell of the spreadsheet, and it would generate and place in that cell a random number from 0 to 1,000:

$$@INT(@RAND*1001).$$

"@RAND" generates a decimal random number between 0 and 1. This random decimal is multiplied by 1,001, resulting in a number that ranges from a low of less than 1.0000 to a high of less than 1,001.0000. The "@INT" command drops the decimal portion and displays the integer portion. Hence, the resulting random number ranges from 0 to 1,000. For example, if the random decimal were 0.0002, the result would be:

$$@INT(0.0002*1001) = @INT(0.2002) = 0.$$

If the random decimal were 0.9999, the result would be:

$$@INT(.9999*1001) = @INT(1000.8999)$$
$$= 1000.$$

The range of random numbers may be changed by substituting a different value for 1,001. To generate a list of random numbers, the formula is copied into as many cells as needed using the "Cell Copy" command. An advantage of using a spreadsheet is that the random numbers can be ordered from lowest to highest easily. This feature is convenient for such tasks as selecting a random sample from an ordered sampling frame. To reorder the numbers, the cell formulas are first converted to values using the "Values" selection from the "Edit" menu. Next, the "Sort" option is selected from the "Database" menu. The result is a list of random numbers in rank order.

Finally, random numbers may also be generated by standard statistical packages, such as SPSS and MINITAB. As with the spreadsheet, SPSS and MINITAB have a sorting procedure available that can be used to order the random numbers.

TABLE B.1   List of Random Numbers

894	920	220	614	090	805	668	331	745	136	071	056	205
493	974	737	304	049	109	097	660	275	036	819	132	807
768	450	669	873	510	712	613	059	924	377	090	315	507
482	379	669	549	746	814	424	217	883	969	246	777	454
621	422	255	762	718	431	883	645	341	148	212	527	557
488	005	671	472	511	746	446	425	889	766	248	860	956
472	782	194	854	272	001	465	118	514	892	258	367	599
554	576	110	002	045	940	724	975	533	401	603	047	221
333	797	019	563	344	349	210	261	204	225	739	730	872
653	346	789	798	616	377	724	625	760	845	430	239	647
018	792	713	967	411	189	654	392	789	308	733	343	168
446	165	992	185	650	158	738	758	284	900	822	217	809
733	098	756	628	982	258	875	694	463	772	162	788	537
324	338	369	374	975	389	657	310	552	951	242	626	135
800	408	564	050	120	844	656	122	270	638	712	442	293
994	349	174	326	424	016	645	595	383	578	393	114	426
776	410	150	051	532	844	219	710	207	763	085	314	858
835	234	461	844	543	475	105	274	191	122	549	991	696
408	051	655	449	318	302	574	581	586	466	123	866	301
356	581	735	113	285	188	235	863	096	585	783	817	030
336	130	491	288	437	351	650	325	673	807	311	844	363
935	737	202	656	201	553	387	933	546	203	930	201	322
975	455	421	422	173	767	163	860	167	939	304	318	227
484	564	624	002	801	589	140	125	059	875	848	345	944
848	669	356	665	029	302	247	804	133	374	407	316	773
836	906	596	608	598	356	481	982	742	757	635	746	967
425	895	530	807	924	685	325	894	571	925	705	559	532
122	251	638	926	678	852	779	707	320	649	809	203	333
034	451	574	656	354	387	913	663	375	079	743	503	635
145	849	295	003	709	118	762	068	784	616	147	959	292
428	232	529	095	487	039	387	957	546	864	107	120	661
755	154	664	651	508	033	915	809	328	137	452	291	539
826	104	222	160	209	051	502	331	146	686	883	400	246
776	604	739	131	166	298	637	123	561	890	701	131	288
407	824	285	927	235	029	020	693	109	638	896	498	486
618	200	842	317	347	457	092	399	155	282	524	001	843
471	229	629	918	141	025	058	833	729	715	300	293	346

(continued)

TABLE B.1   (*continued*)

820	378	250	979	367	537	907	692	685	185	282	276	351
466	615	866	805	239	138	372	292	787	350	852	026	586
694	817	184	101	428	277	646	584	674	582	545	348	245
378	839	626	595	447	107	403	426	421	177	414	308	652
126	857	405	000	284	823	588	927	228	559	376	230	786
728	088	417	036	171	603	988	692	995	285	056	823	211
719	148	527	527	334	371	726	435	651	414	908	170	684

Source   This table of random numbers was generated by the Computer Center at Northern Michigan University with the assistance of John Limback.

Glossary

Accidental Samples: Samples composed of those elements that are readily available or convenient to the researcher. Also called **Availability Samples** or **Convenience Samples**.

Alpha Error: The probability of rejecting a null hypothesis when it is actually true. Also called **Type I Error**.

Alpha Level: The probability at which the null hypothesis will be rejected; the amount of Type I error (probability of rejecting a true null hypothesis) that is acceptable in a research design.

Alphanumeric Variables: Variables consisting only of alphabetical or letter characters, being without the property of quantitative value, and treated as labels or strings by data analysis programs.

Anonymity: A situation in which no one, including the researcher, can link any data to a particular respondent.

Applied Research: Research designed to focus scientific research tools on a practical, real-world problem identified by some client with some practical outcome in mind.

Area Sampling: A multistage sampling technique that involves moving from larger clusters of units to smaller and smaller ones until the unit of analysis, such as *household* or *individual*, is reached. Also called **Cluster Sampling** or **Multistage Sampling**.

Authority: The unquestioned acceptance of someone's leadership or knowledge because of their social position, expertise, or experience.

Available Data Research: Research in which the data used was collected by someone other than the investigator for purposes that differ from the investigator's but that are available to be analyzed.

Availability Samples: Samples composed of those elements that are readily available or convenient to the researcher. Also called **Accidental Samples** or **Convenience Samples**.

Bar Graph: A frequency distribution graph for variables treated as nominal or ordinal, in which each value is plotted on the X axis as a separate bar whose height signifies the frequency of the value.

Basic Research: Research conducted for the purpose of advancing our knowledge about human behavior with little concern for any immediate or practical benefits that might result.

Beta Error: The probability of failing to reject a null hypothesis that is actually false. Also called **Type II Error**.

Bivariate Statistics: Statistics that describe the relationship between two variables.

Blocking: A two-stage system of assigning subjects to experimental and control groups whereby subjects are first aggregated into blocks according to one or more key variables; members of each block are then assigned to experimental and control conditions randomly.

Case Study: A detailed, descriptive account of one individual, situation, organization, group, or other entity.

Categorizing Strategies: Attempts in qualitative data analysis to generalize and abstract by generating concepts and even theories from the raw data.

Causality: The situation where an independent variable is the factor—or one of several factors—that produces variation in a dependent variable.

Central Limit Theorem: If a large number of random samples is taken from any population with a given mean and standard deviation, the distribution of sample statistics will be approximately normally distributed and the mean of the sampling distribution will be equal to the population parameter.

Closed-Ended Questions: Questions that provide respondents with a fixed set of alternatives from which they are to choose.

Cluster Sampling: A multistage sampling technique that involves moving from larger clusters of units to smaller and smaller ones until the unit of analysis, such as *household* or *individual*, is reached. Also called **Area Sampling** or **Multistage Sampling**.

Codebook: An inventory of all the individual items in a data collection instrument together with variable names, variable labels, special codes, and all other information necessary to understand the coding scheme.

Coding: The categorizing of behaviors or elements into a limited number of categories.

Coding Scheme: A plan by which the researcher organizes responses to a variable or an item, together with specifications for how that variable is defined for computerization.

Collaborative Research: Research in which the sponsors, subjects, and other stakeholders

in the research work with the researcher to plan and carry out the research, do the data analysis, prepare the report, and disseminate the results.

Common Sense: Practical judgments based on the experiences, wisdom, and prejudices of a people.

Comparative Research: Research that involves the comparison of two or more events, settings, societies, or cultures in order to determine the similarities and differences among them.

Computer-Assisted Interviewing (CAI): Using computer technology to assist in the completion of questionnaires and interviews.

Computer-Assisted Personal Interviewing (CAPI): Face-to-face interviewing in which the interviewer reads questions off the computer monitor and enters responses directly into the computer.

Computer-Assisted Qualitative Data Analysis (CAQDA): Using computers to do coding, hypothesis testing, and theory building in qualitative data analysis.

Computer-Assisted Self Interviewing (CASI): Respondents read questions from a survey on a computer screen and personally enter their responses at the computer keyboard.

Computer-Assisted Telephone Interviewing (CATI): Conducting an interview via telephone, with the interviewer reading questions from the computer monitor and recording answers directly into a computer file.

Concepts: Mental constructs or images developed to symbolize ideas, persons, things, events, or processes.

Concurrent Validity: A type of criterion validity in which the results of a newly developed measure are correlated with results from an existing measure.

Confidence Level: The probability of being correct when a population value is predicted to fall within a confidence interval.

Confidentiality: Ensuring that information about or data collected from those who participate in a study will not be made public in a way that they can be linked to an individual.

Construct Validity: A complex approach to establishing the validity of measures involving relating the measure to an overall theoretical framework in order to determine whether the instrument confirms a series of hypotheses derived from an existing and at least partially verified theory.

Content Analysis: A method of transforming the content of documents from a qualitative, unsystematic form to a quantitative, systematic form.

Content Validity: The extent to which a measuring device covers the full range of meanings or forms that would be included in a variable that is being measured. Also called **Sampling Validity.**

Context Map: One type of visual display in qualitative data analysis that describes in graphic form the physical or social setting that is the context of the observations.

Contextualizing Strategies: Attempts in qualitative data analysis to treat the data as a coherent whole and retain as much of the raw data as possible in order to capture the whole context.

Contingency Control: A multivariate analysis that examines the relationship between an independent variable and a dependent variable in each of the categories of a third, or test, variable. Also called **Table Elaboration.**

Contingency Question: A question that is answered only if certain responses were given to a previous question.

Continuous Variables: Variables that theoretically have an infinite number of values.

Control Condition: The condition in an experiment that does not receive the experimental stimulus.

Control Group: The subjects in an experiment who have all the same experiences as those in the experimental condition except that they are not exposed to the experimental stimulus.

Control Variables: Variables whose value is held constant in all conditions of an experiment.

Convenience Samples: Samples composed of those elements that are readily available or convenient to the researcher. Also called **Accidental Samples** or **Availability Samples.**

Correlational Research: Research in which the statistical analysis demonstrates that two variables are correlated, or vary together in some systematic way.

Cost–Benefit Analysis: An approach to program evaluation wherein program costs are related to program benefits expressed in dollars.

Cost Effective Analysis: An approach to program evaluation wherein program costs measured in dollars are related to program effects, with effects measured in whichever units can express the outcome or effects of the program.

Cover Letter: A letter that accompanies a mailed questionnaire and serves to introduce and explain it to the recipient.

Criterion Validity: A technique for establishing the validity of measures that involves demonstrating a correlation between the measure and some other standard.

Critical Social Science: The perspective that views society as consisting of various groups that compete over scarce resources, and that scientists are one of those groups, struggling to gain respect, legitimacy, power, and authority.

Critical Thinking: A mode of assessment or a reflective process that helps you assess information and decide on courses of action.

Cross-Sectional Research: Research based on data collected at one point in time.

Data Analysis: The process of deriving some meaning from the observations that have been made during a research project.

Data Archives: Data libraries that lend sets of data, much as ordinary libraries lend books.

Data Display: An organized presentation of data in qualitative data analysis that enables researchers and their audiences to draw some conclusions from the data and to move on to the next stage of the research.

Data Transformation: Preparing data for data analysis by modifying existing variables or generating new ones.

Debriefing: A session following an experiment when subjects are told the complete nature of the experiment, including any deceptions used, and subjects' reactions are assessed.

Deductive Reasoning: Deducing or inferring a specific conclusion from some general or abstract premises or propositions.

Demand Characteristics: Subtle, unprogrammed cues in an experiment that communicate to experimental subjects something about how they should behave.

Dependent Variable: The passive variable in a relationship or the one affected by an independent variable.

Descriptive Research: Research that attempts to discover facts or describe reality.

Descriptive Statistics: Procedures that assist in organizing, summarizing, and interpreting the data from a sample or a population.

Determinism: The belief that there is order in the universe, that there are reasons why everything happens, and that scientists can discover what those reasons are.

Dimensional Sampling: A sampling technique designed to enhance the representativeness of small samples by specifying all important variables and choosing a sample that contains at least one case to represent all possible combinations of variables.

Direct Costs: A proposed program budget or actual program expenditures.

Discrete Variables: Variables with a finite number of distinct and separate values.

Discriminant Validity: A part of construct validity where a valid measure should not show a correlation with variables or measures that are irrelevant to it or that theoretical considerations suggest it should not be correlated.

Discriminatory Power Score: A value calculated during construction of a Likert scale that indicates the degree to which each item discriminates between high scorers and low scorers on the entire scale.

Double-Blind Experiment: An experiment conducted in a way such that neither the subjects nor the experimenters know which groups are in the experimental and which are in the control condition.

Ecological Fallacy: Inferring something about individuals based on data collected at higher units of analysis, such as groups.

Empirical Generalization: A statement about how two variables will be related in most or all situations of a certain type.

Empowerment Research: Collaborative research in which one goal of the research is to develop new knowledge, skills, and capabilities among the various stakeholders in the research—in other words, to empower them.

Ethics: The study of what is proper and improper behavior, of moral duty and obligation.

Ethnographic Interview: Informal and unstructured interviews that can explore a wide range of topics, may last for a long time, and involve a nondirective interviewer. Also called an **In-Depth Interview**.

Ethnography: Like participant observation, it involves observing people in their natural environment and produces a description of a people and the cultural basis of their life.

Evaluability Assessment: A preliminary investigation into a program prior to its evaluation to determine those aspects of the program that are evaluable.

Evaluation Research: The use of scientific research methods to plan and assess social policies, social interventions, and social programs.

Experiential Knowledge: Knowledge gained through firsthand observation of events and based on the assumption that truth can be achieved through personal experience.

Experiment: A controlled method of observation in which the value of one or more independent variables is changed in order to assess its causal effect on one or more dependent variables.

Experimental Condition: The condition in an experiment that receives the experimental stimulus.

Experimental Group: Those subjects who are exposed to the experimental stimulus.

Experimental Research: Research designed such that changes in an independent variable are known to occur before changes are observed in a dependent variable.

Experimental Stimulus: The independent variable in an experiment that is manipulated by the experimenter to assess its effect on a dependent variable. Also called **Experimental Treatment**.

Experimental Treatment: The independent variable in an experiment that is manipulated by the experimenter to assess its effect on a dependent variable. Also called **Experimental Stimulus**.

Experimental Variability: Variation in a dependent variable produced by an independent variable.

Experimenter's Expectations: How experimenters wish an experiment to come out

and how those wishes may be communicated to experimental subjects.

Explanatory Research: Research with the goal to determine why or how something occurs.

External Validity: The extent to which causal inferences made in an experiment can be generalized to other times, settings, or people.

Extraneous Variability: Variation in a dependent variable from any source other than an experimental stimulus.

Face Validity: The degree to which there is a logical relationship between the variable and the proposed measure.

Facts: Some thing or event that has happened or is true.

Field Experiments: Experiments conducted in naturally occurring settings as people go about their everyday affairs.

Field Notes: The detailed recording of the observations made in a field setting.

Field Research: A type of qualitative research that involves observations made of people in their natural settings as they go about their everyday life.

Filter Question: A question whose answer determines which question a survey respondent goes to next.

Focus Group: An interview with a whole group of people at the same time. Also called a **Group Depth Interview**.

Formal Theories: The most abstract theories that deal with general social processes and mechanisms, without being tied to a specific social context.

Formative Program Evaluation: Evaluation research that focuses on the planning, development, and implementation of a program.

Foundation: A nonprofit, legally incorporated entity organized for the purpose of dispersing funds to projects that meet the guidelines of its charter.

Fraud (scientific): The deliberate falsification, misrepresentation, or plagiarizing of data, findings, or the ideas of others.

Frequency Polygon: A frequency distribution graph of an interval or ratio level variable in which interval midpoints are plotted on the X axis, the corresponding frequencies are plotted on the Y axis, and the resulting points are connected by a line.

Grant: The provision of money or other resources to be used for either research or other purposes.

Grounded Theory: A research methodology for developing theory by letting the theory emerge from the data, or be "grounded" in the data.

Group Depth Interview: An interview with a whole group of people at the same time. Also called a **Focus Group.**

Guttman Scale: A measurement scale in which the items have a fixed progressive order or intensity and that has the characteristic of reproducibility.

Histogram: A graph for depicting the frequency distribution of an interval- or ratio-level variable in which intervals of the variable are plotted on the X axis and the height of the bars represent frequencies or percentages.

Historical Research: The examination of the records and other evidence that have survived from the past.

Hypotheses: Testable statements of presumed relationships between two or more concepts.

Idiographic Explanations: Explanations that focus on a single person, event, or situation and attempt to specify all the conditions that helped to produce it.

Independent Variable: The presumed active or causal variable in a relationship.

In-Depth Interview: Informal and unstructured interviews that can explore a wide range of topics, may last for a long time, and involve a nondirective interviewer. Also called an **Ethnographic Interview.**

Index: A measuring technique that combines a number of items into a composite score.

Indicator: An observation assumed to be evidence of the attributes or properties of some phenomenon.

Inductive Reasoning: Inferring something about a whole group or class of objects from knowledge of one or a few members of that group or class.

Inferential Statistics: Procedures that allow us to make generalizations from sample data to the populations from which the samples were drawn.

Informed Consent: Telling potential research participants about all aspects of the research that might reasonably influence their decision to participate.

Internal Validity: An issue in experimentation concerning whether the independent variable actually produces the effect it appears to have on the dependent variable.

Interpretive Approaches: The perspective that social reality has a subjective component to it, that social reality arises out of social interaction and the exchange of social meanings, and that science must gain knowledge of that subjective dimension.

Interval Measures: Measures that classify observations into mutually exclusive categories with an inherent order and equal spacing between the categories.

Interview: A technique in which an interviewer reads questions to respondents and records their answers.

Interviewer Bias: A situation where the characteristics of the interviewer influence how respondents to an interview answer questions.

Interview Schedule: A document, used in interviewing, similar to a questionnaire, that contains instructions for the interviewer, specific questions in a fixed order, and transition phrases for the interviewer.

Item: A single indicator of a variable, such as an answer to a question or an observation of some behavior or characteristic.

Judgmental Sampling: A nonprobability sampling technique in which investigators use their judgment and prior knowledge to choose people for the sample who would best serve the purposes of the study. Also called **Purposive Sampling**.

Laboratory Experiments: Experiments conducted in artificial settings constructed in such a way that selected elements of the natural environment are simulated and features of the investigation are controlled.

Latent Coding: Coding of more abstract or implicit meanings in a document or medium in which the coder must infer whether a representation in a document is an instance of some broader phenomenon.

Levels of Measurement: Rules that define permissible mathematical operations on a given set of numbers produced by a measure.

Likert Scale: A measuring scale consisting of a series of statements followed by a series of response alternatives, often: strongly agree, agree, disagree, or strongly disagree.

Location Measures: Statistics that locate some point, or location, in a distribution.

Longitudinal Research: Research based on data gathered over an extended time period.

Macrolevel Research: Research that focuses on large-scale social structures and the social processes that occur among them.

Macrotheories: Theories that focus on society as a whole, on large social processes such as industrialization or urbanization, or on social institutions such as religion, politics, or the economy.

Manifest Coding: Coding the more objective or surface content of a document or medium.

Matching: A process of assigning subjects to experimental and control groups in which each subject is paired with a similar subject in the other group.

Matrix Display: A visual display in qualitative data analysis that looks very much like a contingency table with its rows and columns but that contains text rather than numerical quantities in the cells of the table.

Matrix Question: A question designed such that response alternatives are listed only once and each question or statement is followed by a box to check or a number or letter to circle.

Measurement: The process of describing abstract concepts in terms of specific indicators by the assignment of numbers or other symbols to these indicants, in accordance with rules.

Measures of Association: Statistics that describe the nature of the relationship between variables, especially the strength and direction of the relationship.

Measures of Central Tendency: Statistics, also known as averages, that summarize distributions of data by locating the "typical" or "average" value, or the center of the distribution of cases.

Measures of Dispersion: Statistics that indicate how dispersed or spread out the values of a distribution are.

Memos: Attempts at theorizing in qualitative data analysis—the researcher writes down ideas about the meanings of the codes that are emerging from the data and the relationships between the various codes.

Microlevel Research: Research that focuses on the face-to-face social interaction and process that occurs among individuals.

Microtheories: Theories that focus on the social interaction and social process that occurs among individuals or in small groups.

Misconduct (scientific): Scientific fraud, plus such activities as carelessness or bias in recording or reporting data, mishandling data, and incomplete reporting of results.

Missing Data: Incomplete data found in available data sets.

Multistage Sampling: A multiple-tiered sampling technique that involves moving from larger clusters of units to smaller and smaller ones until the unit of analysis, such as *household* or *individual*, is reached. Also called **Area Sampling** or **Cluster Sampling**.

Multitrait–Multimethod Approach: A particularly complex form of construct validity involving the simultaneous assessment of numerous measures and numerous concepts through the computation of intercorrelations.

Multivariate Statistics: Statistics that describe the relationships among three or more variables.

Narrative Analysis: Similar to profiles but with interviews, autobiographies, life histories, letters, and other personal materials used to form a descriptive narrative of a person's life or circumstances.

Needs Assessment: The collection of data to determine how many people need particular services and to assess the level of services or personnel that already exist to fill that need.

Network Analysis: A context map that shows connections and relationships between individuals, groups, statuses, or organizations by the use of lines and arrows.

Nominal Definitions: Verbal definitions in which one set of words or symbols is used to represent another set of words or symbols.

Nominal Measures: Measures that classify observations into mutually exclusive categories but with no ordering to the categories.

Nomothetic Explanations: Explanations that focus on a class of events and attempt to specify the conditions that seem common to all those events.

Nonprobability Samples: Samples in which the probability of each population element being included in the sample is unknown.

Nonreactive Observation: Observation in which those under study are not aware that they are being studied and the investigator does not change their behavior by his or her presence. Also called **Unobtrusive Observation.**

Nonreactive Research: Research in which those under study are not aware that they are being studied and the investigator does not change their behavior by his or her presence. Also called **Unobtrusive Research.**

Normal Distribution: A symmetrical, unimodal distribution in which the mode, median, and mean are identical and have the same proportion of cases between the same two ordinates.

Null Hypothesis: In statistical hypothesis testing, a statement that no relationship exists between two variables in the population or that there is no difference between a sample statistic and a population parameter.

Numeric Variable: A variable that has the property of quantitative value; a variable in a computer data file that can be used in computations.

Observations: To notice or record that something or event has occurred or is true.

Open-Ended Questions: Questions without a fixed set of alternatives, to which respondents write their own responses.

Operational Definitions: Definitions that indicate the precise procedures or operations to be followed in measuring a concept.

Opportunity Costs: The value of forgone opportunities incurred by funding one program as opposed to some other program.

Ordinal Measures: Measures that classify observations into mutually exclusive categories that have an inherent order to them.

Panel Study: Research in which data are gathered from the same people at different times.

Paradigms: General ways of thinking about how the world works and how we gain knowledge about the world.

Participant Observation: A method in which the researcher observes people in their natural environment and the researcher is a part of, and participates in, the activities of the people, group, or situation that is being studied.

Peer Review: The process in science whereby scientists with appropriate expertise review and make judgments about the worth of the grant proposals or research reports prepared by other scientists.

Physical Traces: Objects or evidence that result from people's activities that can be used as data to test hypotheses.

Pie Chart: A circular graph depicting the frequency distribution of a variable; the number of degrees of the circle for each section or slice represents the proportionate number of cases for each value of the variable.

Pilot Study: A trial run on a small scale of all procedures planned for a research project.

Policy Research: A type of applied research for policy development that focuses on broad policies that affect communities or society as a whole.

Population: All possible cases of what we are interested in studying.

Positivism: The perspective that the world exists independently of people's perceptions of it and that science uses objective techniques to discover what exists in the world.

Predictive Research: Research that attempts to make projections about what may occur in the future or in other settings.

Predictive Validity: A type of criterion validity wherein scores on a measure are used to predict some future state of affairs.

Preexperimental Designs: Experimental designs that lack the random assignment to conditions and/or the control groups that help control threats to internal validity.

Pretest: A preliminary application of the data-gathering technique for the purpose of determining its adequacy.

Privacy: The ability to control when and under what conditions others will have access to your beliefs, values, or behavior.

Probability Samples: Samples in which each element in the population has a known chance of being selected into the sample.

Probes: Follow-up questions used during an interview to elicit clearer and more complete responses.

Profiles: Vignettes of a person's experience, usually taken from in-depth interviews, that are stated largely in the person's own words, with relatively little interpretation or analysis by the researcher.

Program Evaluation: One kind of evaluation research whose goal is to assess how well social programs or social interventions operate and whether they achieve their goals.

Proportional Reduction in Error (PRE): A property of a measure of association that permits estimation of how much the independent variable contributes to reducing error in predicting values of the dependent variable.

Propositions: Statements linking concepts or variables, often statements about the relationship between two or more variables.

Public Opinion Polling: Measuring the attitudes of some identifiable group on some political or social issues of interest.

Purposive Sampling: A nonprobability sampling technique wherein investigators use their judgment and prior knowledge to choose people for the sample who would best serve the purposes of the study. Also called **Judgmental Sampling**.

Qualitative Research: Research that focuses on data in the form of words, pictures, descriptions, or narratives.

Quantitative Research: Research that uses numbers, counts, and measures of things.

Quasi-Experimental Designs: Special designs that use procedures other than random assignment to experimental and control groups to control many threats to internal validity.

Questionnaire: A set of written questions that people respond to directly on the form itself without the aid of an interviewer.

Quota Sampling: A type of nonprobability sampling that involves dividing the population into various categories and determining the number of elements to be selected from each category.

Random Assignment: A process for assigning subjects to experimental and control groups that uses chance to reduce the variation between experimental and control groups.

Random Errors: Measuring measurement errors that are neither consistent nor patterned.

Ratio Measures: Measures that classify observations into mutually exclusive categories with an inherent order, equal spacing between the categories, and an absolute zero point.

Reactivity: The degree to which the presence of a researcher influences the behavior being observed.

Reductionist Fallacy: Inferring something about groups, or other macrolevels of analysis, based on data collected from individuals.

Reflective Remarks: Reflections, interpretations, connections, or other thoughts that occur to the researcher and that are added to field notes while transcribing field notes or coding the data.

Reliability: The ability of a measure to yield consistent results each time it is applied.

Representative Sample: A sample that reflects the distribution of relevant variables in the target population accurately.

Research Design: A detailed plan outlining how observations will be made.

Research Hypothesis: The alternative statement to the null hypothesis, stating that a relationship is present between variables at the population level, or that there is a difference between a sample statistic and a population parameter.

Respondent-Driven Sampling: A variation of snowball sampling.

Response Bias: Responses to questions that are shaped by factors other than the person's true feelings, intentions, and beliefs.

Response Rate: The percentage of a sample that completes and returns a questionnaire or agrees to be interviewed.

Response Pattern Anxiety: The tendency for people to become anxious if they have to repeat the same response all the time and to change their responses to avoid doing so.

Response Set: The tendency for people to agree or disagree with statements regardless of the content of the statements.

Sample: One or more elements selected from a population.

Sampling Distribution: A distribution of sample statistics; it is a theoretical distribution that is distinguished by being a normal distribution whose mean is the population parameter.

Sampling Error: The extent to which the values of a sample differ from those of the population from which it was drawn.

Sampling Frame: A listing of all the elements in a population.

Sampling Validity: An approach to establishing validity of measures through determining whether a measuring device covers the full range of meanings that should be included in the variable being measured. Also called **Content Validity**.

Scale: A measuring technique, similar to an index, that combines a number of items into a composite score, but a scale has a built in intensity structure, potency, or natural levels of feeling to the items that make it up.

Science: A method of obtaining objective knowledge about the world through systematic observation.

Scientific Community: The people engaged in scientific work along with the values, norms, attitudes, behaviors, and language that guide and direct their work.

Secondary Data Analysis: The reanalysis of data previously collected for some other research project.

Semantic Differential: A scaling technique that involves respondents rating a concept or stimulus on a scale between a series of polar opposite adjectives.

Simple Random Sampling: A sampling technique wherein the target population is treated as a unitary whole and each element has an equal probability of being selected for the sample.

Snowball Sampling: A type of nonprobability sampling characterized by a few cases of the type we wish to study leading to more cases, which, in turn, lead to still more cases until a sufficient sample is achieved.

Social Desirability: Individuals' tendency to give socially desirable, popular answers to questions in order to present themselves in a good light.

Social Impact Assessment (SIA): Procedures for estimating the impact of programs and projects on people, groups, communities, and institutions.

Social Indicators Research: Research whose focus is the development of quantitative measures of important social phenomena.

Social Policy: All those laws, administrative procedures, court rulings, and other formal and informal social practices that shape the distribution of resources and the treatment of people in society.

Social Research: A systematic examination (or reexamination) of empirical data collected by someone firsthand, concerning the social, cultural, or psychological forces operating in a situation.

Sociogram: A context map that shows connections and relationships between individuals, groups, statuses, or organizations by the use of lines and arrows.

Stakeholders: All the people who have an interest in whether a social program operates or how well it does so.

Statistical Estimation: An inferential statistical procedure in which a sample statistic is used to estimate the value of a population parameter.

Statistics: A set of mathematical procedures for assembling, classifying, analyzing, and summarizing numerical data so that some meaning or information is derived.

Stratified Sampling: A sampling technique wherein the population is subdivided into strata with separate subsamples drawn from each strata.

Substantive Theories: Theories that are less abstract and focus on specific forms and contexts of social behavior.

Summated Rating Scales: Scales in which a respondent's score is determined by summing the numbers of questions answered.

Summative Program Evaluation: Program evaluation that centers on assessing the effectiveness and efficiency of programs and the extent to which program effects are generalizable to other settings and populations.

Survey: A data collection technique in which information is gathered from individuals, called respondents, by having them respond to questions or statements.

Systematic Errors: Measurement errors that are consistent and patterned.

Systematic Sampling: A type of simple random sampling wherein every *k*th element of the sampling frame is selected for the sample.

Table Elaboration: A multivariate analysis that examines the relationship between an independent variable and a dependent variable in each of the categories of a third, or test, variable. Also called **Contingency Control.**

Targeted Sampling: A variation on quota and purposive sampling in which procedures are used to ensure that people or groups with specified characteristics will appear in the sample.

Theoretical Sampling: A version of targeted sampling in which a developing theory is used to decide which units or individuals would be most appropriate to select for the sample.

Theoretical Sensitivity: Having data collection and analysis closely guided by emerging theoretical issues.

Theory: A set of interrelated, abstract concepts and statements that offer an explanation of some phenomenon.

Thurstone Scale: A measuring scale consisting of a series of items with a predetermined scale value to which respondents indicate their agreement or disagreement and which has the property of "equal-appearing intervals."

Time Sampling: A sampling technique used in observational research in which observations are made only during specified preselected times.

Traditional Knowledge: Knowledge based on custom, habit, and repetition.

Trend Study: Research in which data are gathered from different people at different times.

True Experimental Designs: Experimental designs that utilize randomization, control groups, and other techniques to control threats to internal validity.

Type I Error: The probability of rejecting a null hypothesis when it is actually true. Also called **Alpha Error.**

Type II Error: The probability of failing to reject a null hypothesis that is actually false. Also called **Beta Error.**

Unidimensional Scale: A multiple-item scale that measures one, and only one, variable.

Units of Analysis: The specific objects or elements whose characteristics we wish to describe or explain and about which data are collected.

Univariate Statistics: Statistics that describe the distribution of a single variable.

Unobtrusive Observation: Observation in which those under study are not aware that they are being studied and the investigator does not change their behavior by his or her presence. Also called **Nonreactive Observation.**

Unobtrusive Research: Research in which those under study are not aware that they are being studied and the investigator does not change their behavior by his or her presence. Also called **Nonreactive Research.**

Validity: The degree to which a measure accurately reflects the theoretical meaning of a variable.

Variables: Concepts that contain a number of categories or values.

Verification: The process of subjecting hypotheses to empirical tests to determine whether a theory is supported or refuted.

Verstehen: Subjective understanding, or the effort to view and understand a situation from the perspective of the people actually in that situation.

References

Achen, C. H. (1986). *The statistical analysis of quasi-experiments.* Berkeley, CA: University of California Press.

Adler, P. A. (1985). *Wheeling and dealing: An ethnography of an upper-level drug dealing and smuggling community.* New York: Columbia University Press.

Agnew, R. A. (1991). Longitudinal test of social control theory and delinquency. *Journal of Research in Crime and Delinquency, 28,* 126–156.

Alford, R. R. (1998). *The craft of inquiry: Theories, methods, evidence.* New York: Oxford University Press.

Alwin, D. F. (1997). Feeling thermometers versus 7-point scales: Which are better? *Sociological Methods and Research, 25,* 318–340.

Amato, P. R. (1987, May). Family processes in one-parent, step-parent, and intact families: The child's point of view. *Journal of Marriage and the Family, 49,* 327–337.

Anderson, A. B., Basilevsky, A., & Hum, D. P. J. (1983). Measurement: Theory and techniques. In P. H. Rossi, J. D. Wright, & A. B. Anderson (Eds.), *Handbook of survey research* (pp. 231–287). New York: Academic Press.

Anderson, B., Silver, B., & Abramson, P. (1988). The effects of the race of the interviewer on race-related attitudes of black respondents in SRC/CPS national election studies. *Public Opinion Quarterly, 52,* 289–324.

Annis, R. C., & Corenblum, B. (1986, December). Effect of test language and experimenter race on Canadian Indian children's racial and self-identity. *Journal of Social Psychology, 126,* 761–773.

Aquilino, W., & LoSciuto, L. (1990). Effects of interview mode on self-reported drug use. *Public Opinion Quarterly, 54,* 362–395.

Aquilino, W. S. (1993). Effects of spouse presence during the interview on survey responses concerning marriage. *Public Opinion Quarterly, 57,* 358–376.

Archer, D., Iritiani, B., Kimes, D. D., & Barrios, M. (1983). Face-ism: Five studies of sex differences in facial prominence. *Journal of Personality and Social Psychology, 45,* 725–735.

Ards, S. (1989). Estimating local child abuse. *Evaluation Research, 13,* 484–515.

Armstrong, G. (1993). . . . Like that Desmond Morris? In D. Hobbs & T. May (Eds.), *Interpreting the field.* Oxford, England: Oxford University Press.

Armstrong, J. S., & Luck, E. J. (1987, Summer). Return postage in mail surveys: A meta-analysis. *Public Opinion Quarterly, 51,* 233–248.

Arnold, D. O. (1970). Dimensional sampling: An approach for studying a small number of cases. *The American Sociologist, 5,* 147–150.

Ashcraft, N., & Scheflen, A. E. (1976). *People space: The making and breaking of human boundaries.* New York: Doubleday.

Bachman, J. G., & O'Malley, P. M. (1984). Yea-saying, nay-saying, and going to extremes: Black–white differences in response styles. *Public Opinion Quarterly, 48,* 491–501.

Backstrom, C. H., & Hursh-Cesar, G. D. (1981). *Survey research* (2nd ed.). New York: Macmillan.

Bailey, K. (1987). *Methods of social research* (3rd ed.). New York: Free Press.

Bailey, R. C., Hser, Y., Hsieh, S., & Anglin, M. D. (1994). Influences affecting maintenance and cessation of narcotics addiction. *Journal of Drug Issues, 24*, 249–272.

Bainbridge, W. S., Brent, E. E., Carley, K. M., Heise, D. R., Macy, M. W., Markovsky, B., & Skvoretz, J. (1994). Artificial social intelligence. In J. Hagan & K. S. Cook (Eds.), *Annual review of sociology: Vol. 20* (pp. 407–436). Palo Alto, CA: Annual Reviews.

Barnes, H. E. (Ed.). (1948). *An introduction to the history of sociology.* Chicago: University of Chicago Press.

Barnett, W. P., & Carroll, G. R. (1995). Modeling internal organizational change. In J. Hagan & K. S. Cook (Eds.), *Annual review of sociology: Volume 21* (pp. 217–236). Palo Alto, CA: Annual Reviews.

Bauer, D. G. (1999). *The "how to" grants manual: Successful grantseeking techniques for obtaining public and private grants* (4th ed.). Phoenix, AZ: Oryx.

Baumrind, D. (1985, February). Research using intentional deception. *American Psychologist, 40*(2), 165–174.

Beauchamp, T. L., Faden, R. R., Wallace Jr., R. J., & Walters, L. (Eds.). (1982). *Ethical issues in social science research.* Baltimore, MD: Johns Hopkins University Press.

Becerra, R. M., & Zambrana, R. E. (1985, Summer). Methodological approaches to research on Hispanics. *Social Work Research and Abstracts, 21*, 42–49.

Becker, H. S. (1953). Becoming a marijuana user. *American Journal of Sociology, 59*, 235–242.

Becker, H. S. (1967). Whose side are we on? *Social Problems, 14*, 239–247.

Bedell, J. R., Ward Jr., J. C., Archer, R. P., & Stokes, M. K. (1985, April 9). An empirical evaluation of a model of knowledge utilization. *Evaluation Review*, 109–126.

Bell, A. P., & Weinberg, M. S. (1978). *Homosexualities: A study of diversity among men and women.* New York: Simon & Schuster.

Benton, T. (1977). *Philosophical foundations of the three sociologies.* Boston: Routledge & Kegan Paul.

Berg, B. L. (1998). *Qualitative research methods for the social sciences* (3rd ed.). Boston: Allyn & Bacon.

Berk, R., & Rossi, P. (1990). *Thinking about program evaluation.* Newbury Park, CA: Sage.

Berk, R. A., Boruch, R. F., Chamber, D. L., Rossi, P. H., & Witte, A. D. (1985, August). Social policy experimentation: A position paper. *Evaluation Review, 9*, 387–430.

Berman, P., & Pauly, E. (1975). *Federal programs supporting educational change: Vol. 2. Factors affecting change agent projects.* Santa Monica, CA: Rand.

Bernard, H. R. (1994). *Research methods in anthropology: Qualitative and quantitative approaches.* Thousand Oaks, CA: Sage.

Berry, S. H., & Kanouse, D. E. (1987, Spring). Physician response to a mailed survey: An experiment in timing of payment. *Public Opinion Quarterly, 51*, 102–114.

Binder, A., Geis, G., & Bruce, D. (1988). *Juvenile delinquency: Historical, cultural, and legal perspectives.* New York: Macmillan.

Blaikie, N. (1993). *Approaches to social enquiry.* Cambridge, MA: Polity.

Bogdan, R., & Taylor, S. J. (1975). *Introduction to qualitative research methods.* New York: John Wiley & Sons.

Bogdan, R. C., & Biklen, S. K. (1992). *Qualitative research for education: An introduction to theory and methods* (2nd ed.). Boston: Allyn & Bacon.

Bohrnstedt, G. W. (1983). Measurement. In P. H. Rossi, J. D. Wright, & A. B. Anderson (Eds.), *Handbook of survey research* (pp. 69–121). New York: Academic Press.

Bollen, K. A., Entwisle, B., & Alderson, A. S. (1993). Macrocomparative research methods. In J. Blake & J. Hagen (Eds.), *Annual review of sociology, Volume 19* (pp. 321–351). Palo Alto, CA: Annual Reviews.

Bollen, K. A., & Paxton, P. (1998). Detection and determinants of bias in subjective measures. *American Sociological Review, 63*, 465–478.

Bonacich, E. (1990). Community forum discussion: What next? *ASA Footnotes, 18*, 7.

Bonjean, C. M., Hill, R. J., & McLemore, S. D. (1967). *Sociological measurement: An inventory of scales and indices.* San Francisco: Chandler.

Borgatta, E., & Bohrnstedt, G. (1981). Levels of measurement: Once over again. In G. Bohrnstedt & E. Borgatta (Eds.), *Social measurement: Current issues* (pp. 23–37). Beverly Hills, CA: Sage.

Boruch, R. F. (1997). *Randomized experiments for planning and evaluation.* Thousand Oaks, CA: Sage.

Bourgois, P. (1995). *In search of respect: Selling crack in El Barrio.* Cambridge, UK: Cambridge University Press.

Bourgois, P., Lettiere, M., & Quesada, J. (1997, May). Social misery and the sanction of substance abuse: Confronting HIV risk among homeless heroin addicts in San Francisco. *Social Problems, 44,* 155–173.

Bowles, R. T. (1981). *Social impact assessment in small communities.* Toronto, Canada: Butterworths.

Brabant, S., & Mooney, L. A. (1997). Sex role stereotyping in the Sunday comics: A twenty year update. *Sex Roles, 37,* 269–281.

Bradburn, N. M. (1983). Response effects. In P. H. Rossi, J. D. Wright, & A. B. Anderson (Eds.), *Handbook of survey research* (pp. 289–328). New York: Academic Press.

Bradburn, N. M., & Sudman, S. (1979). *Improving interview method and questionnaire design.* San Francisco: Jossey-Bass, 1979.

Brajuha, M., & Hallowell, L. (1986).Legal intrusion and the politics of fieldwork. *Urban Life, 14,* 454–479.

Brent, E., & Thompson, A. (1996). *Methodologist's toolchest: User's guide and reference manual.* Columbia, MO: Idea Works.

Bridge, R. G. (1974). *Nonresponse bias in mail surveys. Defense advanced research projects agency R-1501.* Santa Monica, CA: Rand.

Brodsky, S. L., & Smitherman, H. O. (1983). *Handbook of scales for research on crime and delinquency.* New York/London: Plenum.

Bronowski, J. (1978). *The origins of knowledge and imagination.* New Haven, CT: Yale University Press.

Browne, M. N., & Keeley, S. M. (1997). *Asking the right questions: A guide to critical thinking* (5th ed.). Englewood Cliffs, NJ: Prentice-Hall.

Brunner, G. A., & Carroll, S. J. (1967). Effect of prior telephone appointments on completion rates and response content. *Public Opinion Quarterly, 31,* 652–654.

Buckhout, R. (1974). Eyewitness testimony. *Scientific American, 231,* 23–31.

Buetow, S. A., Douglas, R. M., Harris, P., & McCulloch, C. (1996). Computer-assisted personal interviews: Development and experience of an approach in Australian general practice. *Social Science Computer Review, 14,* 205–212.

Burgess, R. G. (1984). *In the field: An introduction to field research.* London: Allen & Unwin.

Burnam, M. A., & Koegel, P. (1988, April). Methodology for obtaining a representative sample of homeless persons: The Los Angeles skid row study. *Evaluation Review, 12,* 117–152.

Cahalan, D. (1989). The Digest poll rides again! *Public Opinion Quarterly, 53,* 129–133.

Campbell, D. T., & Fiske, D. W. (1959). Convergent and discriminant validity by the multitrait–multimethod matrix. *Psychological Bulletin, 56,* 81–105.

Campbell, D. T., & Stanley, J. C. (1963). *Experimental and quasi-experimental designs for research.* Chicago: Rand McNally.

Carroll, L. (1946). *Through the looking glass.* New York: Random House.

Cartwright, A., & Tucker, W. (1967). An attempt to reduce the number of calls on an interview inquiry. *Public Opinion Quarterly, 31,* 299–302.

Catania, J. A., Binson, D., Canchola, J., Pollack, L. M., Hauck, W., & Coates, T. J. (1996). Effects of interviewer gender, interviewer choice, and item wording on responses to questions concerning sexual behavior. *Public Opinion Quarterly, 60,* 345–375.

Catlin, G., & Ingram, S. (1988). The effects of CATI on costs and data quality: A comparison of CATI and paper methods in centralized interviewing. In R. M. Groves, P. P. Blemer, L. E. Lyberg, J. T. Massey, W. L. Nicholls II, & J. Waksberg, (Eds.), *Telephone-survey methodology* (pp. 437–450). New York: JohnWiley & Sons.

Cavan, S. (1966). *Liquor license*. Chicago: Aldine.

Ceci, S. J., Peters D., & Plotkin, J. (1985, September). Human subjects review, personal values, and the regulation of social science research. *American Psychologist, 40,* 994–1002.

Chaiken, M. R., & Chaiken, J. M. (1984, April). Offender types and public policy. *Crime and Delinquency, 30,* 195–226.

Champion, D. J. (1981). *Basic statistics for social research* (2nd ed.). Scranton, PA: Chandler.

Charmaz, K. (1983). The grounded theory method: An explication and interpretation. In R. M. Emerson (Ed.), *Contemporary field research* (pp. 109–126). Prospect Heights, IL: Waveland Press.

Chen, H., & Rossi, P. H. (1980). The multi-goal, theory-driven approach to evaluation: A model linking basic and applied social science. *Social Forces, 59,* 106–122.

Christie, R., & Geis, F. L. (1970). *Studies in Machiavellianism*. New York. Academic Press.

Clandinin, D. J., & Connelly, F. M. (2000). *Narrative inquiry: Experience and story in qualitative research*. San Francisco: Jossey-Bass.

Clark, R., & Maynard, M. (1998). Using online technology for secondary analysis of survey research data—"act globally, think locally." *Social Science Computer Review, 16,* 58–71.

Coelho, R. J. (1983). *An experimental investigation of two multi-component approaches on smoking cessation*. Unpublished doctoral dissertation. East Lansing: Michigan State University.

Coffey, A, & Atkinson, P. (1996). *Making sense of qualitative data: Complementary research strategies*. Thousand Oaks, CA: Sage.

Coleman, J. S., Campbell, J. E., Hobson, L., McPartland, J., Mood A., Weinfield, F., & York, R. (1996). *Equality of educational opportunity*. Washington, DC: U.S. Government Printing Office.

Committee on the Status of Women in Sociology. (1986). *The treatment of gender in research*. Washington, DC: American Sociological Association.

Converse, J. M. (1987). *Survey research in the United States: Roots and emergence, 1890–1960*. Berkeley: University of California Press.

Converse, P. (1970). Attitudes and non-attitudes: Continuation of a Dialogue. In E. R. Tufte (Ed.), *The quantitative analysis of social problems* (pp. 168–189). Reading, MA: Addison-Wesley.

Cook, T. D. (1985). Postpositivist critical multiplism. In R. L. Shotland & M. M. Mark (Eds.), *Social science and social policy* (pp. 21–62). Beverly Hills, CA: Sage.

Cook, T. D., & Campbell, D. T. (1979). *Quasi-experimentation: Design and analysis issues for field settings*. Chicago: Rand McNally.

Cotter, P. R., Cohen, J., & Coulter, P. B. (1982). Race-of-interviewer effects in telephone interviews. *Public Opinion Quarterly, 46,* 278–284.

Coulton, C. J. (1979). Developing an instrument to measure person–environment fit. *Journal of Social Service Research, 3,* 159–174.

Couper, M. P., & Rowe, B. (1996). Evaluation of a computer-assisted self-interview component in a computer-assisted personal interview survey. *Public Opinion Quarterly, 60,* 89–105.

Couvalis, G. (1997). *The philosophy of science: Science and objectivity*. London: Sage.

Cox, D. E., & Sipprelle, C. N. (1971). Coercion in participation as a research subject. *American Psychologist, 26,* 726–728.

Cronbach, L. J. (1951). Coefficient alpha and the internal structure of tests. *Psychometrica, 16,* 197–334.

Cronbach, L. J., & Meehl, P. (1955). Construct validity in psychological tests. *Psychological Bulletin, 52,* 281–302.

Davis, D. W. (1997). Nonrandom measurement error and race of interviewer effects among African Americans. *Public Opinion Quarterly, 61,* 183–207.

Dawson, J. A., & D'Amico, J. J. (1985, April). Involving program staff in evaluation studies: A strategy for increasing information use and enriching the data base. *Evaluation Review, 9,* 173–188.

Decker, S. H., & Van Winkle, B. (1996). *Life in the gang: Family, friends, and violence*. Cambridge, UK: Cambridge University Press.

DeMaio, T. J. (1984). Social desirability and survey measurement: A review. In C. F. Turner &

E. Martin (Eds.), *Surveying subjective phenomena: Vol. 2* (pp. 257–282). New York: Russell Sage Foundation.

DeMartini, J. (1982). Basic and applied sociological work: Divergence, convergence, or peaceful coexistence? *Journal of Applied Behavioral Science, 18*, 203–215.

de Neufville, J. I. (1981). Social Indicators. In M. E. Olsen & M. Micklin (Eds.). *Handbook of applied sociology* (pp. 5–23). New York: Praeger.

Denzin, N. (1989a). *Interpretive interactionism.* Newbury Park, CA: Sage.

Denzin, N. (1989b). *The research act: A theoretical introduction to sociological methods* (3rd ed.). Englewood Cliffs, NJ: Prentice-Hall.

Deutch, S. J., & Alt, F. B. (1977). The effect of Massachusetts's gun control law on gun-related crimes in the city of Boston. *Evaluation Quarterly, 1*, 543–567.

DeVault, M. L. (1996). Talking back to sociology: Distinctive contributions of feminist methodology. In J. Hagan & K. S. Cook (Eds.), *Annual review of sociology: Vol. 22.* Palo Alto, CA: Annual Reviews.

DeVellis, R. F. (1991). *Scale development: Theories and applications.* Newbury Park, CA: Sage.

DeWeese, L. Carroll III. (1983). Social research in industry. In H. E. Freeman, R. R. Dynes, P. H. Rossi, & W. F. Whyte (Eds.), *Applied sociology* (pp. 165–171). San Francisco: Jossey-Bass.

Dignan, M., Michielutte, R., Sharp, P., Bahnson J., Young, L., & Beal P. (1990). The role of focus groups in health education for cervical cancer among minority women. *Journal of Community Health, 15*, 369–375.

Dillman, D. (1991). The design and administration of mail surveys. In W. R. Scott & J. Blake (Eds.), *Annual review of sociology: Volume 17* (pp. 225–249). Palo Alto, CA: Annual Reviews.

Dobratz, B. A., & Shanks-Meile, S. L. (1997). *"White power, white pride!" The white separatist movement in the United States.* New York: Twayne Publishers.

Dohan, D., & Sánchez-Jankowski, M. (1998). Using computers to analyze ethnographic field data: Theoretical and practical considerations. In J. Hagan & K. S. Cook (Eds.), *Annual review*

of sociology: Volume 24 (pp. 477–516). Palo Alto, CA: Annual Reviews.

Dressel, P. L., & Petersen, D. M. (1982, August). Becoming a male stripper: Recruitment, socialization, and ideological development. *Work and Occupations, 9*, 387–406.

Dukes, R. L., Stein, J. A., & Ullman, J. B. Long-term impact of drug abuse resistance education (D.A.R.E). *Evaluation Review, 21*, 483–500.

Dunford, F., Huizinga, D., & Elliott, D. (1989). *The Omaha domestic violence police experiment.* Washington, DC: National Institute of Justice.

Durkheim, E. (1938). *Rules of the sociological method* (S. Solovay and J. Mueller, Trans.). Chicago: University of Chicago Press, 1938.

Edwards, A. L., & Kilpatrick, F. P. (1948). A technique for the construction of attitude scales. *Journal of Applied Psychology, 32*, 374–384.

Eichler, M. (1988). *Nonsexist research methods: A practical guide.* Boston: Allen & Unwin.

Elms, A. C. (1982). Keeping deception honest: Justifying conditions for social scientific research strategies. In T. L. Beauchamp, R. R. Faden, R. J. Wallace, Jr., & L. Walters (Eds.), *Ethical issues in social science research.* Baltimore, MD: Johns Hopkins University Press, 1982.

Engler, R. L., Covell, J. W., Friedman, P. J., Kitcher, P. S., & Peters, R. M. (1987, November 26). Misrepresentation and responsibility in medical research. *New England Journal of Medicine, 317*, 1383–1389.

Erikson, K. T. (1966). *Wayward Puritans.* New York: John Wiley & Sons.

Erikson, K. T. (1967). A comment on disguised observation in sociology. *Social Problems, 14*, 366–373.

Estroff, S. E. (1981). *Making it crazy: An ethnography of psychiatric clients in an American community.* Berkeley: University of California Press.

Evans, W. (1996, Fall). Computer-supported content analysis: Trends, tools, and techniques. *Social Science Computer Review, 14*, 269–279.

Farley, R. (1984). *Blacks and whites: Narrowing the gap.* Cambridge, MA: Harvard University Press.

Fay, B. (1987). *Critical social science: Liberation and its limits.* Ithaca, NY: Cornell University Press.

Feldman, R. A., & Caplinger, T. E. (1977). Social work experience and client behavioral change: A multivariate analysis of process and outcome. *Journal of Social Service Research, 1,* 5–33.

Fernandez, M., & Ruch-Ross, H. S. (1998). Ecological analysis of program impact: A site analysis of programs for pregnant and parenting adolescents in Illinois. *Journal of Applied Sociology, 15,* 104–133.

Ferree, M., & Hall, E. (1990). Visual images of American society: Gender and race in introductory sociology textbooks. *Gender and Society, 4,* 500–533.

Ferrell, J. (1995). Urban graffiti: Crime, control, and resistance. *Youth and Society, 28,* 73–92.

Festinger, L., Riecken, H., & Schachter, S. (1956). *When prophecy fails.* New York: Harper & Row.

Fine, G. A. (1987). *With the boys: Little league baseball and preadolescent culture.* Chicago: University of Chicago Press.

Finsterbusch, K. (1981). Impact assessment. In M. E. Olsen & M. Micklin (Eds.), *Handbook of applied sociology* (pp. 24–47). New York: Praeger.

Finsterbusch, K., & Motz, A. B. (1980). *Social research for policy decisions.* Belmont, CA: Wadsworth.

Fischer, J., & Corcoran, K. (1994). *Measures for clinical practice: A sourcebook* (2nd ed.). New York: Free Press.

Fontana, A., & Frey, J. H. (1994). Interviewing: The art of science. In N. K. Denzin & Y. S. Lincoln (Eds.), *Handbook of qualitative research* (pp. 361–376). Thousand Oaks, CA: Sage.

Fortune, A. E. (1979). Communication in task-centered treatment. *Social Work, 24,* 390–397.

Fowler, F. J. Jr. (1988). *Survey research methods* (rev. ed.). Beverly Hills, CA: Sage.

Fowler, F. Jr., & Mangione, T. (1990). *Standardized survey interviewing.* Newbury Park, CA: Sage.

Fox, R. J., Crask, M. R., & Kim, J. (1988). Mail Survey response rate: A meta-analysis of selected techniques for inducing response. *Public Opinion Quarterly, 52,* 467–491.

Frey, J. H. (1986, Summer). An experiment with a confidentiality reminder in a telephone survey. *Public Opinion Quarterly, 50,* 267–269.

Frey, J. H. (1989). *Survey research by telephone* (2nd ed.). Newbury Park, CA: Sage.

Fulford, K. W. M., & Howse, K. (1993). Ethics of research with psychiatric patients: Principles, problems, and the primary responsibilities of researchers. *Journal of Medical Ethics, 19,* 85–91.

Furnham, A., Abramsky, S., & Gunter, B. A. (1997). Cross-cultural content analysis of children's television advertisements. *Sex Roles, 37,* 91–99.

Galison, P., & Stump, D. (Eds.). (1996). *The disunity of science: Boundaries, contexts, and power.* Stanford, CA: Stanford University Press.

Gallagher, B. J. III. (1987). *The sociology of mental illness* (2nd ed.). Englewood Cliffs, NJ: Prentice-Hall.

Galtung, J. (1967). *Theory and methods of social research.* New York: Columbia University Press.

Geertz, C. (1971). *Islam observed: Religious development in Morocco and Indonesia.* Chicago: University of Chicago Press.

Gelles, R. J. (1978). Methods for studying sensitive family topics. *American Journal of Orthopsychiatry, 48,* 408–424.

Gelles, R. J. (1987). What to learn from cross-cultural and historical research on child abuse and neglect: An overview. In R. J. Gelles & J. B. Lancaster (Eds.), *Child abuse and neglect: Biosocial dimensions* (pp. 15–30). New York: Aldine de Gruyter.

Gibbs, L. (1983, Fall). Evaluation researcher: Scientist or advocate? *Journal of Social Service Research, 7,* 81–92.

Gilligan, C. (1982). *In a different voice: Psychological theory and women's development.* Cambridge, MA: Harvard University Press.

Gilljam, M., & Granberg, D. (1993). Should we take don't know for an answer? *Public Opinion Quarterly, 57,* 348–357.

Gilmore, S., & Crissman, A. (1997). Video games: Analyzing gender identity and violence in this new virtual reality. *Studies in Symbolic Interaction, 21,* 181–199.

Glaser, B. (1978). *Theoretical sensitivity.* Mill Valley, CA: Sociology Press.

Glaser, B., & Strauss, A. (1967). *The discovery of grounded theory.* Chicago: Aldine.

Goduka, I. N. (1990). Ethics and politics of field research in South Africa. *Social Problems, 37,* 329–340.

Goldscheider, F. K., & Waite, L. J. (1991). *New families, no families? The transformation of the American home.* Berkeley: University of California Press.

Goode, E. (1994). *Deviant behavior* (4th ed.). Englewood Cliffs, NJ: Prentice-Hall.

Goode, W. J., & Hatt, P. K. (1952). *Methods in social research.* New York: McGraw-Hill.

Gorden, R. (1992). *Basic interviewing skills.* Itasca, IL: F. E. Peacock.

Gorden, R. L. (1987). *Interviewing: Strategies, techniques, and tactics* (4th ed.). Chicago: Dorsey Press.

Gordon, M. M. (1988). *The scope of sociology.* New York: Oxford University Press.

Gottschalk, L. R. (1969). *Understanding history: A primer of historical method.* New York: A. Knopf.

Gouldner, A. (1976). The dark side of the dialectic: Toward a new objectivity. *Sociological Inquiry, 46,* 3–16.

Goyder, J. (1985, Summer). Face-to-face interviews and mailed questionnaires: The net difference in response rate. *Public Opinion Quarterly, 49,* 234–252.

Graham, K., LaRocque, L., Yetman, R., Ross, T. J., & Guistra, E. (1980). Aggression and barroom environments. *Journal of Studies on Alcohol, 41,* 277–292.

Graham, K. R. (1977). *Psychological research: Controlled interpersonal research.* Monterey, CA: Brooks/Cole.

Gray, B. H. (1982). The regulatory context of social and behavioral research. In T. L. Beauchamp, R. R. Faden, R. J. Wallace, Jr., &. Walters, L. (Eds.), *Ethical issues in social science research* (pp. 329–355). Baltimore, MD: Johns Hopkins University Press.

Greene, J. C. (1994). Qualitative program evaluation: Practice and promise. In N. K. Denzin & Y. S. Lincoln (Eds.), *Handbook of qualitative research* (pp. 530–544). Thousand Oaks, CA: Sage.

Greene, J. G. (1988, April). Stakeholder participation and utilization of program evaluation. *Evaluation Review, 12,* 91–116.

Greenley, J. R., & Schoenherr, R. A. (1981). Organization effects on client satisfaction with humaneness of service. *Journal of Health and Social Behavior, 22,* 2–18.

Groves, R. M. (1989). *Survey errors and survey costs.* New York: John Wiley & Sons.

Groves, R. M., Blemer, P. P., Lyberg, L. E., Massey, J. T., Nicholls II, W. L., & Waksberg, J. (Eds.). (1988). *Telephone survey methodology.* New York: John Wiley & Sons.

Groves, R. M., & Couper, M. P. (1998). *Nonresponse in household interview surveys.* New York: John Wiley & Sons.

Groze, V. (1991). Adoption and single parents: A review. *Child Welfare, 70,* 321–332.

Guba, E. G., & Lincoln, Y. S. (1989). *Fourth generation evaluation.* Newbury Park, CA: Sage.

Guba, E. G., & Lincoln, Y. S. (1994). Competing paradigms in qualitative research. In N. K. Denzin & Y. S. Lincoln (Eds.), *Handbook of qualitative research* (pp. 105–117). Thousand Oaks, CA: Sage.

Gubrium, J. F., & Holstein, J. A. (1997). *The new language of qualitative methods.* New York: Oxford University Press.

Gunther, M. (1993, June 15). Women, minorities get short end on small screen, study says. *Detroit Free Press,* p. 1C.

Guttman, L. (1944). A basis for scaling qualitative data. *American Sociological Review, 9,* 139–150.

Guttman, L. (1950). The basis for scalogram analysis. In S. A. Stouffer, L. Guttman, E. A. Suchman, P. F. Lazarsfeld, S. A. Star, & J. A. Clausen, (Eds.), *Measurement and prediction* (pp. 60–90). Princeton, NJ: Princeton University Press.

Hagstrom, W. (1965). *The scientific community.* New York: Basic Books.

Halfpenny, P. (1982). *Positivism and sociology: Explaining social life.* London: Allen & Unwin.

Harding, S. (1986). *The science question in feminism.* Ithaca, NY: Cornell University Press.

Harvey, L. (1990). *Critical social research*. London: Urwin Hyman.

Hathaway, A. D. (1997). Marijuana and tolerance: Revisiting Becker's sources of control. *Deviant Behavior, 18*, 103–124.

Heckathorn, D. D. (1997, May). Respondent-driven sampling: A new approach to the study of hidden populations. *Social Problems, 44*, 174–199.

Heise, D. R. (1970). The semantic differential and attitude research. In G. F. Summers (Ed.), *Attitude measurement* (pp. 235–253). Chicago: Rand McNally.

Hendershott, A. B., & Norland, S. (1990). Theory based evaluation: An assessment of the implementation and impact of an adolescent parenting program. *Journal of Applied Sociology, 7*, 35–48.

Henry, G. T. (1990). *Practical sampling*. Newbury Park, CA: Sage.

Henry, G. T. (1994). *Graphing data: Techniques for display and analysis*. Newbury Park, CA: Sage.

Higgins, P. C., & Johnson, J. M. (1988). *Personal sociology*. New York: Praeger.

Hirschel, J., Hutchison III, I., & Dean, C. (1992). The failure of arrest to deter spouse abuse. *Journal of Research in Crime and Delinquency, 29*, 7–33.

Hirschi, T. (1969). *Causes of delinquency*. Berkeley: University of California Press.

Holmes, R., & DeBurger, J. (1988). *Serial murder*. Newbury Park, CA: Sage.

Holstein, J. A., & Gubrium, J. F. (1994). Phenomenology, ethnomethodology, and interpretive practice. In N. K. Denzin & Y. S. Lincoln (Eds.), *Handbook of qualitative research* (pp. 262–272). Thousand Oaks, CA: Sage, 1994.

Holstein, J. A., & Gubrium, J. F. (1995). *The active interview*. Thousand Oaks, CA: Sage.

Holsti, O. R. (1969). *Content analysis for the social sciences and humanities*. Reading, MA: Addison-Wesley.

Homans, G. C. (1964). Contemporary Theory in Sociology. In R. E. L. Faris (Ed.), *Handbook of modern sociology* (pp. 951–977). Chicago: Rand McNally.

Honey, M. (1984). *Creating Rosie the riveter: Class, gender, and propaganda*. Amherst: University of Massachusetts Press.

Hooker, E. (1957). The adjustment of the male overt homosexual. *Journal of Projective Techniques, 21*, 18–31.

Horowitz, R. (1987, December). Community tolerance of gang violence. *Social Problems, 34*, 437–450.

Hoshino, G., & Lynch, M. M. (1981). Secondary analysis of existing data. In R. M. Grinnell Jr. (Ed.), *Social work research and evaluation* (pp. 333–347). Itasca, IL: Peacock.

Huber, B. (1981). New human subjects policies announced; exemptions outlined. *ASA Footnotes, 9*, 1.

Hughson, J. (1998). Among the thugs: The "new ethnographies" of football supporting cultures. *International Review for the Sociology of Sport, 33*, 43–57.

Humphreys, L. (1970). *Tearoom trade: Impersonal sex in public places*. Chicago: Aldine-Atherton.

Intrieri, R. C., von Eye, A., & Kelly, J. A. (1995). The aging semantic differential: A confirmatory factor analysis. *The Gerontologist, 35*, 616–621.

Irwin, D. M., & Bushnell, M. M. (1980). *Observational strategies for child study*. New York: Holt, Rinehart & Winston.

Jackson, B. O., & Mohr, L. B. (1986, August). Rent subsidies: An impact evaluation and an application of the random-comparison-group design. *Evaluation Review, 10*, 483–517.

James, J., & Bolstein, R. (1990). The effect of monetary incentives and follow-up mailings on the response rate and response quality in mail surveys. *Public Opinion Quarterly, 54*, 346–361.

Javeline, D. (1999). Response effects in polite cultures: A test of acquiescence in Kazakhstan. *Public Opinion Quarterly, 63*, 1–28.

Jencks, C. (1992). *Rethinking social policy: Race, poverty, and the underclass*. Cambridge, MA: Harvard University Press.

Jenkins-Hall, K., & Osborn, C. A. (1994). The conduct of socially sensitive research: Sex offenders as participants. *Criminal Justice and Behavior, 21*, 325–340.

Johnson, J. M. (1975). *Doing field research*. New York: Free Press.

Johnson, S. M., & Bolstad, O. D. (1973). Methodological issues in naturalistic observation: Some problems and solutions for field research. In L. A. Hamerlynck, L. C. Handy, & E. J. Mash (Eds.), *Behavior change* (pp. 7–67). Champaign, IL: Research Press.

Jones, J. H. (1992). *Bad blood: The Tuskegee syphilis experiment*. New York: Free Press.

Jorgensen, D. (1989). *Participant observation: A methodology for human studies*. Newbury Park, CA: Sage.

Kane, E. W., & Macauley, L. J. (1993). Interviewer gender and gender attitudes. *Public Opinion Quarterly, 57,* 1–28.

Kang, M-E. (1997). The portrayal of women's images in magazine advertisements: Goffman's gender analysis revisited. *Sex Roles, 37,* 979–996.

Karoly, L. A. (1998). *Investing in our children: What we know and don't know about the costs and benefits of early childhood interventions*. Santa Monica, CA: Rand.

Kassebaum, G., Ward, D., & Wilner, D. (1971). *Prison treatment and parole survival*. New York: John Wiley & Sons.

Katz, D. (1949). An analysis of the 1948 polling predictions. *Journal of Applied Psychology, 33,* 15–28.

Katz, J. (1972). *Experimentation with human beings*. New York: Russell Sage Foundation.

Kaye, B. K., & Johnson, T. J. (1999). Research methodology: Taming the cyber frontier: techniques for improving online surveys. *Social Science Computer Review, 17,* 323–337.

Kelle, U. (Ed.). (1995). *Computer-aided qualitative data analysis: Theory, methods, and practice*. Thousand Oaks, CA: Sage.

Kelly, J. R., & McGrath, J. E. (1988). *On time and method*. Beverly Hills, CA: Sage.

Kemeny, J. G. (1959). *A philosopher looks at science*. Princeton, NJ: Van Nostrand.

Kenny, G. K. (1986, June). The metric properties of rating scales employed in evaluation of research: An empirical examination. *Evaluation Review, 10,* 397–408.

Kercher, K. (1992). Quasi-experimental research designs. In E. F. Borgatta & M. L. Borgatta (Eds.), *Encyclopedia of sociology*, Volume 3 (pp. 1595–1613). New York: Macmillan.

Kiesler, S., & Sproull, L. (1986). Response effects in the electronic survey. *Public Opinion Quarterly, 50,* 402–413.

Kimmel, A. (1988). *Ethics and values in applied social research*. Newbury Park, CA: Sage.

Kirk, J., & Miller, M. L. (1986). *Reliability and validity in qualitative research*. Beverly Hills, CA: Sage.

Kirk, R. E. (1982). *Experimental design: Procedures for the behavioral sciences* (2nd ed.). Belmont, CA: Brooks/Cole.

Kish, L. (1965). *Survey sampling*. New York: John Wiley & Sons.

Klockars, C. B. (1979). Dirty hands and deviant subjects. In C. B. Klockars & F. W. O'Connor (Eds.), *Deviance and decency: The ethics of research with human subjects* (pp. 261–282). Beverly Hills, CA: Sage.

Knorr, K. (1981). *The manufacture of knowledge: An essay on the constructivist and contextual nature of science*. Oxford, England: Pergamon Press.

Knudsen, D. D., Pope, H., & Irish, D. P. (1967). Response differences to questions on sexual standards: An interview–questionnaire comparison. *Public Opinion Quarterly, 31,* 290–297.

Kohfeld, C., & Leip, L. (1991, April). Bans on concurrent sale of beer and gas: A California case study. *Sociological Practice Review, 2,* 104–115.

Korn, J. H. (1997). *Illusions of reality: A history of deception in social psychology*. Albany: State University of New York Press.

Krathwohl, D. R. (1988). *How to prepare a research proposal: Guidelines for funding and dissertations in the social and behavioral sciences* (3rd ed.). New York: Distributed by Syracuse University Press.

Krippendorff, K. (1980). *Content analysis: An introduction to its methodology*. Beverly Hills, CA: Sage.

Krueger, R. A. (1994). *Focus groups: A practical guide for applied research* (2nd ed.). Thousand Oaks, CA: Sage.

Kuhn, T. (1970). *The structure of scientific revolutions* (2nd ed.). Chicago: University of Chicago Press.

Kuo, W. H., & Tsai, Y. (1986, June). Social networking, hardiness, and immigrant's mental health. *Journal of Health and Social Behavior, 27,* 133–149.

Lake, D. G., Miles, M. B., & Earle, R. B. (Eds.). (1973). *Measuring human behavior: Tools for the assessment of social functioning.* New York: Teachers College Press.

Lally, J. J. (1977). Social determinants of differential allocation of resources to disease research: A comparative analysis of crib death and cancer research. *Journal of Health and Social Behavior, 18,* 125–138.

Landrine, H. (1985, July). Race and class stereotypes of women. *Sex Roles, 13,* 65–75.

Lantz, H. R., Schmitt, R., Britton, M., & Snyder, E. C. (1968). Pre-industrial patterns in the colonial family in America: A content analysis of colonial magazines. *American Sociological Review, 33,* 413–426.

Larson, C. J. (1995). Theory and applied sociology. *Journal of Applied Sociology, 12,* 13–29.

Lavrakas, P. J. (1987). *Telephone survey methods: Sampling, selection, and supervision.* Beverly Hills, CA: Sage.

Lawless, E. J. (1991). Methodology and research notes: Women's life stories and reciprocal ethnography as feminist and emergent. *Journal of Folklore Research, 28,* 35–60.

Leary, W. E. (1996, October 18). Needle study to get review on questions about ethics. *New York Times,* p. A12.

Lenihan, K. (1977). *Unlocking the second gate.* Department of Labor. Washington, DC: U.S. Government Printing Office. (R & D Monograph 45)

Leo, R. A. (1995, Spring). Trial and tribulations: Courts, ethnography, and the need for an evidentiary privilege for academic researchers. *The American Sociologist, 26,* 113–134.

Leo, R. A. (1996, Spring). The ethics of deceptive research roles reconsidered: A response to Kai Erikson. *The American Sociologist, 27,* 122–128.

Levin, J., & Spates, J. L. (1970). Hippie values: An analysis of the underground press. *Youth and Society, 2,* 59–73.

Lieberson, S., & Silverman, A. R. (1965). The precipitants and underlying conditions of race riots. *American Sociological Review, 30,* 887–898.

Likert, R. A. (1932). Technique for the measurement of attitudes. *Archives of Psychology, 21*(140).

Locke, L. F., Silverman, S. J., & Spirduso, W. W. (1998). *Reading and understanding research.* Thousand Oaks, CA: Sage.

Locke, L. F., Spirduso, W. W., & Silverman, S. J. (2000). *Proposals that work: A guide for planning dissertations and grant proposals* (4th ed.). Thousand Oaks, CA: Sage.

Lockhart, L. L. (1991). Spousal violence: A cross-racial perspective. In R. L. Hampton (Ed.), *Black family violence: Current research and theory* (pp. 85–101). Lexington, MA: Lexington Books.

Lofland, J., & Lofland, L. H. (1995). *Analyzing social settings* (3rd ed.). Belmont, CA: Wadsworth.

Luebke, B. (1989). Out of focus: Images of women and men in newspaper photographs. *Sex Roles, 20,* 121–133.

Lynch, M., & Bogen, D. (1997). Sociology's asociological "core": An examination of textbook sociology in light of the sociology of scientific knowledge. *American Sociological Review, 62,* 481–493.

Mandell, N. (1988). The least-adult role in studying children. *Journal of Contemporary Ethnography, 16,* 433–467.

Manson, S. M. (1986). Recent advances in American Indian mental health research: Implications for clinical research and training. In M. R. Miranda & H. H. L. Kitano (Eds.), *Mental health research and practice in minority communities: Development of culturally sensitive training programs* (pp. 51–89). Rockville, MD: U.S. Department of Health and Human Services. (DHHS Publication No. [ADM] 86–1466)

Marin, G., & Marin, B. (1991). *Research with Hispanic populations.* Newbury Park, CA: Sage.

Marshall, C., & Rossman, G. B. (1995). *Designing qualitative research* (2nd ed.). Thousand Oaks, CA: Sage.

Martin, K. A. (1998). Becoming a gendered body: Practices of preschools. *American Sociological Review, 63,* 494–511.

Martinez, T. A. (1997). Popular culture as oppositional culture: Rap as resistance. *Sociological Perspectives, 40,* 265–286.

Marx, K. (1964). *Selected writings in sociology and philosophy* (T. B. Bottomore and M. Rubel, Eds.; T. B. Bottomore, Trans.). Baltimore, MD: Penguin, 1964. (Original work published 1848).

Marx, K. (1967). *Das kapital.* New York: International. (Original work published 1867–1895.)

Mathios, A., Avery, R., Bisogni, C., & Shanahan, J. (1998). Alcohol portrayal on prime-time television: Manifest and latent messages. *Journal of Studies on Alcohol, 59,* 305–310.

Mavis, B. E., & Brocato, J. J. (1998). Postal surveys versus electronic mail surveys: The tortoise and the hare revisited. *Evaluation & the Health Professions, 21,* 395–408.

Maxwell, J. A. (1996). *Qualitative research design: An interactive approach.* Thousand Oaks, CA: Sage.

Mayo, J. K., Hornick, R. C., & McAnany, E. G. (1976). *Educational reform with television: The El Salvador experience.* Palo Alto, CA: Stanford University Press.

McCord, J. A. (1978). Thirty-year follow-up of treatment effects. *American Psychologist, 33,* 284–289.

McDowell, I., & Newell, C. (1996). *Measuring health: A guide to rating scales and questionnaires* (2nd ed.). New York: Oxford University Press.

McGranahan, D. V., & Wayne, I. (1948). German and American traits reflected in popular drama. *Human Relations, 1,* 429–455.

McKillip, J. (1987). *Need analysis: Tools for human services and education.* Beverly Hills, CA: Sage.

Melton, G. B. (1990). Certificates of confidentiality under the Public Health Service Act: Strong protection but not enough. *Violence and Victims, 5,* 67–71.

Melton, G. B., & Gray, J. N. (1988). Ethical dilemmas in AIDS research: Individual privacy and public health. *American Psychologist, 43,* 60–64.

Merriam, S. B. (1998). *Qualitative research and case study applications in education.* San Francisco: Jossey-Bass.

Merton, R. K. (1973). *The sociology of science.* Chicago: University of Chicago Press.

Miles, M. B., & Huberman, A. M. (1994). *Qualitative data analysis: An expanded sourcebook* (2nd ed.). Thousand Oaks, CA: Sage.

Miller, D. C. (1991). *Handbook of research design and social measurement* (5th ed.). Newbury Park, CA: Sage.

Miller, L. P. (1987, September/October). The application of research to practice: A critique. *American Behavioral Scientist, 30,* 70–80.

Miller, R. W. (1987). *Fact and method: Explanation, confirmation and reality in the natural and social sciences.* Princeton: Princeton University Press.

Miner, J. T., & Miner, L. E. (1999). A guide to proposal planning and writing. In *Directory of research grants.* Phoenix: Oryx Press.

Mitroff, I. (1974). Norms and counter-norms in a select group of the *Apollo* moon scientists: A case study of ambivalence of scientists. *American Sociological Review, 39,* 579–595.

Moody, E. J. (1976). Urban witches. In J. E. Nash & J. P. Spradley (Eds.), *Sociology: A descriptive approach* (pp. 442–453). Chicago: Rand McNally.

Moore Jr., B. (1996). *Social origins of dictatorship and democracy: Lord and peasant in the making of the modern world.* Boston: Beacon Press.

Morash, M., & Greene, J. R. (1986, April). Evaluating women on patrol: A Critique of contemporary wisdom. *Evaluation Review, 10,* 230–255.

Morgan, D. L. (1994). *Focus groups as qualitative research* (2nd ed.). Thousand Oaks, CA: Sage.

Morrison, R. S. (1990). Disreputable science: Definition and detection. *Journal of Advanced Nursing, 15,* 911–913.

Morth, M., & Johnson, P. J. (1999). *Foundation fundamentals: A guide for grantseekers* (6th ed.). New York: The Foundation Center.

Moser, C. A., & Kalton, G. (1972). *Survey methods in social investigation* (2nd ed.). New York: Basic Books.

Mueller, D. J. (1986). *Measuring social attitudes: A handbook for researchers and practitioners.* New York: Teachers College Press.

Murray, L., Donovan, R., Kail, B. L., & Medvene, L. J. (1980). Protecting human subjects during social work research: Researchers' opinions. *Social Work Research and Abstracts, 16,* 25–30.

Myers, D. J. (1997, February). Racial rioting in the 1960s: An event history analysis of local conditions. *American Sociological Review, 62,* 94–112.

Nagel, E. (1961). *The structure of science.* New York: Harcourt, Brace & World.

National Institute of Justice. (1997, September). Criminal justice research under the Crime Act—1995 to 1996. *NIJ research report.* [Online, retrieved 2/5/1999 from (http://www.ncjrs.org/-txtfiles/166142.txt)].

Nesbary, D. (2000). *Survey research and the World Wide Web.* Boston: Allyn & Bacon.

Newman, I., & Benz, C. R. (1998). *Qualitative–quantitative research methodology: Exploring the interactive continuum.* Carbondale: Southern Illinois University Press.

Nielson, J. M. (Ed.). (1989). *Feminist research methods.* Boulder, CO: Westview Press.

Nigro, G. N., Hill, D. E., Gelbein, M. E., & Clark, C. L. (1988, June). Changes in the facial prominence of women and men over the last decade. *Psychology of Women Quarterly, 12,* 225–235.

Nunnally, J. C. (1978). *Psychometric theory.* New York: McGraw-Hill.

Nyden, P., Figert, A., Shibley, M., & Burrows, D. (1997). *Building community: Social science in action.* Thousand Oaks, CA: Pine Forge Press.

Ogilvie, D. M., Stone, P. J., & Shneidman, E. S. (1966). Some characteristics of genuine versus simulated suicide notes. In P. J. Stone, D. C. Dunphy, M. S. Smith, & D. M. Ogilvie (Eds.), *The general inquirer: A computer approach to content analysis* (pp. 527–535). Cambridge, MA: MIT Press, 1966.

O'Hear, A. (1989). *An introduction to the philosophy of science.* New York: Oxford University Press.

Oksenberg, L., Coleman, L., & Cannell, C. F. (1986, Spring). Interviewer's voices and refusal rates in telephone surveys. *Public Opinion Quarterly, 50,* 97–111.

Olesen, V. (1994). Feminism and models of qualitative research. In N. K. Denzin and Y. S. Lincoln (Eds.), *Handbook of Qualitative Research* (pp. 158–174). Thousand Oaks, CA: Sage.

Olzak, S., & Shanahan, S. (1996, March). Deprivation and race riots: An extension of Spilerman's analysis. *Social Forces, 74,* 931–961.

Orne, M. T. (1962). On the social psychology of the psychological experiment: With particular reference to demand characteristics and their implications. *American Psychologist, 17,* 776–783.

Orr, L. L. (1998). *Social experimentation: Evaluating public programs with experimental methods.* Thousand Oaks, CA: Sage.

Ortner, S. B. (1984). The founding of the first Sherpa nunnery and the problem of "women" as an analytic category. In V. Patraka & L. Tilly (Eds.), *Feminist re-visions* (pp. 98–134). Ann Arbor: University of Michigan Women's Studies Program.

Osgood, C. E., Suci, G. J., & Tannenbaum, P. H. (1957). *The measurement of meaning.* Urbana: University of Illinois Press.

Parker, S., & Kleiner, R. J. (1966). *Mental illness in the urban negro community.* New York: The Free Press.

Patton, M. Q. (1981). *Creative evaluation.* Beverly Hills, CA: Sage.

Patton, M. Q. (1982). *Practical evaluation.* Beverly Hills, CA: Sage.

Patton, M. Q. (1987). Evaluation's political inherency: Practical implications for design and use. In D. J. Palumbo (Ed.), *The politics of program evaluation* (pp. 100–145). Newbury Park, CA: Sage.

Patton, M. Q. (1990). *Qualitative evaluation and research methods* (2nd ed.). Newbury Park, CA: Sage.

Paul, R. (1993). *Critical thinking: What every person needs to survive in a rapidly changing world.* Rohnert Park, CA: Foundation for Critical Thinking.

Pearlin, L., Lieberman, M., Menaghan, E., & Mullan, J. (1981). The stress process. *Journal of Health and Social Behavior, 22,* 337–356.

Peirce, K. (1997). Women's magazine fiction: A content analysis of the roles, attributes, and occupations of the main characters. *Sex Roles, 37,* 581–593.

Pepler, D. J., & Craig, W. M. (1995). A peek behind the fence: Naturalistic observations of aggressive children with remote audiovisual recording. *Developmental Psychology, 31,* 548–553.

Pescosolido, B. A., Grauerholz, E., & Milkie, M. A. (1997). Culture and conflict: The portrayal of blacks in U.S. children's picture books through the mid- and late-twentieth century. *American Sociological Review, 62,* 443–464.

Peterson, R. R. (1996). A re-evaluation of the economic consequences of divorce. *American Sociological Review, 61,* 528–536.

Peterson, S., & Kroner, T. (1992). Gender biases in textbooks for introductory psychology and human development. *Psychology of Women Quarterly, 16,* 17–36.

Peterson, S., & Lach, M. (1990). Gender stereotypes in children's books: Their prevalence and influence on cognitive and affective development. *Gender and Education, 2(2),* 185–197.

Peyrot, M. (1985, January). Coerced voluntarism: The micropolitics of drug treatment. *Urban Life, 13,* 343–365.

Phillips, D. P. (1983, August). The impact of mass media violence on U.S. homicides. *American Sociological Review, 48,* 560–568.

Piazza, T. (1993). Meeting the challenge of answering machines. *Public Opinion Quarterly, 57,* 219–231.

Piliavin, I., & Briar, S. (1964). Police encounters with juveniles. *American Journal of Sociology, 70,* 206–214.

Piliavin, I., Masters, S., & Corbett, T. (1977). Factors influencing errors in AFDC payments. *Social Work Research and Abstracts, 15,* 3–17.

Pollner, M., & Adams, R. E. (1997). The effect of spouse presence on appraisals of emotional support and household strain. *Public Opinion Quarterly, 61,* 615–626.

Polsky, N. (1967). *Hustlers, beats, and others.* Chicago: Aldine.

Powdermaker, H. (1966). *Stranger and friend: The way of an anthropologist.* New York: Norton.

Prothro, J. W. (1956). Verbal shifts in the American presidency: A content analysis. *American Political Science Review, 50,* 726–739.

Purcell, P., & Stewart, L. (1990). Dick and Jane in 1989. *Sex Roles, 22(3–4),* 177–185.

Rabinow, P., & Sullivan, W. M. (Eds.). (1987). *Interpretive social science: A second look.* Berkeley: University of California Press.

Rainwater, L., & Pittman, D. J. (1967). Ethical problems in studying a politically sensitive and deviant community. *Social Problems, 14,* 357–366.

Rank, M. R. (1994). *Living on the edge: The realities of welfare in America.* New York: Columbia University Press.

Rathje, W., & Murphy, C. (1992). *Rubbish! The archaeology of garbage.* New York: HarperCollins.

Ratzan, R. M. (1981). The experiment that wasn't: A case report in clinical geriatric research. *The Gerontologist, 21,* 297–302.

Rea, L., & Parker, R. (1992). *Designing and conducting survey research.* San Francisco: Jossey-Bass.

Reece, R., & Siegal, H. (1986). *Studying people: A primer in the ethics of social research.* Macon, GA: Mercer University Press.

Reese, H. W., & Fremouw, W. J. (1984). Normal and normative ethics in behavioral science. *American Psychologist, 39,* 863–876.

Reichardt, C. S., Trochim, W. M. K., & Cappelleri, J. C. (1995). Reports of the death of regression-discontinuity analysis are greatly exaggerated. *Evaluation Review, 19,* 39–63.

Reid, P., & Finchilescu, G. (1995). The disempowering effects of media violence against women on college women. *Psychology of Women Quarterly, 19,* 397–411.

Reinharz, S. (1992). *Feminist methods in social research.* New York: Oxford University Press.

Reiss, A. K., & Rhodes, L. (1967). An empirical test of differential association theory. *Journal of Research in Crime and Delinquency, 4,* 28–42.

Rennison, C. M. (1999). Crime victimization 1998: Changes 1997–98 with trends 1993–98. *National crime victimization survey.* Washington, DC: U.S. Department of Justice, Bureau of Justice Statistics.

Reynolds, P. D. (1979). *Ethical dilemmas and social science research*. San Francisco: Jossey-Bass.

Roberts, C. W. (Ed.). (1997). *Text analysis for the social sciences: Methods for drawing inferences from texts and transcripts*. Mahwah, NJ: Lawrence Erlbaum.

Robinson, J., Shaver, P., & Wrightsman, L. (Eds.). (1991). *Measures of personality and social psychological attitudes*. San Diego, CA: Academic Press.

Roscoe, P. B. (1995). The perils of "positivism" in cultural anthropology. *American Anthropologist, 97*, 492–504.

Rosenberg, M. (1968). *The logic of survey analysis*. New York: Basic Books.

Rosenhan, D. L. (1973). On being sane in insane places. *Science, 179*, 250–258.

Rosenthal, R. (1967). Covert communication in the psychological experiment. *Psychological Bulletin, 67*, 356–367.

Rosenthal, R. (1991). Replication in behavioral research. In J. W. Neuliep (Ed.), *Replication research in the social sciences* (pp. 1–30). Newbury Park, CA: Sage.

Rosenthal, R., & Rosnow, R. (1975). *The volunteer subject*. New York: John Wiley & Sons.

Rossi, P. H. (1987, November/December). No good applied social research goes unpunished. *Society, 25*, 74–79.

Rossi, P. H., Berk, R. A., & Lenihan, K. J. (1980). *Money, work, and crime: Experimental evidence*. New York: Academic Press.

Rossi, P. H., Freeman, H. E., & Lipsey, M. W. (1999). *Evaluation: A systematic approach* (6th ed.). Thousand Oaks, CA: Sage.

Rossi, P. H., Wright, J. D., & Anderson, A. B. (1983). Sample surveys: History, current practice, and future prospects. In P. H. Rossi, J. D. Wright, & A. B. Anderson (Eds.), *Handbook of survey research* (pp. 1–20). New York: Academic Press.

Rossi, P. H., Wright, J. D., Fisher, G. A., & Willis, G. (1987, March 13). The urban homeless: estimating composition and size. *Science, 235*, 1336–1341.

Roth, D., Bean, J., Lust N., & Saveanu, T. (1985, February). *Homelessness in Ohio: A study of people in need*. Columbus: Ohio Department of Mental Health, Office of Program Evaluation and Research.

Rudd, M. D., Rajab, M. H., Orman D. T., Stulman, D. A., Joiner, T., & Dixon, W. (1996). Effectiveness of an outpatient intervention targeting suicidal young adults: Preliminary results. *Journal of Consulting and Clinical Psychology, 64*, 179–190.

Rule, J. B. (1978). *Insight and social betterment: A preface to applied social science*. New York: Oxford University Press.

Runcie, J. F. (1980). *Experiencing social research* (rev. ed.). Homewood, IL: Dorsey Press.

Rutman, L. (1984). Introduction. In L. Rutman (Ed.), *Evaluation research methods: A basic guide* (2nd ed.) (pp. 9–25). Beverly Hills, CA: Sage.

Scarce, R. (1994). (No) trial (but) tribulations: When courts and ethnography conflict. *Journal of Contemporary Ethnography, 23*, 123–149.

Schaeffer, N. C. (1980). Evaluating race-of-interviewer effects in a national survey. *Sociological Methods and Research, 8*, 400–419.

Scheaffer, R. L., Mendenhall, W., & Ott, L. (1996). *Elementary survey sampling* (5th ed.). Belmont, CA: Wadsworth.

Scheper-Hughes, N. (1992). *Death without weeping: The violence of everyday life in Brazil*. Berkeley: University of California Press.

Schiebinger, L. L. (1999). *Has feminism changed science?* Boston: Harvard University Press.

Schuman, H., & Presser, S. (1979). The open and closed question. *American Sociological Review, 44*, 692–712.

Schutte, N. S., & Malouff, J. M. (1995). *Sourcebook of adult assessment strategies (applied clinical psychology)*. New York: Plenum Press.

Scott, C. (1961). Research on mail surveys. *Journal of the Royal Statistical Society, 124*, 143–195. (Series A)

Scott, J. A. (1990). *Matter of record: Documentary sources in social research*. Oxford, England: Polity Press.

Scriven, M. (1991). *Evaluation thesaurus* (4th ed.). Newbury Park, CA: Sage.

Sechrest, L., & Belew, J. (1983). Nonreactive measures of social attitudes. *Applied Social Psychology Annual: 4*. Beverly Hills, CA: Sage.

Seidman, I. E. (1991). *Interviewing as qualitative research*. New York: Teachers College Press.

Seidman, I. E. (1998). *Interviewing as qualitative research* (2nd ed.). New York: Teachers College Press.

Seiler, L. H., & Hough, R. L. (1970). Empirical comparisons of the Thurstone and Likert techniques. In G. F. Summers (Ed.), *Attitude measurement* (pp. 159–173). Chicago: Rand McNally.

Shapin, S. (1995). Here and everywhere: Sociology of scientific knowledge. In J. Hagan & J. S. Cook (Eds.), *Annual Review of Sociology, Volume 21* (pp. 289–321). Palo Alto, CA: Annual Reviews Inc.

Shearing, C. D. (1973). How to make theories untestable: A guide to theorists. *The American Sociologist, 8*, 33–37.

Sheatsley, P. B. (1983). Questionnaire construction and item writing. In P. H. Rossi, J. D. Wright, & A. B. Anderson (Eds.), *Handbook of survey research* (pp. 195–230). New York: Academic Press.

Sheley, J. F. (1976). A study in self-defeat: The public health venereal disease clinic. *Journal of Sociology and Social Welfare, 4*, 114–124.

Sherman, L. (1992). *Policing domestic violence: Experiments and dilemmas*. New York: Free Press.

Sherman, L., & Berk, R. A. (1984). The specific deterrent effects of arrest for domestic assault. *American Sociological Review, 49*, 261–271.

Shilts, R. (1987). *And the band played on: Politics, people, and the AIDS epidemic*. New York: St. Martin's Press.

Shupe Jr., A. D., & Bromley, D. G. (1980). Walking a tightrope: Dilemmas of participant observation of groups in conflict. *Qualitative Sociology, 2*, 3–21.

Singer, E., VonThurn, D. R., & Miller, E. R. (1995). Confidentiality and response: A quantitative review of the experimental literature. *Public Opinion Quarterly, 59*, 446–459.

Skidmore, W. (1979). *Theoretical thinking in sociology*. Cambridge, England: Cambridge University Press.

Skipper, Jr., J. K. (1979). Stripteasers: A six-year history of public reaction to a study. In L. Carga & J. H. Ballantine (Eds.), *Sociological footprint* (pp. 12–18). Boston: Houghton Mifflin.

Skocpol, T. (1984). Emerging agendas and recurrent strategies in historical sociology. In T. Skocpol (Ed.), *Vision and method in historical sociology* (pp. 356–391). Cambridge, England: Cambridge University Press.

Smart, B. (1976). *Sociology, phenomenology, and Marxian analysis: A critical discussion of the theory and practice of a science of society*. Boston: Routledge & Kegan Paul.

Smith, A. W. (1987, March/April). Problems and progress in the measurement of black public opinion. *American Behavioral Scientist, 30*, 441–455.

Smith, D. (1984). Discovering facts and values: The historical sociology of Barrington Moore. In T. Skocpol (Ed.), *Vision and method in historical sociology* (pp. 313–355). Cambridge, England: Cambridge University Press.

Smith, D. (1997, May 1). Study looks at portrayal of women in media. *New York Times*, p. A17.

Smith, H. W. (1981). *Strategies of social research* (2nd ed.). Englewood Cliffs, NJ: Prentice-Hall.

Smith, J. K. (1989). *The nature of social and educational inquiry: Empiricism versus interpretation*. Norwood, NJ: Ablex.

Smith, S. H., & McLean, D. D. (1988). *ABC's of grantsmanship*. Reston, VA: American Alliance for Health, Physical Education, Recreation, and Dance.

Smith, S. S., & Richardson, D. (1983). Amelioration of deception and harm in psychological research: The important role of debriefing. *Journal of Personality and Social Psychology, 44*(5), 1075–1082.

Smith, T. W. (1987, Spring). That which we call welfare by any other name would smell sweeter: An analysis of the impact of question wording on response patterns. *Public Opinion Quarterly, 51*, 75–83.

Smith, T. W. (1997). The impact of the presence of others on a respondent's answers to questions. *International Journal of Public Opinion Research, 9*, 33–47.

ocolar, M. J. (1981). *Greater use of exemplary education programs could improve education for disadvantaged children.* Washington, DC: U.S. Government Printing Office. (GAO Report to Congress).

Spilerman, S. (1976, October). Structural characteristics of cities and the severity of racial disorders. *American Sociological Review, 41,* 771–793.

Squire, P. (1988). Why the 1936 *Literary Digest* poll failed. *Public Opinion Quarterly, 52,* 125–133.

Stein, J. (1964). *Fiddler on the roof.* New York: Crown.

Stevens, R. E., Wrenn, B., Ruddick, M. E., & Sherwood, P. K. (1997). *The marketing research guide.* New York: Haworth Press.

Stevens, S. S. (Ed.). (1951). *Handbook of experimental psychology.* New York: John Wiley & Sons.

Stine, G. J. (1998). *Acquired immune deficiency syndrome: Biological, medical, social, and legal issues* (3rd ed.). Upper Saddle River, NJ: Prentice-Hall.

Stoecker, R. (1999). Making connections: Community organizing, empowerment planning, and participatory research in participatory evaluation. *Sociological Practice: A Journal of Clinical and Applied Sociology, 1,* 209–231.

Stott, C., & Reicher, S. (1998). How conflict escalates: The Inter-group dynamics of collective football crowd "violence." *Sociology, 32,* 353–377.

Straus, M. A. (1969). *Family measurement techniques.* Minneapolis: University of Minnesota Press.

Straus, M. A., & Gelles, R. J. (1988). How violent are American families? Estimates from the national family violence resurvey and other studies. In G. T. Hotaling, D. Finkelhor, J. T. Kirkpatrick, & M. A. Straus (Eds.), *Family abuse and its consequences: New directions in research* (pp. 14–36). Newbury Park, CA: Sage.

Straus, M. A., Hamby, S. L., Boney-McCoy, S., & Sugarman, D. B. (1996, May). The revised conflict tactics scales (CTS2): Development and preliminary data. *Journal of Family Issues, 17,* 283–316.

Strauss, A., & Corbin, J. (1994). Grounded theory methodology: An overview. In N. K. Denzin &

Y. S. Lincoln (Eds.), *Handbook of qualitative research* (pp. 273–285). Thousand Oaks, CA: Sage.

Strauss, A. L. (1987). *Qualitative analysis for social scientists.* Cambridge (England)/New York: Cambridge University Press.

Street, D., Vinter, R. D., & Perrow, C. (1966). *Organizations for treatment: A comparative study of institutions for delinquents.* New York: Free Press of Glencoe.

Strickland, S. P. (1972). *Politics, science, and dread disease.* Cambridge, MA: Harvard University Press.

Sudman, S. (1965). Time allocation on survey interviews and other field occupations. *Public Opinion Quarterly, 29,* 638–648.

Sudman, S. (1976). *Applied sampling.* New York: Academic Press.

Sudman, S. (1985, June). Mail surveys of reluctant professionals. *Evaluation Research, 9,* 349–360.

Sudman, S., & Bradburn, N. M. (1982). *Asking questions.* San Francisco: Jossey-Bass.

Sullivan, C. (1991). The provision of advocacy services to women leaving abusive partners. *Journal of Interpersonal Violence, 6,* 41–54.

Sullivan, T. J. (1992). *Applied sociology: Research and critical thinking.* Boston: Allyn & Bacon.

Sullivan, T. J. (2000). *Introduction to social problems* (5th ed.). Boston: Allyn & Bacon.

Taylor, S. (1982). *Durkheim and the study of suicide.* New York: St. Martin's Press.

Teich, A., & Frankel, M. (1992). *Good science and responsible scientists: Meeting the challenge of fraud and misconduct in science.* Washington, DC: American Association for the Advancement of Science.

Thorne, B. (1993). *Gender play: Girls and boys in school.* New Brunswick, NJ: Rutgers University Press.

Thurner, A. W. (1984). *Rebels on the range: The Michigan copper miners' strike of 1913–1914.* Lake Linden, MI: John H. Forster Press.

Thurstone, L. L., & Chave, E. J. (1929). *The measurement of attitudes.* Chicago: University of Chicago Press.

Tourangeau, R., & Smith, T. W. (1996). Asking sensitive questions: The impact of data collection mode, question format, and question context. *Public Opinion Quarterly, 60,* 275–304.

Treas, J., & VanHilst, A. (1976). Marriage and re-marriage rates among older Americans. *The Gerontologist, 16,* 132–140.

U.S. Bureau of the Census. (1998). *Statistical abstract of the United States: 1998* (118th ed.). Washington, DC: U.S. Government Printing Office.

Vidich, A. J., & Bensman, J. (1958). *Small town in mass society.* Princeton, NJ: Princeton University Press.

Vidich, A. J., & Lyman, S. M. (1994). Qualitative methods: Their history in sociology and anthropology. In N. K. Denzin & Y. S. Lincoln (Eds.), *Handbook of qualitative research* (pp. 23–59). Thousand Oaks, CA: Sage.

Wallace, W. (1971). *The logic of science in sociology.* New York: Aldine de Gruyter.

Ward, D. A., & Kassebaum, G. G. (1972). On biting the hand that feeds: Some implications of sociological evaluations of correctional effectiveness. In C. H. Weiss (Ed.), *Evaluating action programs: Readings in social action and education* (pp. 300–310). Boston: Allyn & Bacon.

Warriner, K., Goyder, J., Gjertsen, H., Hohner, P., & McSpurren, K. (1996). Charities, no; lotteries, no; cash, yes: Main effects and interactions in a Canadian incentives experiment. *Public Opinion Quarterly, 60,* 542–562.

Warwick, D. P., & Lininger, C. (1975). *The sample survey: Theory and practice.* New York: McGraw-Hill.

Watters, J. K., & Biernacki, P. (1980). Targeted sampling: Options for the study of hidden populations. *Social Problems, 36,* 416–430.

Webb, E. J., Campbell, D. T., Schwartz, R. D., Sechrest, L., & Grove, J. B. (1981). *Nonreactive measures in the social sciences.* Boston: Houghton Mifflin.

Weber, M. (1946). Science as a vocation. In H. H. Gerth & C. W. Mills (Eds.), *Max Weber: Essays in sociology.* New York: Free Press. (Original work published 1922).

Weber, M. (1957). *The theory of social and economic organization* (A. M. Henderson and T. Parsons, Trans.). New York: Free Press. (Original work published 1925).

Weber, R. P. (1990). *Basic content analysis* (2nd ed.). Beverly Hills, CA: Sage.

Weeks, M. F. (1988). Call scheduling with CATI: Current capabilities and methods. In R. G. Groves, P. P. Blemer, L. E. Lyberg, J. T. Massey, W. L. Nicholls II, & J. Waksberg, (Eds.), *Telephone survey methodology* (pp. 403–420). New York: John Wiley & Sons.

Weinberg, M. S. (1968). Sexual modesty, social meanings, and the nudist camp. In M. Truzzi (Ed.), *Sociology and everyday life* (pp. 212–220). Englewood Cliffs, NJ: Prentice-Hall.

Weinstein, R. M. (1982). The mental hospital from the patient's point of view. In W. R. Gove (Ed.), *Deviance and mental illness* (pp. 121–146). Newbury Park, CA: Sage.

Weiss, C. H. (1997). How can theory-based evaluation make greater headway? *Evaluation Review, 21,* 501–524.

Weiss, C. H. (1998). *Evaluation: Method for studying programs and policies* (2nd ed.). Upper Saddle River, NJ: Prentice-Hall.

Weitzman, E. B., & Miles, M. B. (1995). *Computer programs for qualitative data analysis.* Thousand Oaks, CA: Sage.

Weitzman, L. J. (1985). *The divorce revolution: The unexpected social and economic consequences for women and children in America.* New York: Free Press.

Welch, M., Fenwick, M., & Roberts, M. (1997). Primary definitions of crime and moral panic: A content analysis of experts' quotes in feature newspaper articles on crime. *Journal of Research in Crime and Delinquency, 34,* 474–494.

Whalen, J., & Zimmerman, D. H. (1998). Observations on the display and management of emotion in naturally occurring activities: The case of "hysteria" in calls to 9-1-1. *Social Psychology Quarterly, 61,* 141–159.

Whyte, W. F. (1958). Freedom and responsibility in research: The Springdale case. *Human Organization, 17,* 1–2.

Wilkinson, S. (1998). Focus groups in feminist research: Power, interaction, and the co-construction of meaning. *Women's Studies International Forum, 21,* 111–125.

Williams Jr., J. A., Vernon J., Williams, M., & Malecha, K. (1987, March). Sex role socialization in picture books: An update. *Social Science Quarterly, 68*, 148–156.

Williams, T. (1989). *The cocaine kids: The inside story of a teenage drug ring*. Reading, MA: Addison-Wesley.

Wilson, T. (1970). Normative and interpretive paradigms in sociology. In J. Douglas (Ed.), *Understanding everyday life: Toward the reconstruction of sociological knowledge* (pp. 57–79). New York: Aldine.

Wolf, D. L. (Ed.). (1996). *Feminist dilemmas in fieldwork*. Boulder, CO: Westview Press.

Wolfgang, M. E. (1981). Confidentiality in criminological research and other ethical issues. *Journal of Criminal Law and Criminology, 72*, 345–361.

Wuebben, P. L., Straits, B. C., & Schulman, G. I. (1974). *The experiment as a social occasion*. Berkeley, CA: Glendessary Press.

Wysocki, D. K. (1999, April). *Feminist methods: The use of the Internet to find difficult to reach populations*. Paper presented at the annual meetings of the Midwest Sociological Society, Minneapolis.

Wysong, E., & Wright, D. W. (1995). A decade of DARE: Efficacy, politics and drug education. *Sociological Focus, 28*, 283–311.

Xu, M., Bates, B. J., & Schweitzer, J. C. (1993). The impact of messages on survey participation in answering machine households. *Public Opinion Quarterly, 57*, 232–237.

Yinger, J. (1995). *Closed doors, opportunities lost: The continuing costs of housing discrimination*. New York: Russell Sage Foundation.

Young, C., Savola, K., & Phelps, E. (1991). *Inventory of longitudinal studies in the social sciences*. Newbury Park, CA: Sage.

Zeller, R. A., & Carmines, E. G. (1980). *Measurement in the social sciences: The link between theory and data*. Cambridge, MA: Cambridge University Press.

Name Index

Subject Index